Buoyant Convection in Geophysical Flows

NATO ASI Series

Advanced Science Institute Series

A Series presenting the results of activities sponsored by the NATO Science Committee, which aims at the dissemination of advanced scientific and technological knowledge, with a view to strengthening links between scientific communities.

The Series is published by an international board of publishers in conjunction with the NATO Scientific Affairs Division

A Life Sciences **B Physics**	Plenum Publishing Corporation London and New York
C Mathematical and Physical Sciences **D Behavioural and Social Sciences** **E Applied Sciences**	Kluwer Academic Publishers Dordrecht, Boston and London
F Computer and Systems Sciences **G Ecological Sciences** **H Cell Biology** **I Global Environment Change**	Springer-Verlag Berlin, Heidelberg, New York, London, Paris and Tokyo

PARTNERSHIP SUB-SERIES

1. Disarmament Technologies	Kluwer Academic Publishers
2. Environment	Springer-Verlag / Kluwer Academic Publishers
3. High Technology	Kluwer Academic Publishers
4. Science and Technology Policy	Kluwer Academic Publishers
5. Computer Networking	Kluwer Academic Publishers

The Partnership Sub-Series incorporates activities undertaken in collaboration with NATO's Cooperation Partners, the countries of the CIS and Central and Eastern Europe, in Priority Areas of concern to those countries.

NATO-PCO-DATA BASE

The electronic index to the NATO ASI Series provides full bibliographical references (with keywords and/or abstracts) to about 50,000 contributions from international scientists published in all sections of the NATO ASI Series. Access to the NATO-PCO-DATA BASE is possible via a CD-ROM "NATO Science and Technology Disk" with user-friendly retrieval software in English, French, and German (©WTV GmbH and DATAWARE Technologies, Inc. 1989). The CD-ROM contains the AGARD Aerospace Database.

The CD-ROM can be ordered through any member of the Board of Publishers or through NATO-PCO, Overijse, Belgium.

Series C: Mathematical and Physical Sciences – Vol. 513

Buoyant Convection in Geophysical Flows

edited by

E. J. Plate

and

E. E. Fedorovich
University of Karlsruhe,
Germany

D. X. Viegas
University of Coimbra,
Portugal

and

J. C. Wyngaard
Pennsylvania State University,
U.S.A.

SPRINGER-SCIENCE+BUSINESS MEDIA, B.V.

Proceedings of the NATO Advanced Study Institute on
Buoyant Convection in Geophysical Flows
Pforzheim, Baden-Württemberg, Germany
17-27 March 1997

A C.I.P. Catalogue record for this book is available from the Library of Congress.

ISBN 978-94-010-6125-4 ISBN 978-94-011-5058-3 (eBook)
DOI 10.1007/978-94-011-5058-3

Printed on acid-free paper

TABLE OF CONTENTS

PREFACE

Buoyant convection is of interest in many fields of geophysical fluid mechanics, in particular in atmospheric and oceanic dynamics, where buoyancy-driven processes play important roles on a variety of scales of motion.

Although the importance of buoyant convection has been recognised for many decades, only recently have the tools become available for effective theoretical and experimental analysis of convective flows. Starting in the 1960s with the pioneering laboratory investigations of atmospheric convection by J. Deardorff, D. Lilly and colleagues and the introduction of bulk models of convectively mixed layers by F. Ball and D. Lilly, convective flows have become a testing ground for numerical models and for sophisticated experimental techniques. The 1960s and 1970s saw a series of field experiments carried out in different countries and aimed at better understanding of the peculiar properties of geophysical convection. The famous Kansas experiment of 1968 allowed fundamental new insights into the nature of convective turbulence in the atmospheric surface layer and has provided a unique data set for two generations of boundary-layer meteorologists. In the 1970s, the first numerical experiments by J. Deardorff on large-eddy simulation of the atmospheric boundary layer raised theoretical studies of geophysical convection to a new level.

Today, buoyant convection in geophysical flows is an advanced and still-developing area of research relevant to problems of the natural environment. During the last decade, significant progress has been achieved through experimental studies, both in nature and in the laboratory, and through large-eddy and direct numerical simulations. Coherent structures have been found to play a key role in geophysical boundary layers and in larger-scale atmospheric and hydrospheric circulations driven by buoyant forcing. New aspects of the interaction between convective motions and rotation have recently been discovered in nature and have been investigated numerically. Extensive experimental data have been collected on the role of convection in cloud dynamics and microphysics. New theoretical concepts and approaches have been outlined regarding scaling and parameterizations of physical processes in buoyancy-driven geophysical flows.

In several different countries technically advanced laboratory facilities have been constructed for experimental studies of geophysical convection. Efforts to simulate the atmospheric convective boundary layer in a wind tunnel have been made at the Institute of Hydrology and Water Resources Planning (IHW), University of Karlsruhe. The construction of their stratified wind tunnel, which was financed by the German Science Foundation (DFG) and built by M. Rau, was finished in the early 1990s. The first results from the tunnel display a plethora of fascinating flow phenomena closely resembling regimes of turbulent convection in the atmospheric boundary layer.

While proceeding with the model studies of buoyant convection, the IHW group established scientific contacts with other research teams over the world dealing with geophysical convection studies. A natural result was the idea of bringing people from these teams together in order to exchange their knowledge and deliver it to the community of young researchers interested in buoyant convection studies. The

Advanced Study Institute (ASI) Programme of the NATO Science Committee was a natural vehicle for implementing this idea.

From the beginning the concept of the ASI on Buoyant Convection in Geophysical Flows was supported by a number of experts in the area of convection research and modelling. About twenty of them agreed to contribute to the ASI as invited lecturers. The Programme of the ASI "Buoyant Convection in Geophysical Flows" drafted in 1995-1996 by the editors of this volume (formerly – members of the ASI Organising Committee) was approved and accepted for funding by the NATO Science Committee.

The response to our invitation to participate in the ASI was gratifying: we received more than 150 applications from potential participants. Unfortunately, only about 80 could be accepted.

The ASI took place during the period from 17 to 27 March 1997 in Pforzheim, a small town at the northern edge of Schwarzwald (Black Forest), Baden-Württemberg, Germany.

The programme of the Institute included buoyancy effects in different media: atmosphere, hydrosphere, and the Earth's mantle; on a wide range of scales: from small-scale phenomena in unstably stratified and convectively mixed layers to deep convection in the atmosphere and the ocean; by different methods of research: field measurements, laboratory simulations, theoretical analysis, and numerical modelling, and within diverse application areas: dispersion of pollutants, parameterization of convection in applied geophysical models, and hazardous phenomena associated with convection. Much of the ASI lecture programme was devoted to fundamentals of convection as a physical phenomenon.

We believe that the present volume, which contains focused versions of the invited lectures, will be a useful compendium on the subject for years to come. The volume falls naturally into four parts.

The first part contains a collection of introductory lectures focusing on fundamental and phenomenological aspects of geophysical convection, and presenting historical and conceptual overviews of convection studies (chapters by E. J. Plate, J. C. Wyngaard, J. C. R. Hunt, S. S. Zilitinkevich *et al.*, R. B. Stull, G. S. Golitsyn, R. H. Käse, D. H. Lenschow, and C. Kiemle *et al.*).

The second part of the volume comprises lecture material on convection modelling and parameterization (chapters by J. C. Wyngaard, R. B. Stull, E. Fedorovich, and C.-H. Moeng).

In the third part, the lectures of R. N. Meroney, and E. Fedorovich and R. Kaiser on experimental studies of geophysical convection in the laboratory are presented.

Overview of applied aspects of convection studies and convective cloud dynamics is given in the fourth part of the volume (chapters by F. T. M. Nieuwstadt, D. X. Viegas, C.-H. Moeng, and A. P. Siebesma).

We hope that the ASI succeeded in filling the gap between fundamental studies of convective geophysical flows and applied modelling of natural phenomena associated with buoyant forcing. Lecturers of the ASI represented both scientific and engineering communities. Their treatment of a variety of the buoyancy-driven natural processes within a common methodological framework should foster links between the theoretical and applied branches of convection research and modelling.

We are grateful to all lecturers of the ASI for their contributions, especially to those who gave their time to prepare their lecture material for publication in this volume. Thanks are due also to the ASI students, whose active participation in lectures and discussions made the ASI a creative and lively scientific meeting.

Our special thanks go to the Local Arrangement Committee members Susanne Rau and Klaus Ammer for their vital help in organising the ASI and in attending to the needs of lecturers and participants. We also extend our gratitude to the administration and personnel of MARITIM Hotel "Goldene Pforte" in Pforzheim, who provided a very comfortable venue for the ASI.

We gratefully acknowledge the financial support of the NATO Science Committee. The grant issued by NATO covered the principal portion of the ASI organisation costs. We are also thankful for donations to the ASI by the University of Karlsruhe, Gemeinschaftskernkraftwerk Neckar GmbH, and Neckarwerke Elektrizitätsversorgugs-AG.

Finally we would like to thank Robert Kolotilo and Dmitrii Mironov for their assistance in preparing the ASI book for publication.

Erich Plate, Evgeni Fedorovich, Domingos Viegas, and John Wyngaard
March 1998

CONVECTIVE BOUNDARY LAYER: A HISTORICAL INTRODUCTION

E. J. PLATE
Institute of Hydrology and Water Resources Planning
Karlsruhe University
Kaiserstrasse 12, 76128 Karlsruhe, Germany

Abstract

A review is given of early research and concepts on the convective boundary layer, which set the stage for all subsequent developments that have been possible via numerical calculations. The historical development proceeded first with inquiries into the stationary turbulent boundary layer. Starting from concepts developed for aeronautical applications of aerodynamics, early research on the atmospheric boundary layer was concerned almost exclusively with stationary flows. Only in the sixties, was the non-stationarity of the planetary boundary layer considered for the first time, with results that left a number of questions open. The first approaches, which are summarized in this paper, were only concerned with obtaining profiles of mean velocity, mean shear, mean temperature, and mean heat flux, which were governed by conservation equations of mass, momentum, and energy.

1. Introduction

As a lower boundary condition for atmospheric motions, the planetary boundary layer is of major interest to meteorological modelling, in particular when one considers processes which take place near the earth's surface, for example in agriculture, where estimates for evaporation and transpiration are needed, or environmental processes, such as emissions from chimneys and exhaust gases from automobiles. Recently, the interest of city planners has also been directed toward such processes, and it is to be expected that environmental issues in city planning will require models which also need parameterizations of the lower part of the atmosphere. The wide interest in the latter issues has been the reason for a previous NATO Advanced Study Institute, conducted by the Institute of Hydrology and Water Resources Planning (Cermak *et al.* [7]).

In response to this interest, scientists all over the world have created a body of knowledge that forms the subject of micrometeorology: the study of the processes in the planetary boundary layer. They started the investigations by considering stationary turbulent boundary layers, which were first considered between 1920 and 1930, notably by G. Taylor, Th. von Kármán, and by L. Prandtl and his students, and used basic concepts from the theory of turbulence, as developed by G. Taylor and A. Kolmogorov.

Their tool used to overcome the non-linearity of the dominant equations of fluid mechanics was the idea of similarity. The dominant concept which evolved from these

E.J. Plate et al. (eds.), Buoyant Convection in Geophysical Flows, 1–22.

early studies was the logarithmic law for the neutral boundary-layer flow. This law was originally obtained based on semi-empirical concepts such as the mixing length, but in later years it has been derived strictly by means of dimensional arguments, notably by Th. von Kármán. A. Kolmogorov predicted the -5/3 power law for the turbulence spectrum.

The initial concepts were transferred in the fifties from aerodynamics to the atmospheric boundary layer by a group of scientists which included H. Lettau, W. Swinbank, C. Priestley, and H. Panofsky. They first considered neutrally stratified flow, but soon the studies were extended to stratified flows, after A. Monin and A. Obukhov derived the basic scaling length for such flows. Much of the research in the sixties and seventies was concerned with experimental investigations made to extend the data basis first provided by "Project Prairie Grass". Field studies were conducted in Australia, England, Russia, and USA for finding the shape of the functions representing mean velocities, turbulent intensities and spectra, which were made dimensionless by the similarity law of A. Monin and A. Obukhov, or by similar quantities based on the scaling lengths for the neutral boundary layer. Such studies were greatly aided when wind tunnels became available in which the (neutrally stratified) boundary layer could be modelled, and when the first proofs that wind-tunnel modelling indeed yielded scaled down versions of atmospheric boundary layer flows were given by J. Cermak and A. Davenport, (see also the summary by Plate [22]). Most of the knowledge available on stationary atmospheric boundary layers that we have today has been available from these studies and has been summarized in a series of books on the planetary boundary layer, starting with books by Plate [21] and Haugen [17]. Recent books are by Stull [29], Arya [2], and Garratt [15], where early literature is extensively reviewed.

2. Analytical Description of the Atmospheric Boundary Layer

We split up the turbulent quantities into two parts. For example, $\theta\,(x,z,t)$ is the potential temperature, which is a turbulent scalar quantity defined as:

$$\theta(x,z,t) = \overline{\theta}(x,z,t) + \theta'(x,z,t),$$

where the term $\overline{\theta}$ is a slowly varying function of t, averaged over a time interval Δt of many minutes, whereas the deviation $\theta'(x,z,t)$ varies rapidly and has a mean value of zero during any Δt. The potential temperature is the temperature corrected for the adiabatic lapse rate. If its mean value $\overline{\theta}(x,z,t)$ increases with height, the air is stably stratified, if it decreases, the fluid is unstable, or super-adiabatic. We consider the general case of a boundary-layer flow according to Fig. 1.

An analytical description of boundary-layer flows must start with the basic equations for the fluxes of heat (the energy equation) and of momentum (the momentum equation), whose derivations can be found, for example, in Schlichting [28]. The boundary-layer assumptions are made, *e.g.* no pressure gradient apart from the hydrostatic one exists in the vertical, and the Reynolds number is assumed to be so high that direct effects of viscosity can be neglected. Furthermore, we consider only two-dimensional flows along a flat surface. Under these assumptions, the equation for the x component of momentum becomes:

$$\frac{\partial \overline{u}}{\partial t}+\overline{u}\frac{\partial \overline{u}}{\partial x}+\overline{w}\frac{\partial \overline{u}}{\partial z}=-\frac{1}{\rho}\frac{d\overline{p}}{dx}+\frac{\partial}{\partial z}\left[-\overline{u'w'}\right], \tag{1}$$

where the expression in brackets is the kinematic momentum flux related to the turbulent shear stress τ by the expression:

$$\tau=-\rho\cdot\overline{u'w'}. \tag{2}$$

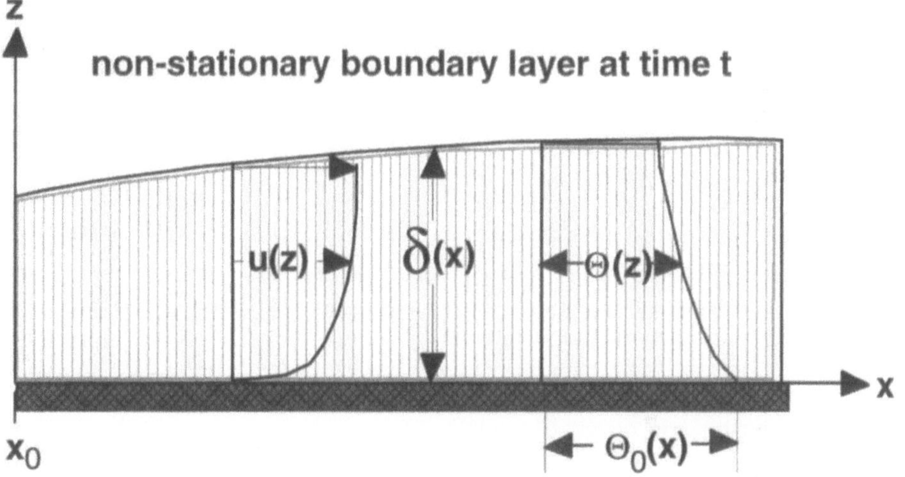

Figure 1. Definitions for a boundary layer flow along a heated plate.

According to boundary-layer theory, the terms in the equation for the vertical momentum are smaller by one order of magnitude, and can therefore be neglected.

The second basic equation is the continuity equation:

$$\frac{\partial \overline{u}}{\partial x}+\frac{\partial \overline{w}}{\partial z}=0. \tag{3}$$

In thermally stratified flow, one must also consider the thermal energy equation:

$$\frac{\partial \overline{\theta}}{\partial t}+\overline{u}\frac{\partial \overline{\theta}}{\partial x}+\overline{w}\frac{\partial \overline{\theta}}{\theta z}=-\frac{\partial \overline{w'\theta'}}{\partial z}+\frac{\varepsilon}{c_p}, \tag{4}$$

where ε is the energy dissipation per unit mass, and c_p is the specific heat of air at the constant atmospheric pressure. The expression $\overline{w'\theta'}$ is identified as the kinematic heat flux $H(z,t)$, *i.e.*

$$H(z,t) = Q_h(z,t)/\rho \cdot c_p = \overline{w'\theta'} . \tag{5}$$

To this set of equations, we add as another important equation for this discussion the equation of conservation of turbulent kinetic energy per unit mass q^2:

$$\frac{\partial \overline{q^2}}{\partial t} + \frac{\partial}{\partial z}\overline{w' \cdot \left(\frac{p'}{\rho} + q^2\right)} = -\overline{u'w'} \cdot \frac{\partial u}{\partial z} + \frac{g}{T} \cdot \overline{w'\theta'} - \varepsilon , \tag{6}$$

where:

$$q^2 = \frac{1}{2}\left(\overline{u'^2} + \overline{v'^2} + \overline{w'^2}\right). \tag{7}$$

With (2)-(7), we have a set of mathematical equations for the thermally stratified boundary layer. It is interesting to note that in these equations the buoyancy is reflected only in the heat flux term $\overline{w'\theta'}$. The equations are essentially non-linear and thus cannot be solved in closed form. Early meteorologists therefore reduced the considerations by concentrating on special cases.

3. Constant Shear Stress Layer: Stationary case

The basic result of the study of the stationary stratified boundary layer is the concept of a multi-layer structure, as shown in Fig. 2. It consists of a bottom part, very close to the lower boundary, which is governed by friction alone. Above this layer is the surface layer, in which wind shear and buoyancy effects in combination determine the flow. The structure of this layer ranges from stable, corresponding to an air density that decreases with height, to neutral, and to unstable (or super-adiabatic, characterized by the density growth with height), which is the most interesting state.

The governing equations for the stationary, constant-thickness layer pictured by early meteorologists requires that the basic equations reduce to:

$$0 = -\frac{1}{\rho}\frac{d\overline{p}}{dx} - \frac{\partial}{\partial z}\overline{u'w'} \quad \text{for momentum,} \tag{8}$$

$$0 = -\frac{\partial}{\partial z}\overline{w'\theta'} + \frac{\varepsilon}{c_p} \quad \text{for energy,} \tag{9}$$

and:

$$0 = -\overline{u'w'} \cdot \frac{\partial u}{\partial z} + \frac{g}{T} \cdot \overline{w'\theta'} - R \text{ for turbulent energy,} \tag{10}$$

where:

$$R = \varepsilon - \frac{\partial}{\partial z} \overline{w' \cdot \left(\frac{p'}{\rho} + q^2 \right)}. \tag{11}$$

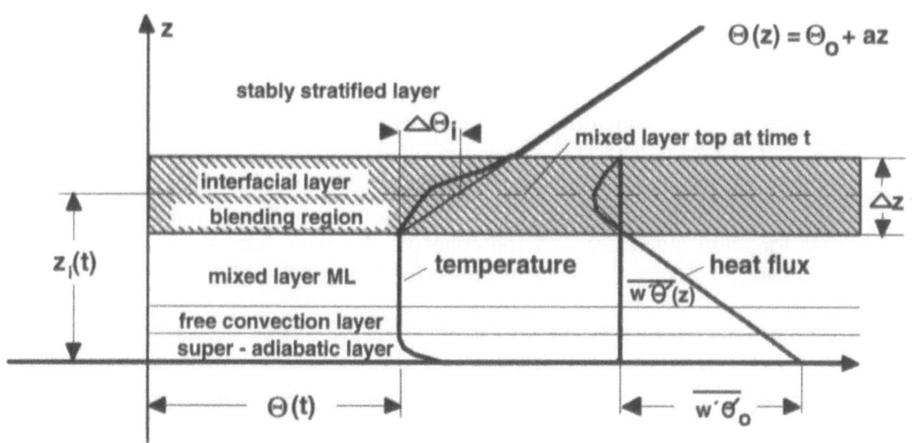

Figure 2. Definitions of the inversion-capped planetary boundary layer.

If the pressure gradient in the x direction is neglected (it is imposed in atmospheric flows by cyclonic gradients balanced by the Coriolis forces), one immediately obtains from Eqs. (8) and (9):

$$\tau = \tau_0 = -\rho \overline{w'u'} = \rho \cdot u_*^2 , \tag{12}$$

where u_* is the shear velocity, *i.e.* this type of layer is a constant stress layer. For the heat flux we obtain:

$$H = H(z) = H_0 + \frac{1}{c_p} \int_0^z \varepsilon(z) dz . \tag{13}$$

For strong heat fluxes the effect of dissipation - which is the mechanical energy dissipated into heat - can be neglected, and it is found that the constant stress layer is also a layer of constant heat flux.

The main characteristics of the flow are the profiles of mean velocity $\overline{u}\,(z)$ and potential temperature $\overline{\theta}\,(z)$. The set of equations given above is not sufficient to determine these profiles, because a relationship is missing between these quantities and the turbulence terms, such as $\overline{w'u'}$ and $\overline{w'\theta'}$. The traditional first-order approach for providing a link between turbulence and mean quantities is obtained by introducing an eddy viscosity ν_T (respectively, a turbulent thermal diffusivity D_θ) and by writing:

$$\tau_0 = \rho \cdot \nu_T \frac{\partial \overline{u}}{\partial z}, \tag{14}$$

$$H(z) = -D_\theta \cdot \frac{\partial \overline{\theta}}{\partial z}. \tag{15}$$

A solution of these equations requires a specification for the diffusivities of momentum and of heat, which are assumed to be determined by overall, rather than by local, features of the flow field. Most of the basic considerations of early fluid mechanists were directed at finding physically reasonable and mathematically tractable expressions for these quantities. For the concepts that have been used, a reference is made to standard textbooks on turbulent flows (for instance, Rodi [26]).

3.1. NEUTRALLY STRATIFIED LAYER

When applying the above equations to the layers indicated in Fig. 2, one must make different assumptions for different layers. For the neutral stratification case, the logarithmic profile:

$$\frac{u(z)}{u_*} = \frac{1}{\kappa} \cdot \ln\left(\frac{z-d}{z_0}\right), \tag{16}$$

derived by dimensional analysis is valid, assuming

$$\frac{\kappa \cdot z}{u_*} \cdot \frac{\partial \overline{u}}{\partial z} = const \tag{17}$$

for the dimensionless shear.

In Eq. (17), the basic scaling quantities are the shear velocity u_* as a velocity scale, and the length scale z, indicating that the gradient depends on the distance from the floor only. The quantity κ is the von Kármán constant, set equal to 0.4. The length z_0, which is the roughness height, is determined by the boundary conditions, as is the zero-plane displacement *d*. Both arise from the fact that near the surface the logarithmic profile is no longer valid: velocity and turbulence structure are governed by the aerodynamic properties of roughness elements – trees, plants, or buildings – which form the lower boundary of the atmospheric sublayer. Extensive data on values of *d* and z_0 have been summarized by Plate [21] and Arya [1], and for urban environments by Theurer [31].

Although temperature differences usually exist in all types of atmospheric boundary layers, the thermal effects are only dynamically effective if the thermal turbulence is larger or of the same order of magnitude as the mechanically (by shear stresses) generated turbulence. A measure for the relative importance is the length L, where L is the Monin-Obukhov length [20] defined by:

$$L = -\frac{u_*^3}{\kappa \cdot \frac{g}{T} \cdot H_0}. \tag{18}$$

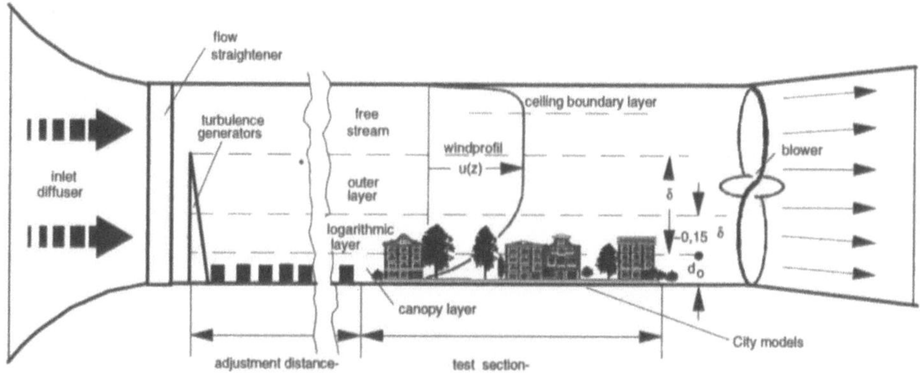

Figure 3. Typical wind tunnel with flow profiles indicated.

A neutrally stratified wind profile also occurs when cloud cover prevents solar radiation from penetrating to the ground, and as a result surface and air temperatures are equal, and no uplift exists. Also, in strong wind conditions the stratification effect is not important, and for practical purposes one can assume that neutral stratification prevails. Because of the practical importance of strong wind flows for wind forces on structures, the equations for neutrally stratified boundary layers form the basis for modelling the atmospheric boundary layer in a wind tunnel. A typical wind tunnel for atmospheric modelling in neutral stratification is shown in Fig. 3. The assumptions made are that the Reynolds number is not important (drag coefficients being independent of Reynolds number), so that u_* is a free scaling velocity that is only used for scaling the turbulence spectrum, and the geometric ratio l/z_0 is the only scaling ratio, where l is a characteristic length scale for the surface elements. Thus, if the logarithmic profile is the same for model and prototype, modelling of the neutral surface layer is assured as has been proved by numerous comparisons of model and field studies (see, for example, Plate [22]). On this basis, the technique of wind-tunnel modelling of atmospheric boundary layers has been well developed, and many wind tunnels have been constructed for determining wind forces on structures and other effects when atmospheric stability is not important.

Based on dimensional grounds, the thickness of the planetary boundary layer for neutral stratification is

$$\delta \approx \beta \frac{u_*}{|f|}. \tag{19}$$

where f is the Coriolis parameter and β is a constant factor between 0.2 and 0.3.

3.2. THE SUPER-ADIABATIC LAYER

For strongly unstable flows, the flux of heat must be considered as a dynamic factor. The gradient of the potential temperature $\theta(x,z,t)$, (which we shall use henceforth instead of the actual temperature), is negative and induces an upward heat flux Q_h. (Also, a flux of latent heat develops through the process of evaporation, but we shall not consider this effect here). For the situation shown in Fig. 2, the heat flux is generated at the ground and is equal to H_0. In the super-adiabatic layer buoyancy induced turbulence arises and interacts with the mechanically generated turbulence. The degree of interaction depends on the relative strength of the two generating mechanisms. The ratio of buoyant forces to shear forces is expressed through L according to Eq. (18), and by introducing a scaling temperature:

$$\theta_* = \frac{-H_0}{\rho c_p u_*}, \tag{20}$$

which yields scales by means of which velocity and temperature distributions in the atmospheric boundary layer are scaled. The application of scaling principles leads to the assumption that the dimensionless gradients of temperature and velocity are dependent only on z/L. The empirical functions $f(z/L)$ given by Businger *et al.* [5] are typical, (see also Arya [2]):

$$\left.\begin{aligned} \frac{\kappa \cdot z}{u_*}\frac{\partial \bar{u}}{\partial z} &= \left(1-15\frac{z}{L}\right)^{-0.25} \\ \frac{\kappa \cdot z}{\theta_*}\frac{\partial \bar{\theta}}{\partial z} &= \left(1-15\frac{z}{L}\right)^{-0.5} \end{aligned}\right\} \quad \text{for} \quad -5 \le \frac{z}{L} \le 0. \tag{21}$$

These equations can be integrated numerically to yield velocity and temperature distributions in the super-adiabatic layer. There are other ways of expressing the stratification effect. A frequently used modification for the velocity distribution is the use of an additive function of z/L to the logarithmic law.

3.3. LAYER OF SHEARLESS CONVECTION

Significant is that from a critical value of $z/L = z/L_{crit} \approx -5$ onward, the effect of instability results in a layer where neither the temperature $\bar{\theta}$ nor the velocity $\bar{u}$ change with height. A layer of shearless convection is created, where the wind field turbulence velocity is scaled by the quantity:

$$w(z)_* = \sqrt[3]{\frac{g}{T} \cdot H_0 \cdot z}\ , \tag{22}$$

which reflects the fact that near the ground the maximum size of the eddies is determined by the distance from the ground. Most likely this is only the case in the region where the buoyant thermals form. Once the thermals have been formed, experimental evidence suggests that they are limited in size not by the height z, but by the extent of the whole layer in which thermals can develop – much like the outer flow in a neutrally stratified boundary layer where the eddy size is determined by the thickness δ of the boundary layer. Therefore, when the thermals are fully developed, the free-convection layer becomes the mixed layer. All flow quantities become independent of height and the size of the turbulence elements is determined by the thickness z_i of the mixed layer below the inversion. For this layer, Deardorff [8] suggested the scaling velocity

$$w_* = \sqrt[3]{\frac{g}{T} \cdot H_0 \cdot z_i}\ . \tag{23}$$

The border between the super-adiabatic layer and the free-convection / mixed layer is not well defined, but it is likely that it is determined by the ratio w_* / u_* This ratio is a function of z_i /L, as can be seen if Eq. (18) is solved for u_* to yield:

$$u_* = -\left(\kappa \cdot \frac{g}{T} \cdot H_0 \cdot L\right)^{1/3}, \tag{24}$$

so that:

$$\frac{w_*}{u_*} = -\left(\frac{z_i}{\kappa \cdot L}\right)^{1/3}. \tag{25}$$

Above a certain critical value of this ratio, roughly identified as 0.4, only buoyancy is important for generating turbulence. Below a certain lower critical value, the mechanical production of turbulence prevail, whereas in between the flow is determined by both factors.

3.4. STABLE LAYER

Early morning conditions for the state of the atmosphere are governed by stable conditions, and the thickness of the stable layer is approximately given by the Eq. (26) (Zilitinkevich [34]):

$$\frac{h \cdot f}{u_*} = \alpha\left(\frac{u_*}{f \cdot L}\right), \tag{26}$$

where α is of the order 1. The expression

$$\frac{h \cdot f}{u_*} = \beta\sqrt{\frac{u_*}{f \cdot L}}, \tag{27}$$

with ß ~ 0.4 has also been recommended (Arya [1]). Other authors prefer to express the thickness of the layer with the dimensionless ratio:

$$\frac{h \cdot f}{u_*} = const, \tag{28}$$

where the constant value is about 0.1 to 0.2 (Grant [16]).

Velocity and temperature distributions in the stable layer can be approximated by the gradient formulation (Arya [2]):

$$\frac{\kappa \cdot z}{u_*}\frac{\partial \bar{u}}{\partial z} = \frac{\kappa \cdot z}{\theta^*}\frac{\partial \bar{\theta}}{\partial z} = \left(1 + 5\frac{z}{L}\right) \quad \text{for } 1 > z/L > 0, \tag{29}$$

from which the profiles of temperature and velocity may be obtained by numerical integration.

4. Non-Stationary Layer Without Changes in *x* Direction

New insight into the nature of the boundary layer was gained when the diurnal cycle of the boundary layer development was considered. If I am not mistaken, the first who investigated the diurnal development of the atmospheric boundary layer were Priestley [23] and Ball [3], who initiated the study of the free-convection layer, which later on was studied extensively by Lilly [19], and in a water tank by Deardorff *et al.* [9] and Deardorff and Willis [10]. Diurnal evolution of the temperature profile in the atmospheric boundary layer is shown in Fig. 4. It is described in detail by Plate [21]:

Fig. 4a: "The diurnal changes in solar radiation set up a cycle of cooling and heating of the planetary boundary layer which is strongly reflected in the wind field. Early in the morning, before sunrise, the stratification of the air layer is stable because the ground is cooler than the air masses above."

Fig. 4b: "With the rising of the sun, on a clear day, solar radiation makes the ground heat up much faster than the overlying air. When the ground is warmer than the air, heat moves from it to the air in contact with the ground. The warm air expands and becomes lighter than the air above it. Parcels of heated air tend to move up into the cooler air above, while cooler air moves down, *i.e.* the air layer is unstable."

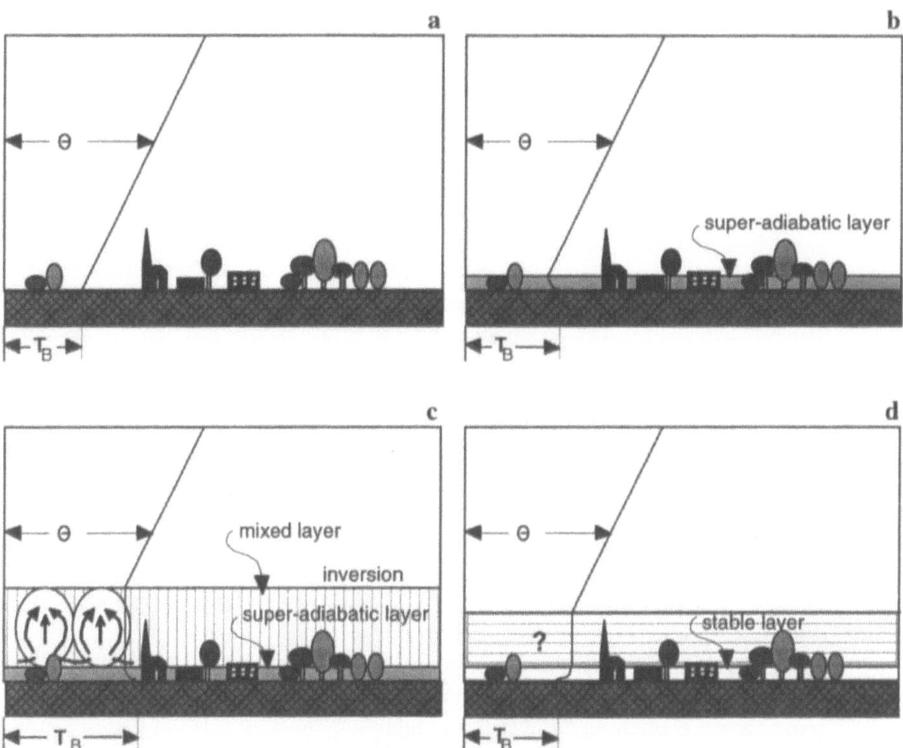

Figure 4. Development of the day time temperature profile for clear sky situation. a - before sunrise: stable stratification; b - shortly after sunrise: T_B increases, super-adiabatic layer is formed; c - late morning: mixed layer forms, thermals rise from super-adiabatic layer; d - late afternoon: ground cools faster than air, what happens to mixed layer?

Fig. 4c: "If the air could adjust instantaneously to the buoyancy that is generated by the heating, a layer of constant (potential) temperature equal to that at the ground would be maintained at all times. Actually, however, the heating of the air lags behind heating of the ground. Clear and dry air over dry ground obtains almost all its heat from the ground. Since directly at the ground no velocity can exist, the transfer of heat from the ground takes place by molecular conduction, down a very strong gradient. The molecular conduction boundary layer is very thin and changes into the forced-convection layer, in which heat is transported by a turbulence that is almost entirely caused by mechanical, *i.e.* shear stress action. At larger distances from the ground, turbulent shear stresses become progressively weaker and gradients both in temperature and velocity decrease. The temperature distribution remains super-adiabatic; but with increasing height less turbulence is maintained by shear stress, and an increasing amount of turbulence is buoyancy induced. Out of the super-adiabatic layer, turbulent parcels of warmer air rise into colder regions aloft and mix with the surroundings in the free-convection layer, which is vigorously agitated because the warm thermal plumes or thermals have significant vertical uplift velocities, and the mass transport associated with them creates a downward motion of colder, but also turbulent air. The result is a mixing of the air which is so strong that in the free convection layer the (potential) temperature is almost constant. The layer in which the temperature field is influenced by the heat flux from the ground is the mixed layer. It is capped by an inversion, *i.e.* by the stably stratified air aloft, which begins near the elevation, where the undisturbed stable potential temperature is equal to the mean temperature of the convective layer. In the course of the day, when the ground gets warmer and the air above it heats up, the convective layer increases in thickness, at a rate that is determined by the heat flux from the ground, which in turn depends on the temperature gradient across the super-adiabatic layer.

Late in the afternoon, an instant of quasi-equilibrium is reached, and the ground is at nearly the same temperature as the layer above it. Since then there are almost no temperature gradients near the ground, the conductive heat flow from the ground ceases to exists, and in the absence of fluxes all temperature fluctuations must die out, so that the neutral stratified boundary layer is asymptotically free of fluctuating temperatures everywhere. It is this condition to which the considerations of the neutrally stratified boundary layer apply, and it is apparent that (under clear sky conditions and for weak winds) it is a rather exceptional state of the atmosphere."

Fig. 4d: Later in the day, the temperature of the air exceeds that of the ground, and the ensuing heat flux down to the ground results in a stably stratified temperature gradient layer near the ground and causes a gradual cooling of the convective layer. The height at which the upper inversion occurs is lowered, while a ground based inversion builds up, particularly after the sun sets. If the conditions for cooling of the surface persist long enough, then eventually the upper and lower inversions destroy the constant temperature between them and join, and further cooling steepens the temperature gradient, until the next morning, when the cycle starts again."

The dynamics of this development was at that time only partially understood, notably the development of the mixed layer above the super-adiabatic layer generated by heating of the surface. The structure of the surface layer had been studied extensively, but only for stationary conditions, and the analysis of the flow profiles of both mean flow and turbulence variances was based on stationary conditions. The transitional conditions between the different stages were little understood, and much work must still be done to develop an understanding of these conditions. Numerical models may help to provide data for further analysis, because experiments in nature or in the laboratory are extremely difficult for these conditions (but see, for example, Grant [16] for recent observations on the development of the nocturnal stable boundary layer just after the evening transition).

Equations (1) to (3) are not sufficient to determine the characteristics of the non-stationary convective layer. A breakthrough for the analysis of the mixed layer was the assumption of Ball [3] that due to the rapid mixing by the thermals the temperature profile is constant in the mixed layer (as is well documented by later studies, as well as by the early data of Warner and Telford [32]), and that the dissipation (which is the energy due to turbulence per unit mass being converted to heat) is so small that it can be neglected. In this manner, Eq. (4) becomes:

$$\frac{d\overline{\theta}}{dt} = -\frac{\partial H}{\partial z}, \tag{30}$$

and:

$$\frac{\partial}{\partial z}\left[\overline{w'\left(\frac{p'}{\rho} + q^2\right)}\right] = -\overline{u'w'} \cdot \frac{\partial \overline{u}}{\partial z} + \frac{g}{T}\overline{w'\theta'}. \tag{31}$$

In these equations, the averages are not taken only over time, but also over a horizontal plane of sufficient width in order to average out the temperature over many thermals and quiescent regions. Then, if Eq. (30) is integrated over the height z (with a very thin super-adiabatic layer assumed), one obtains:

$$\frac{d\overline{\theta}}{dt} \cdot z = H_0 - H(z), \tag{32}$$

i.e. in this non-stationary case the heat flux decreases linearly with height.

Equation (32) has two unknowns, H and $\overline{\theta}$, and therefore a second equation is needed for closure. Equation (31) cannot be used without adding other unknowns. Therefore, a second equation is obtained through an empirical balance equation. In his original study, Ball closed the set of equations by assuming that the net heat flux is zero across the mixed layer. This assumption did not agree well with observations. Later studies have improved the result by using different closure assumptions. A simple one, which is in good agreement with experimental data, was given by Plate [21], who accounted for a temperature overshoot into the inversion layer. The entrainment of the thermals into the upper stable layer displaces warm air downward and thus contributes to the heating of the mixed layer. He assumed that the overshooting distance was proportional to the height z_i. This assumption yields a geometry of the temperature profiles as shown in Fig. 5, and results in:

$$z_i^2 = \frac{2H_0 \cdot t}{2b - a}, \tag{33}$$

which gives the important formulation of the convective layer growth rate. The derivation of this equation is presented in the Appendix.

Equation (33) can be slightly modified to better illustrate its meaning. Heidt [18] introduced the ratio $\eta = b/a$ and obtained Eq. (33) in the form:

$$z_i^2 = \frac{2H_0 \cdot t}{a(2\eta - 1)}. \tag{34}$$

For this expression, the condition $0.5 < \eta \leq 1$ must hold. Heidt confirmed the validity of this equation by means of experiments in a water tank, and by evaluation of experimental data from other sources. He found η in the range from 0.84 to 0.90 for both laboratory and field experiments.

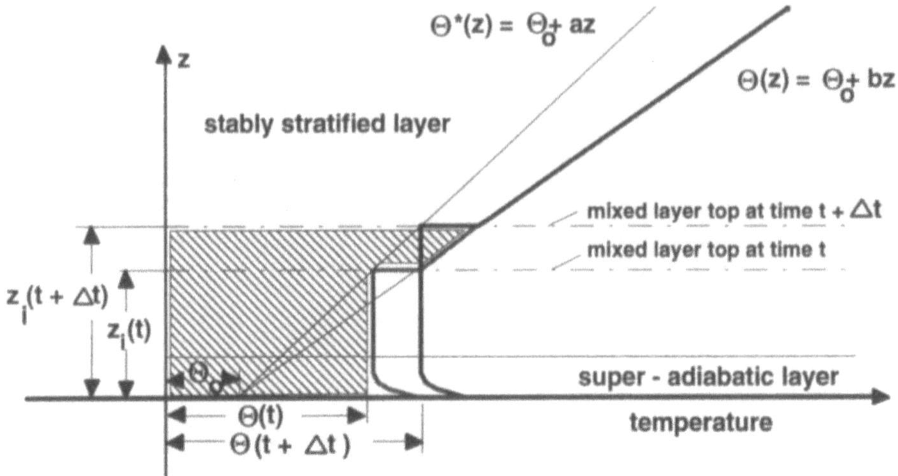

Figure 5. Geometric representation of an overshoot temperature proportional to the height for obtaining the heat balance of the non-stationary mixed layer growth: two-layer model.

From Fig. 5, one obtains for the temperature θ^* at z_i :

$$\theta^*(z_i) = T_0 + b\sqrt{\frac{2H_0 \cdot t}{a(2\eta - 1)}} . \tag{35}$$

It is interesting to note that Eq. (34) implies (as shown in the Appendix) that

$$H(z_i) = -A \cdot H_0 , \tag{36}$$

where A is a constant with a best value of about 0.2.

The assumption A=*const* has been used by a number of scientists (for example, by Tennekes [30], and Carson [6]) in lieu of the basic assumptions of Plate [21]. A first approach for an explanation of the constancy of A is based on dimensional analysis, described by means of an entrainment function P_E :

$$P_E = \frac{1}{w_{zi}} \cdot \frac{dz_i}{dt} = f_s , \tag{37}$$

where w_{zi} is a characteristic velocity and f_s is the stability-dependent parameter. On dimensional grounds, the logical choice for w_{zi} is the velocity scale w_* for the mixed layer defined by Eq. (23), so that:

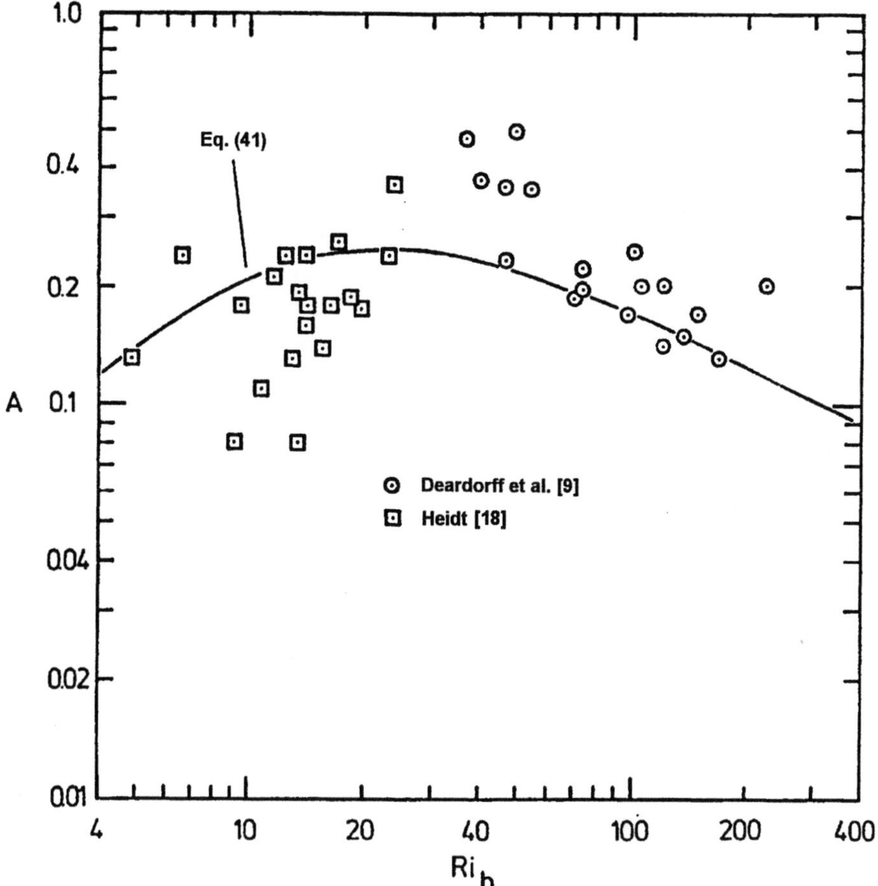

Figure 6. Relationship between the entrainment coefficient A and the Richardson number Ri_h (Faust [12]).

$$P_E = \frac{1}{w_*} \cdot \frac{dz_i}{dt}. \tag{38}$$

This function depends on the strength of the stability, which can be expressed through the Richardson number:

$$\mathrm{Ri}_h = \sqrt[3]{\frac{g \cdot z_i^4 \cdot a^3}{T_0 \cdot H_0^2}}. \tag{39}$$

With these expressions, it can be shown that the assumption $A{=}const$ is identical to the result:

$$P_E = (1+2A)\cdot \mathrm{Ri}_h^{-1}. \tag{40}$$

The similarity approach cannot be used to find out what physically happens at the interface between the mixed layer and inversion. To cast more light on the entrainment process, a number of studies have been conducted (Wyngaard [33], Driedonks and Tennekes [11], see also the paper by Fedorovich [13] in this volume) whose authors have tried to refine this result on the basis of the energy equation of turbulent motion, Eq. (6) or Eq. (31). They attempted to show that A is not a constant, but depends on the Richardson number expressed by Eq. (39).

Plotting A against the Richardson number for many experimental data (collected by Faust [12]) yielded a best fit curve:

$$A = \frac{0.033\cdot \mathrm{Ri}_h}{1+0.0814\cdot \mathrm{Ri}_h^{3/2}}, \tag{41}$$

which is shown in Fig. 6

However, as Faust (and many others) concluded: as long as a two layer model is used for describing the convective boundary layer, a constant value for A is as good approximation for the observed data as are the curves obtained from different models. Recent data indicate that the thickness of the interfacial layer between convective flow and the stable upper layer is an important parameter, resulting in a second determining Richardson number based on the thickness of this layer. It may be argued nevertheless that this layer is a result of the interaction of the two adjacent layers, and therefore it should be described, in a dimensional analysis, by the parameters of the other two layers.

5. A Parallel Example of Stratified Flow Mixing

An interesting analogy to the mixed layer growth problem, which casts light on the process of entrainment, may be obtained from the stirred tank results of Rouse and Dodu [27]. They performed experiments in the water tank shown in Fig. 7, where initially a layer of low density fluid (fresh water with density ρ_1) was placed on top of a layer with a higher density fluid (salt water with initial density ρ_2). Then the higher density fluid was stirred by means of a mixer, and artificial turbulence was induced. Due to the stirring, the dense fluid migrated into the lighter fluid at a rate dz_i / dt, while the entrained lighter fluid mixed with the lower fluid to create a density $\rho(t)$. This process can be described by considering the conservation equations of fluid mechanics as applied to the model of Fig. 7.

The equation for mass continuity applied to the time intervals t and $t+\Delta t$ yields

$$\rho_1\cdot \Delta z_i + \rho(t)\cdot z_i = \left[\rho(t) + \frac{\partial \rho}{\partial t}\Delta t\right]\cdot (z_i + \Delta z_i)$$

for the mass change, so that:

$$\frac{\partial\rho}{\partial t} = \frac{[\rho_1 - \rho(t)]}{z_i}\frac{\Delta z_i}{\Delta t}. \tag{42}$$

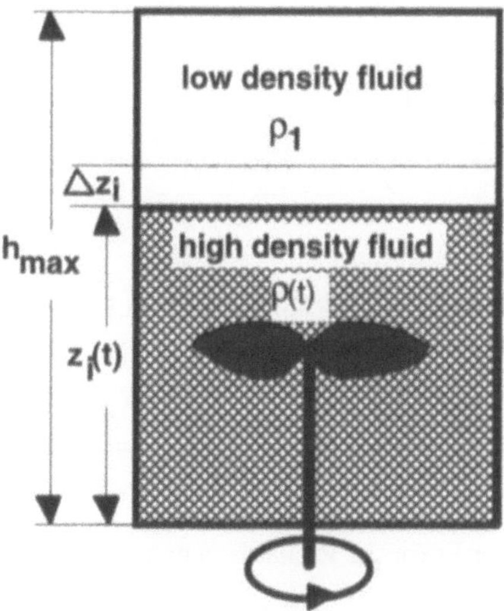

Figure 7. Mixed layer analogy of Rouse and Dodu [27].

The equation expressing the change in mechanical energy for this case can be stated as:

$$\frac{\Delta E}{\Delta t} = \frac{\Delta W}{\Delta t} - D, \tag{43}$$

where $\Delta E / \Delta t$ is the change in mechanical energy per unit area (it is approximately equal to the change in potential energy), D is the energy dissipation, and $\Delta W/\Delta t$ is the work done by the stirrer, per unit area. These quantities can be written as:

$$\frac{\Delta W}{\Delta t} = \rho_2 \cdot u_{ch}^3, \tag{44}$$

and (see the Appendix):

$$\frac{\Delta E}{\Delta t} = \frac{1}{2}[\rho_1 - \rho(t)]z_i \cdot g \cdot \frac{\Delta z_i}{\Delta t}. \tag{45}$$

Let the volume of the fluid below z_i be very large, so that $\rho(t) \approx \rho_2$, where ρ_2 is the initial density of the lower fluid. If one makes the additional assumption that the dissipation during the time interval Δt is proportional to the work done by the stirrer during Δt, with a factor of proportionality $1-\alpha$, then one obtains for Eq. (43):

$$\frac{1}{2}(\rho_2 - \rho_1) \cdot g \cdot z_i(t) \cdot \frac{\Delta z_i}{\Delta t} = \alpha \cdot \frac{\Delta W}{\Delta t}, \tag{46}$$

Consequently, analogous to Eq. (34),

$$z_i^2 = \frac{4\alpha \cdot (\Delta W/\Delta t) \cdot t}{(\rho_2 - \rho_1) \cdot g}. \tag{47}$$

Equation (46) can also be written in terms of an entrainment factor P_E. By using Eq. (44) and rearranging, one obtains:

$$P_E = \frac{\Delta z_i}{\Delta t} \frac{1}{u_{ch}} = \alpha \cdot \frac{2\rho_2 \cdot u_{ch}^2}{(\rho_2 - \rho_1) \cdot g \cdot z_i} = 2\alpha \cdot \mathrm{Fr}^2, \tag{48}$$

where the Froude number Fr (or the Richardson number Ri) is given by:

$$\mathrm{Fr}^2 = \mathrm{Ri}^{-1} = \frac{u_{ch}^2}{[(\rho_2 - \rho_1)/\rho_2] \cdot g \cdot z_i}, \tag{49}$$

which is in agreement with the considerations on mixing by thermals. Note that the characteristic velocity in this case is defined through the energy input by the stirrer which can be measured directly. The coefficient α was determined experimentally by Rouse and Dodu [27], and includes the efficiency of the stirrer. For details see the original paper.

6. Further Convective-Layer Studies at Karlsruhe University

At this point I want to conclude the historical review. I have presented the state of the knowledge on the convective layer as it existed in 1970-1975. Research at the Institute of Hydrology and Water Resources Planning at the University of Karlsruhe was essentially conducted to investigate problems which were left unsolved by the analyses just presented. The German Science Foundation through the Sonderforschungsbereich 80 gave us the opportunity to look deeper into the structure of the convective boundary layer, and in a series of three dissertations we further explored its structure. The first one was a study by Heidt [18], who essentially duplicated the experiments of Deardorff *et al.* [9], but with the purpose of finding out what happens at the interface between the mixed layer and inversion. His findings have already been discussed: they did not yield any additional insights into the fluid mechanics of the entrainment process. He basically developed a technique on how to very accurately determine the coefficients, in particular the coefficient A. By his introduction of the factor η he made a good contribution in systematizing the dimensionless presentation.

The research of Heidt was continued by Faust [12]. We believed that more information could be gained by studying single thermals. Experiments on this issue were conducted by Faust in a water tank, in which initially the water was stably stratified with a constant temperature gradient of about 0.2 to 1.0 °C/cm. The bottom of the tank had a circular plate arranged in the center which could be heated. At time t=0 heating of the plate was started by an electrical energy step-function input. A bubble of heated water formed above the plate and evolved into a thermal plume, whose growth was monitored by means of measuring the developing temperature profile in the center of the plume. Although the experiments were conducted in a tank of limited size which resulted in a substantial downward flow induced by the rising thermals, Faust succeeded in developing a compensating model, so that his results could be applied to thermals in any size container. To describe the thermal, Faust developed a model of a periodic bubble growing out of the super-adiabatic layer. The special feature of this bubble model was that no entrainment was allowed in the lower part of the plume, so that with an increasing distance from the floor and increasing velocity the diameter of the plume had to decrease first. Only at greater heights did the plume assume a structure that could be described by standard thermal plume models based on similarity assumptions. A most fascinating observation was that the generation and decay of the thermal plume was almost perfectly periodic.

In order to allow the diameter of the hot plate to enter as a dimensional quantity, Faust tried to relate his results to an urban heat island – an application, which was only partially successful. His main contribution was a rather excellently documented thermal plume model. What could not be achieved was a model by which the entrainment process into the stable layer could be described: Faust assumed this to be a passive process, the plumes losing their momentum and displacing fluid from the inversion. Basically, it led to an equation for the height h of the individual thermal:

$$h^2 = \frac{2}{a(2\eta - 1)} \cdot \frac{D^2}{L^2} H_0 t\,, \tag{50}$$

which shows that only the factor D/L, where D is the diameter of the hot plate and L the diameter of the container, modifies Eq. (34).

The questions which cannot be answered by water tank models are associated with the occurrence of the wind shear. What is the critical value of w_* / u_*? Is the value of 0.4 given above indeed correct? How does the shear influence the thermals? Our next experiments were directed towards answering such questions. We first built a small wind tunnel with neutrally stratified flow, and used a hot plate as a heat source for generating a single thermal in a cross flow. The experiments were conducted by Blöchl [4]. They showed that under the effect of even a small amount of cross wind, the thermal plume attached to the ground and formed a heated plume parallel to the bottom. This approach therefore did not appear very relevant for the atmosphere, and so we decided to build a large recirculating wind tunnel for studying convection with shear. This facility was designed and made operational by M. Rau (Rau *et al.* [24], and Rau and Plate [25]), and results obtained by means of this facility are presented by Fedorovich and Kaiser [14] in this volume. The tunnel is now available for studying many of the open questions involving shear and including problems associated with the study of pollution transport in convective layers. The tunnel can serve as a standard data source for verification of numerical models.

Appendix

Derivation of Eq. (33). The heat balance of the mixed layer is obtained from the simple geometry of Fig. 5 by assuming that the change in temperature is proportional to the change in height, *i.e.* it is assumed that the overshoot falls on a straight line passing through T_0. We define: $\Delta z + z_i(t) = z_i(t+\Delta t)$. Then we can write the equation for the heat content in the layer below $z_i(t+\Delta t)$ at time *t*, which is the heat content of the diagonally marked volume in Fig. 5, and obtain:

$$S(t) = z_i \cdot (T_0 + b \cdot z_i) + \frac{1}{2}\Delta z(2T_0 + 2a \cdot z_i + a \cdot \Delta z) \approx z_i \cdot T_0 + b \cdot z_i^2 + T_0 \cdot \Delta z + a \cdot z_i \cdot \Delta z \, .$$

For the heat content at time $t + \Delta t$, we obtain for the layer below $z_i(t+\Delta t)$:

$$S(t+\Delta t) = (z_i + \Delta z_i)(T_0 + b \cdot z_i + b \cdot \Delta z_i) \approx z_i \cdot T_0 + b \cdot z_i^2 + 2b \cdot z_i \cdot \Delta z_i + \Delta z_i \cdot T_0 \, .$$

The change in heat during time Δt is equal to the heat flux $H_0 \cdot \Delta t$ from below, consequently the conservation of heat yields: $H_0 \cdot \Delta t = S(t+\Delta t) - S(t)$, and

$$H_0 = \frac{S(t+\Delta t) - S(t)}{\Delta t} = (2b - a) \cdot z_i \cdot \frac{\Delta z_i}{\Delta t} \, ,$$

from which Eq. (33) follows.

Derivation of Eq. (36). From the temperature profile assumption:

$$\theta^*(z_i) = T_0 + b \cdot z \quad \rightarrow \quad \frac{d\theta^*}{dt} = b \cdot \frac{dz_i}{dt}$$

we obtain, in conjunction with the heat balance Eq. (32) at $z = z_i$:

$$b\frac{dz_i}{dt} \cdot z_i = H_0 - H(z_i) = \frac{b}{2}\frac{dz_i^2}{dt} \, .$$

Inserting Eq. (33) yields:

$$H(z_i) = -\frac{1-\eta}{2\eta-1} \cdot H_0 \quad \rightarrow \quad A = \frac{1-\eta}{2\eta-1}.$$

Derivation of Eq. (45). The change in potential energy per unit area is given by $\Delta E = P(t+\Delta t) - P(t)$, which can be obtained by considering the potential energy in the marked volume of Fig. 7 at time *t*:

$$P(t) = \frac{1}{2} z_i \cdot \rho(t) g z_i + \left(z_i + \frac{1}{2} \Delta z_i \right) \cdot \rho_1 \Delta z_i,$$

and at time $t+\Delta t$:

$$P(t+\Delta t) = \frac{1}{2} z_i \cdot \left(\rho(t) + \frac{\partial \rho}{\partial t} \Delta t \right) g z_i + \left(z_i + \frac{1}{2} \Delta z_i \right) \left(\rho(t) + \frac{\partial \rho}{\partial t} \Delta t \right) g \Delta z_i.$$

Therefore,

$$\frac{\Delta E}{\Delta t} = \frac{1}{2} g \cdot z_i^2 \frac{\partial \rho}{\partial t} \cdot \left(1 + 2 \frac{\Delta z_i}{z_i} \right),$$

and in conjunction with Eq. (42) one obtains Eq. (45).

References

1. Arya, S. P. (1982) Atmospheric boundary layers over homogeneous terrain, in E. J. Plate (ed.), *Engineering Meteorology*, Elsevier, Amsterdam, pp. 233-267.
2. Arya, S. P. (1988) *Introduction to Micrometeorology*, Academic Press, New York.
3. Ball, F. K. (1960) Control of inversion height by surface heating, *Quarterly Journal of the Royal Meteorological Society* **86**, 483-494.
4. Blöchl, B. (1988) Windkanalmodellierung von Wärmeinseln in turbulenter Grenzschicht, *Mitteilungen Heft 34*, Institut für Hydrologie und Wasserwirtschaft, Universität Karlsruhe.
5. Businger, J. A., Wyngaard, J. C., Izumi, Y., and Bradley, E. F. (1971) Flux-profile relationships in the atmospheric surface layer, *Journal of the Atmospheric Sciences* **28**, 181-189.
6. Carson, D. J. (1973) The development of a dry inversion-capped convectively unstable boundary layer, *Quarterly Journal of the Royal Meteorological Society* **99**, 450-467.
7. Cermak, J. E., Davenport, A. G., Plate, E. J. and Viegas, D. X., (eds.) (1995) *Wind Climate in Cities*, Kluwer Academic Publishers, Dordrecht.
8. Deardorff, J. W. (1970) Convective velocity and temperature scales for the unstable boundary layer and for Rayleigh convection, *Journal of the Atmospheric Sciences* **27**, 1211-1213.
9. Deardorff, J. W., Willis, G. E., and Lilly, D. K. (1969) Laboratory investigation of non-steady penetrative convection, *Journal of Fluid Mechanics* **35**, 7-31.
10. Deardorff, J. W. and Willis G. E. (1985) Further results from a laboratory model of the convective planetary boundary layer, *Boundary-Layer Meteorology* **32**, 205-236.
11. Driedonks, A. G. M. and Tennekes, H. (1984) Entrainment effects in the well mixed atmospheric boundary layer, *Boundary-Layer Meteorology* **30**, 75-105.
12. Faust, K. (1981) Modelldarstellung von Wärmeinselströmungen durch Konvektionsstrahlen, *Mitteilungen Heft 19*, Institut für Hydrologie und Wasserwirtschaft, Universität Karlsruhe.
13. Fedorovich, E. (1998) Bulk models of the atmospheric convective boundary layer, see this volume.
14. Fedorovich, E. and Kaiser, R. (1998) Wind tunnel model study of turbulence regime in the atmospheric convective boundary layer, see this volume.

15. Garratt, J. R. (1992) *The Atmospheric Boundary Layer*, Cambridge University Press, Cambridge.
16. Grant, A. L. M. (1997) An observational study of the evening transition boundary layer, *Quarterly Journal of the Royal Meteorological Society* **123**, 657-677.
17. Haugen, D. A., (ed.) (1973) *Workshop in Micrometeorology*, American Meteorological Society, Boston.
18. Heidt, F. D. (1975) Zeitlicher Abbau der stabilen Schichtung eines Fluides durch freie Konvektion, *Mitteilungen Heft 6*, Institut für Hydrologie und Wasserwirtschaft, Universität Karlsruhe.
19. Lilly, D.K. (1968) Models of cloud-topped mixed layers under a strong inversion *Quarterly Journal of the Royal Meteorological Society* **94**, 292-309.
20. Monin, A. S. and Yaglom, A. M. (1971) *Statistical Fluid Mechanics* **1**, MIT Press, Cambridge, Massachusetts.
21. Plate, E. J. (1971) *Aerodynamic Characteristics of Atmospheric Boundary Layers*, US Atomic Energy Commission, Oak Ridge, Tennessee.
22. Plate, E. J. (1982) Wind tunnel modelling of wind effects in engineering, in E. J. Plate (ed.), *Engineering Meteorology*, Elsevier, Amsterdam, pp. 573-639.
23. Priestley, C. H. B. (1959) *Turbulent Transfer in the Lower Atmosphere*, University of Chicago Press, Chicago.
24. Rau, M., Bächlin, W., and Plate, E. (1991) Detailed design features of a new wind tunnel for studying the effects of thermal stratification, *Atmospheric Environment*, **25A**, 1258-1263.
25. Rau, M. and Plate, E. (1995) Wind tunnel modelling of convective boundary layers, in J. E. Cermak *et al.* (eds.), *Wind Climate in Cities*, Kluwer Academic Publishers, Dordrecht, pp. 431-456.
26. Rodi, W. (1980) Turbulence models and their application in hydraulics - a state of the art review, *IAHR Publication*, Delft.
27. Rouse, H. and Dodu, J. (1955) Diffusion turbulente a travers une continuité de densité, *La Houille Blanche* **4**, 522-532.
28. Schlichting, H. (1968) *Boundary Layer Theory*, MacGraw Hill, New York.
29. Stull, R. (1988) *An Introduction to Boundary Layer Meteorology*, Kluwer Academic Publishers, Dordrecht.
30. Tennekes, H. (1973) A model for the dynamics of the inversion above a convective boundary layer, *Journal of the Atmospheric Sciences* **30**, 558-567.
31. Theurer, W. (1995) Point sources in urban areas: modelling of neutral gas clouds with semi-empirical models, in J. E. Cermak *et al.*, (eds.) *Wind climate in cities*, Kluwer Academic Publishers, Dordrecht, pp. 485-502.
32. Warner, J. and Telford, J. W. (1967) Convection below cloud base, *Journal of the Atmospheric Sciences* **24**, 374-382.
33. Wyngaard, J. C. (1971) Local free convection, similarity, and the budget of shear stress and heat flux. *Journal of the Atmospheric Sciences* **28**, 1171-1182.
34. Zilitinkevich, S. S. (1972) On the determination of the height of the Ekman boundary layer, *Boundary-Layer Meteorology* **3**, 141-145.

CONVECTION VIEWED FROM A TURBULENCE PERSPECTIVE

J. C. WYNGAARD
Penn State University
University Park 16802, Pennsylvania, USA

1. Introduction

Turbulence is known as one of the outstanding unsolved problems in physics. In his 1979 paper on the evolution of ideas in turbulence Liepmann [7] wrote "The scientific study of turbulent flow spans approximately one hundred years; during that time some of the greatest names in physics, mechanics, and engineering have at one time or another taken a crack at the problem. Progress in many directions has been made, indeed significant progress. However, the 'turbulence problem' ... remains."

In turbulence research one sees periods of feverish activity in what can become blind alleys, punctuated by occasional leaps forward. The long view reveals slow, steady progress in the physical understanding of the structure of turbulent flows in engineering and geophysics, accumulating evidence of the futility of head-on analytical attacks on turbulence, and repeated indications that experiment paces the growth in our understanding of turbulence.

Some people study turbulence because "it is there"; most are probably motivated by the need to deal with turbulence in applications. Applications areas today tend to lie at the boundaries of disciplines—e.g., at the boundary between turbulence and climate or mesoscale or severe-storm dynamics, between turbulence and acoustics, or between turbulence and atmospheric dispersion. Those in the atmospheric or oceanic sciences who need to learn about convection are apt to be newcomers to the field of turbulence; those in engineering turbulence who need to learn about turbulence in the atmosphere or in the ocean are apt to be newcomers to convection. I thought, therefore, that I should devote this first lecture to a view of convection from a turbulence perspective. My reference point is convection in the atmospheric boundary layer.

2. The Surface Layer: The Kansas Experiment and Its Surprises

In 1963 the Boundary-Layer Branch of the Air Force Cambridge Research Laboratories in Bedford, Massachusetts began experimental study of the atmospheric surface layer. In the decade to follow they had an active and leading role in the development of a new generation of turbulence sensors (such as the sonic anemometer) and carried out extensive measurement programs at homogeneous field sites. One high point of this activity was their 1968 expedition to Kansas. As the newest member of that research group (I had joined it in 1967) I participated in that 1968 field program, perhaps the most ambitious and complete surface-layer measurement program up to that time. The Kansas experiment yielded a

E.J. Plate et al. (eds.), Buoyant Convection in Geophysical Flows, 23–39.

substantial body of new knowledge of the structure of turbulence in the atmospheric surface layer, including some surprising new findings about convective turbulence. I will summarize some of these surprises here.

2.1. PRESSURE TRANSPORT AND THE TKE BUDGET

In the traditional decomposition of all flow variables into ensemble-mean and fluctuating parts (velocity = $U_i + u_i$, potential temperature = $\Theta + \theta$, pressure = $P + p$), the fluctuating velocity equation becomes in the Boussinesq approximation

$$\frac{\partial u_i}{\partial t} = -u_j \frac{\partial U_i}{\partial x_j} - U_j \frac{\partial u_i}{\partial x_j} - \frac{\partial}{\partial x_j}\left(u_i u_j - \overline{u_i u_j}\right) - \frac{1}{\rho_0}\frac{\partial p}{\partial x_i}$$

$$-2\epsilon_{ijk}\Omega_j u_k + \frac{g}{T_0}\theta\delta_{3i} + \nu\frac{\partial^2 u_i}{\partial x_j \partial x_j}. \quad (1)$$

with the overbar representing an ensemble average. Dotting this equation with u_i, ensemble averaging, and rearranging, using the consequences of incompressibility, local isotropy, and large Reynolds number, yields the turbulent kinetic energy (TKE) budget

$$\frac{\partial}{\partial t}\frac{\overline{u_i u_i}}{2} + U_j \frac{\partial}{\partial x_j}\frac{\overline{u_i u_i}}{2} =$$

$$-\overline{u_j u_i}\frac{\partial U_i}{\partial x_j} \quad \text{(shear production)}$$

$$-\frac{\partial}{\partial x_j}\frac{\overline{u_i u_i u_j}}{2} \quad \text{(turbulent transport)}$$

$$-\frac{1}{\rho_0}\left(\frac{\partial \overline{p u_i}}{\partial x_i}\right) \quad \text{(pressure transport)}$$

$$+\frac{g}{T_0}\left(\overline{\theta u_3}\right) \quad \text{(buoyant production)}$$

$$-\epsilon. \quad \text{(viscous dissipation)} \quad (2)$$

In a quasi-steady, horizontally homogeneous atmospheric surface layer with coordinates aligned with the mean wind and in standard notation $[u_i = (u, v, w),\ \ x_i = (x, y, z)]$ this reduces to

$$0 = -\overline{uw}\frac{\partial U}{\partial z} \quad \text{(shear production)}$$

$$-\frac{\partial}{\partial z}\frac{\overline{u_i u_i w}}{2} \quad \text{(turbulent transport)}$$

$$-\frac{1}{\rho_0}\left(\frac{\partial \overline{pw}}{\partial z}\right) \quad \text{(pressure transport)}$$

$$+\frac{g}{T_0}\left(\overline{\theta w}\right) \quad \text{(buoyant production)}$$

$$-\epsilon. \quad \text{(viscous dissipation)} \quad (3)$$

Equation (3) expresses the balance of mechanisms that maintain the mean kinetic energy per unit mass of the turbulence in the quasi-steady, locally homogeneous near-surface region of the atmospheric boundary layer. Lumley and Panofsky [8] summarized our understanding of this balance as of the early 1960s. The few measurements that existed at that time seemed to indicate that under convective conditions the turbulent transport term represents a rate of loss that occurs physically through export of TKE out of the surface layer. That loss rate was believed to be essentially balanced by the rate of gain through buoyant production, although this was more an assumption than a quantitative inference from data. Pressure transport, which had not yet been measured, was assumed to be small, perhaps because it had been found to be generally small in various canonical laboratory flows. Thus, it was concluded that in the convective surface layer the shear-production term was balanced by viscous dissipation.

The 1968 Kansas experiment was the first in which all terms but pressure transport in the surface-layer TKE budget were measured directly. The data base from this experiment was quite extensive, consisting of about 30 one-hour runs made over a two-week period under nominally steady conditions. The fetch was remarkably uniform, with no prominent obstacles to wind flow for many kilometers. The instrumentation was the most up-to-date and reliable available, including sonic and hot-wire anemometry at three levels on the 32-m tower (Haugen *et al.*, [4]).

The TKE budget results, Figure 1, indicated a large imbalance, a gain term. After considering various interpretations of this imbalance, the Kansas team judged it most likely due to pressure transport. Since the covariance $\overline{pw}$ vanishes at the surface, the finding that pressure transport is a gain term in the unstable surface layer implies that $\overline{pw}$ becomes negative away from the surface in unstable conditions.

The Kansas results were unsettling in that they refuted the contemporary interpretation of the TKE budget. The implication that the rate of TKE gain through pressure transport, through a mechanism that is still not physically clear, was as large as any other term in the budget not only disagreed with the conventional wisdom, but also indicated that the unstable surface layer behaved like no other known turbulent flow. Despite careful and skeptical scrutiny of these results by many observers in the years since the Kansas experiment, the picture they portray slowly gained support. By the mid 1970s direct measurements of pressure fluctuations with a specially designed sensor [9] soon confirmed that $\overline{pw}$ was indeed negative and that pressure transport was a substantial gain term in the unstable surface layer.

Figure 2 shows a "consensus" TKE budget ca 1990, about 20 years after the publication of the Kansas results. It reflects input from a number of scientists reporting on several experiments. It is consistent with direct measurements of pressure transport, including recent measurements [10] with a pressure sensor of design different from that used by [9]. This consensus budget agrees well with the Kansas findings. Any concerns that the Kansas TKE budget results were anomalous should now be erased. There remain debates ([14], [18]) about the effects of probe- and tower-induced flow distortion on turbulence measurements, but such are part of the fabric of micrometeorological life. In the broader picture, it is clear that the convective surface layer is a very unusual turbulent flow.

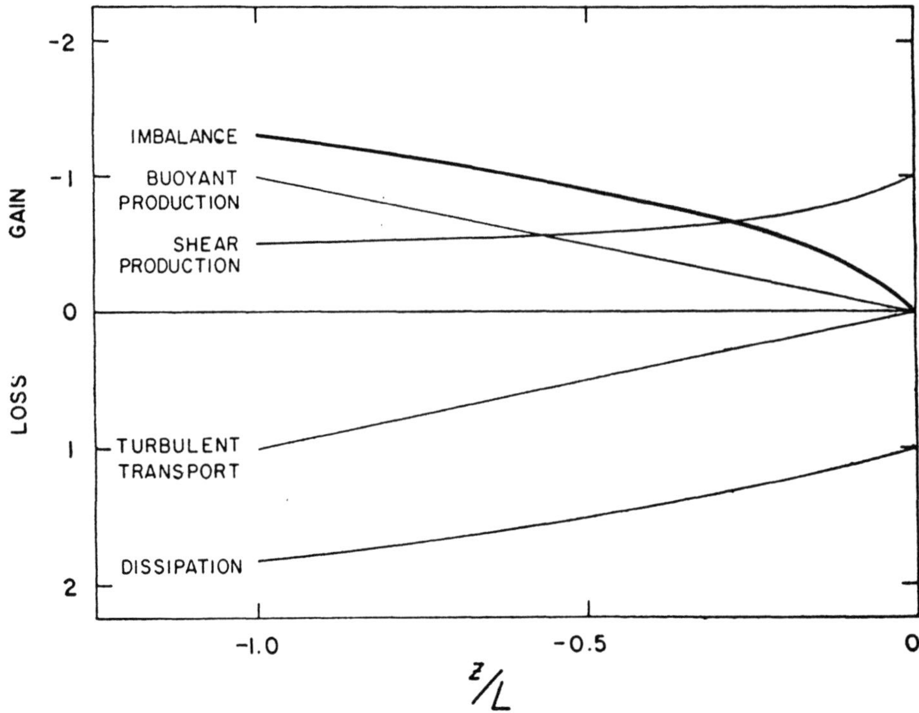

Figure 1. The TKE budget in the convective surface layer as deduced in the 1968 Kansas experiment. Buoyant production, shear production, turbulent transport, and dissipation were measured directly; their imbalance, which was presumed to be due to pressure transport, is indicated. From [22].

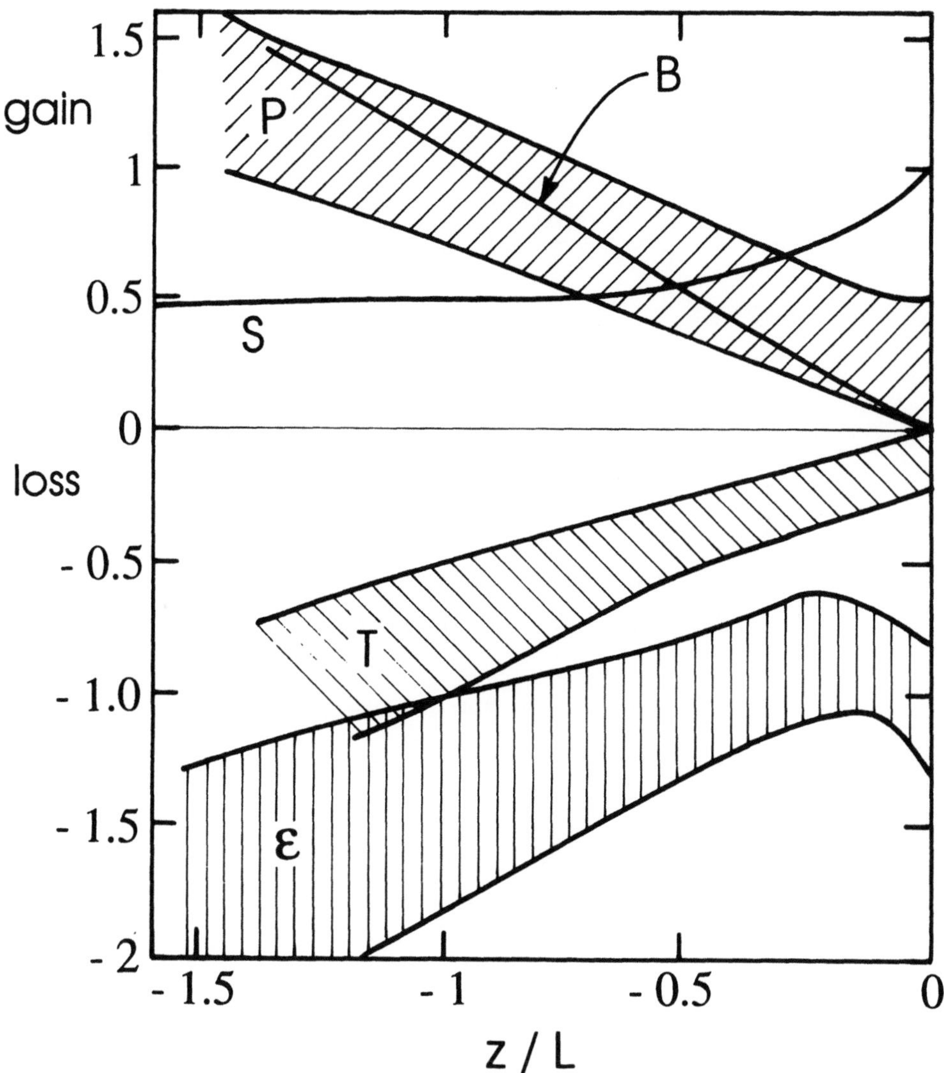

Figure 2. A consensus TKE budget for the convective surface layer. The shaded areas represent the spread of measurements from experiments summarized by Hogstrom [5]. B is buoyant production; P, pressure transport; S, shear production; T, turbulent transport; and ϵ, viscous dissipation. From [21].

2.2. REYNOLDS FLUX BUDGETS

Although they are a natural extension of the TKE budget, Reynolds flux budgets were not discussed in the literature until much later. Corrsin [2] credits Reynolds [12] with the first exposition of the TKE balance, and Chou [1] with the first shear-stress balance.

The Reynolds stress budget is

$$
\begin{aligned}
\frac{\partial \overline{u_i u_k}}{\partial t} + U_j \frac{\partial \overline{u_i u_k}}{\partial x_j} = \\
-\overline{u_j u_k}\frac{\partial U_i}{\partial x_j} - \overline{u_j u_i}\frac{\partial U_k}{\partial x_j} \quad &\text{(shear production)} \\
-\frac{\partial \overline{u_i u_k u_j}}{\partial x_j} \quad &\text{(turbulent transport)} \\
-\frac{1}{\rho_0}\left(\overline{u_k \frac{\partial p}{\partial x_i}} + \overline{u_i \frac{\partial p}{\partial x_k}}\right) \quad &\text{(pressure covariance)} \\
-2\epsilon_{ijm}\Omega_j \overline{u_m u_k} - 2\epsilon_{kjm}\Omega_j \overline{u_m u_i} \quad &\text{(Coriolis)} \\
+\frac{g}{T_0}\left(\overline{\theta u_k}\delta_{3i} + \overline{\theta u_i}\delta_{3k}\right) \quad &\text{(buoyant production)} \\
-\frac{2\epsilon}{3}\delta_{ik}. \quad &\text{(viscous dissipation)}
\end{aligned}
\tag{4}
$$

The temperature flux equation is

$$
\begin{aligned}
\frac{\partial \overline{\theta u_i}}{\partial t} + U_j \frac{\partial \overline{\theta u_i}}{\partial x_j} = \\
-\overline{u_j u_i}\frac{\partial \Theta}{\partial x_j} - \overline{\theta u_j}\frac{\partial U_i}{\partial x_j} \quad &\text{(gradient production)} \\
-\frac{\partial \overline{\theta u_i u_j}}{\partial x_j} \quad &\text{(turbulent transport)} \\
-\frac{1}{\rho_0}\left(\overline{\theta \frac{\partial p}{\partial x_i}}\right) \quad &\text{(pressure covariance)} \\
+\frac{g}{T_0}\overline{\theta^2}\,\delta_{i3} \quad &\text{(buoyant production)} \\
-2\epsilon_{ijk}\Omega_j \overline{u_k \theta}. \quad &\text{(Coriolis)}
\end{aligned}
\tag{5}
$$

In deriving these equations we expressed the molecular term in each as the sum of diffusion and destruction components [19]. The diffusion term is negligible on the grounds of large Reynolds and Peclet numbers; the destruction term has been simplified through the concept of *local isotropy*, or isotropy of the finest-scale spatial structure of turbulence. We will return to this concept later.

In a quasi-steady, horizontally homogeneous surface layer and in micrometeorological coordinates and notation these yield the budget of Reynolds shear stress $\overline{uw}$:

$$
\begin{aligned}
0 = &-\overline{w^2}\frac{\partial U}{\partial z} \quad \text{(gradient production)} \\
&-\frac{\partial \overline{uw^2}}{\partial z} \quad \text{(turbulent transport)} \\
&-\frac{1}{\rho_0}\left(\overline{w\frac{\partial p}{\partial x}} + \overline{u\frac{\partial p}{\partial z}}\right) \quad \text{(pressure covariance)} \\
&+\frac{g}{T_0}\overline{\theta u}. \quad \text{(buoyant production)}
\end{aligned}
\tag{6}
$$

The budget of vertical temperature flux $\overline{w\theta}$ under these conditions reduces to

$$
\begin{aligned}
0 = &-\overline{w^2}\frac{\partial \Theta}{\partial z} \quad \text{(gradient production)} \\
&-\frac{\partial \overline{\theta w^2}}{\partial z} \quad \text{(turbulent transport)} \\
&-\frac{1}{\rho_0}\left(\overline{\theta\frac{\partial p}{\partial z}}\right) \quad \text{(pressure covariance)} \\
&+\frac{g}{T_0}\overline{\theta^2}, \quad \text{(buoyant production)}
\end{aligned}
\tag{7}
$$

the Coriolis term being negligible in the surface layer.

To my knowledge the first measurements of the shear-stress balance (6) and the temperature flux balance (7) were made in the 1968 Kansas experiment. These results were also surprising. Perhaps the biggest surprise was that despite the tremendous intermittency displayed by time series of $uw(t)$ and $w\theta(t)$ in the convective surface layer [4]—these time series showed that most of the fluxes were contained in large but infrequent "bursts"—the turbulent transport terms are small. This implies that the flux budgets are in a state of local balance: in each case the gradient production term is essentially balanced by the loss through the pressure term (Figure 3). One might not have expected this behavior in this convective flow, particularly since its TKE balance does have a large turbulent transport term. The state of local balance of the Reynolds fluxes is consistent with the observed existence of well-behaved eddy diffusivities for momentum and heat in the unstable surface layer.

That these flux budgets were balanced by pressure destruction was also a surprise, if a lesser one. It was once assumed by some workers that they were balanced by molecular destruction, as is the TKE budget, with pressure effects playing a secondary role. As we mentioned earlier, the molecular terms in the flux budgets represent two processes. One is molecular diffusion, which is negligible at the huge Reynolds and Peclet numbers of the

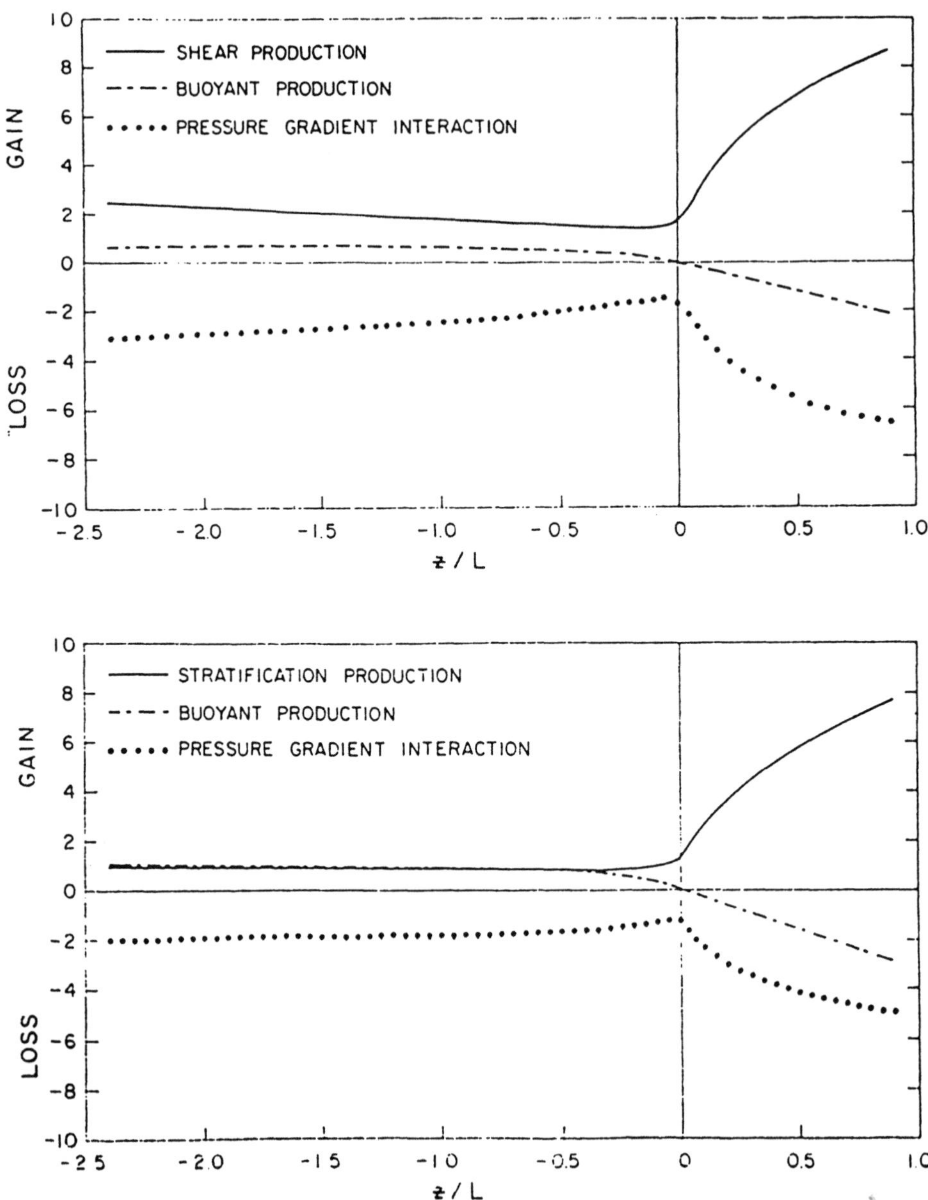

Figure 3. The budgets of $\overline{uw}$ (top) and $\overline{w\theta}$ (bottom) as measured in Kansas. Turbulent transport is negligible. From [23].

surface layer; the other is molecular destruction. In the case of the $\overline{uw}$ budget, for example, the molecular destruction term can be written

$$\text{rate of molecular destruction} = 2\nu \overline{\frac{\partial u}{\partial x_j}\frac{\partial w}{\partial x_j}} = 2\nu \int_0^\infty k^2 \Phi_{13} dk_1 dk_2 dk_3, \quad (8)$$

where Φ_{ij} is the spectral density tensor. Scaling arguments in [23] show that molecular destruction of $\overline{uw}$ can be significant only if the integral in (8) receives its main contributions from the highest wavenumbers—i.e., those past the inertial range. The one-dimensional stress cospectra measured in Kansas and elsewhere fall as $k^{-7/3}$ in the inertial subrange, however, so that the integrand in (8) peaks before the inertial range and molecular destruction cannot be significant. The local isotropy we assumed in writing (6) and (7) also implies that molecular destruction of shear stress is negligible.

These flux budgets reinforce an old message in turbulence: there is no simple, direct link between local behavior (e.g., the wildly variable momentum and heat flux signals from the convective surface layer) and ensemble-mean behavior (e.g., the locally balanced momentum and heat flux budgets there). Local behavior and ensemble-mean behavior are different things. If momentum and scalar constituents are carried across the entire flow by a single eddy event, as in the convective boundary layer, for example, it does not follow that the eddy diffusivity is not well behaved.

The Kansas results have also focused attention on the vitally important but physically elusive roles of pressure fluctuations in turbulent convection. Now that pressure fluctuations are being measured directly, the flux budgets are playing an important role in field testing of the pressure sensor (Oncley et al. [10]). The production terms in the budget of horizontal temperature flux are so well documented that if, as is believed, the budget is in essentially local balance, the magnitude of the pressure covariance term is known fairly accurately.

2.3. MONIN-OBUKHOV SIMILARITY: ITS SUCCESSES AND FAILURES

We approached the analysis of the Kansas data with the expectation that they would follow Monin-Obukhov (M-O) similarity. As we hand-plotted the results on graph paper we were elated. The mean wind shear and mean temperature gradient profiles were beautifully M-O similar; the vertical velocity and temperature variances behaved equally well (Figure 4). Scaling arguments for the asymptotic behavior of the M-O functions under very stable and very unstable conditions gave predictions that agreed well with the data. Overall, these successes were quite spectacular.

Virtually all of the Kansas results were presented in the M-O framework. In some cases the M-O plots had considerable scatter, but usually the trend of the M-O function was clear. In a few cases, however, there was large scatter and no evidence of an underlying, universal M-O function. These cases invariably involved statistics of horizontal velocity fluctuations.

Figure 5, for example, shows a M-O plot of Kansas data for the standard deviation of lateral velocity σ_v. The top panel shows data from three different runs, each having a different value of the M-O length L. We measured at heights of 5.66, 11.3, and 22.6 m so there are three data points for each run. The top plot makes it clear that in a given convective

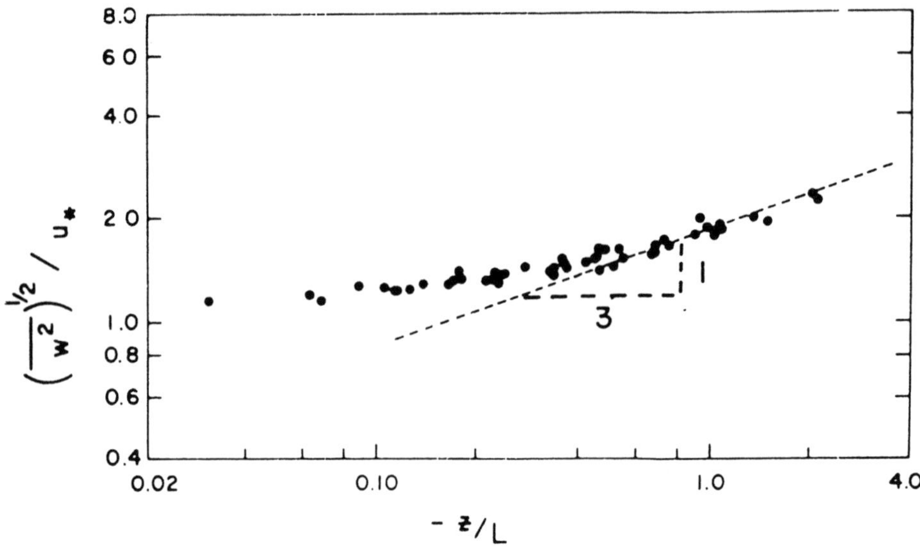

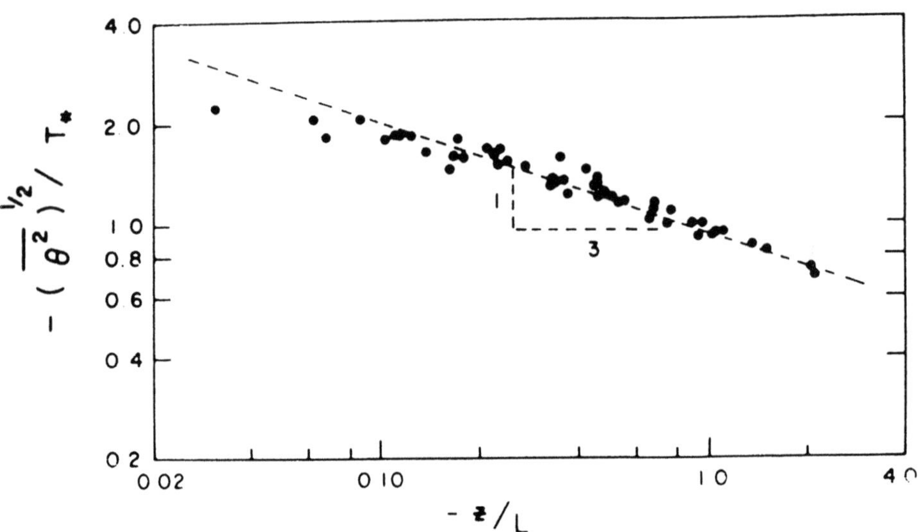

Figure 4. The spectacular M-O similarity of the fluctuations of vertical velocity (top) and temperature (bottom) in Kansas. From [23].

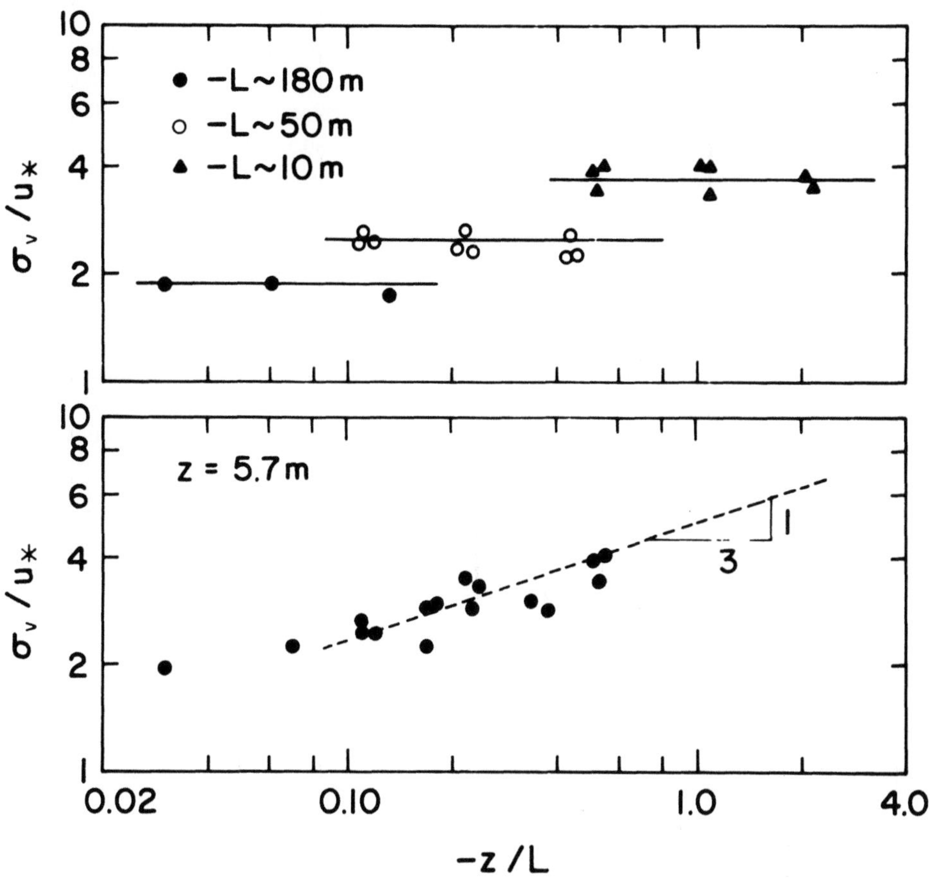

Figure 5. Two plots of σ_v, the standard deviation of lateral wind fluctuations. The bottom plot, from one height, may give the false appearance of M-O similarity, but the top plot, from three heights, shows it is not M-O similar. From [20].

surface layer σ_v does not vary with height z. The bottom panel shows that if we hold z fixed and plot data from different runs according to their z/L value, the result gives the false appearance of following M-O similarity.

There were early doubts about the applicability of the M-O hypothesis to horizontal velocity fluctuations. Lumley and Panofsky [8] wrote: "... it might be tempting to follow similarity theory and write $\sigma_v = Cu_* f(z/L)$ .... However, this formulation is plainly incorrect." It is not clear that much heed was paid to this warning, perhaps because the physical basis of the "incorrectness" of applying M-O scaling to horizontal velocity fluctuations was not clear. But after Deardorff [3] contended, on the basis of his first LES results for the convective boundary layer, that the velocity scale in the convective boundary layer is $w_* = (gQ_0 z_i/T_0)^{1/3}$, where Q_0 is the surface temperature flux and z_i is the boundary-layer depth, it was recognized that these large, convective eddies drive the large-scale horizontal wind fluctuations in the surface layer as well. Soon it became evident that this explained why M-O similarity does not apply to horizontal velocity statistics.

Peltier *et al.* [11] showed that even though the horizontal scale of horizontal velocity in the unstable surface layer is z_i, continuity requires a smaller length scale for vertical velocity. To sketch out their development, we start from the continuity constraint in its simplest form, zero velocity divergence:

$$\frac{\partial u}{\partial x} + \frac{\partial v}{\partial y} + \frac{\partial w}{\partial z} = 0. \tag{9}$$

From (9) we obtain the vertical velocity w as

$$w(z) = -\int_0^z \left(\frac{\partial u}{\partial x} + \frac{\partial v}{\partial y}\right) dz'. \tag{10}$$

If we use a Fourier-Stieltjes representation in the horizontal directions, e.g.,

$$u(\mathbf{x}, z) = \int e^{i\mathbf{k}\cdot\mathbf{x}} dU(\mathbf{k}, z), \;\; w(\mathbf{x}, z) = \int e^{i\mathbf{k}\cdot\mathbf{x}} dW(\mathbf{k}, z), \;\; \mathbf{k} = (k_x, k_y), \;\; \mathbf{x} = (x, y), \tag{11}$$

then Eq. (10) says the Fourier-Stieltjes coefficients are related by continuity:

$$dW(\mathbf{k}, z) = -\int_0^z \Big(ik_x dU(\mathbf{k}, z') + ik_y dV(\mathbf{k}, z')\Big) dz'. \tag{12}$$

Through Deardorff's finding that w_* is the important velocity scale in the convective boundary layer we expect the magnitude $k = |\mathbf{k}|$ of the energy-containing wavenumbers of the horizontal velocity to be of order z_i^{-1}, because the large convective eddies that fill the boundary layer extend right to the surface. At these small wavenumbers the dependence of the Fourier coefficients on z near the surface should be much like that of the mean velocity profile. In that case we can write the integral (12) over z as

$$dW(\mathbf{k}, z) \simeq -z\Big(ik_x dU(\mathbf{k}, z) + ik_y dV(\mathbf{k}, z)\Big), \quad kz_i \sim 1, \tag{13}$$

within a proportionality factor of order 1. Multiplying (13) by its complex conjugate and ensemble averaging then shows that ϕ_w, the 2-D spectrum of w, behaves at small wavenumbers as

$$\phi_w(k, z) \sim (kz)^2 \phi_h(k, z) \sim \left(\frac{z}{z_i}\right)^2 \phi_h(k, z) \ll \phi_h(k), \tag{14}$$

where ϕ_h is the two-dimensional spectrum of horizontal velocity. The result (14) implies that w fluctuations have negligible energy at horizontal scales of order z_i where the horizontal energy $\overline{u^2} \sim w_*^2$ resides.

Equation (14) indicates that the w fluctuations have their energy at horizontal wavenumbers of order z^{-1}, or at length scales of order z. If we then assume that the free-convection scales $g/T_0, Q_0$, and z determine the w-energy, then $\overline{w^2} \sim \left(\frac{g}{T_0}Q_0 z\right)^{2/3} \sim u_f^2$. Thus, the vertical and horizontal velocities in the convective surface layer scale quite differently:

$$u, v : \text{velocity scale } w_*, \text{ horizontal length scale } z_i;$$

$$w : \text{velocity scale } u_f \sim \left(\frac{g}{T_0}Q_0 z\right)^{1/3} \sim w_* \left(\frac{z}{z_i}\right)^{1/3}, \text{ horizontal length scale } z. \quad (15)$$

This suggests that w statistics, but not u and v statistics, ought to be M-O similar. This is broadly consistent with the evidence.

There has been speculation that because the horizontal velocity fluctuations are not M-O similar, the TKE budget in the surface layer is also not M-O similar. One argument goes as follows: the large, slow convective eddies establish transient, local wind profiles near the surface and thereby generate small-scale turbulence through the local rate of shear production. This local production rate depends on the local surface stress generated by these convective eddies, but that in turn depends on the convective velocity scale w_* and the surface roughness length z_0. Thus, the local rate of shear production (and, through the energy cascade, the local rate of viscous dissipation) involve more than the M-O governing parameters, so that the TKE budget ought not to be M-O similar. Given the lack of supporting experimental evidence, however, it seems that any such effects could be below the present level of detectability.

3. The Convective Mixed Layer and Its Peculiarities

The mixed layer, the middle 50–75% of the PBL, is not nearly as accessible to experimental study as the surface layer—not only because of the difficulty and expense of making measurements there, but also because of the very much poorer statistical convergence properties of turbulence data there. Much of what we know about the mixed layer was learned through large-eddy simulation (LES) and measurements in laboratory flows such as convection tanks and unstable boundary layers.

Among the strongest peculiarities of the convectively mixed layer are its dispersion characteristics. R. Lamb, an early worker in this area, summarized some of these characteristics in a review paper [6]:

> "... Deardorff and Willis ... studied diffusion phenomena from a simulated ground-level cross-wind line source in a water tank model of the convective mixed layer. An important result of this study was that beginning at a travel time of about $0.5h/w_*$, the centerline of a ground level plume begins to rise rather rapidly until by a travel time of the order of h/w_*, it has reached an elevation of about $0.75h$. This finding was so grossly out of line with the conventional Gaussian model of diffusion that speculation was raised, e.g., by Pasquill ..., that inasmuch as the ascent of the centerline of

the surface-source plume is a manifestation of free convection, it is inapplicable to atmospheric diffusion except under very low wind-speed conditions."

Lamb used scaling arguments to contend that these results ought to be applicable to the PBL under fairly typical convective conditions.

Laboratory convection tank diffusion results are widely used today to predict diffusion in the atmospheric boundary layer under convective conditions. Figure 6 shows some measurements made by Willis and Deardorff of mean concentrations downwind of elevated sources. These measurements have caused the Gaussian plume model, long the standard tool for routine predictions of short-range dispersion in the atmosphere, to be gradually replaced by improved models that incorporate the decidely non-Gaussian nature of dispersing plumes in the convective boundary layer. These and other implications of the convection-tank results are discussed by Weil [13].

The unusual properties of convective turbulence are also shown by the different ways it diffuses "top-down" and "bottom-up" scalars. These refer to conservative constituents undergoing horizontally homogeneous diffusion through the boundary layer with a zero flux at one boundary and a non-zero flux at the other. Examples include potential temperature and the mixing ratio of water vapor or nonreactive trace constituents. A top-down scalar (with the non-zero flux at the top) has a much smaller eddy diffusivity than a bottom-up scalar diffusing in the same convective boundary layer (Figure 7).

4. Summary and Implications

The convective boundary layer is truly an unusual turbulent flow:

- Its TKE budget in the surface layer has a large source from pressure transport, something not seen in shear flows, to my knowledge.
- The *local* behavior of stress and heat flux in the surface layer is wildly intermittent. However, their budgets are in a state of local balance, with negligible turbulent transport—despite the existence of boundary-layer spanning, highly energetic convective eddies.
- Monin-Obukhov similarity apparently works spectacularly for vertical velocity and fails spectacularly for horizontal velocity.
- The convective boundary layer as a whole has diffusion characteristics that are quite unlike those of other turbulent flows.

In short, convective turbulence is a different animal from shear-flow turbulence.

Convective boundary layer turbulence owes these unusual properties to the forcing by buoyancy. This forcing is one-sided: heating from below. Our convective boundary layer would be different in important ways if it were also cooled from above at the same rate.

The message here is clear. We cannot reliably infer the energy-containing properties or characteristics of a turbulent flow, even qualitatively, from those of another flow of different geometry and forcing. In particular, if one has need for knowledge of the energy-containing-range structure of convection, or for closures for a model of convection, then one needs to obtain this information from studies of convection.

Atmospheric boundary layer turbulence research in general has been limited by the

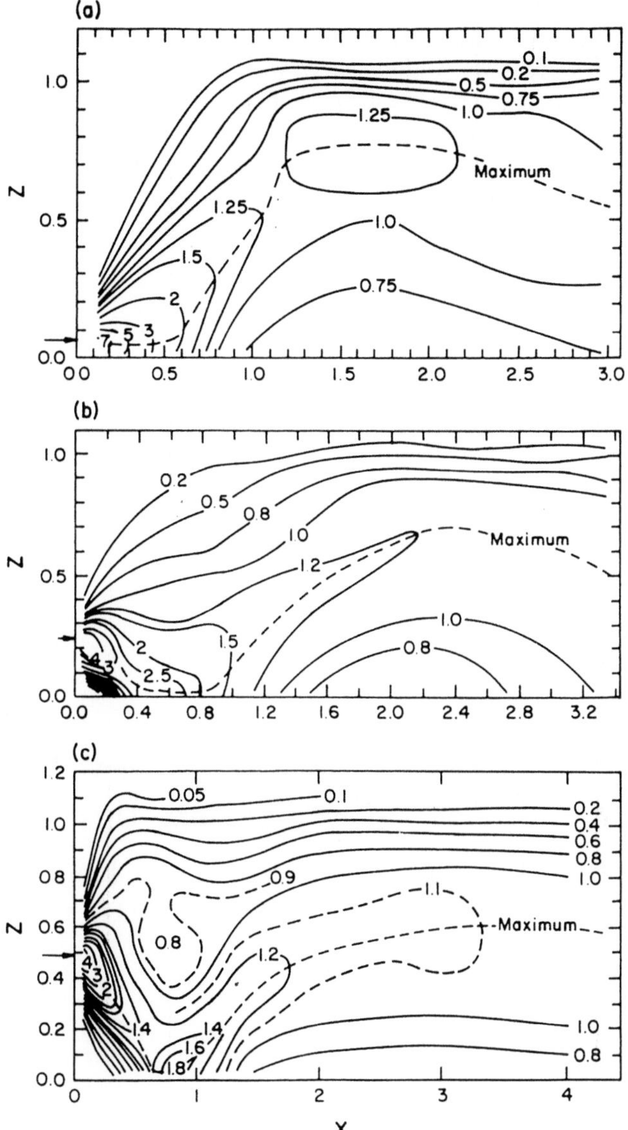

Figure 6. Laboratory convection tank results showing contours of nondimensional crosswind-integrated concentration as a function of dimensionless height Z and downwind distance X for sources at three release heights in a convective boundary layer. These results by Willis and Deardorff [15–17] show that the Gaussian-plume model is not a good representation of diffusion in the convective boundary layer. From Weil [13].

great difficulty in making the measurements we need. Because of the resulting lack of data, much of the atmospheric turbulence modeling of the past two or three decades has used closures originating in shear flows. Unfortunately, we often find those closures to be inadequate as the data needed to test and refine them do appear.

With the emergence of numerical simulation and laboratory models of the atmospheric boundary layer, our data deficiences are being somewhat alleviated. I will discuss these aspects in my second lecture.

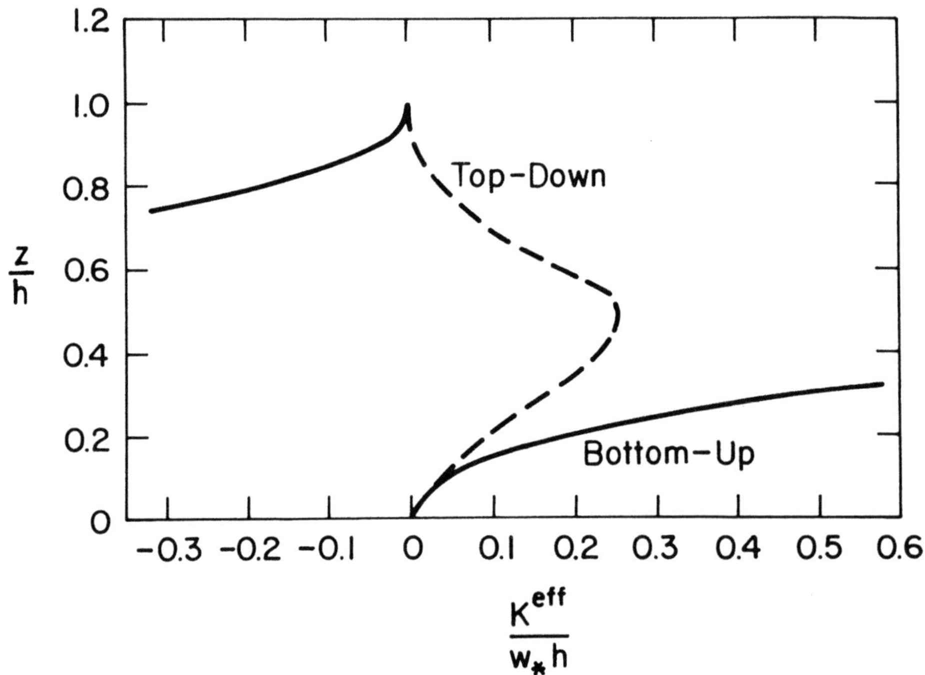

Figure 7. The eddy diffusivities for top-down and bottom-up scalars are quite different. From [24].

References

1. Chou, P. Y. (1945) Pressure flow of a turbulent fluid between two infinite parallel planes, *Quart. Appl. Math.* **3**, 198–209.
2. Corrsin, S. (1974) Limitations of gradient transport models in random walks and in turbulence, *Adv. Geophys.* **18A**, 25–60.
3. Deardorff, J. W. (1970) Convective velocity and temperature scales for the unstable planetary boundary layer and for Rayleigh convection, *J. Atmos. Sci.* **27**, 1211–1213.
4. Haugen, D. A., Kaimal, J. C., and Bradley, E. F. (1971) An experimental study of Reynolds stress and heat flux in the atmospheric surface layer. *Quart. J. Roy. Meteorol. Soc.*, **97**, 168–180.
5. Hogstrom, U. (1990) Analysis of turbulence structure in the surface layer with a modified similarity formulation for near-neutral conditions, *J. Atmos. Sci.* **47**, 1949–1972.
6. Lamb, R. G. (1982) Diffusion in the CBL, in F.T.M. Nieuwstadt and H. Van Dop (eds.), *Atmospheric Turbulence and Air Pollution Modelling*, Reidel, 159–229.
7. Liepmann, H. W. (1979) The rise and fall of ideas in turbulence, *Am. Scientist*, **67**, 221–228.
8. Lumley, J. L. and Panofsky, H. A.(1964) *The Structure of Atmospheric Turbulence*, Wiley.
9. McBean, G. A. and Elliott, J. A. (1975) The vertical transports of kinetic energy by turbulence and pressure in the boundary layer. *J. Atmos. Sci.*, **32**, 753–766.
10. Oncley, S. P., Horst, T., Praskovsky, A., and Wilczak, J. (1995) The TKE budget from the FLAT experiment, *11th Symp. on Boundary Layers and Turbulence*, Charlotte, NC. AMS, Boston, 5–8.
11. Peltier, L. J., Wyngaard, J. C., Khanna, S., and Brasseur, J. G. (1996) Spectra in the unstable surface layer, *J. Atmos. Sci.* **53**, 49–61.
12. Reynolds, O. (1895) On the dynamical theory of incompressible viscous fluids and the determination of the criterion, *Phil. Trans. Roy. Soc. London, Ser. A* **186**, 123–164.
13. Weil, J. C. (1988) Dispersion in the convective boundary layer, in A. Venkatram and J. Wyngaard (eds.), *Lectures on Air Pollution Modeling*, AMS, 167–227.
14. Wieringa, J. (1980) A reevaluation of the Kansas mast influence on measurements of stress and cup anemometer overspeeding, *Bound.-Layer Mereor.*, **18**, 411–430.
15. Willis, G. E. and Deardorff, J. W. (1976) A laboratory model of diffusion into the convective boundary layer, *Quart. J. Roy. Meteor. Soc.* **102**, 427–445.
16. Willis, G. E. and Deardorff, J. W. (1978) A laboratory study of dispersion from an elevated source within a modeled convective boundary layer, *Atmos. Environ.* **12**, 1305–1312.
17. Willis, G. E. and Deardorff, J. W. (1981) A laboratory study of dispersion from a source in the middle of the convectively mixed layer, *Atmos. Environ.* **15**, 109–117.
18. Wyngaard, J. C. (1981) The effects of probe-induced flow distortion on atmospheric turbulence measurements, *J. Appl. Meteor.*, **20**, 784–794.
19. Wyngaard, J. C. (1982) Boundary-layer modeling, in F.T.M. Nieuwstadt and H. Van Dop (eds.), *Atmospheric Turbulence and Air Pollution Modelling*, Reidel, 69–106.
20. Wyngaard, J. C. (1988) Structure of the PBL, in A. Venkatram and J. Wyngaard (eds.), *Lectures on Air Pollution Modeling*, AMS, 9–61.
21. Wyngaard, J. C. (1992) Atmospheric turbulence, *Ann. Review of Fluid Mech.* **24**, 205–233.
22. Wyngaard, J. C. and Coté, O. R. (1971) The budgets of turbulent kinetic energy and temperature variance in the atmospheric surface layer, *J. Atmos. Sci.* **28**, 190–201.
23. Wyngaard, J. C., Coté, O. R., and Izumi, Y. (1971) Local free convection, similarity, and the budgets of shear stress and heat flux, *J. Atmos. Sci.* **28**, 1171–1182.
24. Wyngaard, J. C. and Weil, J. C. (1991) Transport asymmetry in skewed turbulence, *Phys. Fluids* **A 3**, 155–162.

EDDY DYNAMICS AND KINEMATICS OF CONVECTIVE TURBULENCE

J. C. R. HUNT
Environmental Fluid Dynamics Program,
Arizona State University, Tempe, AZ, USA
Department of Applied Mathematics and Theoretical Physics
University of Cambridge, Silver Street, Cambridge, CB3 9EW, UK

Abstract

This review of 'convective' turbulence focusses on mechanisms that determine how eddies are formed, grow and interact with each other. It is shown how these are related to the interactions of the turbulence with the rigid surfaces and other regions of fluid motion bounding the convective layer. The effects of weak cross winds and non-uniformity in surface conditions are also discussed.

1. Introduction

Convective turbulence forms one of the subclasses of turbulent motion driven by random body forces [34]. In this case they are buoyancy forces arising from density fluctuations advected by the turbulent flow itself, so unlike some cases the forcing and the flow are strongly coupled. The density fluctuations are usually generated by temperature variation, though they can be caused by variation in mass concentration, as they are in geological flows [35] and in clouds [56]. These variations result from the initial variations in the density field or from those on the boundaries of the flow, particularly heating and cooling.

Convective turbulence occurs in a wide variety of natural flows; in the atmosphere, oceans, molten rocks, and surface/subterranean water. Other kinds of turbulence are likely to be present at the same time, particularly those caused by shear gradients in the large scale and mean velocity fields. Other dynamical influences arise from particles and variations in fluid composition such as evaporating cloud droplets, very hot particles in volcanic

E.J. Plate et al. (eds.), Buoyant Convection in Geophysical Flows, 41–82.

flows and the boundary effect of breaking waves in density interfaces. For the mechanics of the situation reviewed here, only convective turbulence is present at the large scale; and in these situations it plays an essential role in the overall behaviour of the fluid and its surroundings (or 'system'), for example in determining the thermal equilibrium and the rates of mass and heat transfer between the fluid and the surroundings.

In our consideration of the atmosphere and oceans, we make use of the simplifying fact that the large scale regions containing the turbulence are approximately 'flat' layers bounded on one side by a relatively rigid surface (the ground or the sea surface) and on the other by a thin 'inversion' layer that is relatively impermeable to the transfer of heat and matter is mobile on the time scale of the large scale of the convective turbulence.

Recent research on geological flows [35] has shown how the formation of rocks arose from fluid processes where convective turbulence occurred in separate horizontal layers above and below a density interface – another kind of inversion layer. In some kinds of geological convection, as in smoke fires within buildings, most of the heat flux may be introduced at one level surface while the upper surface is rigid and not thermally conducting. However as the whole volume of fluid heats up, with the regions near both surfaces heating up fastest (for reasons given in section 4), the surfaces may dissolve or disintegrate leading to fluxes of matter into the convection that may diminish or strengthen the convection depending on whether they are positively or negatively buoyant [36].

In these as in most naturally occurring convective flows the convection is not reflexionally symmetric about horizontal planes because more heat flux or other source of buoyancy flux is provided at one bounding surface than the other. This generally leads to strong asymmetry but a simplification in the structure of the turbulence.

Despite this 'imperfection' in symmetry we can generally assume (cf. [2]) that the bounding surfaces are approximately flat and that the heat and buoyancy fluxes at the surface, or sources of buoyancy generation in the interior, are approximately uniform.

Criteria are discussed in section 5 to show when small departures from flatness of the surface and from uniformity of the surface conditions only have small effects on the statistics of of the turbulence [41]. This lack of sensitivity does not appear to extend to the forms of the large eddies, which can change quite markedly without appreciably changing many of the macroscopic features of the flow [47]. Stably stratified turbulent flows, including boundary layers are very much more sensitive to surface changes [9].

The main reason for this relative insensitivity to surface inhomogeneity is that in turbulent convection the components of velocity with the longest correlation distances are the *vertical* rather than the *horizontal*, as in shear flow turbulence. An important consequence is that turbulent convection has a broadly similar structure even in volumes which are bounded laterally by resistive surfaces or by regions of strong inhomogeneity. So the concepts, especially the statistical results, of this review of horizontally unbounded flows can be approximately applied to more complex situations such as tall cumulus clouds whose strong turbulent convection is very inhomogeneous.

In turbulent shear flows recent research [39] has shown that the turbulence is insensitive to boundaries even when they are separated only a few layer depths apart.

As in most turbulent flows the structure of convective turbulence is strongly affected by its initiation and by the boundary conditions. No tendency is observed towards an asymptotic universal state (as suggested by Landau and Lifshitz [49]). For a review of such turbulence questions see [73]. However different types of convective flow do have features in common both in terms of the eddy structure and the statistics of the temperature and velocity fields. In different zones of convective flow there are also similarities with other types of turbulent flow, notably wall effects and wave-turbulence interactions at density interfaces.

This review focusses on the two main types of eddy structure and how they relate to the statistical properties of the turbulence and to the surface conditions, near rigid surfaces and inversion layers respectively. The thermal field below rigid bottom surfaces significantly affects the turbulent eddy structure and thence the spatial/temporal variation of temperatures in the fluid, a point that seems to have been overlooked in most previous discussions of this subject. The generation of convection by internally produced buoyancy forces are not discussed (e.g. the phase changes in clouds or radiation near cloud boundaries which certainly produce some special effects [26]).

The initiation of unstable density gradients which cause turbulent convection can occur in many ways and on many scales; even in stably stratified flows when oscillations grow to large amplitudes. The eddy motions resulting from such events may be quite similar to those we focus on here where continuous convection is produced by steady input of energy and buoyancy flux [86].

In many geophysical flows this occurs in spatially varying flows, such as when cool air is advected over a warm sea surface and the vertical extent of the convection may be largely determined by the time (or distance) taken for fluid to enter the convective region. This upward 'encroachment' of a convective region is similar to the spreading of a local region of turbulence in non-buoyant flows [82] but is not reviewed here.

This review is primarily addressed to those who need to use the ideas and the calculations of the specialist for various practical and scientific purposes but for whom the literature is generally too technical. It may also encourage more specialists in this field to observe birds circling in convective eddies and bubbles in the cooking pot and generally sharpen the scientific questions about convection. This should help in the search for more accurate or efficient numerical models. The results of new developments in research methods are made full use of in this review, notably the uses of numerical simulations on the one hand [68] and remote sensing on the other [25]. But these do not displace the value of many earlier studies (including those of bird flight) which can now be better explained with our new understanding.

2. General Aspects of Convective Eddies

2.1. INITIALISATION OF CONVECTIVE TURBULENCE

Turbulence is driven by natural convection when the area average of the density at each level $z, \overline{\rho}(z)$, decreases in the direction parallel to the local gravitational (or other equivalent) acceleration g (assumed to be vertical hereafter) in a volume of fluid of sufficiently large scale that molecular affects are negligible small. This means that the transport of vorticity by molecular viscosity ($\rho\nu$) and of heat (and other variables that affect the density) by molecular thermal conductivity ($\rho\kappa$) are small compared to their transport by the velocity field. As in other potentially unstable flows a non-dimensional number can be defined whose magnitude indicates the presence of unstable motion and ultimately turbulence, and which can be interpreted as a ratio of forces that amplify and damp-out eddying motion. Between level surfaces a distance h apart which are heated and cooled, uniformly with initially constant temperature gradient $\partial T/\partial z$, and density gradient $\rho_0\beta\partial T/\partial z$, where ρ_0 is the density at a reference level ($z = 0$), and β is the expansion coefficient, convective motions are initiated when the non-dimensional Rayleigh number, Ra, that characteristics unstable buoyancy forces is large enough. To apply this analysis to compressible fluid, T is replaced by θ, the potential temperature, i.e. the temperature of a fluid element displaced adiabatically to a reference level (e.g. the ground for the atmosphere). At the reference level by definition $T = \theta$ [24].

For this flow system a typical vertical buoyancy force is $g(\beta\partial T/\partial z)(hl^3)$ acting on a small volume of scale l moving upwards a distance of order l, in a characteristic time period τ, Ra is the ratio of this force to the viscous drag $\rho\nu l(l/\tau)$ where the τ is the critical time for the temperature perturbation in the volume to be diffused by thermal conductivity. Since $\tau \sim l^2/\kappa$, the ratio is independent of τ and is given by

$$\mathrm{Ra}_l = \frac{g\beta\partial T/\partial z.l^4}{\rho\nu.\kappa}$$

Normally Ra is defined in terms of the fixed scale h, i.e.

$$\mathrm{Ra} = \frac{g\beta \partial T/\partial z . h^4}{\rho \nu \kappa}$$

Since the most unstable motions have length scales that are only about a (1/5) of the distance h between the plates, if $\mathrm{Ra}_l \sim 1$ at its critical value for the initiation of convection $\mathrm{Ra} = \mathrm{Ra}^* \simeq 10^3$. Understanding transition is helpful when considering the more complex processes affecting eddies in strong turbulent convection.

At the onset of natural convection many types of flow pattern are observed(e.g. [12], [48]). However, the initial patterns of motion usually span the vertical space of the region (we shall exclude the effects of strong magnetic fields acting on conducting fluids and strong rotation) and have regular spatio-temporal patterns that persist, even though its initial phase or location can not usually be predicted in advance. Thus, as in other forms of 'slow' transition to turbulence, for values of Ra just above Ra*, natural convection is not generally ergodic. As $\Delta\mathrm{Ra} = (\mathrm{Ra} - \mathrm{Ra}^*)$ increases, the strength of the convection increases and new patterns emerge with their own spatial-temporal time scales; at some value of ΔRa the flow in any given realisation becomes unpredictable, chaotic and generally ergodic [67]. The spectra and probability distribution progressively tend to the smooth forms found in the random multi-scale motions characteristic of all fully turbulent flows. Although these motions are chaotic, they are not independent of influences specific to the particular flow; in open flow without boundaries they are correlated, statistically, with motion at earlier times or where there are boundaries the statistics of large scale motions are affected by their presence. For these reasons turbulent flows do not have a universal structure, although they all have certain common features especially for small scale motions. Turbulent flows can be classified into types corresponding to the basic instabilities that drive them, such as shear flows or rotating flows. This leads to common features in the structure and certain statistical properties of the large eddies, e.g. [33].

2.2. DEFINING CHARACTERISTIC EDDIES

Even in the restricted class of turbulent flows in wide horizontal layers driven by heating and/or cooling at the bottom and top level boundaries ($z = 0, h$ respectively) (with no other forcing, whether by shear, rotation or body forces), the forms of the large scale turbulence motions can differ significantly. The differences largely arise because of the coupling between the convection and the dynamic and thermodynamic conditions of these boundaries, which may be fluid or solid, and in either case may act as rigid or flexible surfaces. Depending on the interactions at these surfaces they may act as thermal conducting or non conducting boundaries, or some combination of both. For natural convection at very high values of Ra it is a useful hypothesis to regard the main eddy motions as being made up of two distinct characteristic types (Fig. 1).

The most energetic eddies are significant firstly because they contain most of the kinetic potential energy. Secondly because they occupy a sizable proportion of the space (but usually less than 50%). Thirdly they also dominate many of the statistical measures because of their temporal and/or spatial persistence (or correlation), whether defined in a fixed frame of reference or one moving with the eddies [31].

The first of these two types are puff eddies and generally move unsteadily upwards or downwards, while the other type are elongated plume-eddies that originate at either or both of the upper and lower boundaries, depending on the heat transfer and the dynamics as will be discussed later.

In both types (denoted by e, or where the distinction is significant by pfe and ple) the vortical motions ($w^{(e)}$) and the density perturbations $\Delta\rho^{(e)}$ are located within the eddies' bounding surfaces S, with a typical diameter across the eddy of l. Thus from the definition of vorticity the typical variation in velocity across the eddy is $W^{(e)} \sim lw^{(e)}$. This surface is open. It moves with a velocity $\mathbf{u}^{(S)}$; its unit **outward** normal vector is $\mathbf{n}^{(S)}$ so that the external flow (denoted by X) whose velocity is $\mathbf{u}^{(X)}$ may have a normal component of velocity $\mathbf{u}^{(X)}.\mathbf{n}$. This leads to a net flow across the surface with a normal relative velocity $\mathbf{u}^{(n)} = (\mathbf{u}^{(X)} - \mathbf{u}^{(S)}).\mathbf{n}^{(S)}$. This is related to the 'entrainment' and 'detrainment' velocities, which are defined later. Note that for the randomly moving or puff eddies the internal flow has a significant component normal to the surface i.e. $\mathbf{u}^{(S)}.\mathbf{n}^{(S)} \neq 0$. Whereas for the elongated plumes the internal motion $\mathbf{W}^{(e)}$ is largely parallel to (S) and there is a smaller entrainment velocity $\mathbf{u}^{(en)} = \mathbf{u}^{(n)}$ normal to (S). When there is a weak external cross wind $(\overline{\mathbf{u}})$ so that $\mathbf{u}^{(X)} = \overline{\mathbf{u}}$, in strong convection plumes move with a velocity $\mathbf{u}^{(s)}$, which approximately equals $\overline{\mathbf{u}}$. Then the normal velocity $\mathbf{u}^{(en)}$ is defined relative to moving plumes (see section 5.2).

In convective turbulence the dominant eddies generate their own kinetic energy driven by the horizontal pressure differences caused by horizontal variations in the hydrostatic pressure differences over the scale which are of the order of $g\Delta\rho^{(e)}l$. The vorticity equation shows how the horizontal component of vorticity grows (i.e. $\frac{d}{dt}|\mathbf{w} \wedge \hat{\mathbf{z}}| > 0$) as a result of the term $(\nabla\rho \wedge \mathbf{g})$ which is of the order of $(\frac{\Delta\rho}{\rho_0}^{(e)} g/l)$. This effect tends to increase the vertical movement and the horizontal acceleration.

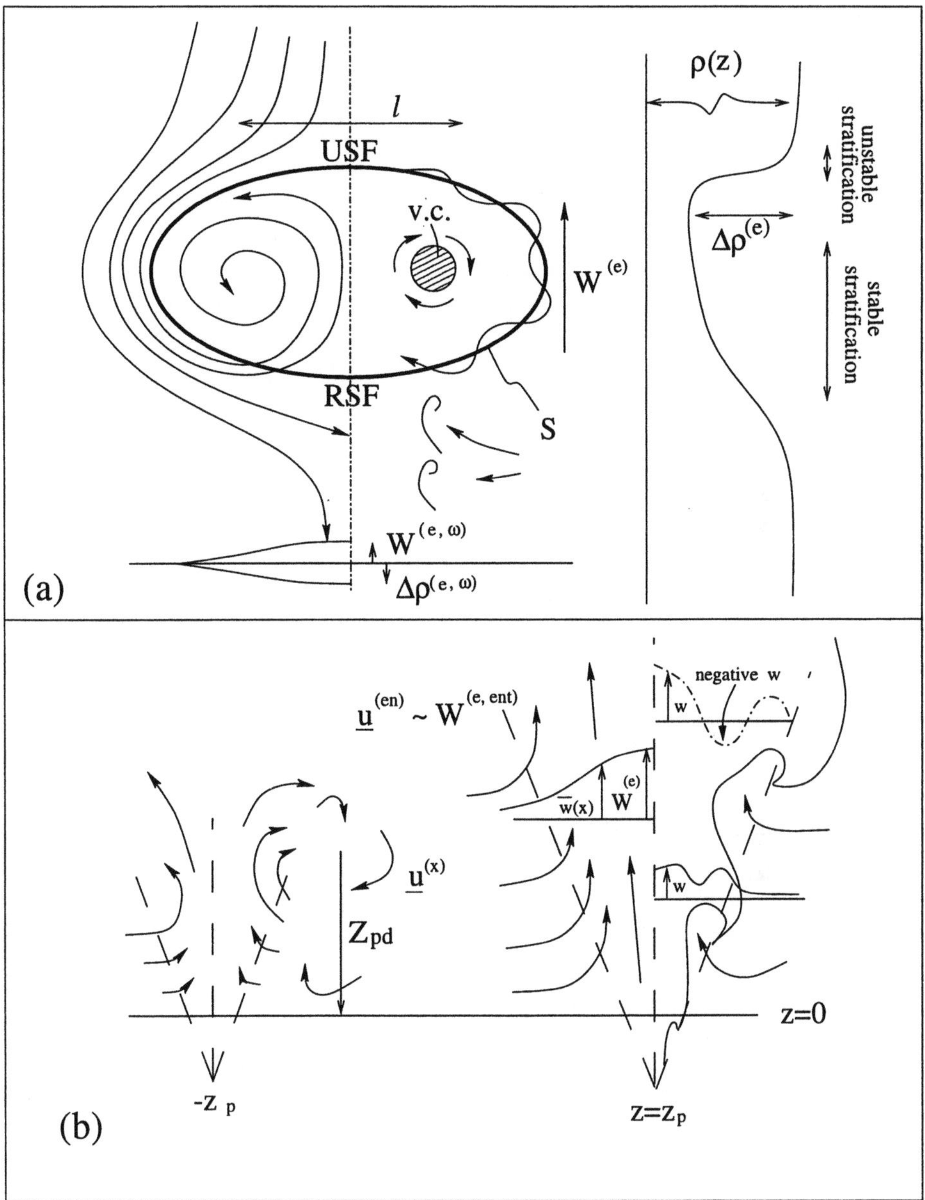

Figure 1. Typical large scale eddies in convective turbulence. (a) Puff eddies (*pfe*) showing either side of the centre line, the streamlines relative to the eddy and small scale vertical and turbulent motion and also a vertical profile of average density on the centre line. A coherent vortex core (v.c.) can exist. (b) Plume eddies (*ple*) showing either side of the centre line the mean and fluctuating outline and velocity profiles. The profile of $w(x)$ — · — ·—, with larger, even negative fluctuations, occurs with phase changes in clouds. A smaller plume is shown that disintegrates under the influence of the external turbulent/straining velocity field $\mathbf{u}^{(x)}$ (after [15]).

2.3. PUFF-EDDIES (pfe)

The actual velocity field $\mathbf{W}^{(e)}$ within these regions is also determined by their geometric form. In the puff eddies where the horizontal and vertical dimensions are both of the same order l, so that the net buoyancy force $(-\Delta\rho^{(e)}g/\mathcal{V})$ is balanced by the small rate of increase in the eddy momentum $(\rho W^{(e)}\mathcal{V})$ and the change in the pressure force and momentum flux $(\sim \rho W^{(e)2}l^2)$. The latter is often referred to loosely as a drag force. For an open surface that is changing with time has to be defined carefully.

Below and above the upper stagnation region of S fluid from the interior of the eddy and the more dense ambient fluid are advected first into the upper stagnation flow (USF) and then around (S) towards the rear stagnation flow (RSF). Some mixing between these two streams takes place in RSF, where small scale turbulence in the external flow and internal flows $u^{(E)}, w^{(E)}$ are amplified by the straining motion, e.g. [82]. But the main source of local turbulence and mixing are the density and velocity fluctuations growing in the unstable density gradient on the upper surface. The reason why the top of the thermal (or a gas bubble) is relatively smooth is that the the flow is accelerating and suppressing the growth of fluctuations [42]. The time for the reduced disturbances to grow to a significant amplitude comparable with l is $(g/\Delta\rho^{(e)}/l)^{1/2}$ which is of the same order as the time for them to be advected to the rear of the eddy [4].

The overall effects of these fluctuations is to diffuse the vorticity outwards, at the same time as some external flow being entrained inwards with an average speed $W^{(e,en)}$ (relative to the frame of reference moving with the eddy), leading to a growth of the volume $\mathcal{V}^{(e)}$ [84], and to mixing of heat and density between the external fluid. The turbulence and the divergence of streamlines in the region (RSF) leads to vorticity being transported across mean streamlines and shed downstream in the eddy's wake e.g. [16]. The wake diameter is smaller than that of a solid body because the shedding of vorticity is inhibited by the inward entrainment and is only about 30% of l. As with other momentum wakes, the typical velocity defect $W^{(e,w)}$ is less than 30% of $W^{(e)}$.

Note that in RSF vertical mixing is inhibited by the stable density gradient (because $\partial\rho/\partial z < 0$) and therefore the heat (and buoyancy) concentrated in the central of the eddy is detrained into the wake at a lower rate than the vorticity, whose eddy diffusivity in stable stratification in general greater than that of a scalar, e.g. [82].

A quantitative approximate model for the dynamics of these eddies begins by considering the rate of change of the momentum inside and outside the eddy defined in terms of an equivalent impulse $I^{(e)}$. This is used in place of the mathematically improper momentum integral [3]. In this case $I^{(e)}$ is defined to allow for the growth of the volume and entrainment into it. This must balance the integral (over the volume $\mathcal{V}^{(e)}$ of the eddy) of the buoyancy force $B^{(e)} = \int g\Delta\rho^{(e)} dv$, and the net downstream flow (or drag) force $D^{(e)} = \rho W^{(e)} W^{(e,w)} l_w^2 = \lambda_D W^{(e)2} l^2$, where the wake drag coefficient λ_D is about 0.1. It follows that

$$\frac{dI^{(e)}}{dt} = B^{(e)} - D^{(e)}. \tag{1}$$

The growth of the volume $\mathcal{V}(\propto l^3)$ by lateral entrainment is modelled by the equation

$$(d/dt)l^3 = W^{(e,\mathrm{ent})} l^2 = \lambda_{\mathrm{ent}} W^{(e)} l^2, \tag{2}$$

where the coefficient $\lambda_{\mathrm{ent}} \cong 0.1$. The relatively slow detrainment of buoyancy implies that

$$dB^{(e)}/dt = d/dt(l^3 \Delta\rho^{(e)}) \simeq 0. \tag{3}$$

Combining these three equations, (1) to (3), shows that the thermal's width l increases as a result of entrainment in proportion to $t^{1/2}$ while the velocity $W^{(e)}$ decreases as $t^{1/2}$. There is a sub-class of more coherent puffs where the vortex lines form vortex rings with the circulation of the $\oint$ proportional to $W^{(e)} l$ that is approximately constant. Since l increases, $w^{(e)} \sim W^{(e)}/l$ decreases, this internal destruction of vorticity by mixing of fluid element having vorticity of the opposite sign just balances its rate of generation by buoyancy forces. Much of the work of these forces is generating small motions and viscous dissipation (their heat generation in the atmosphere has negligible effect on the buoyancy for vertical motions that are small compared to the depth of the atmosphere). Only about 1/3 of the work is converted to the kinetic energy of the mean motion [84], whereas in stably stratified shear flows (where buoyancy forces reduce the kinetic energy) this factor is even lower; so puffs are relatively more efficient because of a less distorted structure of small scale turbulence.

2.4. ELONGATED PLUME EDDIES (*ple*)

The second type of eddy tends to be nearly vertical and to extend upwards and/or downwards from the lower and upper boundaries respectively. External flow is entrained into them both close to the boundaries where they are formed and for a considerable distance above (and/or below). Naturally the entrainment ceases if they extend as far as the other boundary or if they disintegrate at some intermediate level (Fig. 1). At that level $z = z_{pd}$, the buoyant fluid in the plume detrains across the eddy surface and disperses into the surrounding flow. See Fig. 1.

These plume eddies tend to be formed near boundaries from the natural coalescence of puffs or smaller plumes. Also entrainment into the plume helps enforce this coalescence [72]. This steady train of puffs and smaller plumes produces a mean upward (or downward) velocity with some fluctuations (but not usually enough to cause reverse flow – at least in dry convection). Once they are formed these internal fluctuations are sustained by the large scale eddies driven by shear at their outer surface. They significantly affect the flow and mixing at the boundaries when the plumes impact on the opposite surface [70].

Some of these plumes are quite small with diameters $l \sim \frac{h}{100}$ while for the large plumes, l reaches as large as $h/4$ in the interior. Some of the smaller plumes may detrain at a low level $z_{pd} \sim \frac{h}{5}$, but most are entrained into the larger plumes. These dominant plumes are observed to be quasi steady because their time scale τ_p is equivalent to the time it takes for a parcel of fluid to recirculate several (about 4) times around the main convection cells (up in one plume and down in another or on the down drafts between the plumes), i.e. $\tau_p \sim 4h/W^{(e)}$.

The flows in these large plumes have some similarities (and some significant differences) with those formed above isolated steady heat sources, whose quasi steady analysis we now describe, following [64],[84]. The vertical flow is driven by the buoyancy force $g\Delta\rho^{(e)}l^2/\rho_0$ per unit height, leading to an increase in momentum flux per unit height of $\rho {W^{(e)}}^2 l^2$ (for upward plumes). The difference between the upward velocity in the plume and the surroundings generate Kelvin-Helmholz shear instabilities and turbulent eddies on the scale of l. These lead to a linear growth of l with a height (z) (or with depth for downward plumes). Since the square root of $W^{(e)2}l^2$ increases, it follows that $W^{(e)}l^2$ and therefore the net volume flux along the plume increase with z. This requires that there is a net entrainment flux into the plume of $q^{(\mathrm{ent})} \sim W^{(e)}l$ per unit height, which increases as a result of buoyancy forces. Thus an external radial 'entrainment' velocity $W^{(e,\mathrm{ent})} \sim q^{(\mathrm{ent}}/l$, is proportional to $W^{(e)}$. Since there is no net flow leaving these plumes until they reach their detraining level, along their length the heat and buoyancy fluxes are constant (i.e. $Q_B^{(e)} = \frac{g\Delta\rho^{(e)}}{\rho}W^{(e)}l^2$).

Thence model equations for such a plume (in the central region of a convective zone where there is no temperature gradient outside the plume [64]). Firstly the change in the momentum force can be stated

$$M = \rho(W^{(e)2} + \sigma_w^{(e)2} + \Delta p^{(e)}/\rho)l^2 \quad \text{and} \quad \partial M/\partial z = g\Delta\rho l^2. \tag{4}$$

Here $\sigma_w^{(e)2}$ is the variance in vertical velocity caused by turbulence and $\Delta p^{(e)}$ is the pressure difference between the plume and the external region which to the first approximation does not vary with z. Secondly the continuity equation yields

$$\frac{\partial}{\partial z}(W^{(e)}l^2) = l^2 W^{(e,\mathrm{ent})} = \alpha l^2 W^{(e)}, \tag{5}$$

where α is the entrainment coefficient. The buoyancy flux

$$Q_B^{(e)} = \frac{g\Delta\rho}{\rho_0} W^{(e)} l^2 = \text{const.} \tag{6}$$

Also if the internal turbulence in a plume is driven by the shear at its outer surface, it is proportional to the average velocity in the plume i.e. $(\sigma_w^{(e)})^2 = \lambda_\sigma^2 W^{(e)2}$. Given this assumption is invalid, M in (4) can be expressed as $(1+\lambda_\sigma^2)\rho W^{(e)2} l^2$. Variations in λ_σ can significantly affect $q^{(\mathrm{ent})}$ [7].

Thus for an isolated upward plume

$$l_H \simeq \alpha z \quad \text{and} \quad W^{(e)} \simeq [Q_B^{(e)}/f]^{1/3}. \tag{7}$$

The results in (8) imply that $q^{(\mathrm{ent})}$ is proportional to $f^{2/3}$ and therefore that the entrainment rate into the plumes increases along their length, unlike jets where it is constant (or decreases for round and planar geometries respectively). Where the dominant large eddies in convective turbulence can be approximated by flows similar to these plumes, it is found that they are the main means of transporting the average **heat** flux F_θ, between the level boundaries separated by a height h. Therefore the characteristic vertical velocity in the large eddies w_* can be approximated by the plume velocity i.e.

$$w_* \sim W^{(e)}(z \sim h) \simeq [F_\theta g\beta h]^{1/3}, \tag{8}$$

where β is the expansion coefficient [18]. Whether plumes are the dominant forms of eddy and the determination of F_θ depends on the conditions at the boundaries considered in sections 3 and 4.

2.5. INTERACTIONS OF CHARACTERISTIC EDDIES

We have already noted that because the large eddies in their active and growing phase are entraining external fluid, these eddies tend to merge when they are near to each other, as demonstrated in studies of merging turbulent plumes (e.g. [72]). Pairs of 'puffs', as with other three dimensional vortical eddies, can also merge [72]. They are most likely just to move around each other but in some cases eddies of equal strength interact strongly so that their velocity fields strain each other's vortical field destructively leaving only a disorganised tangle of vortex filaments (e.g. [44]). As with the behaviour of all large eddy structures in turbulence it tends to be quite dependent on how they are formed, which in this case is usually near the boundaries.

There are limitations to the growth of buoyant turbulent plumes through merging because the boundaries ensure equal upward and downward flow.

Frictional processes there prevent the upward and downward zones of motion from being separated by a distance L_c much greater than the depth h (see section 3). In both experiments and numerical simulations of convective turbulence, the horizontal scale for convection cells reaches a limiting value L_c. In the atmospheric boundary layer, where the release of latent heat in a moist atmosphere also affects convection, 'super-cells' occur where convective plumes (which usually include up and down drafts **within** them) can be concentrated into regions with horizontal scales of more than 100 km, which is about 10 times the height of the troposphere. Here rotational effects are probably important and the entrainment into the plumes are less significant for interaction than the mean horizontal flows set up by the convection [58].

In the external flow region (X) outside the large characteristic eddies, flow visualisation studies show how smaller scale, mainly incoherent eddying exists whose amplitude is denoted by $u^{(X)}$ and integral scale $L^{(X)}$. This turbulence is generated in the wakes of rising/falling puff eddies, or near elongated plumes when they move or interact. Numerical and radar pictures show plumes sometimes at 30deg or from the vertical e.g. [70] (see also section 5). Puffs and plumes in this region may disintegrate at critical levels $(z_p^{(e)})$ where their buoyant motions $W^{(e)}$ are less than those of surrounding eddying motions $u^{(X)}$ (e.g. [15]).

This turbulence in (X)may also be advected from boundary regions and bring with it significant density perturbations. Their effect on the local turbulence in (X) can be estimated by considering the ratio of buoyancy forces across an eddy to its momentum drag which is an eddy Richardson number $\mathrm{Ri}^{(e,X)} \simeq g\frac{\Delta\rho^{(X)}}{\rho}L^{(X)}/(u^{(X)})^2$. Typically $\mathrm{Ri}^{(e,X)} \lesssim 1$ where $L^{(X)}$ is of the order of h_s, typically the thickness of the layer which released the eddies. Studies of such turbulence density fluctuations in an unconfined region shows that this potential energy is gradually converted into kinetic energy so that $\mathrm{Ri}^{(e,X)}$ is of order unity [5].

In the numerical simulation of [5], the Reynolds number was probably too low, only about 200, for the eddy motions to form into strong thermal eddies. Perhaps therefore the mixing of the eddies buoyancy with the surroundings was too rapid. This occured on a time scale of $L^{(X)}/u^{(X)}$ which is certainly less than the time on which the coherent vertical motions of thermals and puffs decay.

2.6. STATISTICAL ANALYSIS OF EDDY STRUCTURE

The description of the characteristic large scale eddy structure for these and other turbulent flows is based on the results of field, laboratory and numerical experiments and flow visualisations. In the earlier studies data analysed simply in terms of the magnitudes and direction of the instantaneous velocity field one in some cases using tracers and Lagrangian tracking (e.g. [64],[84]). This method worked well for identifying and quantifying the structure of elongated plumes by mapping regions of coherent upward and downward vertical velocity (e.g. [51]). Data from the former plume studies could be used to explain and quantify main features of the probability distributions of vertical velocity and the statistics of the component, such as the skewness and the mode-quantities of importance for air pollution dispersion [31]. This data explains some of the dynamics we shall refer to in sections 3 and 4 [70].

But to study the *time-dependence* of the large plume-like structures more elaborate techniques are needed, firstly to define key features of the eddy structure as they change and secondly to describe quantitatively and conceptually these changes. The features are usually 'modes' such as those obtained by proper orthogonal decomposition [28]. Here each component of the random velocity field $u_i(\mathbf{x}, t)$ is represented in terms of a series of orthogonal functions $\phi_{n_i}(\mathbf{x})$ which are determined empirically from the measured (or computed) cross correlation $R_{ij}(\mathbf{x}, \mathbf{x}')$ at pairs of points $\mathbf{x}, \mathbf{x}'$. The statistical analysis of the data provides the variances $\overline{a_n^2}$ of the random amplitudes $a_n(t)$, so that

$$u_i(\mathbf{x}, t) = \sum a_{n_i}(t)\phi_{n_i}(\mathbf{x}). \tag{9}$$

When most of the energy is in the leading term (i.e. $\overline{a_1^2} \gg \overline{a_n^2}, n \geq 2$) the leading mode $\phi_n(\mathbf{x})$ is usually found to be very similar to the form of the eddy structure derived by more direct means such as flow visualisation or conditional sampling (as in the updraft/downdraft studies). But when several modes have similar levels of energy, e.g. $\overline{a_1^2} \sim \overline{a_2^2} \sim a_3^2$, the method can be difficult to interpret and may even be misleading in describing typical eddy structures [8]. The main value of this approach is that it leads to coupled equations for a_1, a_2, a_3 etc. whose solutions demonstrate the time dependent behaviour of the modes and of the eddy structure.

This 'output' is best analysed using the techniques of dynamical systems especially for low Rayleigh number convection, where plotting the 'trajectory' on phase planes demonstrates qualitative and statistical behaviour.

So far this approach has only been applied to numerical simulation of convection where there is sufficient data to compute $R_{ij}(\mathbf{x}, \mathbf{x}')$ and thence $\phi_n(\mathbf{x}_1), a_n(t)$. Therefore it has been restricted to moderate values of Ra $< 10^4$ [73]. These methods have not yet been applied to eddies that move randomly through the flow such as puff eddies. For these eddy structures the 'best' method currently available is to map the instantaneous velocity field and identify the regions of coherent velocity. These are the (pfe) eddies.

The objectivity of the method is to define these vortical eddies in terms of the magnitude of the invariants of the strain tensor. The local pressure can also be factored in. The method has so far only been applied to moving eddies in shear flow turbulence (e.g. [23], [38]), but could well be applied to convective turbulence.

To analyse the small scale turbulence and how it is advected and interacts with larger scales, most studies have so far been based on second and third moment spectra. This method provided a measure of the small scale 'eddies' (or at least the small scale Fourier components, see [79]) and demonstrated that these have the same Kolmogorov energy distribution as for other kinds of turbulence. Atmospheric observations provided this information over many scales of motion from 100 m to 0.1 m. However, measurement at a fixed location (in a crossflow) show that the spectra of the *third* moments $\Phi_{w,w^2}(\kappa_1) \propto \epsilon\kappa_1^{-2}$, where ϵ is the mean energy dissipation rate per unit mass. This is calculated from the vertical velocity w at one time t and its square w^2 at another time $t + \Delta t$ in a cross wind. The resulting form does not correspond to that found in other kinds of turbulence, where the small scales are isotropic. Numerical simulations are consistent with this result by showing how small scales contribute to the third movements – an important result for vertical dispersion near the surface. A heuristic explanation has been proposed in terms of interactions between small and large scale turbulence, [31]. This is a notable example of one type of highly non-Gaussian and anisotropic turbulence, of which there are others where the universality of the Kolmogorov spectra is only observed in second order spectra [22].

Numerical simulations and laboratory experiments show how in convection, as in other turbulent flows, small scale eddies are randomly advected by those with larger scales. In convection this occurs in the movement of small eddies around the outside or in the wake of a puff eddy or shear-generated ring-like vortices in plumes [70] or small scale axial plumes rising within an elongated plume eddy. Typically these coherent eddies have scales of about 1/3 to 1/10 of that of the larger eddies. Following Tennekes [78], by considering the frequencies associated with the random large scale advection, the Eulerian frequency spectrum can be deduced. This enables estimates to be made for how the large eddy structures impinge on the inversion layer and generate local waves and turbulence (see section 4). For experimental confirmation see the spectral measurements in [43] for an analogous form of turbulence near an interface in a laboratory mixing box. A practical application of such modelling would be to calculate how turbulence affects electromagnetic propagation.

These smaller scales of course are not only advected but are stretched and distorted leading to mechanical dissipation at a rate ϵ per unit mass per unit time. As in other free shear flows it is found that its magnitude in the eddy is given by $\epsilon^{(e)} \sim W^{(e)3}/l$ and therefore for **plume** eddies (given the result (7) for the variation of $W^{(e)}$ and l with z) $\epsilon^{(e)}l^2$ is constant with height. When convection is dominated by the turbulence in these plumes, which tends to happen when it is asymmetric, if the plume eddies behave in the isolated form described in (1), the horizontal averaged dissipation $\overline{\epsilon}$ would be approximately constant with height.

In the region (X) between the plumes the smaller scale less coherent eddies are being strained by the divergence of the large scale flow between the plumes descending to their energy by dissipation. One observes from radar derived velocity vectors how the entrainment into rising plumes produces regions of high strain rate between them [25].

It is clear that mechanical energy must be dissipated in natural convection between planes because the thermal energy input produces no mean motion (when averaged over the whole flow) and no mechanical energy output. Its magnitude can be understood using the turbulence kinetic energy equation. In this case there is no mean motion and the conditions are statistically steady. If the mean buoyancy flux is gF_θ (normalised on the temperature), then

$$0 = gF_\theta - \frac{\partial}{\partial z}(\overline{wq^2} + \overline{wp}) - \overline{\epsilon}(z), \tag{10}$$

where $q^2 = \frac{1}{2}(u^2+v^2+w^2)$. The time mean values denoted by $\overline{(\ )}$ equal ensemble means and averages across all 'cells' because this very high Reynolds number turbulence is ergodic and horizontally homogeneous. In bottom upwards' asymmetric convection where F_θ decreases with height, in the downdrafts between the upward plumes, w is negative and the transport term $(-\frac{\partial}{\partial z}(wq^2))$ is positive and comparable in magnitude to $\overline{\epsilon}$. This can also be interpreted as a local production showing term because $\partial w/\partial z < 0$. In symmetric convection where F_θ is approximately constant in the interior, there must be equally strong upward and downward plumes so that the average dissipation in the plumes is constant with height. But entrainment into the plumes is still strong leading to a significant smaller scale turbulence in region (X) between the plumes and a significant level of dissipation (although lower than in the plumes). At locations where plumes or puffs detrain, locally the upward buoyancy flux $gW^{(e)}\Delta\theta^{(e)}$ tends to zero, but the dissipation is still large because of the local straining processes. This process also involves considerable mixing and thence damping of temperature fluctuations [87].

The variation of the mean temperature field $\overline{\theta}$, as well as the fluctuations in temperature (θ') and in buoyancy $g\theta'$ can be understood by considering the eddy structure and the different nature of micro-mixing between fluid

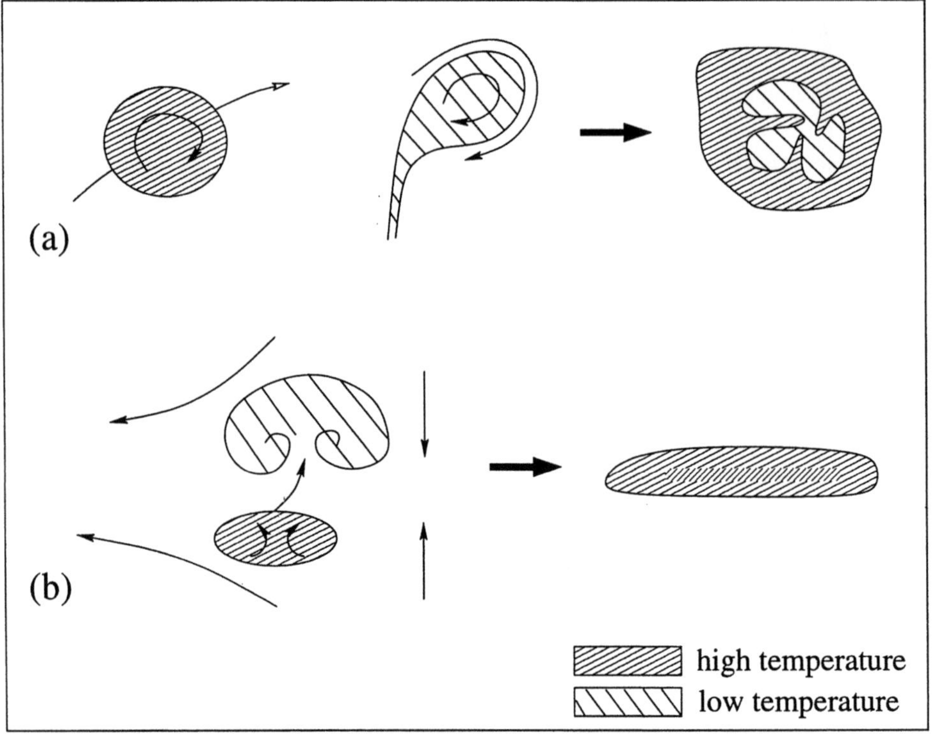

Figure 2. Temperature fluctuations in interacting eddies with (a) weak small scale mixing, with the variance $\langle\theta^2\rangle_\nu$ approximately unchanged (b) strong small scale mixing with the variance decreased.

elements in turbulent flows (cf. [61] and Fig. 2). Consider the rate of change of the temperature of a fluid element $d\theta/dt$ in the well mixed central part of a natural convection flow (symmetric or non-symmetric) where there are no mean temperature gradients, so that

$$d\theta/dt \equiv \frac{\partial\overline{\theta}}{\partial t} + \frac{d\theta'}{dt} = \kappa\nabla^2\theta, \tag{11}$$

where (except very close to the boundary)

$$\partial\overline{\theta}/\partial t = -\partial F_\theta/\partial z\ , \tag{12}$$

$$\frac{d\theta'}{dt} \equiv \partial\theta'/\partial t + u\frac{\partial\theta'}{\partial x} + v\frac{\partial\theta'}{\partial y} + w\frac{\partial\theta'}{\partial z}, \tag{13}$$

and where κ is the molecular diffusivity of the fluid.

The Eulerian mean value of $d\theta'/dt$ in a flow which is statistically stationary is $\frac{\partial}{\partial z}\overline{w\theta'}$. In the interior of the convective region (at a very high

Peclet number where $W^{(e)}l/\kappa \sim 10^7$ for air), averaging (11) shows that the mean effects of molecular diffusion are negligible and that the Eulerian and Lagrangian (denoted by the superscript L) means of $d\theta'/dt$ are determined by the gradient of the heat flux (and therefore the asymmetry of the convection) i.e.

$$\partial\overline{\theta}/\partial t = -\frac{\partial}{\partial z}(\overline{w\theta'}) \equiv -\overline{d\theta'/dt} = -\overline{d\theta'/dt^{(L)}}. \qquad (14)$$

This means that on average the temperature of fluid elements are not changing, i.e. $d\overline{\theta}/dt = 0$. If this were interpreted to mean that micro-mixing was negligible everywhere, i.e. $d\theta/dt \simeq 0$ (see Fig. 2a), it would imply (following Chatwin and Sullivan [14]) that defining a mean $\langle\theta\rangle_{\mathcal{V}}$ and fluctuation $\theta'_{\mathcal{V}}$ over a fluid volume $\mathcal{V}$ and deriving statistics it would follow that the variance of the total temperature would be conserved i.e. $\langle\theta^2\rangle = \langle\theta\rangle^2_{\mathcal{V}} + \langle\theta'^2_{\mathcal{V}}\rangle$.

This is a reasonable approximation to the mixing process in the plumes where high levels of fluctuations are observed comparable to the average change in θ between the plume and the external region, i.e. $\theta' \sim \Delta\theta^{(e)}$ (Fig. 2a). However when plumes or puff eddies disintegrate and discharge their heat and buoyancy directly, or indirectly via the boundary regions, into the exterior flow turbulence, although $\langle\theta\rangle$ is approximately conserved, the same result would suggest equally high values of $\langle\theta'^2\rangle$ in the external region (or downdrafts in asymmetric convection). This is not observed, because in these detrainment regions there is strong straining and shear of the turbulence, which always leads to a high level of mixing and molecular dissipation [87]. In fact $\overline{d\theta'^2/dt}$ can locally be as large as $\Delta\theta^{(e)2}W^{(e)}/l^{(e)}$.

This process is analogous to that in stably stratified flows where the total mean flux is equal to the eddy flux $(\overline{w\theta})$ but its magnitude depends on micro-mixing. The rise in the mean temperature (12) and having a steady value for the ratio of temperature standard derivation σ_θ to surface F_{θ_s} heat flux (given a constant velocity fluctuation w_*) are dependent on some micro-mixing occurring especially at the detrainment zones of buoyant eddies.

3. Interaction between Convective Eddies and Rigid Boundaries

3.1. KINEMATIC AND THERMAL BOUNDARY CONDITIONS

Rigid boundaries affect convective motions near them through kinematic and thermal boundary conditions applied to the velocity and temperature fields at the interface between the fluid and solid. The former conditions are that the normal and parallel components of the mean and fluctuating velocity are zero at the boundary itself, say at $z = 0$, or if the surface is rough (for example with sand of 100 μ m or buildings of 100m) the flow above the roughness is found to be effectively zero at some level where $z = z_0 + d$, z_0 is the roughness length (typically 1/30 of the heights of the actual roughness elements) and d is the displacement height (that depends on the spacing and height of these elements, e.g. [37]). Here we consider natural surfaces that are flat (slopes are left to section 5) and assume that $d = 0$. But we should note that the mechanical and thermal effects of the flow *between* large roughness elements is likely to affect convective eddies above them; a point that needs study. The dimensionless parameters that determine how these conditions affect the flow are the roughness parameter z_0/h and the Reynolds number Re $= w_* h/\nu$. For geophysical flows effectively only the former is significant.

The thermal conditions depend on how the heating and cooling is driving the convection. It is useful to consider the two main types (i) where the heating/cooling is provided at the boundaries, and (ii) in the interior with a volume heating rate V_θ. In the former cases the defining quantities are the surface fluxes (per unit area) F_{θ_S} and difference between the surface temperature T_s and the mean values in the centre of the convective region, i.e. $T_s - \overline{\theta}^{(CL)}$. Note that on a rough surface the temperature near the surface above the roughness elements is determined by the thermal roughness height z_{0_T} which can be significantly smaller than z_0 [27].

In some situations these parameters are fixed or their relations are fixed by non-convective processes, for example the surface temperature can be fixed if there is a very high thermal conductivity of the ground (so that the fluid-ground convection Peclet number $\mathrm{Pe}_g = (w_* h)/\kappa_g << 1$. Alternatively the flux maybe fixed at the surface, for example by radiation, or at a level 'in the ground' well below the surface,denoted by F_{θ_g}. Then the distributions of θ and F_{θ_s} over the surface depend on the coupling between the thermal fields in the fluid and the 'ground'. Adrian *et al.* [1] briefly and non-quantitatively discussed these sub-surface problems; for earlier theoretical studies see [13]. The uniformity of these thermal boundary conditions are particularly critical in convection; in fact non-uniformity is often cited as the reason for differences in turbulence and eddy structure (e.g. [82]); this may not always be correct as we shall show.

3.2. IMPACTING EDDIES

It is instructive first to review how eddies and random fluctuations generally impact onto a rigid surface when buoyancy forces are neglected . It will be shown that even though the eddies are generated by such forces the effects of boundaries on them are determined mainly by inertial and frictional forces. This idealised analysis provides useful quantitative estimates for eddy motions and statistical quantities. See Fig. 3.

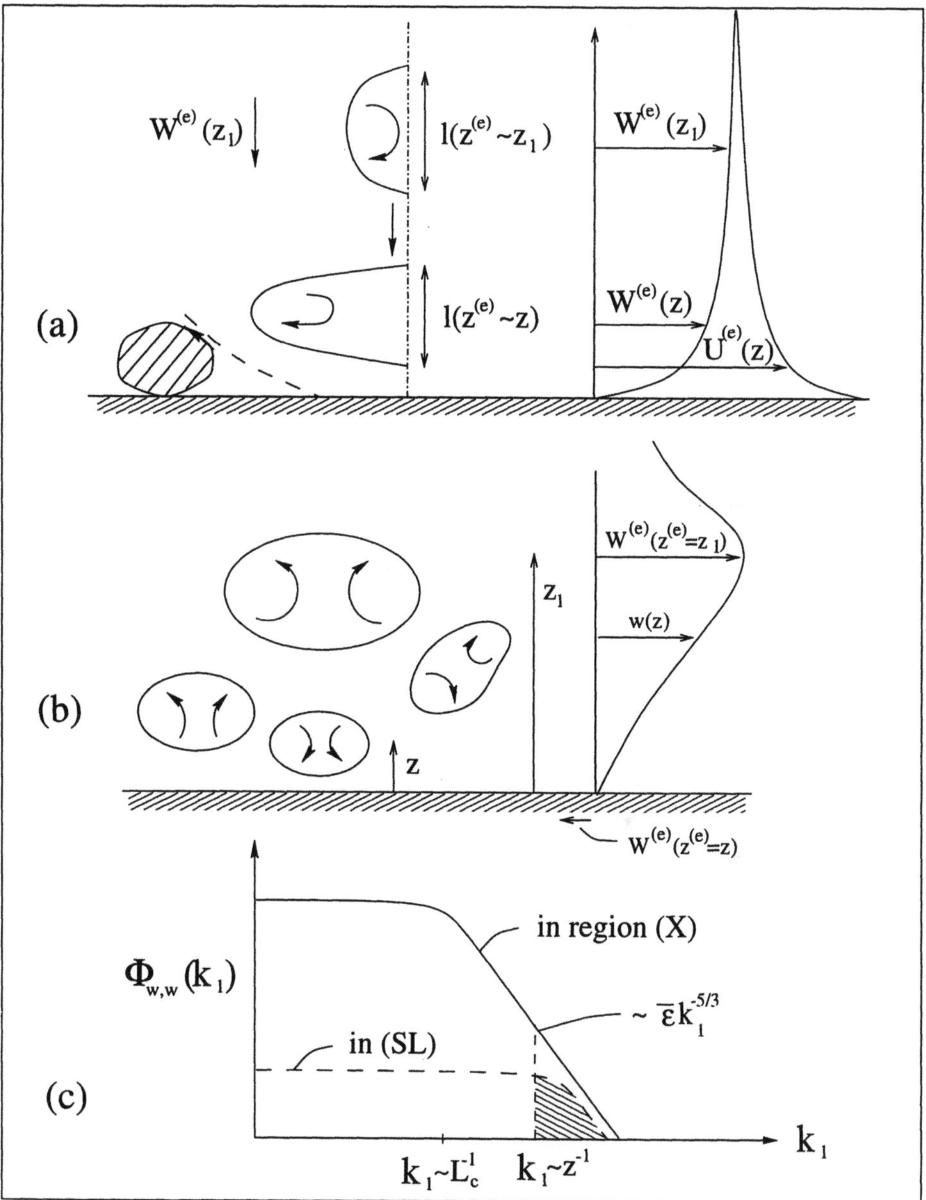

Figure 3. Non-buoyant effects of a rigid boundaries on eddies. (a) The variation of the peak velocity components within an eddy $(U^{(e)}, W^{(e)})(z^{(e)} \simeq z)$ as it approaches a surface. Note the reverse vorticity induced by the boundary (denoted by ////). (b) Showing profiles of the vertical velocity fields of actual eddies at one moment in time (——). (c) The one-dimensional spectrum of vertical velocity (— in the region (X) above the surface layer (SL); - - - in (SL) at height z).

3.2.1. *Non-Buoyant Effects*

Consider an eddy with length l, such as a vortex ring at a level $z^{(e)}$ above a rigid plane moving towards it with a velocity $W^{(e)} < 0$. Firstly this implies that at $z = z^{(e)}, |w| \sim |W^{(e)}|$. Secondly the condition of zero normal velocity means that $w = 0$ at $z = 0$. This distorts the whole velocity field around the eddy, for example by reducing the vertical velocity approximately in proportion to $z/z^{(e)}$ and increasing the horizontal velocity components u, v. u also induces a pressure gradient at the eddy that slows it down. The velocity of the eddy $w^{(e)}(z)$ approaches zero rather abruptly depending on the eddy shape. (Fig. 3a,b)

A dynamical analysis of how a full spectrum of eddies, with an inertial range isotropic spectrum for the small scales [29] is distorted by a rigid surface shows the vertical component is dominated by the small scales of the impacting turbulence, and leads to some quite general scaling laws for the variance and correlations of w, viz.;

$$\sigma_w \sim (\bar{\epsilon} z)^{(1/3)}, \tag{15}$$

and

$$\overline{w(z)w(z_1)}/\overline{w^2(z_1)} \simeq z/z_1 \,. \tag{16}$$

The former result in (15) and also calculations of the spectra (Fig. 3c), show that the variance of w is approximately equal to that of the small scale eddies in the inertial range spectrum,having wave number greater than $1/z$. This is because vertical velocities on scales greater than z are reduced by the surface. This very rapid decrease at the surface is consistent with the abrupt 'braking' of finite size eddies as they reach the surface. The second statistical result (16) can also be explained in terms of the velocity field around finite eddies; the vertical velocity at level $z, w(z)$, can be considered as caused by the sum of the contribution of eddies located at level z (i.e. $z^{(e)} = z$) having a length scale of the order of z, and of the damped velocity field of the larger eddies at greater heights where $z^{(e)} = z_1 > z$, viz. $w(z) \simeq W^{(e)}(z^{(e)} = z) + (z/z_1)W^{(e)}(z^{(e)} = z_1)$. Only the latter component is correlated with the fluctuating velocity at height z_1 since $\overline{W^{(e)}(z)W^{(e)}(z = z_1)} \simeq 0$. Therefore $\overline{w(z)w(z_1)} \sim (z/z_1)(W^{(e)})^2(z = z_1)$ which implies (16). For extensive verification (including ejecting eddies) see [32], and for extension to higher moments in convective turbulence see [70].

The same linear calculation based on the hypothesis that eddies respond rapidly to the surface when they approach it, is less accurate in its prediction for the horizontal fluctuations i.e. $\sigma_u^2 + \sigma_v^2$. But by estimating the self induced stretching of the vortex lines in the eddy the extra amplification, which is about 20%, can be accounted for, as recent laboratory experiments have demonstrated [43]. (An even greater increase in $u^{(e)}$ is found when isolated vortices impact on to surfaces.) The effect of the frictional stresses at the boundary is to generate vorticity in the opposite direction to that of the impinging vortex (see Fig. 3a) [62]; firstly this slows down the tangential components (i.e. σ_u, σ_v) – a local fluctuating logarithmic velocity profile for $u(z), v(z)$ has been proposed ([46], [54]) – and secondly that can lead to the vorticity being ejected from the wall. These processes are considered later in the context of ejecting eddies.

3.2.2. *Buoyant Effects*

In most turbulent flows driven by natural convection there are large unstable temperature gradients in their 'surface' layers of thickness h_s at the boundaries, so that impacting eddies are usually more dense than the fluid near a bottom boundary ,or less dense than at the top. In either case the impacting eddy of scale l experiences an 'Archimedes' buoyancy force, which is of the order of $g(d\rho/dz)(h_s l^3)/\rho_o$. For this to affect significantly the trajectory of the eddy as it approaches the boundary it would have to be greater than that caused by the pressure gradient associated with the image eddy i.e. $\rho W^{(e)^2}.l^3/z^3$. This does not happen in the surface layer for typical cases of natural geophysical convection. Thus the main dynamics affecting impacting eddies onto rigid surfaces are not directly buoyancy related.

3.3. EJECTING EDDIES

In layers near rigid level surfaces in natural convection usually the fluid is unstable because there is a positive heat flux (i.e. $F_\theta > 0$), either at the bottom or top surface. However there are situations when one of the layers is stably stratified as a result of a negative or zero heat flux at the boundary, whether applied steadily or on time scales even as short as those of the largest eddies. The instability causes eddying on the scale of the surface layer, whose thickness h_s is derived below. But the progressive interactions and coalescence of these eddies lead to large scale plumes on the scale of the whole depth h. These are approximately steady (on their large time scale) or the interactions can induce intrinsically unsteady puff eddies. An analysis of the former naturally leads on to the latter.

Within the surface layer the small scales l of the local plume or puff eddies produced by the unstable temperature gradient grow as they rise in the surface layer h_s so that $l \sim z$. Following [63], most authors argue on dimensional grounds that their velocity at height z should only depend on z and the local surface (bottom) heat flux F_{θ_s} (normalised on the absolute temperature T_s at $z = 0$) so that

$$\sigma_w \sim (gF_{\theta_s})^{1/3} z^{1/3} . \tag{17}$$

This statistical result is not easy to explain in terms of local plumes or puff eddies. However a dynamical (and realistic!) explanation is possible, based firstly on the interactions of these surface eddies, which leads to the energy dissipation rate ϵ being comparable to gF_{θ_s} and secondly on the fact that the kinematical relations (15), (16) and their physical interpretation also apply to these ejecting eddies. Thence (17) follows from (15) and (10). The horizontal fluctuations are mainly produced by the larger scale motions to be discussed.

The eddies in this layer are intrinsically unsteady, as they grow and interact, on a Lagrangian time scale of

$$\tau_L \sim z/\sigma_w(z) \sim z^{2/3}/(gF_{\theta_s})^{1/3} \sim z^{2/3}/\bar{\epsilon}^{1/3}. \qquad (18)$$

The mutual entrainment of the buoyant eddies in the surface layer and their formation (in certain circumstances to be determined) are such that over a horizontal radius of about L_c a surface flow with velocity u_s is entrained into them at all levels but about half of the total volume flux of the fully developed plumes (at $z = h$) is drawn in within the surface layer. (Fig. 4) *All* the buoyancy flux Q_b in the plume [as defined in (6)] enters through this layer (since in the plume, where $z > h_s$ the entrained fluid only has the buoyancy of the external flow which is defined to be zero!). Thus $Q_b \simeq gF_{\theta_s}.L_c^2$. The diameter of and initial velocity of the plume must be comparable with h_s and u_s – a consequence of continuity, and balance between inertia and local pressure gradients. This ensures that the effective 'origin' for the plume (7) is at a level $z_p \simeq -\frac{h_s}{\alpha}$; which explains why the width of updrafts grow slowly with z and not really as rapidly as plumes located on the surface [51]. If additional entrainment fluid is fed into the plumes near the surface, the plume origin rises and the surface inflow u_s decreases as in some natural ventilation schemes [53].

The local shear stresses generated by the vertical velocity gradient $\partial u/\partial z$ of the surface flow into the plume is sufficiently great that on a time scale τ_p of the plume a spatially varying logarithmic profile is set up (cf. [46], [54], [76]). This is confined to a thin inner layer of depth h_{ss} where the surface shear stress is $u_{*_c}(x)$ (where x is the horizontal distance from the plume centre) where the vertical component of turbulence associated with u_{*_c} is greater than that driven by the unstable heat flux F_{θ_s}, even though it is the overall effect of heat flux that drives the flow. This layer's depth h_{ss} is comparable to the local length scale $\widetilde{L}_{MO}$ which is analogous to the Monin-Obukhov length scale (normally defined for surface layer turbulence with significant mean wind speeds comparable to the turbulence levels). $\widetilde{L}_{MO}$ may be defined as the height z where the contribution by the shear stress to the average dissipation rate in the convection 'cell' $\langle\epsilon\rangle(z)$ i.e. $u_{*_c}^3/\kappa z$ exceeds that caused by heat flux gF_θ. Thus

$$\widetilde{L}_{MO} \sim u_{*_c}^3/(gF_{\theta_s}\kappa) \, , \qquad (19)$$

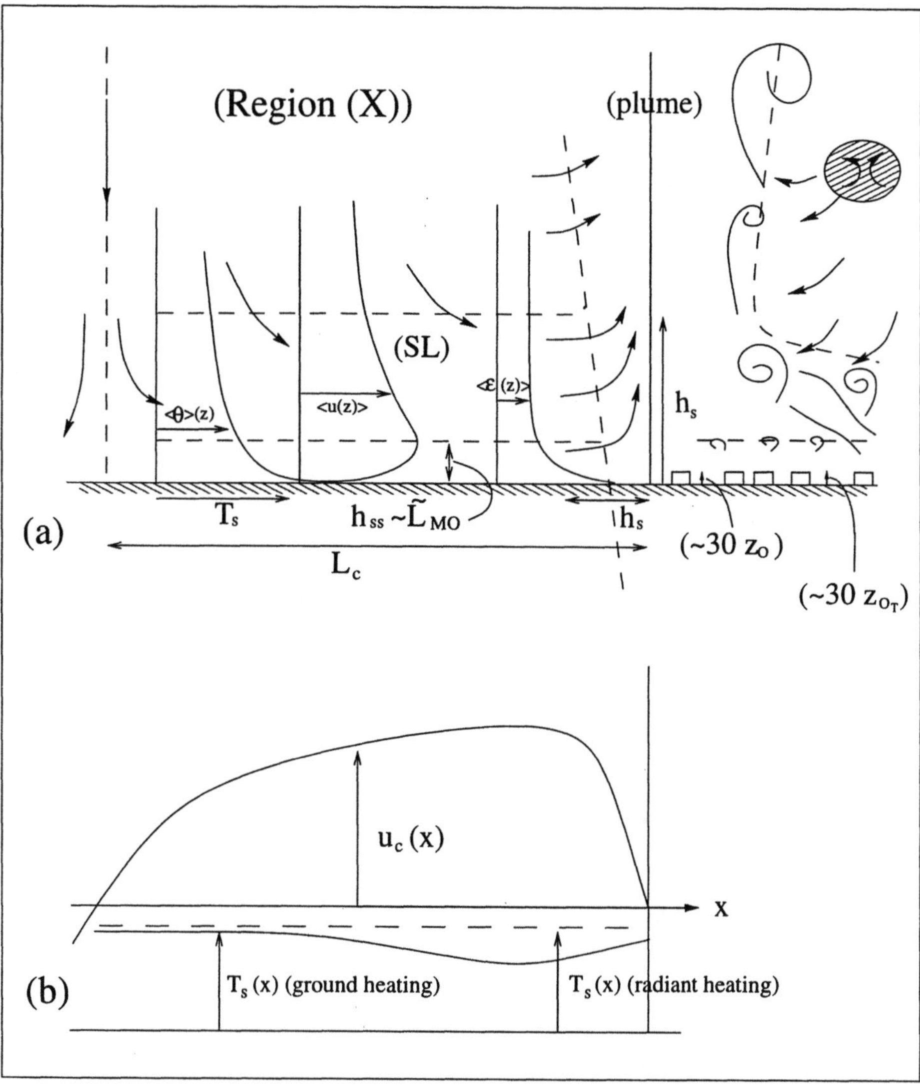

Figure 4. Large scale plume eddy above a heated rigid rough surface [90]. (a) Flow structure and typical profiles. (b) Horizontal variations in the convection cell of surface friction velocity $U_{*_c}(x)$ and of the surface temperature T_s depending on whether the heating is provided by strong radiation at the surface or by the ground below the surface.

where κ is von Karman's constant. The actual level of u_{*_c} is determined by matching the average shear stress velocity at the level $\tilde{L}_{MO}$ with the flow in the upper part of the surface layer $\tilde{L}_{MO} < z < h_s$ where convective eddies determine the eddy diffusivity and the depth of h_s. Thence from the works [76] and [90]

$$u_{*c} \sim \kappa w_*/\ln(h_s/z_o), \tag{20}$$

where

$$h_s/L_c \sim \sigma_w(z \sim h_s)/u_s(z \sim h_s) \sim (h_s/h)^{1/3} . \tag{21}$$

Thus the depth of the inner surface layer is related to the scale of the convection cell L_c. For the typical value of L_c/h of about 1/5, it follows that $(h_s/h) \sim (L_c/h)^{3/2} \sim 1/10$. Similar laminar flow sub layer calculations are consistent with bulk measurements for convection smooth surfaces [11].

This shows that the typical vertical velocities of eddies at the top of the surface layer are less than about 50% of the horizontal velocity. This level of average horizontal velocity causes almost all the buoyant plumes or puffs to be bent over and transported laterally into the plumes. Thus in the surface layer, *between the plumes*, the vertical heat flux $F_\theta(x,z)$ decreases to zero at height h_s.

Therefore over the central part of the convective region, in the external region outside the plumes, heat or any matter originating from the surface has mostly had to come down from the upper surface layer. This shows how material from the surface may only reach parts of the atmosphere above the the surface layer quite slowly (on a time scale of above $2h/w_* \sim 2 \times 10^3$secs in the atmospheric convective boundary layer). But material (e.g. pollution from elevated sources) entering the boundary layer from intermediate heights or through its upper surface may be transported downwards considerably faster [89].

The lateral velocities $u_s(z \sim h_s)$ in the surface layer and thence sweep eddies of scale $l \sim z$ towards the plumes. It follows that the Eulerian time scale $\tau_E^{(w)}$ of the vertical fluctuation for a fixed observer at an average height in the surface layer is of the order of $\tau_E^{(w)} \sim h_s/w_*$, which is smaller than the Lagrangian time scale τ_L by a factor of the order of $(h/h_s)^{1/3}$.

This difference is a consequence of the great anisotropy of the turbulence near the surface in convective turbulence [83]. It is not found in most types of turbulence where all the velocity components are of the same order. For example in the neutral boundary layer these two scales are of the same order for each component of velocity.

The different scales of motion in the surface layer need to be taken into account in defining, modelling and measuring the vertical diffusivity for momentum (or eddy viscosity ν_e) and heat/mass transfer K_z. We recall the distinction between turbulence defined relative to average motions $\langle u\rangle, \langle w\rangle$ of the convection 'cell' e.g. the Reynolds stress $\langle uw\rangle$ and the mean velocity and temperature fluctuations at a height z averaged over the **whole flow** area on which scale $\overline{\langle u\rangle} = \overline{\langle w\rangle} = 0$ but $\overline{\langle\theta\rangle} \neq 0$. Because the mean velocity

$\overline{u}$ is zero for convection with no mean flow $\overline{\nu_e}$ cannot be defined in this situation. In each convection cell

$$\langle \nu_e \rangle = \langle uw \rangle / (\partial \langle u \rangle / \partial z) \sim z^{4/3} (gF_\theta)^{4/3} \text{ for } z > h_{ss} \sim \tilde{L}_{MO} \,. \tag{22}$$

However for the thermal eddy diffusivity, where $\langle \theta' w' \rangle$ denote the correlation of the fluctuations relative to the 'cell average' velocity and temperature fields $\langle w \rangle$ and $\langle \theta \rangle$, so that $\overline{\theta w} = \overline{\langle \theta \rangle \langle w \rangle} + \overline{\langle \theta' w' \rangle}$, the mean value can be defined over the whole plane

$$\overline{K_z} = -(\overline{\theta w})/(\partial \overline{\theta} / \partial z) \,. \tag{23}$$

Whereas for the 'cell' average

$$\overline{\langle K_z \rangle} = -\langle \theta' w' \rangle / (\partial \langle \theta \rangle / \partial z) \,. \tag{24}$$

Thus the mean is greater than the cell average, i.e. $\overline{K_z} > \langle K_z \rangle$ since $\overline{\langle \theta \rangle \langle w \rangle}$ is positive and $\overline{\theta} = \overline{\langle \theta \rangle}$.

Since $\langle w \rangle$ is proportional to z where $\langle \theta \rangle$ is greatest under the plumes, (23) suggests that the mean value $\overline{K_z}$ increases more rapidly with z than the cell average $\langle K_z \rangle$. This is consistent with the unexplained field and numerical result that $K_z \propto z^{3/2}$, which is a slightly faster increase with height than the $z^{4/3}$ result of 'free-convective' scaling. If this explanation is correct, it would support the continued use of free convection scaling for 'cell average' modelling, as in [90].

3.4. THERMAL STRUCTURE OF THE SURFACE LAYER AND THE SUB-SURFACE LAYER

In the surface shear stress layer $z_o < z \leq \tilde{L}_{MO}$ the potential temperature profile (averaged on the time scale τ_p) is given by

$$\Delta\theta(x) = \theta(x,z) - T_s(x) = -\frac{F_{\theta s}(x)}{\kappa u_{*_c}(x)} \ln(z/z_{oT}). \tag{25}$$

The surface temperature $T_s(x)$ and surface flux $F_{\theta_s}(x)$ vary with distance x from the plume centre line in the convection cell. This depends on the variations of the local surface shear stress u_{*_c} and the coupling of the thermal field in the fluid to that in the adjacent solid ($z < 0$). The average value of $u_{*_c}(x)$ in terms of w_* is given by (20). The variation of u_{*_c} derived from the analysis of the surface and free convective layers shows that u_{*_c} increases on radial lines towards the plume but reaches a maximum about one radius from the plume centre line, where $x \sim h_s \sim l$. It decreases to zero at the centre of the plume where $z = 0$ (Fig. 4).

As explained in section 3.1, if the heating/cooling is transferred through the rigid boundary (or 'ground') and the ground's thermal conductivity is large enough (in relation to convective transport by the flow) i.e. $\mathrm{Pe}_g = w_* L/\kappa_g \ll 1$, then the surface temperature T_s is effectively constant.

In this case the air flow and the (fixed) nature of the surface determine F_{θ_s}. This condition can also occur when the ground temperature is determined by a local radiation balance at the surface. In dry desert conditions the incoming radiation flux F_R (which is mainly a function of the solar radiation and cloud cover) is balanced by the outgoing radiation proportional to σT_s^4.

However on surfaces on which there are roughness elements with thermal properties, such as vegetation and/or buildings, and below which there is material with complex thermal properties, the surface temperature T_s is determined by a complex balance of radiation and conduction into the ground; latent heat effects, soil thermal capacity, resistance to the air flow (or roughness length) and local friction velocity u_{*_c}. The flux into the air flow F_θ is also determined by the same factors, so that T_s and F_{θ_s} are in general linked implicitly. An interesting situation can occur in the evening when F_R is reduced and the ground is warm. Then F_θ is essentially determined by the heat flux from the ground. Note that $\overline{F_{\theta_s}} = \overline{F_{\theta_g}}$ where, in the ground, $F_\theta = \kappa_g \frac{\partial \theta}{\partial z}$ and $f_\theta \to +F_{\theta_g}$ as $z \to -\infty$. This also corresponds to laboratory experiments such as that of [81].

These conditions relating F_{θ_s} and F_{θ_g} are necessary, because F_{θ_s} varies as a result of the variations in heat transfer caused by the variation in surface friction $u_{*_c}(x)$. If it is *assumed* that F_{θ_s} is constant (25) shows that T_s decreases by say $\Delta\theta_s \sim (F_{\theta_s} \ln(h_s/z_{oT})/u_{*_c})$ as a result of these variations. This causes steady state temperature gradients in the ground, which change $F_{\theta s}$ by $O(\frac{\kappa_g \Delta\theta_s}{\sqrt{\tau_p \kappa_g}})$. Thence $\mathrm{Pe}_g = w_* L/\kappa_g$ must be larger than $\ln^4(h_s/z_{o_T})$ (i.e. 10 to 100) for the changes in $F_{\theta s}/F_{\theta g}$ to be negligible so that the original assumption of small changes in F_{θ_S} is invalid.

Furthermore the variations of T_s along the surface layer can have dynamical effects which can change the pattern of thermal convection. The lateral pressure gradient $\partial p/\partial x$ set up by the cooling of the surface can be sufficiently great to counteract the inflow caused by the plume if $\ln^4(h_s/z_o) > (h/h_s)$.

Thus plumes form part of a dynamically and thermally consistent system both in the fluid and in the solid boundary, either if the surface temperature T_s is fixed (in which case the flux F_{θ_s} will vary in proportion to u_{*_c}) or if F_{θ_s} (rather than F_{θ_g}) is fixed along with some constraint to limit the magnitude of $\Delta\theta_s$ (as in the experiments of [74], [50]).

But if $\mathrm{Pe}_g \gtrsim 1$ and the surface thermal flux is driven by F_{θ_g}, then plumes can only exist for a period of time τ_p it takes for the variation in surface temperature to penetrate into the solid boundary and thence change the surface flux. This essentially is a mechanism for puff eddies to be formed. The size l and time scale τ_{pfe} of these eddies depend on how much heat can be extracted from the ground in the time it takes to cool the ground by the fluid flow before the heat is restored by the heat flux. Thus as Pe_g increases (or κ_g decreases) l and τ_{pfe} decrease in proportion to the thickness of the thermal layer in the ground and therefore to $\mathrm{Pe}_g^{-1/2}$.

Experimental evidence is incomplete; Scorer [72] observed that atmospheric puff eddies certainly occur in the evening. Cooks note that by placing special low thermal conductivity plates below a cooking pan, the location of bubbling is random and quite small scale. Townsend's [81] experiment on convection showed a strong tendency to form unsteady puffs, whereas the fixed surface heat flux experiments all show plume eddies.

4. Interactions of Convective Regions with Inversion Layers

4.1. GENERAL QUESTIONS

Many regions of natural convection are confined above or below by regions [or inversion layers (I.L.)] of sufficiently stable density profiles that their structure persists for a long time under the impact of turbulent eddies from the convective region (Fig. 5). These bounding inversion layers h_s respond to their continuous impingement in a number of ways that affect motions outside them as well as in the convective layers (C.L.). Note that I.L. may adjoin on its other side either another region ($\mathcal{U}$) of fluid or a solid boundary.

(i) The mean height $h(t)$ above the lower surface of the I.L. varies. This not only determines the dynamics of the convective eddies which depend on z but also the depth to which pollutants are diffused. Mean concentrations are generally proportional to h^{-1}. Both in the upper oceans and atmosphere, h varies with time because of diurnal solar heat flux. Therefore 'mean' properties are those defined on a time scale large compared to that of the large eddies in the convective cell (i.e. $\tau_p \simeq h/w_*$), but less than that on which the overall flow is changing [e.g. $h/(dh/dt)$].

It is conventional to refer to the vertical velocity at which this interface moves as a 'boundary' entrainment velocity E_b, [85], but usually only when dh/dt is positive. For most situations of convective turbulence below an inversion layer, when $dh/dt > 0$,

$$E_b \equiv dh/dt \ll w_* \quad \text{or in dimensionless form} \quad \hat{E}_b = E_b/w_* \ll 1 \,. \tag{26}$$

We also assume here that when dh/dt is *negative*,

$$dh/dt \ll w_* \,. \tag{27}$$

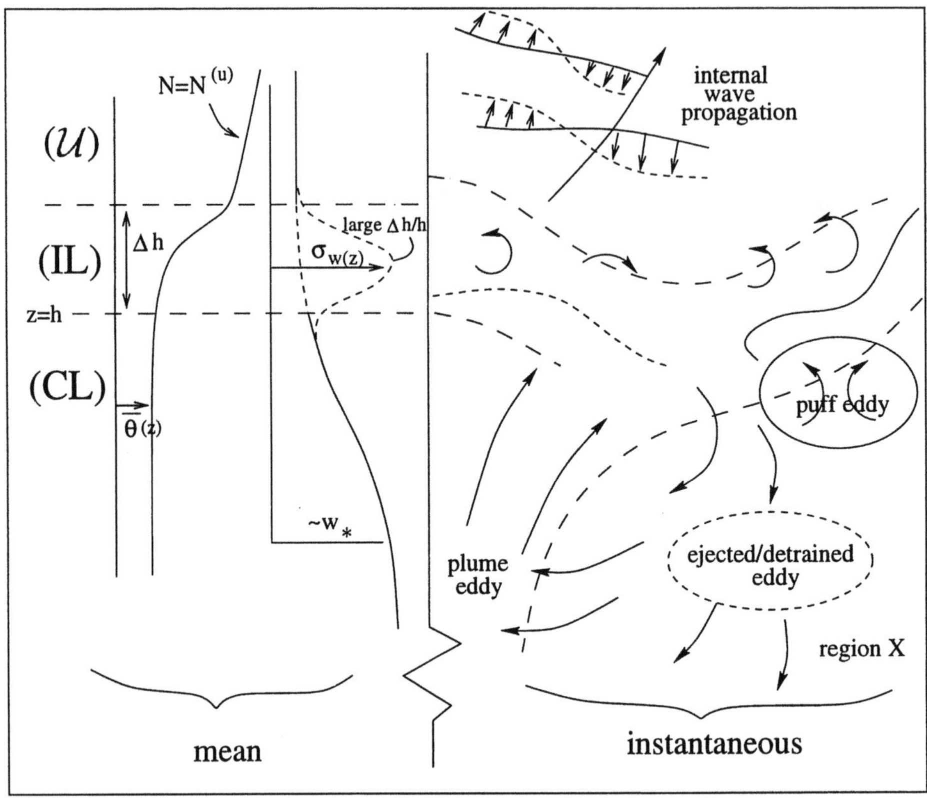

Figure 5. Interaction of convective turbulence with inversion layer showing the different possible profiles of vertical velocity variance $\sigma_w(z)$ (large and small $\Delta h/h$), impact of plumes and puff eddies, detrained eddies and re-entrainment into plumes. Note internal wave generation in $\mathcal{U}$.

But where the stable stratification in the region $(\mathcal{U})$ is weak dh/dt may be comparable to w_* and the latter assumption is invalid. The ratio $\hat{E}_b$ is a critical parameter for modelling how a convective layer evolves. It depends greatly on the mechanics within the I.L.; (on this at least most investigators agree!).

(ii) The I.L. is not at a fixed height and the thickness of the I.L. also varies with time. The mean thickness is denoted by Δh. Radar pictures, images of cloud tops and laboratory model studies show how there are large scale undulations on the length scale h, and on a scale Δh smaller scale eddies lead to mixing events [25]. In most geophysical situations, the ratio $\Delta h/h$ is small, but this ratio varies considerably depending on the flows above and below I.L. The mixing processes within I.L. also determine $\hat{E}_b$. It is convenient to normalise $|d\Delta h/dt|$ on w_* and express it as $\widehat{\Delta E_B}$.

(iii) Whether the convective region moves or not, there are may be fluxes of heat e.g. F_{θ_I} and convection F_{c_I} across the interface at $z = h$, caused by the turbulent mixing motions in the inversion layer and by differences in temperature and concentration across the layer $\Delta\theta_I, \Delta c_I$. There are geophysical and other practical situations where $dh/dt = 0$, and where vigorous convection occurs in two regions separated by a stable interface. Clearly following the usual approach in heat transfer studies this transport process is quantified (perhaps confusingly) by the flux entrainment velocity

$$E_{f(\theta)} = F_{\theta_I}/\Delta\theta_{\mathrm{I}} \; ; \; E_{f(c)} = F_{c_I}/\Delta c_{\mathrm{I}} \, . \tag{28}$$

It cannot be assumed that these two parameters are identical because temperature is a scalar with dynamic properties, whereas the weak concentrations of matter is not [45].

The relevant dimensionless parameters are $\hat{E}_{F(\theta)} = E_{f(\theta)}/w_*$ and $\hat{E}_{F(s)} = E_{F(c)}/w_*$. These ratios can be related to the ratio $\hat{E}_b$ by analysis of comparable flows, since clearly the same mixing processes affect both ratios [57]. When the condition (23) is satisfied, it is also found that the fluid entrainment velocities are small compared to w_* (i.e. $\hat{E}_{F(\theta)} \ll 1$).

(iv) When the upper region $(\mathcal{U})$ above the I.L. is stably stratified the eddying and wave motion within I.L. generates internal waves in $(\mathcal{U})$. If the density gradient is uniform with Brunt Vaisala frequency N_u the waves propagate away; if it is non-uniform or there are significant velocity gradients these waves can be trapped. Either mechanism leads to flux of wave energy from $F_{(WE)}$ (per unit area per unit time) from I.L. and ultimately from the convective layer. It is natural to normalise $F_{(WE)}$ on the main cause of energy loss, namely the total mechanical dissipation in the convective layer $\bar{\epsilon}h \sim w_*^3$. The ratio $\hat{F}_{(WE)} = F_{(WE)}/w_*^3$ is found to be small (typically 10 to 20% in the atmospheric boundary layer) but its magnitude depends on the mechanisms in I.L.

(v) The main question from a Lagrangian point of view about I.L. is where do the fluid elements come from in the layer and where do they go to? See Fig. 5. If large eddies are mainly in the form of plumes which are strong enough to reach I.L., then these transport most of the fluid entering I.L. and much of this fluid will have come from the other surface (for example the heated and polluted bottom of the convective layer, so that I.L. may be hotter and more polluted than region (X) between the plumes). Because $\hat{E}_b \ll 1$ it follows that this fluid must then leave I.L. into the external region (X). About 50% reaches the surface layer at the other boundary. These entrainment requirements and the mixing processes in IL determine the buoyancy and momentum of this departing fluid. However if the convection is in the form of puffs, these impinge randomly on the I.L. and equally the fluid leaves randomly as it is entrained into downward puffs and as instability and waves develop. In this case most fluid particles cannot travel directly from the other surface; rather, through inter-eddy mixing, they may be transported on several different eddies (Fig. 2b). Therefore the rate of transport of heat and matter are slower and the temperature and concentration in I.L. are less closely linked to that near the other (e.g. lower) surface.

The variables that determine the motions in I.L. and its interactions with the convective layer on a time scale comparable with the largest eddy motions, e.g. $\tau^{(e)}(\sim h/w_*)$, include the temperature T_s, flux F_{θ_s} and the roughness lengths z_o, z_{oT} (at the other boundary), the depth h, the buoyancy frequency N_u in region $(\mathcal{U})$, the temperature jump across $\Delta\theta_I$ and (in most circumstances) the depth Δh of I.L. Models can be constructed to describe the main features of the velocity field given these variables [20]. Alternatively computations can simulate the initiation of convection at say $t = 0$ so that I.L. develops along with the structure of the whole convective layer. Then $T_s, F_{\theta_s}, \theta_u(z), h$ would be specified at $t = 0$.

Typically this would be the approach of a large eddy or numerical simulation. Provided eddy viscosity models reflect more closely the intrinsically inhomogeneous nature of the turbulence (discussed in the next section) in the convective layer and in I.L., it is also possible to use this traditional approach to study the evolution of convective layers [21], [80].

4.2. MECHANISMS

The instantaneous height $Z_I(x, y, t)$ of the inversion layer (I.L.) is displaced and distorted by the adjacent turbulent eddies near the top of the convective layer (C.L.) (Fig. 5a). Z_I has the same vertical displacement $Z(t)$ as a fluid element lying in this surface. By definition $Z(t)$ moves with the fluid. Its motion and the temperature of the fluid element $\theta(t)$ are affected by moving in the stable density gradient of I.L.

If there was no mixing the magnitude of Z would be limited because increasing Z increases the potential energy of the fluid elements and therefore its variance is determined by their average kinetic energy, i.e.

$$\overline{Z^2} \sim \min[w_I^2/N_u^2 \, , \frac{w_I^4}{\Delta h^2 N_I^4}], \tag{29}$$

where $N_I^2 = g\frac{\Delta\theta}{T_s}/\Delta h$ and w_I^2 is the typical variance in the I.L. The typical time to reach this height is $(Z^2)^{1/2}/w_I$. This would imply that dh/dt is zero for a given level of turbulence in C.L. However fluid elements do mix in stable layers often as a result of overturning internal waves so that a fluid element displaced upwards changes its temperature by a small amount $\gamma Z\partial\theta/\partial z$ where the parameter $\gamma \ll 1$ [61]. By considering the oscillations of particles in the stable stratification this implies that a first estimate for the boundary entrainment velocity is

$$E_b = \frac{dh}{dt} \sim \frac{\overline{dZ_i}}{dt} \sim \frac{\overline{dZ^{(L)}}}{dt}, \tag{30}$$

whence from (29),(30)

$$\hat{E}_b \simeq \gamma w_I / w_*. \tag{31}$$

A close study of the mechanisms in I.L. is necessary to estimate these two dimensionless variables γ and (w_I/w_*) in terms of relevant dimensionless variables of the I.L., C.L. and u. These are $\mathrm{Ri}_{(I)} = \frac{g\Delta\theta h}{T_s W_*^2}, \frac{N_t}{N_u}$ and $\Delta h/h$. The latter ratio determines the relative size of the large eddies ($\sim h/5$) in C.L. to the depth of I.L. The physical interpretation of the former ratio or 'bulk Richardson number for I.L.' depends on the kind of motion that is taking place in I.L. Laboratory experiments [57] and numerical simulations [68] indicate that when coherent eddies approach I.L. and travel parallel to it so that a strong local shear is set up the shear instability leads to Kelvin-Helmholtz 'billows' forming; the criterion is that $\mathrm{Ri}_{(I)} \lesssim 1$ [84]. It is not clear whether this mechanism is significant when the large scales are not coherent. When puff-like eddies impinge with velocity of the order of w_* directly on I.L., their kinetic energy is sufficient to deflect the I.L. by an amount approximately given by (29) where $w_I \sim w_*$. On the 'rebound' some mixing takes place at a rate such that γ is small. This mechanism is most significant when $\mathrm{Ri}_I \sim 1$ [52]. Essentially this concept was the basis of the first to model I.L. [19]. But it was not wholly successful because when I.L. is even more stable, impinging eddies do not have the energy to deflect the interface directly. However by moving unsteadily parallel to it pressure fluctuations generate waves. These moving eddies might be vortices rising within a plume which are some point along the large scale motion [68] or may be puff-like eddies of different scales advecting each other.

For thin layers where $\Delta h \lesssim h/10$, the significant wave lengths are much greater than Δh, whereas when $\Delta h/h$ is larger (of the order of 1/5) internal wave modes within the I.L. are set up by these eddy motions [20] and [10]. In both cases waves break which leads to weak mixing so that $E_b \sim \mathrm{Ri}^{-(1+r')}$ where $r' \sim 0.2\pm0.1$. Note also that in the former case the vertical turbulence tends to decrease near $z = h$, whereas in the latter case it increases and leads to the well known bumpiness sometimes felt by aircraft passengers as they pass through inversion layers. Surprisingly, detailed calculations show and laboratory experiments confirm, that in the latter case the trapped waves have a longer period and cause less mixing than the former one. The difference in E_b may be as large as a factor of 10 when $\mathrm{Ri}_I \sim 20$. This suggests that the precise modelling and forecasting of the growth of the main inversion layers [i.e. $h(t)$] are quite sensitive to the value of $\Delta h/h$ when Ri_I is large *and* there is no mean shear. In fact in most atmospheric flows Ri_I is close to unity and shear is significant, so that for many situations this great sensitivity is not of practical importance. For radiative and phase change processes see [56].

5. Departures from Ideal Convection

5.1. GENERAL CONSIDERATIONS

The statistics of turbulent convection in any given situation are characterised by the typical vertical velocity w_* and the scale of the largest eddies typically $h/5$ and characteristic heat flux F_{θ_s}, so the effect of any perturbation to the surface geometry and the initial or boundary conditions can be assessed in relation to these scales. However these scales may give a better guide to the sensitivity to perturbations for some variables and not for others, especially the eddy structure which has been shown even in ideal conditions to be quite sensitive to the homogeneous parameters. Any perturbation inhomogeneity is likely to be even more significant for eddy structure and its effect on the flow may need to be considered in relation to some assumed eddy structure.

Even more than for studying ideal laterally homogeneous, steady convection, when one examines departures from ideal one needs to consider specific cases. We shall focus here on atmospheric and oceanic boundary/mixed layers.

5.2. CROSS WIND EFFECTS NEAR THE GROUND

Field observations (including a terrifying glider trip by the author of this review) show that in conditions of very strong convection in very dry conditions even when a synoptically driven cross wind $\overline{u}$ exists of the order of w_*, a plume structure continues to exist in just the same form as when there is no wind [64], [40]. For these plume eddies to keep moving over the ground requires the surface layer to keep heating up the air entering from the external region and then providing a continuing flow into the plumes. Figure 6 shows the likely pattern of the average flow in the local coordinate frame of the moving plume whose speed $u^{(S)}$ is defined at height $z = h_s$. Note how the mean (averaged on a typical plume scale) velocity $\langle u\rangle(z)$ (in the absence of the wind) reverses at a critical height z_c defined by $\langle u\rangle_o(z_c) \simeq u^{(S)}$ where $\langle u\rangle_o(z)$ is the surface layer profile in the absence of the cross wind. The study of perturbed turbulent boundary layers over moving waves [6] showed that a reversal of the mean flow of $z = z_c$ in the shear stress layer (or inner surface layer in their notation) does not disrupt the overall flow structure. Its main effect is likely to raise the effective

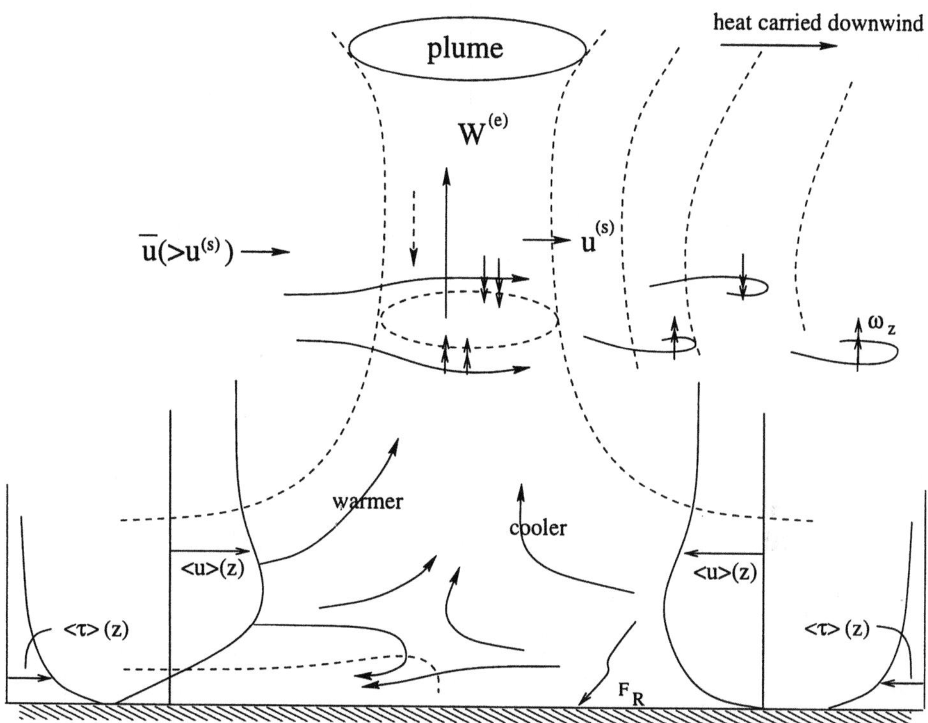

Figure 6. Flow structure in coordinates moving with a plume at speed $u^{(s)}$ drive by a cross wind $\overline{u}$. Note the reverse flow near the ground in these coordinates and the vorticity and some heat carried downwind.

roughness length for the outer flow from the value z_o to $z_o \exp(\kappa u^{(S)}/u_{*_c})$. From the discussion of the structure in the surface layer and equations (20) and (25, it follows that. for given F_{θ_s} and ω_z. I_{*_c} decreases/increase far on the upwind/downwind side of the plume. Thence the temperature reaching the plume would increase/decrease. Thus as $\overline{u}/u_{*_c}$ increases, this mechanism helps maintain the stability of the plumes' structure in a cross wind.

The other effects of a cross wind on plumes can be analysed by considering the horizontal velocity of the plume $u^{(S)}$ in relation to that of the cross wind $\overline{u}(z \gtrsim h_s)$. Let the difference be $\Delta u^{(S)}$. $u^{(S)}$ is determined by the acceleration of the fluid elements as they rise through the surface boundary layer where over the distance $z > h_{ss}$, the difference in the mean velocity is of the order of u_*. Therefore $\Delta u^{(S)} \sim u_{*_c}$. As with a plume leaving a chimney, this velocity difference means that there is a vortex 'sheet' around the plume with vorticity ω_z. Some of this vorticity is advected downwind (causing plumes to be elongated in the direction of the cross wind). The interaction between the entrainment into the plume and ω_z causes the plume to move downwind and to bend over [16]. Such studies show that these distortion effects are only significant when $\Delta u^{(S)} \gtrsim 1/3$ of the plume speed (i.e. $W^{(e)} \sim w_*$) so that we can expect significant distortion to the plume structure when $u_* \gtrsim 1/3 w_*$.

The effects of weak cross flow ($u_{*_c} \ll w_*$) do not change the statistical properties of the turbulence provided the *mean* shear stress $\overline{\tau} = \overline{u_*^2}$ generated by mean flow is very weak compared to w_*^2. Surface layer research in the Indian desert by Rao *et al.* [65] shows that even when $\overline{u}(z = z_{\text{ref}})$ (where $z_{\text{ref}} \sim h_s \sim 10m$) is of the order of 2-3 times as large as w_*, the shear stress layer eddy viscosity $\kappa u_{*_c} z$ is determined by the convective turbulence (i.e. u_{*_c} is proportional to w_*). This is why the surface shear stress τ is proportional to w_* and to $\overline{u}(z \simeq z_{\text{ref}})$. These experiments and the large eddy simulation results of [55] (which have rather desert-like boundary conditions) confirm that only when $\overline{\tau}^{1/2} \gtrsim 0.3 w_*$ is the structure of the plume-dominated convective turbulence destroyed by a cross wind.

No work has been done on how cross winds affect the structure of the puffs leaving the boundary. If the main mechanism for puff generation involves the sub-surface unsteady thermal field, it is clear that this will be quite changed. The puffs would then come from the basic instability of the temperature profile in the surface layer.

There are other, more complex kinds of cross wind, that are less well understood but which can have strong effects on the turbulence. Krishnamurti and Howard [48] have shown in the laboratory (what is observable in a saucepan of soup) that once a mean motion is initiated convection can maintain it despite frictional effects. This can appear locally as a crossflow. Typically these self-induced velocities are a fraction of w_*, so that the basic structure of convection is not changed.

5.3. SOME CROSS WIND EFFECTS AT THE INVERSION LAYER

The existence of a mean cross wind $\overline{u}^{(u)}$ in the upper region outside a convective layer is likely to drive a mean velocity $u^{(CL)}$ in the well mixed region below the inversion layer. In general there is a velocity jump across the layer of $\Delta \overline{u}^{(I)} = \overline{u}^{(u)} - \overline{u}^{(CL)}$. On geophysical scales Coriolis forces lead to a change in flow direction of $\Delta \phi^{(I)}$, (Fig. 7), that depends on the relation between mean shear stress $\tau_{xz}, \tau_{y,z}$ and $\overline{u}^{(CL)}, \overline{v}^{(CL)}$. Given the jump $|\Delta \overline{u}^{(I)}| \lesssim w_*$ these shear stress gradients do not significantly alter the statistics of the structure of the turbulence in the *interior* of the convective layer.

The main changes to the turbulence occur in I.L. where the gradient of the mean velocity is of the order of $\Delta \overline{u}^{(I)} / \Delta h$ which is comparable to the fluctuating gradients $w_* / \Delta h$ when $\Delta \overline{u}^{(I)}$ is comparable to w_*.

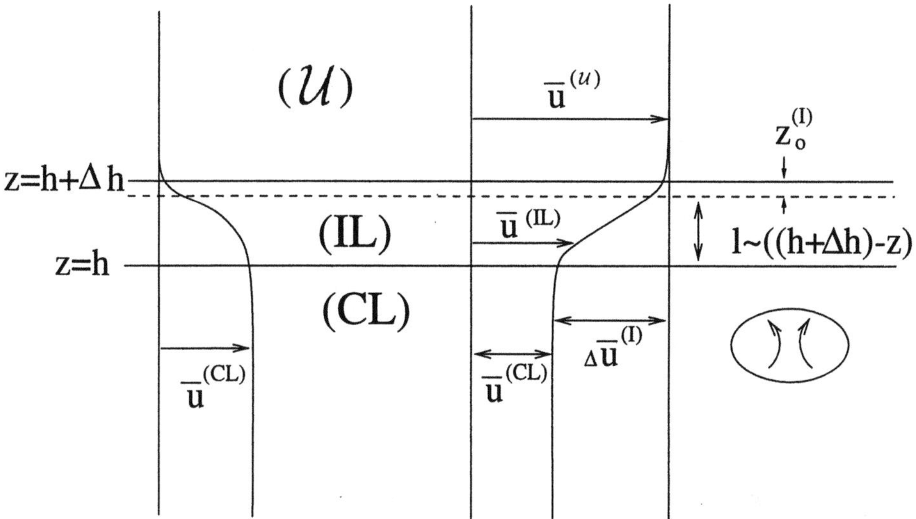

Figure 7. Inversion layer when there is a cross wind showing how a turbulent shear layer develops and how eddy motion and the mean profile can be similar to a turbulent stratified boundary layer below a rigid rough surface at $z = h + \Delta h$ (after [17, 74]).

In the discussion of section 4.3 we noted how the strong density gradient in I.L. inhibits vertical motion and therefore acts like a rigid surface on eddies a distance $z(\lesssim L_x)$ below the layer. (The condition is that $w_*/N^{(I)} \ll L_x$ where $L_x \sim h/5$ is the integral scale of eddying motion in the convective layer.) The effect of mean shear $\partial\overline{u}/\partial z$ is to strain the eddies and increase dissipation. Numerical simulation and field data show that it is the controlling factor for the scale of turbulence for stable flows when $\mathrm{Ri} = [N^{(I)}/(\partial\overline{u}/\partial t)]^2)$ is less than unity [69],[31]. It also acts to block large scale eddies from influencing the turbulence above I.L. The distorting and localising effect of shear means that the dynamics and the structure of I.L. are changed. Now local processes of production and dissipation of energy and scalar fluctuations, dominate more than those associated with the interactions of the convective eddies and the I.L. Trapped waves are changed by shear but their role appears to be less critical.

When this same argument is applied to the boundary layer over a rigid surface (where eddies are also 'blocked' at the surface and the turbulence is in local equilibrium) it leads to Prandtl's concept of a logarithmic shear boundary layer. Csanady [17] showed on these grounds that the mean velocity profiles $\overline{u}(h+\Delta h-z)$ are also logarithmic near stable inversion layers, with the local roughness length being proportional to the vertical displacements on the layer ($\sim \sigma_W/N^{(I)}$). Recent laboratory studies [75] support this idea and also show how the large gradients in the logarithmic layer must adjust smoothly to the mean velocity $\overline{u}^{(u)}$ outside the I.L. This concept is quite consistent with the growing practice in modelling convective layers of specifying the length scales near the inversion layer in terms of the distance from the layer (when $\sigma_W/N^{(I)} \ll 1$) – so-called 'non-local' scaling [80]. A criticism of these models is that they do not adequately differentiate between situations dominated by the longer range wave-eddy actions; and a particularly striking feature is that in the shear-free I.L. case Δh increases in proportion to h while with shear Δh remains approximately constant even though h is varying (which is implied by the log profile).

5.4. EFFECTS OF SURFACE UNDULATIONS

When the fluid is heated above a large surface, that is sloping at a small angle α_s, the difference between the reduced hydrostatic vertical pressure gradient in the surface layer and above it causes a horizontal pressure gradient which drives a mean motion up the slope which is of the order of $w_*\alpha_s$. The same mechanism can drive local recirculating flows above an undulating surface when it is heated (or cooled). This is why the field experiments and numerical studies confirm that when α_s is small the eddy structure is unchanged [41],[47].

But what happens near an undulating but unheated surface, with the energy being supplied internally or through the other surface? It is found that mechanically stirred turbulence with with integral scale L_x produces mean recirculating motions over vertical scale $\lambda \gg L_x$ above a rigid undulating surface with wave length λ [88]. Therefore the same effect can be expected in convective driven turbulence. Indeed even when the surface is thermally active this mechanical process may contribute to the mean motion. Both processes may lead to deformation of the surface whether it is sand or molten rock, by mechanical and/or thermal processes.

5.5. UNSTEADY BOUNDARY CONDITIONS

The last non-ideal effect to be considered is that of temporal variations in the energy sources that drive convective turbulence. The most rapid changes in the atmospheric turbulence occur when a positive surface heat flux F_{θ_s} is turned on (on a time scale less than τ_p for fully developed convection); puffs or thin plumes form in the unstable temperature profile; eventually they are absorbed into larger eddy structures. As F_{θ_s} increases in the atmosphere (and ocean) the depth h of the convective layer grows at a rate $E_b = dh/dt$ that is about 1/10 of the typical value of the turbulent velocities w_*. Therefore the vertical structure of the turbulence can adjust to the changing scale of the convective layer, at time t, $h(t)$ (when averaged over the eddy time h/w_*). Thus expressions for the vertical variance such as (15), can be applied during this growth phase of the layer.

Note that if the flux F_{θ_s} is constant with time dh/dt can still increase as a result of mixing in the inversion layer. But it is much slower.

The main reason why this growth process is not reversible is because if F_{θ_s} is steadily reduced from time t_1 so that $dh/dt < 0$ for $t > t_1$, the temperature (and density) distribution $\theta(z)$ in the upper region $(\mathcal{U})$ does not revert to its previous stable profile; $\theta(z)$ is now nearly constant with height above the new value of $h(t_1)$ and below the previous value $h(t_1)$. This is why experiments and numerical simulation show the 'collapse' of convective layers is not a self-similar process (as is the growth). Another reason for this complexity is because the thermal behaviour at and below the bottom surface ($z = 0$) plays a greater role which is not determined by the time scales of convection. In the case of the atmospheric boundary layer, when radiant heating diminishes in the afternoon, this leads to a reduction in temperature T_s. In desert conditions T_s may drop to a value below $\theta(z \simeq h_s)$ and the surface layer may become stably stratified and quite still even before sunset. Meanwhile turbulence persists and slowly decays above this layer. By contrast over vegetated and urban areas, the main effect is a gradual decline in heat flux F_{θ_s} into the atmosphere being supplied directly from the stored heat of surface roughness or from the ground (F_{θ_g}). In the latter case the conditions are appropriate for the formation of puff eddies [71]. Numerical simulations [77] show that the density profile in the surface layer continues to be unstable, at least in part, until the solar radiation ceases. As in the desert conditions the turbulence declines in the central part of the convective layer, as is correlated with ever weakening entrainment processes in the upper part of the convective layers which ceases to be 'capped' by a well defined inversion layer.

The turbulence in the central part of the convective layer decays more slowly than simple neutrally buoyant turbulence (e.g. in a wind tunnel) because the buoyant eddies in this region contain potential energy (equal to the work required to displace them downwards $\propto g\Delta\rho^{(e)}l^3h \sim w_*^2l^3$ as well as as kinetic energy $(\propto \rho w_*^2 l^3)$). If the decay were only associated with the dissipation of kinetic energy, as in homogeneous turbulence, then the typical eddy velocity in the central region would vary with time t as

$$u^{(e)}(t) \sim u^{(e)}(t_1)(\frac{t}{(h/w_*)(t_1)})^{-p/2},$$

where the exponent $p \simeq 1.1 \pm 0.1$ [60]. However if the small eddies of scale $l(\sim \Delta h^{(I)} \sim h/10)$ have even a very slight level of buoyancy (say 10% of the rms value), they will have a small but finite rise velocity of say 10% of w_*, that will persist on a time of the order of $10h/w_*$ until they reach the top. This is consistent with the images of decaying cumulus clouds and with numerical simulations [77] but not necessarily with the theory in [5].

6. Concluding Remarks

This review has shown how recent research has explained many of the key phenomena in convective turbulence in many different geophysical situations. Much insight comes from using local analyses drawing on wider ideas and models of turbulence, and from matching these together. However there are several important aspects where there is considerable uncertainty, such as the decay of convective turbulence, the entrainment of fluid from the inversion layer, the roles of roughness elements at the surface and the processes that determine the different forms of eddy structure. The matching of the different sub-regions also needs more detailed study.

Acknowledgements

This work at A.S.U. was supported by N.S.F. I am grateful to many recent and earlier collaborations and colleagues with especial thanks to H.J.S. Fernando, I. Oroud, H.E. Huppert, A. Grachev, J.C. Kaimal, P.J. Mason, J.C. Wyngaard and S. Zilitinkevich. Thanks also to the organisers of the workshop Erich Plate and Evgeni Fedorovich. Many thanks to Alison Harrison for an excellent job of preparing this paper.

References

1. Adrian, R.J., Ferreira, R.T.D.S. and Boberg, T. (1986) Turbulent thermal convection in wide horizontal fluid layers, *Experiments in Fluids* **4**, 121-141.
2. Batchelor, G. K. (1954) Heat convection and buoyancy effects in fluids, *Q. J. Roy. Met. Soc.* **80**, 339-358.
3. Batchelor, G. K. (1967) *Introduction to Fluid Dynamics*, Cambridge University Press, Cambridge.
4. Batchelor, G.K. (1987) Stability of a large gas bubble rising through liquid, *J. Fluid Mech.* **184**, 399-422.
5. Batchelor, G.K., Canuto, V.M. and Chasnov, J.R. (1991) Homogeneous buoyancy-generated turbulence, *J. Fluid Mech.* **235**, 349-378.
6. Belcher, S.E. and Hunt, J.C.R. (1993) Turbulent shear flow over slowly moving waves, *J. Fluid Mech.* **251**, 109-148.
7. Bhat, G.S. and Narasimha, R. (1996) A volumetrically heated jet: large-eddy structure and entrainment characteristics, *J. Fluid Mech.* **325**, 303-330.
8. Bonnet, J.P. and Glauser, M.N. (eds.) (1993) *Eddy structure identification in free turbulent shear flows*, Kluwer Academic Publishers, Dordrecht.
9. Brost, R.A. and Wyngaard, J.C. (1978) A model study of the stably stratified planetary boundary layer, *J. Atmos. Sci.* **25**, 1427-1440.
10. Carruthers, D.J. and Moeng, C.-H. (1987) Waves in the overlying inversion of the convective boundary layer, *J. Atmos. Sci.* **44**, 1801-1808.
11. Castaing, B., Gunaratne, G., Heslot, F., Kadanoff, L., Libchaber, A., Thomae, S., Wu, X., Zaleski, S. and Zanetti, G. (1989) Scaling of hard thermal turbulence in Rayleigh-Bénard convection, *J. Fluid Mech.* **204**, 1-30.
12. Chandrasekhar, S. (1961) *Hydrodynamic and hydromagnetic stability*, Clarendon, Oxford.
13. Chapman, C.J. and Proctor, M.R.E. (1980) Nonlinear Rayleigh-Bénard convection between poorly conducting boundaries, *J. Fluid Mech.* **101**, 759-782.

14. Chatwin, P.C. and Sullivan, P.J. (1979) The relative diffusion of a cloud of passive contaminant in incompressible turbulent flow *J. Fluid Mech.* **91**, 337-355.
15. Ching, C.Y., Fernando, H.J.S., and Robles, A. (1995) Breakdown of line plumes in turbulent environments, *J. Geophys. Res.* **100**, 4707-4713.
16. Coelho, S.L.V. and Hunt, J.C.R. (1989) Vorticity dynamics of the near field of strong jets in crossflows, *J. Fluid Mech.* **200**, 95-120, also Proc. IMA/SMAI.
17. Csanady, G.T. (1978) Turbulent interface layers, *J. Geophys. Res.* **83**, 2329-2342.
18. Deardorff, J.W. and Willis, G.E. (1967) Investigation of turbulent thermal convection between horizontal plates, *J. Fluid Mech.* **28**, 675-704.
19. Deardorff, J.W., Willis, G.E. and Stockton, B.H. (1980) Laboratory studies of the entrainment zone of a convectively mixed layer, *J. Fluid Mech.* **100**, 441-64.
20. Fernando, H.J.S, Hunt, J.C.R. and Carruthers, D.J. (1994) Turbulence, waves and mixing at stratified density interfaces; modelling and experiments, in N. Rockliff and I. Castro (eds.) *Proc. 4th IMA Conf. on 'Stratified flows ...'*, Clarendon, Oxford, pp. 175-200.
21. Frech, M. and Mahrt, L. (1995) A two scale mixing formulation for the atmospheric boundary layer, *Bound. Layer Met.* **23**, 91-104.
22. Frisch, U. (1996) *Turbulence*, Cambridge University Press, Cambridge.
23. Fung, J.C.H., Hunt, J.C.R., Perkins, R.J., Wray, A.A. & Stretch, D. (1991), Defining the zonal structure of turbulence using the pressure and invariants of the deformation tensor, in A.V. Johansson and P.H. Alfredsson (eds.), *Advances in Turbulence 3*, Springer Verlag, Berlin, pp. 395-404.
24. Gill, A. E. (1982), *Atmosphere-Ocean Dynamics*, Academic Press.
25. Gossard, E. E. and Strauch, R. G. (1983) *Radar observations of clear air and clouds.* Elsevier, Amsterdam.
26. Gregory, D. and Miller, M.J. (1989) A numerical study of deep tropical convection, *Q. J. Roy. Met. Soc.* **490**, 1209-1241.
27. Hewer, F.E. and Wood, N. (1997), The effective roughness length for scalar transfer in neutral conditions over hilly terrain, *Quart. J. Roy. Met. Soc.* (In press).
28. Holmes, P. J., Berkooz, G. & Lumley, J. L. (1996), *Turbulence, coherent structures, symmetry and dynamical systems*, Cambridge University Press, Cambridge.
29. Hunt, J.C.R. (1984) Turbulence structure in thermal convection and shear-free boundary layers, *J.Fluid Mech.* **138**, 161-184.
30. Hunt, J.C.R., Kaimal, J.C. & Gaynor, J.E. (1985) Some observations of turbulence structure in stable layers, Q.J. Roy. Met. Soc. **111**, 793-815.
31. Hunt, J.C.R., Kaimal, J.C. and Gaynor, J.E. (1988) Eddy structure in the convective boundary layer - new measurements and new concepts, *Q.J. Roy. Met. Soc.* **114**, 821-858.
32. Hunt, J.C.R., Moin, P., Lee, M, Moser, R.D., Spalart, P., Mansour, N.N., Kaimal, J.C. and Gaynor, E. (1989) Cross correlation and length scales in turbulent flows near surfaces, in H.-H. Fernholz and H.E. Fiedler (eds.) *Advances in Turbulence* **2**, Springer-Verlag, Berlin, pp 128-134.
33. Hunt, J.C.R., Vassilicos, J.C. & Kevlahan, N.K.R. (1994). Turbulence: a state of nature or a collection of phenomena?, in H. Branover and Y. Unger (eds.) *Progress in Turbulence Research*, Progress in Astronautics and Aeronautics Series vol. 162, pp. 1–18.
34. Hunt, J. C. R. (1995) Effects of body forces on turbulence, in R. Benzi (ed.) *Advances in Turbulence V*, Kluwer, pp. 229-235.
35. Huppert, H.E. (1986) The intrusion of fluid mechanics into geology, *J. Fluid Mech.* **173**, 557-594.
36. Huppert, H.E. and Sparks, R.S.J. (1988) Melting the roof of a chamber containing a hot, turbulently convecting fluid, *J. Fluid Mech.* **188**, 107-131.
37. Jackson, P.S. (1981) On the displacement height in the logarithmic velocity profile, *J. Fluid Mech.* **111**, 15-25.
38. Jeong, J. Hussain, F., Schoppa, W. and Kim, J. (1997) Coherent structures near

the wall in a turbulent channel flow, *J. Fluid Mech.* **332**, 185-214.
39. Jimenez, J. and Moin, P (1991) The minimal flow unit in near-wall turbulence, *J. Fluid Mech.* **225**, 213-240.
40. Kaimal and Hougen, D.A. (1967) Characteristics of vertical velocity fluctuations observed on a 430 m. tower, *Q. J. Roy. Met. Soc.* **93**, 305-317.
41. Kaimal, J.C., Eversole, R.A., Lenschow, D.H., Stankov, B.B., Kahn, P.H. and Businger, J.A. (1982) Spectral characteristics of the convective boundary layer over uneven terrain, *J. Atmos. Sci.* **39**, 1098-1114.
42. Keffer, J.F., Kawall, J.G., Hunt, J.C.R. and Maxey, M.R. (1978) The uniform distortion of thermal and velocity mixing layers, *J. Fluid Mech.* **86**, 465-490.
43. Kit, E.L.G., Strang, E.J. and Fernando, H.J.S. (1997) Measurement of turbulence near shear-free density interfaces, *J. Fluid Mech.* **334**, 293-314.
44. Kiya, M., Ohyama, M. and Hunt, J.C.R. (1986) Vortex pairs and rings interacting with shear layer vortices, *J. Fluid Mech.* **172**, 1-15.
45. Komori, S. and Nagata, K. (1996) Effects of molecular diffusivities on counter-gradient scalar and momentum transfer in strongly stable stratification, *J. Fluid Mech.* **326**, 205-237.
46. Kraichnan, R. (1962) Turbulent thermal convection at arbitrary Prandtl number, *Phys. Fluids* **5**, 1374-1389.
47. Krettenauer, K. and Schumann, U. (1992) Numerical simulation of turbulent convection over wavy terrain, *J. Fluid Mech.* **237**, 261-299.
48. Krishnamurti, R. and Howard, L.N. (1981) Mean flow set up by tilted plumes in a confined space, *Proc. Nat. Acad. of Sci. U.S.A.* **78**.
49. Landau, L.E. and Lifshitz, L.M. (1959) *Fluid Mechanics*, Pergamon, Oxford.
50. Lawson, R.E., Snyder, W.H., Shipman, M.S. (1998) A laboratory model of diffusion in the convective boundary layer, 10th Joint Conference on Applications of Air Pollution, American Meteorological Society, January 1998.
51. Lenschow, D.H. and Stephens, P.L. (1980) The role of thermals in the convective boundary layer, *Bound. Layer Met.* **19**, 509-532.
52. Linden, P.F. (1973) The interaction of a vortex ring with a sharp density interface: a model for turbulent entrainment, *J. Fluid Mech.* **60**, 467-480.
53. Linden, P.F. (1996) Natural ventilation of buildings, International Congress of Theoretical and Applied Mathematics, Kyoto.
54. Long, R.R. and Chen, T-C. (1981) Experimental evidence for the existence of the 'mesolayer' in turbulent systems, *J. Fluid Mech.* **105**, 19-59.
55. Mason, P. J. (1992) Large eddy simulation of dispersion in convective boundary layers. *J. Atmos. Sci.* **41**, 1561-1571.
56. MacVean, M.K. and Mason, P.J. (1990) Cloud top entrainment instability through small scale mixing, *J. Atmos. Sci.* **47**, 1012-1030.
57. McGrath, J.L., Fernando, H.J.S. and Hunt, J C R (1997), Turbulence, waves and mixing at shear-free density interfaces. Part II - laboratory experiments, *J. Fluid Mech.* **347**, 235-261.
58. Moncrieff, M. W. (1992) Organised convective system; archetypal dynamical models, mass and momentum flux theory, and parametrization, *Q. J. Roy. Met. Soc.* **118**, 819-850.
59. Mory, M. (1991) A model of turbulent mixing across a density interface including the effect of rotation, *J. Fluid Mech.* **223**, 193-207.
60. Nieuwstadt, F.T.M. and Brost, R.A. (1986) The decay of convective turbulence, *J. Atmos. Sci.* **43**, 532-546.
61. Pearson, H.J., Puttock, J.S. and Hunt, J.C.R. (1983), A statistical model of particle motions and vertical diffusion in a homogeneous stratified turbulent flow, *J. Fluid Mech.* **129**, 219–249.
62. Perot, B. and Moin, P. (1995) Shear-free turbulent boundary layers. Part 1. Physical insights into near-wall turbulence, *J. Fluid Mech.* **295**, 199-227.
63. Prandtl, L. (1932) Meteorologische Anwendung der Stromungslehre, *Beitr. Phys.*

Atmos. **19**, 188-202.

64. Priestley, C.H.B. (1959) *Turbulent transfer in the lower atmosphere*, University of Chicago Press, Chicago.
65. Rao, K.G., Narasimha, R. and Prabhu, A. (1996) Estimation of drag coefficient at low wind speeds over the monsoon trough land region during MONTBLEX-90, *Geophys. Res. Let.* **23**, 2617-2620.
66. Rodi, W. (ed.) (1982) *Turbulent buoyant jets and plumes*, Pergamon, Oxford.
67. Ruelle, D. and Takens, F. (1971) On the nature of turbulence, *Commun. Math. Phys.* **20**, 167-192.
68. Schmidt, H. and Schumann, U. (1989) Coherent structure of the convective boundary layer derived from large-eddy simulations, *J. Fluid Mech.* **200**,, 511-562.
69. Schumann, U. (1996) Direct and large eddy simulation of stratified homogeneous shear flows, *Dyn. Atmos. Oceans* **23**, 81-98.
70. Schumann, U. and Schmidt, H. (1989) Heat transfer by thermals in the convective boundary layer, in H.H. Fernholz and H.F. Fiedler (eds.), *Advances in Turbulence 2*, Springer, Berlin, pp. 210-215.
71. Scorer, R.S. (1954) The nature of convection as revealed by soaring birds and dragonflies, *Q. J. Roy. Met. Soc.* **80**, 68-77.
72. Scorer, R.S. (1978) *Environmental Aerodynamics*, Ellis Horwood, Chichester.
73. Sirovich, L. (ed.) (1991) *New Perspectives in Turbulence*, Springer-Verlag, Berlin.
74. Sparrow, E.M., Husar, R.B. and Goldstein, R. J. (1970) Observations and other characteristics of thermals, *J. Fluid Mech.* **41**, 793-800.
75. Strang, E. and Fernando, H.J.S. Entrainment and mixing in a stratified layer, submitted to *J. Fluid Mech.*
76. Sykes, R.I., Henn, D.I. and Lewellen, D.S. (1993) Surface layer description under free-convection conditions, *Q. J. Roy. Met. Soc.* **119**, 409-421.
77. Sorbjan, Z. (1997) Decay of convective turbulence revisited, *Bound. Layer Met.* **82**, 501-515.
78. Tennekes, H. (1975) Eulerian and Lagrangian time microscales in isotropic turbulence, *J. Fluid Mech.* **67**, 561-567.
79. Tennekes, H. and Lumley, J.L. (1971) *First Course in Turbulence*, M.I.T. Press, Cambridge, Mass.
80. Therry, G. and Lacarrère, P. (1983) Improving the eddy kinetic energy model for planetary boundary layer description. *Bound. Layer Met.* **25**, 63-88.
81. Townsend, A.A. (1959) Temperature fluctuations over a heated horizontal surface, *J. Fluid Mech.* **5**, 209-241,
82. Townsend, A.A. (1976) *The Structure of Turbulent Shear Flow*, Cambridge University Press, Cambridge.
83. Turfus, C. and Hunt, J.C.R. (1987), A stochastic analysis of the displacements of fluid elements in inhomogeneous turbulence using Kraichnan's method of random modes, in G. Comte-Bellot and J. Mathieu (eds.) *Advances in Turbulence* **1**, (Proc. European Turbulence Conf., Lyon) Springer Verlag, Berlin, pp.191-203.
84. Turner, J.S. (1973) *Buoyancy Effects in Fluids*, Cambridge University Press, Cambridge.
85. Turner, J.S. (1986) Turbulent entrainment: the development of the entrainment assumption, and its application to geophysical flows, *J. Fluid Mech.* **173**, 431-471.
86. Voropayev, S.I., Afanasyev, Y.D. and van Heist, G.J.F. (1993) Experiments on the evolution of gravitational instability of an overturned, initially stably stratified fluid, *Phys. Fluid A* **5**, 2461-2466.
87. Warhaft, Z. (1980) An experimental study of the effect of uniform strain on thermal fluctuations in grid-generated turbulence, *J. Fluid Mech.* **99**, 545-573.
88. Wong, H.Y.W. (1985) *Shear free turbulence and secondary flow near angled and curved surfaces*, Ph.D. Thesis, University of Cambridge.
89. Wyngaard, J.C. and Brost, R.A. (1984) Top-down and bottom-up diffusion in the convective boundary layer, *J. Atmos. Sci.* **41**, 102-112.

90. Zilitinkevich, S., Grachev, A., and Hunt J.C.R. (1998) Surface frictional processes and non-local heat/mass transfer in the shear-free convective boundary layer. See this volume.

SURFACE FRICTIONAL PROCESSES AND NON-LOCAL HEAT/MASS TRANSFER IN THE SHEAR-FREE CONVECTIVE BOUNDARY LAYER

S. ZILITINKEVICH[1,2], A. GRACHEV[2], AND J. C. R. HUNT[3]

[1]*Institute for Hydrophysics, GKSS Research Centre*
21502 Geesthacht, Germany

[2]*A. M. Obukhov Institute of Atmospheric Physics, RAS*
109017 Moscow, Russia

[3]*Department of Applied Mathematics and Theoretical Physics,*
University of Cambridge
Cambridge, CB3 9EW, UK

Abstract

During the last several decades the near-surface frictional processes in the convective boundary layer (CBL) were considered in the spirit of the Prandtl [49] theory of free convection, implying the concepts of (i) universal chaotic turbulence and (ii) local correspondence between turbulent fluxes and mean gradients. Accordingly the fluxes of heat and water vapour in the atmospheric surface layer are parameterized disregarding gross features of the CBL. Conventional practical tools are the Monin-Obukhov similarity theory and downgradient turbulence closure models. In strong convection, however, local-transport models break down. This is because of the buoyancy-driven semi-organised structures embracing the CBL, namely, comparatively narrow uprising plumes surrounded by wider downdraughts. The surface layer feeds the plumes through the CBL-scale convergence flow patterns. These can be treated as internal boundary layers of radial geometry strongly affected by the buoyancy forces. Generally, convergence flows superimpose on mean wind. In calm weather they yield their own velocity shears characterised by the "minimum friction velocity". As a result, turbulent mixing and heat/mass transfer are facilitated depending on gross features of the CBL, such as the CBL depth and the surface roughness length. Only the first step has been taken in analysing the above mechanism. In the present paper an advanced theoretical model for the non-local frictional processes is proposed for the roughness length to the CBL depth ratios less than 0.0001. Theoretical predictions are compared with new experimental data. It is shown that data from large-eddy simulation (LES), which served as an "empirical" basis in early models, diverge drastically from field data. It is conceivable that LES, although efficient in other respects, has limited capability in treating surface roughness effects. The proposed model is in excellent agreement with atmospheric data.

E.J. Plate et al. (eds.), Buoyant Convection in Geophysical Flows, 83–113.

1. Introduction

Traditional theoretical models for the surface-layer turbulence, namely, the Monin and Obukhov [42] similarity theory and lower-order downgradient turbulence closures (*e.g.*, those classified by Mellor and Yamada [37]), imply two concepts: (i) turbulent flows are universally composed of fully organised mean components and fully chaotic components, and (ii) turbulent fluxes are linked uniquely to local mean gradients,

turbulent flux = - (eddy diffusivity)×(mean gradient).

Here, a coefficient called eddy viscosity/conductivity/diffusivity in its turn depends on local properties of the mean flow.

During the last decade it has been understood that in general terms the above concepts break down. In addition to fully organised and fully chaotic motions, a further type of motion has been recognised, namely, large-scale coherent structures which are semi-organised, *i.e.* neither chaotic nor deterministic. Such structures play an important role in the convective boundary layer (CBL). They strongly contribute to the CBL turbulence and result in essentially non-local vertical turbulent transport in both the CBL interior and the surface layer.

The physical nature of non-local near-surface transport is especially clear in the shear-free convection regime. Here, at very high Rayleigh numbers, coherent structures are the buoyancy-driven 3-dimensional Benard-type convective cells composed of narrow uprising plumes and wide downdraughts. They embrace the entire CBL (~2 km in height) and include pronounced large-scale (~5 km in diameter) convergence flow patterns close to the surface. Roughly speaking, the letter represent internal boundary layers (IBLs) of topologically radial geometry. Due to convective turbulence, the IBL depth in a typical cell grows up towards the plume axis, so that IBLs feed the plumes. Furthermore, the IBL flows yield horizontal velocity shears that generate mechanical turbulence, and consequently, facilitate vertical exchange of heat, water vapour, and other scalar admixtures at the surface. Hence the surface-layer turbulence reveals an essential dependence on gross features of the CBL, such as the CBL depth, h, and the surface roughness length, z_{0u}. Traditional models overlook this mechanism and underestimate the surface fluxes.

Leaving detailed discussion of early non-local models to Subsection 2.3, we notice here that even the most advanced models oversimplified the nature of the surface layer. The near-surface convergence flow patterns were treated as either unstably stratified but horizontally homogeneous flows (Schumann [52]) or heterogeneous but neutrally stratified internal boundary layers (Sykes *et al.* [57], Zilitinkevich [71,72]). However, the effects of heterogeneity and unstable stratification are both important.

In addition, no reliable field data on the dimensionless minimum friction velocity and the heat/mass transfer coefficients dependent on z_{0u}/h were available until present. The early models, therefore, were calibrated using data from numerical large-eddy simulation (LES) studies. However, these data diverge drastically from new atmospheric data analysed in the present paper (Figs. 8 and 9).

In the framework of a new theoretical model for non-local surface friction and vertical transport processes, this divergence is explained. Moreover, it is clearly shown that

LES still has limited capability of treatment of surface roughness effects. The other shortcomings of the early models indicated above are also clarified.

The new model is developed for strong convection regimes over surfaces that are not very rough, $z_{0u}/h < 10^{-4}$. Verification of the model using measurements in atmospheric boundary layers on land and sea shows excellent agreement between model predictions and experimental data. The model opens the way to practical calculation of the surface fluxes in conditions of calm weather convection, when traditional parameterizations fail.

2. Overview

2.1. TRADITIONAL LINE

The commonly accepted approach in modelling the transport properties of the atmospheric surface layer is the Monin-Obukhov similarity theory. This theory holds that, except within close vicinity of the surface, the governing parameters of the surface-layer turbulence are the near-surface fluxes of momentum and potential temperature, τ_s and Q_s, and the buoyancy parameter, β,

$$\tau_s = \rho_0 \left| \overline{\mathbf{u}'w'} \right|_{z=0}, \quad Q_s = \overline{\theta' w'}\Big|_{z=0}, \quad \beta = g / \bar{\theta}\Big|_{z=0}, \tag{1}$$

where z is the height over the surface, g is the gravitational acceleration, ρ_0 is the reference air density, $\mathbf{u}=\mathbf{i}u+\mathbf{j}v$ is the horizontal velocity vector (with the streamwise, u, and transverse, v, components), w is vertical velocity, θ is potential temperature, primes denote turbulent fluctuations, and the over-bar denotes the ensemble or time mean value. The velocity, length, and temperature scales composed from these parameters are

$$u_* = (\tau_s / \rho_0)^{1/2}, \quad L = u_*^3 / \beta Q_s, \quad T_* = Q_s / u_*, \tag{2}$$

where u_* is the friction velocity, and L is the Monin-Obukhov length (we skip minus, traditionally put on the r.h.s. of the expression of L, so that we may deal with a positive scale). Using these scales the theory holds that dimensionless vertical gradients of the mean wind velocity and potential temperature are universal functions of the dimensionless height z/L,

$$\partial u / \partial z = (u_* / k_u z)\Phi_u(z/L), \quad \partial \theta / \partial z = -(T_* / k_T z)\Phi_T(z/L), \tag{3}$$

where $k_u \approx k_T \approx 0.4$ are the von Kármán constants, and $\Phi_u(z/L)$ and $\Phi_T(z/L)$ are known empirical functions that satisfy the conditions $\Phi_u(0)=1$ and $\Phi_T(0)=1$. Integrating Eq. (3) over z and employing the roughness lengths, z_{0u} for wind and z_{0T} for temperature, to parameterize the flow-surface interaction, the expressions of the vertical profiles are derived,

$$u(z) = k_u^{-1} u_* \left[\ln(z / z_{0u}) + \Psi_u(z/L) - \Psi_u(z_{0u}/L) \right], \tag{4a}$$

$$\theta(z) - \theta_s = -k_T^{-1} T_* \left[\ln(z / z_{0T}) + \Psi_T(z/L) - \Psi_T(z_{0T}/L) \right]. \tag{4b}$$

Here, subscript "s" stands for the value of a variable at the surface (at z=0), and the functions Ψ_u and Ψ_T represent integrals,

$$\Psi_{\{u,T\}}(\xi) = \int \left[\Phi_{\{u,T\}}(\xi) - 1 \right] \xi^{-1} d\xi. \tag{5}$$

Given the left hand side (l.h.s.) of Eqs. (4a,b) from measurements or from a numerical model, with knowledge of the roughness parameters z_{0u} and z_{0T}, Eqs. (2), (4a,b) are solved for u_*, Q_s, and L. Then the fluxes of momentum, $\tau_s = \rho_0 u_*^2$, and heat, $H_s = \rho_0 c_p Q_s$, are found, where c_p is the specific heat of air at constant pressure (see Monin and Obukhov [42], Monin and Yaglom [43]).

This approach is successfully applied when the wind is strong enough. In particular Eqs. (4a,b) yield well grounded logarithmic profiles in neutral stratification (when u_* is high or Q_s is small, so that $L = u_*^3 / \beta Q_s$ >100m).

In conditions of vanishing mean wind, the momentum flux τ_s, as defined by Eq. (1a), and the friction velocity $u_* = \sqrt{\tau_s}$ should also be taken as approaching zero. Then $L \to 0$, $T_* \to \infty$, and Eq. (4b), in contrast to the second equation in Eq. (3b), has no physical significance. Indeed, with small values of L the concept of the roughness length is no longer consistent with the similarity theory under consideration. This concept implies the existence of well-developed logarithmic intervals in the velocity and temperature profiles immediately above a thin roughness/viscous layer adjacent to the rough/smooth surface (see, *e.g.*, Section 5.4, Monin and Yaglom [43]). In particular, z_{0T} is defined as a conventional height at which the logarithmic temperature profile, extended logarithmically towards the surface, achieves a value equal to the surface temperature. The above similarity theory, however, does not incorporate any logarithmic portions in the profiles when L is small enough, *i.e.* $L < 30\, z_{0u}$.

The classical Prandtl [49], Obukhov [45,46], and Monin and Obukhov [42] free convection formulation, $\partial\theta / \partial z = -C_{fc} Q_s^{2/3} \beta^{-1/3} z^{-4/3}$, yields no logarithmic interval in the temperature profile,

$$\theta(z) = \theta_a + 3 C_{fc} Q_s^{2/3} (\beta z)^{-1/3}, \tag{6}$$

where θ_a is the mean potential temperature in the well-mixed interior of the CBL, and C_{fc} is a dimensionless constant (C_{fc} =1, after Kader and Yaglom [26]). It is not surprising, therefore, that the heat transfer law that seemingly follows from Eq. (6), substituting the roughness length z_{0T} for z and the surface temperature θ_s for $\theta(z_{0T})$, *viz.*,

$$Q_s = (3C_{fc})^{-3/2}(\beta z_{0T})^{1/2}\Delta\theta^{3/2}, \quad \Delta\theta \equiv \theta_s - \theta_a, \tag{7}$$

was never proposed by the authors of the classical theory. They apparently understood the inconsistency of the similarity theory of free convection with the use of the roughness length concept. The heat transfer law Eq. (7) was, nevertheless, employed for practical purposes (*e.g.*, Louis [32]), even though it was poorly founded.

A more logical free-convection heat transfer model, although also based on the idea of local flux-gradient correspondence, was proposed simultaneously by Malkus [35] and Priestley [50]. Here, the near-surface viscous sub-layer is supposed to be thick ($\propto \nu / u_*$) as the friction velocity is very small or even equal to zero. As a result the effect of the flow-roughness interaction is neglected, and the dynamically-smooth-surface model is applied. Then molecular viscosity, ν, and molecular heat conductivity, κ, together with Q_s and β, are assumed to be the controlling factors of the temperature profile close to the surface, while far from the surface the free-convection regime is adopted. With these assumptions the heat transfer law is

$$Q_s = C_0(\beta\kappa / \mathrm{Pr})^{1/3}\Delta\theta^{4/3}, \tag{8}$$

where C_0 is a dimensionless coefficient, $\mathrm{Pr}=\nu/\kappa$ is the Prandtl number, and $\Delta\theta$ is the temperature difference between the surface and the convective layer interior.

In engineering applications, Eq. (8) is often written in terms of the Nusselt, Nu, and Rayleigh, Ra, numbers, *viz.*,

$$\mathrm{Nu} = C_0\,\mathrm{Ra}^{1/3}, \tag{9}$$

where

$$\mathrm{Nu} \equiv Q_s h / \kappa\Delta\theta, \quad \mathrm{Ra} \equiv \beta\Delta\theta h^3 / \kappa\nu, \tag{10}$$

and h is the depth scale (in the atmosphere, the CBL depth). Notice that h drops out from Eq. (9), which indicates that the heat transfer law under consideration is a local one independent of gross features of the CBL.

Golitsyn and Grachev [17,18,19] verified Eq. (8) and the similar water-vapour transfer law using atmospheric measurements over placid water. They found that the above formulation is suitable for geophysical applications with C_0=0.14 (which is close to many of laboratory estimates). The same parameterization of the heat and water-vapour fluxes over smooth water/land surfaces was suggested earlier by Deardorff [11] and Liu *et al.* [31]. Howard [23] extended this analysis to include mean shear. It stands to reason that Eqs. (8) and (9) can hardly be applied over rough surfaces.

2.2. NON-LOCAL EFFECTS: GENERAL CONCEPT AND EMPIRICAL EVIDENCE

The primary drawback of the above traditional convective heat transfer laws lies in the fact that in both the Monin-Obukhov similarity theory, Eqs. (2)-(6), and the Malkus and

Priestley formulation, Eqs. (8) and (9), horizontal velocities in the surface layer are factually scaled with the free-convection velocity scale,

$$W_c = (\beta Q_s z)^{1/3}. \tag{11}$$

This assumption, although reasonable at first sight, does not reflect the role of the CBL-scale buoyancy-driven cells. In the cells, vertical velocities are at their maximum in the middle of the layer and become small (of the order of W_c) close to the surface. On the contrary, horizontal velocities associated with cells are negligibly small in the middle of the layer and become pronounced near the surface where the flow converges to the plume base (clearly they are also pronounced at the CBL upper boundary where the flow diverges from the plume top). Hence, horizontal velocities in the surface layer should be scaled with the Deardorff [9-12] global convection velocity scale,

$$W_* = (\beta Q_s h)^{1/3}, \tag{12}$$

rather than with W_c. When the depth scale of the near-surface thermal boundary layer, h_s, differs from h by only one order of magnitude (which might be more or less plausible in shallow convective layers typical of laboratory experiments), the difference between W_c and W_* may not be very significant. This probably explains detectable but not unduly large scatter in laboratory estimates of the coefficient C_0 in Eq. (9), $0.1<C_0<0.2$ (*e.g.*, Table 1 in Zilitinkevich [69]). In typical atmospheric conditions however, when h_s~1m and h~10^3m, the traditional heat transfer law, Eq. (9), should be revised. Qualitatively the above features of convection were already noticed by Prandtl [49] and subsequently by Kraichnan [27] and Plate [48].

Businger [5,6] was the first who clearly pointed out the importance of the CBL-scale convergence flow patterns in the surface-layer scaling. He introduced the fruitful concept of the "minimum friction velocity", namely, the concept of the friction velocity U_* generated by the aggregate effect of variable CBL-scale horizontal velocity shears regardless of their orientation. Here, U_* is determined from measurements through scalar averaging, $U_*^4 =< (u'w')^2 > + < (v'w')^2 >$ (firstly take the square and then average), where angle brackets designate the time average. Notice essential difference with the traditional vector averaging, $U_*^4 =< u'w' >^2 + < v'w' >^2$ (firstly average and then take the square), that implies $U_* \to 0$ when the mean wind vanishes. Businger postulated that U_* depends on the bulk governing parameters of the CBL turbulence, β, Q_s, and h, and on the roughness length of the underlying surface, z_{0u},

$$U_* / W_* = \Phi_*(h / z_{0u}), \tag{13}$$

where Φ_* is an unknown universal function expected to be monotonously decreasing with the increasing argument.

Wyngaard and Cote [67] and Panofsky *et al.* [47] revealed basically the same physics in their analysis of the horizontal velocity variances, σ_u^2 and σ_v^2, in the convective surface layer. They discovered that σ_u^2 and σ_v^2 do not follow the Prandtl [49], Obukhov [45,46] and Monin and Obukhov [42] similarity theory. In the free-convection limit the above variances are scaled with the Deardorff velocity scale,

$$\sigma_u^2 \equiv \overline{u'^2} = \sigma_v^2 \equiv \overline{v'^2} \sim 0.5W_*^2, \tag{14}$$

in contrast to the vertical velocity variance that follows the Prandtl theory very well,

$$\sigma_w^2 \equiv \overline{w'^2} = 1.1W_c^2. \tag{15}$$

Here, the empirical estimate, 1.1, of the coefficient on the r.h.s. of Eq. (15) is quite trustworthy. It follows from both field observations (Lenschow *et al.* [30]) and LES studies (Moeng and Wyngaard [41]). The estimates of the coefficient on the r.h.s. of Eq. (14) are much less stable, occurring in the interval between 0.2 and 0.6. A comprehensive discussion of the problem is given by Yaglom [68]. See also the Zilitinkevich [71,72] theoretical model for σ_u / W_* as dependent on z_{0u} / h [Eq. (25) below], and a summary diagram of σ_u / W_* versus z/h including atmospheric, laboratory, and LES data in Fig. 8 of Hibberd and Sawford [21].

It is evident that large-scale structures, although contributing to σ_u, σ_v, U_* and to the heat/mass transfer coefficients in calm weather, scarcely affect the mean flux of momentum when it is well pronounced. This is why the structures can be referred to as "inactive turbulence", as shown by Townsend [1961], Bradshaw [4], and Högström [22].

Coherent structures in the atmospheric CBL were comprehensively examined by Lenschow and Stephens [29], Williams and Hacker [61,62], and Williams *et al.* [63,64]. Wyngaard [65,66] proposed the concept of bottom-up and top-down diffusion associated with large-scale updraughts and downdraughts. Hunt [24] and Hunt *et al.* [25] provided scaling analysis of both chaotic and semi-organised CBL turbulence. Mahrt [33] has found that even weak shears systematically organise the convective structures into asymmetric convection. A large number of LES numerical experiments resolving coherent structures in the CBL interior were carried out (Moeng and Wyngaard [40,41], Schmidt and Schumann [51], Sykes and Henn [56], Nieuwstadt *et al.* [44], to mention a few).

2.3. EARLY THEORETICAL MODELS AND LES STUDIES

Presumably the first quantitative theoretical model for the minimum friction velocity and the associated heat transfer was developed by Schumann [52]. In his model, the vertical temperature difference $\Delta\theta$, Eq. (7), was related to the near-surface internal boundary layer (IBL) controlled by the horizontal convergence flow pattern within a large-scale convective cell (Fig. 1). The layer depth, h_s, and horizontal scale, R, were both taken to be proportional to the CBL depth,

$$h_s \propto h, \quad R \propto h, \tag{16}$$

and the proportionality factors were estimated indirectly with the aid of the atmospheric data available at the time (from Caughey [8]). Vertical profiles of the horizontal velocity and potential temperature within the IBL were modelled neglecting the horizontal heterogeneity of the flow and employing the Monin-Obukhov similarity theory, Eqs. (4) and (5), with the minimum friction velocity, U_*, substituted for the conventional friction velocity, u_*.

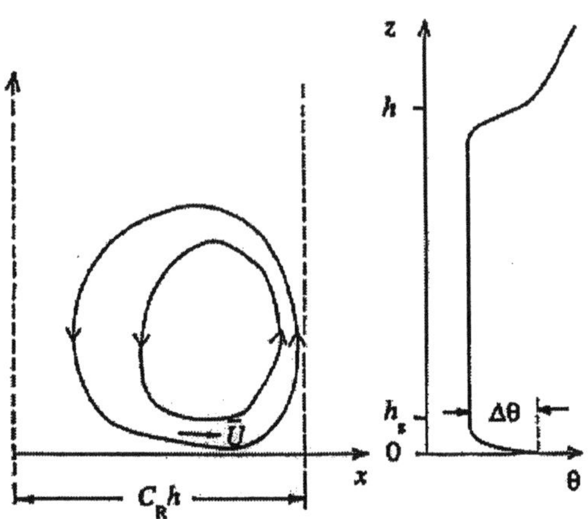

Figure 1. Schematic illustration of the Schumann [52] model.

For simplicity, no distinction was made between the roughness lengths for wind, z_{0u}, and for temperature, z_{0T}. Then, employing the hydrostatic balance to determine the horizontal pressure difference in the near-surface layer and using simplified heat and momentum balance equations therein, the following expressions were eventually derived:

$$U_* / W_* = A_1 (z_{0u} / h)^{1/6}, \quad Q_s = A_2 (\beta z_{0u})^{1/2} \Delta\theta^{3/2}, \tag{17}$$

where A_1=0.52 and A_2 =2.15 are empirical dimensionless coefficients.

In Eq. (17), the first formula specifies the Businger function Φ_*, Eq. (13). Here, the second formula is principally identical to Eq. (8) but provides a tenfold higher heat transfer due to the difference in empirical coefficients, A_2 =2.15 whereas $(3C_{fc})^{-3/2} = 0.2$. As is seen from Figs. 5 and 6, Eqs. (17) agree closely with the LES data of Schmidt and Schumann [51] and Sykes *et al.* [57]. As applied to the atmosphere, z_{0u} in the second formula in Eq. (17) should be replaced by z_{0T} (see, *e.g.*, Table 1 in Beljaars and Holtslag [2]).

At first sight, the similarity between the heat transfer laws Eq. (17) and Eq. (8) is surprising. Indeed, Eq. (17) follows from a non-local minimum-friction-velocity model, whereas Eq. (8) has obviously nothing to do with any non-local transport. The point is that the IBL depth, h_s, taken quite reasonably to be proportional to the CBL depth, h, drops out from the heat transfer formulation because of the accepted horizontal homogeneity of the IBL and simplified versions of the heat balance and momentum balance equations.

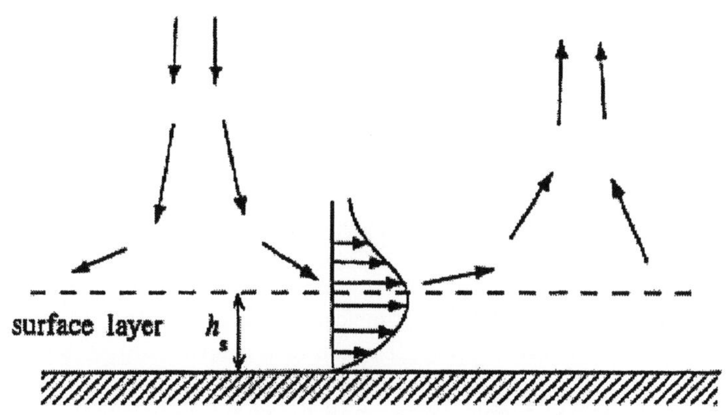

Figure 2. Schematic illustration of the Sykes *et al.* [57] model.

Sykes *et al.* [57] developed an alternative scaling analysis for the heterogeneous convergence flows in the IBL, to some extent in the spirit of the "blending layer" concept of Mason [36]. They determined h_s from a horizontal momentum equation simplified by neglecting the pressure gradient term (which seems to be an oversimplification, as the neglected term presents the driving force). Then the balance between the inertial term, $\sim W_*^2 / h$, and the stress-gradient term, $\sim U_*^2 / h_s$, yielded

$$h_s \propto h(U_* / W_*)^2. \tag{18}$$

Within the IBL, Sykes *et al.* employed the logarithmic velocity and temperature profiles, neglecting the effect of stratification (which was taken into account in the earlier Schumann model). This yielded the expressions of the required parameters, namely, the surface layer depth, h_s (identified as a typical height of the near surface wind maximum, Fig. 2), the minimum friction velocity, U_*, and the potential temperature flux, Q_s,

$$(h_s / h)\ln^2(h_s / z_{0u}) = k_u^2 A_3^2 / A_4, \tag{19}$$

$$(U_* / W_*)[2\ln(U_* / W_*) + \ln(A_4 h / z_{0u})] = k_u A_3, \tag{20}$$

$$Q_s = (A_5 \Delta\theta h_s)^{3/2} \beta^{1/2} / h. \tag{21}$$

Here, k_u=0.4 is the von Kármán constant, and A_3=1.0, A_4=2.0, and A_5=4.4 are dimensionless coefficients evaluated from LES studies of the vertical structure of the CBL close to the surface using the LES code of Sykes and Henn [56].

Although Eq. (18) can hardly be considered as a step forward compared to the Schumann Eq. (16), the heat transfer law Eq. (21), in contrast to the Schumann Eq. (17), implies a non-local vertical transport dependent on the CBL depth, h.

Stull [55] employed the Deardorff velocity scale, Eq. (12), to characterise the non-local convective transport, neglected the role of the surface roughness length, and provided similarity arguments in support of very simple heat transfer formulation,

$$Q_s = A_6 W_* \Delta\theta \equiv (A_6 \Delta\theta)^{3/2} (\beta h)^{1/2}. \tag{22}$$

He estimated the coefficient A_6 in Eq. (22) as A_6=0.0063, using data from the BLX83 field experiment in Oklahoma (roughness length 0.05 m). For practical applications he proposed an empirical interpolation between Eq. (22) and the traditional flux-gradient relation for forced convection.

It is worth mentioning that the Stull empirical estimate of the heat transfer differs dramatically from the above LES estimates (see Fig. 6). The divergence between atmospheric and LES data with respect to the heat transfer will be discussed in more detail below in the context of the new theoretical model.

Another reasonable and simple extension of the conventional surface-layer similarity theory was proposed by Deardorff [11] and further developed by Benoit [3], Miller [39] and Beljaars [1].

Bearing in mind that the Deardorff velocity scale W_* is a natural scale for horizontal velocity in the near-surface flow patterns, the above authors merely replaced the mean wind velocity $u(z)$ on the l.h.s. of Eq. (4a) by the sum $u(z)+A_7 W_*$, and adopted the dimensionless coefficient A_7 as a constant (generally it should depend on z_{0u}/h). Then in the free convection limit the generalised Eq. (4a) [substituting $A_7 W_*$ for $u(z)$ on the l.h.s. of the equation] served to determine the minimum friction velocity U_* (substituted everywhere for u_*). A similarly modified version of Eq. (4b) provided the heat transfer law. Fitting this model with LES data from Sykes *et al.* [57], Beljaars [1] obtained $A_7 \approx 1.2$.

Zilitinkevich [71,72] made an attempt to assemble the large-scale convective turbulence, considering axi-symmetric convective cells as basic large-scale semi-organised structures and using analytical approximation of the circulation pattern in a typical cell.

In his model the CBL-scale vertical velocity was parameterized by extending the Sorbjan z-dependent convective scaling (Section 4.4.2 in Sorbjan [54], further discussion in Zilitinkevich [70]), whereas the horizontal velocity was derived by integrating the continuity equation.

The modelled circulation in the cell consisted of a comparatively narrow and strong updraught, wider and weaker downdraught, and the convergence/divergence horizontal velocity field (with the centripetal movement in the lower half of the CBL and the centrifugal movement in its upper half). It is shown in Figs. 3 and 4, together with the

Williams and Hacker [61] airborne data presenting a well-pronounced large-scale convective structure in the atmosphere.

With the given basic flow pattern, a neutral-static-stability model for the near surface internal boundary layer was employed. Then the area-averaged resistance and heat/mass transfer laws immediately provided the expressions of the minimum friction velocity;

$$U_* = A_* W_* \left[\ln(h / z_{0u}) - B_s\right]^{-1/3}, \tag{23}$$

and the potential temperature flux,

$$Q_s = A_T (\beta h)^{1/2} \Delta\theta^{3/2} \left[\ln(h / z_{0u}) - B_s\right]^{-1/2} \left[\ln(h / z_{0T}) - B_T\right]^{-3/2}. \tag{24}$$

Similarly, the area-average square horizontal velocity in the internal boundary layer provided the expression of the horizontal velocity variance,

$$\sigma_u^2 = \sigma_v^2 = A_U W_*^2 \left[\ln(h / z_{0u}) - B_s\right]^{4/3}. \tag{25}$$

Here, $A_* = 0.14$, $B_s = 5.7$, $A_U = 0.25$, $A_T = 0.04$, and $B_T \approx B_s = 5.7$ are dimensionless empirical coefficients estimated from the LES data of Schumann [52] and Sykes *et al.* [57].

In the above model [71,72], a neutrally stratified IBL module is employed. Moreover analytical approximation of the circulation pattern in a convective cell, although supported by scaling analysis and LES for the CBL interior, is questionable close to the surface. In particular the universally adopted dependence of the vertical velocity on height in the free-convection layer, $w \sim z^{1/3}$, is not justified at the very surface (cf. detailed treatment of large scale convective eddies by Wilczak and Businger [60]).

All the above theoretical models relied on LES studies rather than atmospheric measurements. Figures 5 and 6 show theoretical curves for the dimensionless minimum friction velocity, U_* / W_*, and the reciprocal of the heat transfer coefficient, $\Delta\theta w_* / Q_s$, versus dimensionless roughness length, z_{0u} / h, together with LES data [51,57]. Equation (22) represents nothing other than the atmospheric estimate of

$$\Delta\theta W_* / Q_s = A_6^{-1} = 160, \tag{26}$$

assigned to $z_{0u} \sim 0.05$m, $W_* \sim 1.7$m/s, and $h \sim 1.5$km, and thus to $z_{0u} / h \sim 3 \times 10^{-5}$. This diverges dramatically from the corresponding LES estimate, $\Delta\theta W_* / Q_s = 60$, chiefly due to the fact that $z_{0T} << z_{0u}$. A systematic and considerable discrepancy between the atmospheric and LES data on the minimum friction velocity has been determined more recently by Grachev *et al.* [20] (see also Figs. 8 and 9).

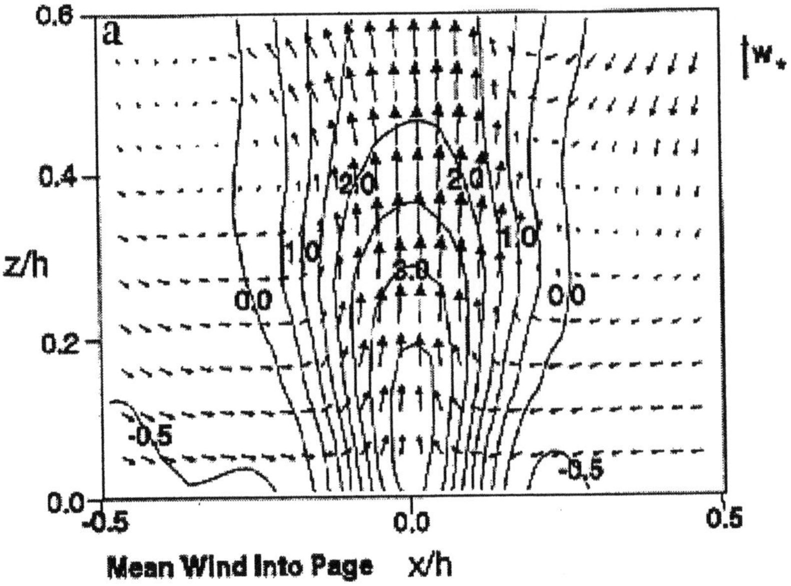

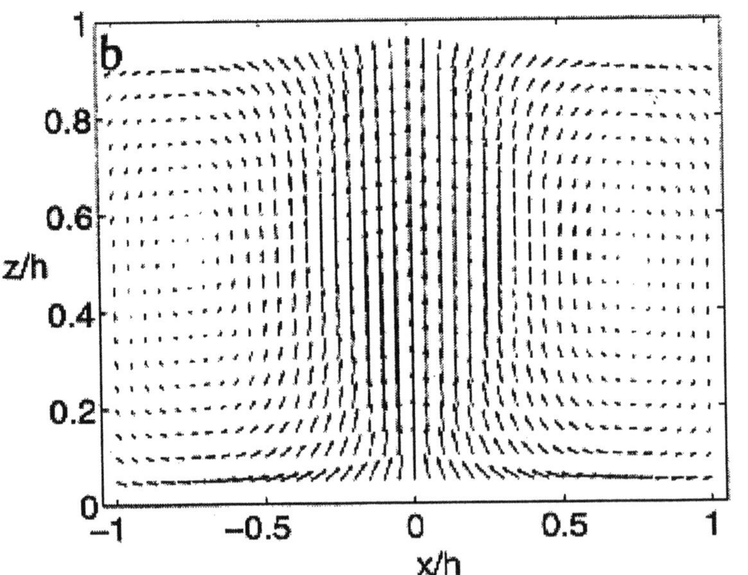

Figure 3. Schematic illustration of the Zilitinkevich [71,72] model for the vertical cross section of typical circulation cell in the CBL: *a*, airborne measurements (Williams and Hacker [61]); *b*, model approximation.

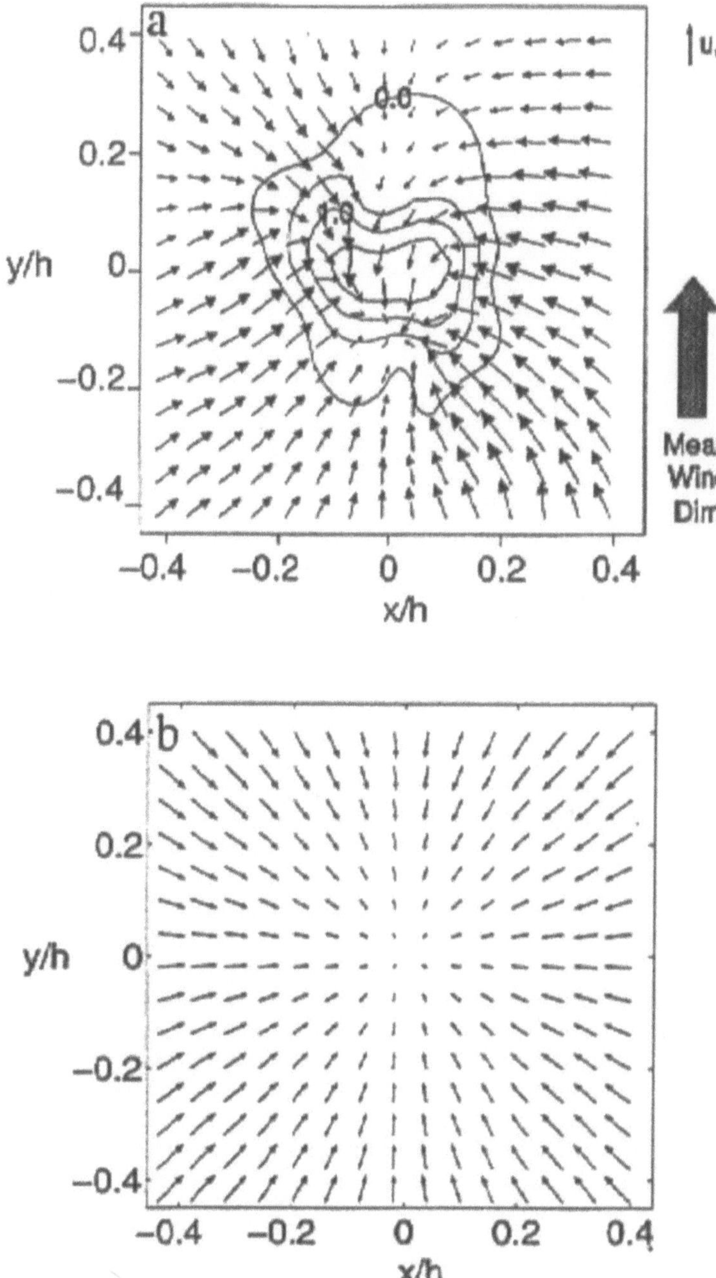

Figure 4. Same as for Fig. 3, but for the horizontal cross section of typical convergence flow pattern in the near-surface layer: *a*, airborne measurements (Williams and Hacker [61]); *b*, model approximation.

These findings, as well as the above critical comments on the early models, served as a motivation for the development of the new convective minimum-friction-velocity and heat/mass-transfer model given below.

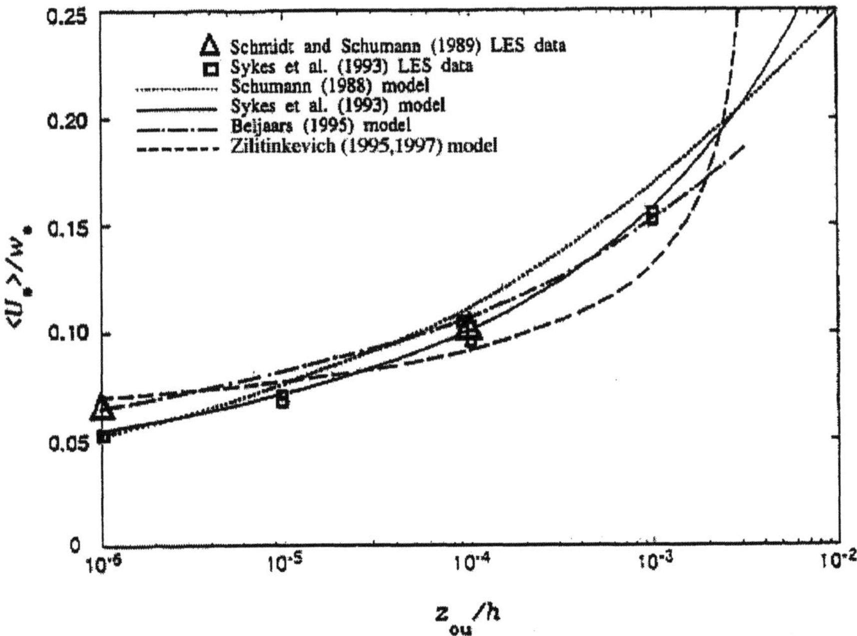

Figure 5. Dimensionless minimum friction velocity, U_* / W_*, versus the roughness length to the CBL depth ratio, z_{0u}/h. Symbols are LES data from [51,57]. Theoretical curves are after the Schumann [52] Eq. (17a), Sykes *et al.* [57] Eqs. (19) and (20), Beljaars [1] generalised Eq. (4a), and Zilitinkevich [71,72] Eq. (23).

In this paper we restrict ourselves to atmospheric applications. The physical nature of the convective heat transfer in the atmospheric boundary layers and the engineering convective flows can be different. In a large number of engineering problems and in laboratory experiments, the vertical extension of convective layers is several orders of magnitude smaller than in atmospheric CBLs. In this case the near-surface internal boundary layers, driven by large-scale semi-organised structures, are basically viscous rather than turbulent. This makes an essential difference in theoretical modelling of the minimum friction velocity and heat/mass transfer.

In the present paper we do not consider the engineering-fluid-mechanical aspect of this problem. The state of the art in this field, as of the year 1994, is given by Siggia [53]. We refer only to the heat transfer law,

$$\mathrm{Nu} \propto \mathrm{Ra}^{2/7}\,\mathrm{Pr}^{-1/7}, \tag{27}$$

reported by Siggia with reference to the Castaing *et al.* [7] laboratory experiments. The theoretical model underlying Eq. (27) accounts for non-local transport due to large-scale buoyancy-driven structures.

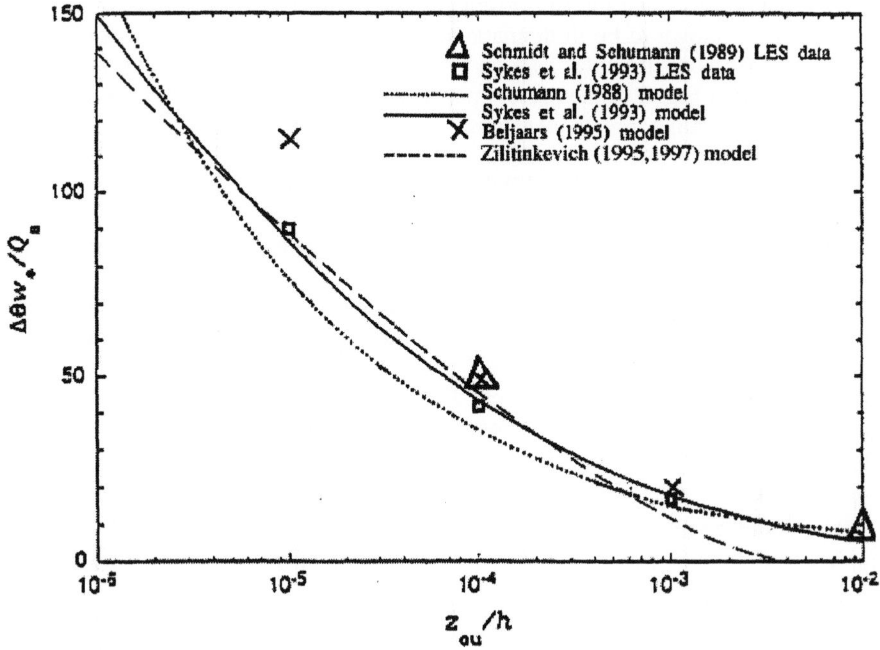

Figure 6. Same as for Fig. 5, but for the reciprocal of the heat transfer coefficient, $\Delta\theta W_* / Q_s$, neglecting the difference between z_{0T} and z_{0u}. Theoretical curves are after the Schumann [52] Eq. (26a), Sykes *et al.* [57] Eqs. (26b), Beljaars [1] Eq. (26d) and Zilitinkevich [71,72] Eq. (26e).

Notice that Eq. (27), contrary to the traditional formulation, Eqs. (9)-(10), involves the convective zone depth, h. Thus revision of the traditional local heat transfer formulations is undertaken in both geophysical and engineering fluid mechanics.

3. Revised Non-Local Formulation

3.1. MOTIVATION TO REVISION

As seen from the overview, all recent theoretical models suffer from disadvantages.

The Schumann [52] model employs a horizontally homogeneous IBL module and suggests a virtually local heat transfer law, Eq. (17).

The Sykes *et al.* [57] model neglects the pressure gradient in the horizontal momentum balance equation [that results in a doubtful expression of the IBL depth scale, Eq. (18)] and ignores the buoyancy effect on the IBL vertical structure.

The Zilitinkevich [71,72] model employs an analytical approximation of the typical convection cell that is not justified close to the surface. It also ignores the buoyancy forces in the IBL.

The problem obviously calls for further investigation.

An important point is that the LES results used as empirical background in the above three models are shown to be in dramatic contradiction with results from atmospheric measurements [see experimental data on the minimum friction velocity in Grachev *et al.* [20] and also in Fig. 8 of the present paper, and data on the heat transfer in Fig. 9]. This contradiction calls for an explanation.

3.2. WHAT IS WRONG: FIELD MEASUREMENTS OR LES?

First we shell clarify the limits of applicability of the roughness length concept as such, in the problem under consideration.

The roughness length formulation is introduced so as to avoid consideration of the "roughness layer", *i.e.*, the layer partially occupied by obstacles (roughness elements) where turbulence is immediately affected by flow-obstacle interactions. It focuses on parameterization of the area-averaged flow-surface interaction in terms of the logarithmic boundary layer model. Obviously it looses physical significance when no logarithmic portions in the velocity and temperature profiles are observed.

The roughness length with respect to wind, z_{0u}, is known to be approximately $0.1 \div 0.03$ of the height of typical roughness elements. In this case a reasonable estimate of the roughness layer depth is $h_0 \sim 30\, z_{0u}$. Obviously, logarithmic profiles can apply only at heights $z > 30\, z_{0u}$.

Alternatively, it has been empirically established, in the framework of the Monin-Obukhov similarity theory for sheared convection, that logarithmic profiles are observed only below the height $z \sim 0.1L$, where L is the Monin-Obukhov length, Eq. (2). Then, regarding the minimum friction velocity, U_*, and the "minimum Monin-Obukhov length",

$$L_* = U_*^3 / \beta Q_s = (U_* / W_*)^3 h, \tag{28}$$

logarithmic profiles in the shear-free convection regime can be observed only at heights $z < 0.1 L_*$.

As a result, the existence of logarithmic portions in the profiles and, hence, the very concept of the roughness length in the shear-free convection over rough surfaces, are justified only when

$$300 z_{0u} / h << (U_* / W_*)^3. \tag{29}$$

In the Schmidt and Schumann [51] and Sykes *et al.* [57] LES studies, z_{0u} / h and U_* / W_* were varied in the intervals $10^{-6} < z_{0u} / h < 10^{-3}$ and $0.05 < U_* / W_* < 0.15$ (see Fig. 5). Even in the most favourable case, when $z_{0u} / h = 10^{-6}$ and, according to the above LES estimates, $U_* / W_* = 0.05$, the condition Eq. (29) is hardly obeyed. It would therefore be unreasonable to expect logarithmic velocity and temperature profiles close to the surface. Hence the Monin-Obukhov type parameterization of the surface fluxes employed in the above LES codes is not justified, and consequently, the LES estimates for the minimum friction velocity and the heat transfer at the surface are not reliable.

This is the reason why LES results shown in Figs. 5 and 6 diverge drastically from atmospheric measurements (see Figs. 8 and 9 below). Needless to say, under these circumstances we should rely on field measurements.

3.3. FIELD DATA

We make use of data sets on the atmospheric CBL turbulence and mean structure over the sea from (i) the San Clemente Ocean Probing Experiment (SCOPE) and (ii) the Tropical Ocean - Global Atmosphere Coupled Ocean - Atmosphere Response Experiment (TOGA COARE), and over land from (iii) the BOREX-95 Experiment.

Marine data were collected by the NOAA Environmental Technology Laboratory (Boulder, CO). It includes 50-min average covariance and inertial-dissipation estimates of turbulent fluxes, mean meteorological variables, radiation fluxes, turbulence spectra, sea-surface temperature, and the CBL height. The measurement system is described by Fairall *et al.* [15].

The experiment SCOPE was performed to investigate air-sea interactions with remote and in situ measurements with the Scripps Institute Floating Instrument Platform (FLIP) off southern California in September 1993. FLIP was moored about 15 km NW off the NW point of San Clemente Island with good open ocean exposure from NW winds (Kropfli and Clifford [28]).

A sonic anemometer/thermometer and a high-speed infrared hygrometer were used to make measurements of turbulent fluxes of momentum, potential temperature, and specific humidity. The instruments were deployed at the end of a 20m long boom. Inversion heights were determined from the radiosonde temperature and humidity profiles using sondes launched from the R/V Titan and the U.S. Navy installation nearby San Nicolas Island. A dual inertial navigation system was used to correct for ship motions. The sea surface temperature was determined from the floating thermistor measurements at 5cm depth, with corrections for the cool skin effect. A conventional aspirated T/RH sensor and infrared hygrometer were used to determine mean temperature and humidity.

Excluding a few days of fog, the surface layer was unstable during the experiment. Winds were not strong, averaging around 4 m/s. A moderate NW swell was almost always present. Several episodes of weakening and turning winds occurred in which the wind direction and sea wave direction were out of balance. For details see [14].

In the experiment TOGA COARE, measurements were carried out from R/V Moana Wave for three cruise legs in the Tropical Western Pacific warm pool area during 1992-1993. The ship was positioned at the centre of the intensive flux array (2° S latitude and 156° E longitude; see Webster and Lukas [59]) for a total of 70 days of measurements. She usually operated in a "drift" mode, meaning wind-speed measurements were relatively unaffected by errors induced by the mean ship motion.

A total of 1622 turbulence-flux and mean-profile measurements (nominally 50-min average) were made in the wind-speed interval 0.2 to 13m/s. The CBL height (always close to 600 m) was determined from radiosonde measurements and from the backscatter intensity of a wind profiler. Editing the data for unfavourable relative wind directions, precipitation, and salt contamination of signals yielded 860 usable flux values. Wind velocity was derived from a sonic anemometer after appropriate corrections for the ship motion. Winds were referenced to the sea surface using near-surface currents from the nearby buoy. Small corrections were applied to the mean-profile measurements based on ship-to-ship intercomparisons and post-calibrations (Fairall *et al.* [13,15]).

The RISØ National Laboratory experiment BOREX-95 was carried out over a flat, homogeneous surface near the Borris site in the western part of Denmark in July 1995 (Mikkelsen *et al.* [38]). A sonic anemometer/thermometer was used to make turbulent measurements of the fluxes of momentum and sensible heat at the reference height 7 m above the surface. Both heat flux and air temperature exhibited a normal diurnal cycle with the deep daytime convective boundary layer and the shallow nocturnal stable-static-stability boundary layer.

In the present paper, data from the above experiments are analysed, and the runs exhibiting strong unstable stratification with wind velocities below 1 m/s are selected as representative of the shear-free convection regime. The surface roughness length with respect to wind, z_{0u}, is calculated by extending the traditional formulation to conditions of strong convection with due regard for the gustiness corrections to the mean wind profile, where the contribution from gusts is scaled with the Deardorff velocity W_* (Fairall and Grachev [16]). Selected data are used to determine the minimum friction velocity (from measured absolute values of the momentum flux) and the aerodynamic heat and mass transfer coefficients as dependent on the CBL governing parameters. The resulting empirical dependencies are shown in Figs. 8-13. When these results are compared with those from LES (also shown in Figs. 8 and 9), it is apparent that the latter are unusable in the problem under investigation.

3.4. BASIC SCALES OF TYPICAL CONVECTION CELL

In general we distinguish between three types of motion: first, mean flow (which is not present in the velocity field as we consider the shear-free convection); second, large-scale semi-organised structures (in our case convective cells); and third, small-scale random fluctuations representing chaotic background turbulence generated by both velocity shears and buoyancy forces. The life-times of the semi-organised structures are much longer than the turbulence time scales. Therefore, the structures can be treated as comparatively stable, quasi-stationary and approximately axi-symmetric circulation cells, playing the same role as mean flow with respect to small-scale turbulence.

Leaving aside practical difficulties in partitioning semi-organised structures and chaotic eddies, the large-scale flow pattern in a typical cell can be modelled to a first approximation as a two-dimensional flow, dependent on vertical, z, and radial, r, coordinates, varying in the intervals $0<z<h$ and $0<r<R$, where R is the cell radius.

We consider the convective boundary layer affected by the surface fluxes of potential temperature, $Q_s = \overline{\theta' w'}|_{z=0}$, and water vapour, $E_s = \overline{q' w'}|_{z=0}$, where q' is the fluctuation of specific humidity. Then the buoyancy fluctuation, b', and the buoyancy flux, B_s, are

$$b' \equiv -g\rho' / \rho_0 = \beta\theta' + 0.61 g q', \quad B_s = \beta Q_s + 0.61 g E_s, \tag{30}$$

where ρ' is the density fluctuation and ρ_0 is the reference density.

The Deardorff velocity scale, the Monin-Obukhov length scale, and also the buoyancy scale are determined as

$$W_* = (B_s h)^{1/3}, \quad L_* = U_*^3 / B_s, \quad b_* = B_s / W_* . \tag{31}$$

Here, the scale b_* characterises the buoyancy difference between uprising plume and ambient fluid in a typical convection cell.

We keep the chief terms in the vertical momentum equation *applied to the plume*,

$$w\frac{\partial w}{\partial z} \sim -\frac{g}{\rho_0}\Delta\rho \sim b_*, \tag{32}$$

where Δp is the global density difference in the cell. Then, using the CBL depth, h, to measure the vertical distance, yields the vertical velocity scale $w \sim (b_* h)^{1/2} = W_*$.

Similarly, we keep the chief terms in the horizontal momentum equation for the near-surface convergence flow,

$$u\frac{\partial u}{\partial r} \sim -\frac{1}{\rho_0}\frac{\partial p}{\partial r} \sim \frac{\Delta p}{\rho_o R}. \tag{33}$$

Then, estimating the horizontal increment of the surface pressure, Δp, with the aid of the hydrostatic balance equation and *applying this to the downdraught*,

$$\frac{\Delta p}{\rho_0} \sim \frac{g}{\rho_0}\Delta\rho h \sim b_* h, \tag{34}$$

yields exactly the same scale for the horizontal velocity, $u \sim (b_* h)^{1/2} \sim W_*$.

Finally, the continuity equation,

$$\frac{1}{r}\frac{\partial}{\partial r}(ru) = -\frac{\partial w}{\partial z}, \tag{35}$$

immediately yields $R \sim h$.

3.5. REVISED MODEL FOR THE SURFACE LAYER

We employ the idea that the small-scale turbulence always have time to adjust to the large-scale semi-organised eddies, *i.e.*, in other words, that the inflow of a large eddy can be treated as the near-surface internal boundary layer (IBL). The key points in the problem under consideration are, then, the depth and the vertical structure of the IBL linked with the horizontal convergence branch of the convective cell (Fig. 7).

The growth rate equation for the IBL depth, h_I, is derived from the turbulent kinetic energy (TKE) equation, which treats the flow in the cell in a certain sense as a mean flow, *i.e.*, it applies the term "turbulence" only to small-scale chaotic fluctuations within the IBL. The latter is a convective sheared boundary layer evolving against non-stratified fluid. The TKE equation vertically averaged over the IBL depth then yields

$$u_I \frac{dh_I}{dr} = w_I + C_e (B_s h_I)^{1/3}, \tag{36}$$

where $u_I = u(h_I)$ and $w_I = w(h_I)$ are horizontal (radial) and vertical velocities at the IBL upper boundary, and C_e is a dimensionless constant (a reference value of $C_e \approx 0.24$ is taken from laboratory experiments in the regime of free entrainment, see Section 2.4 in [69]).

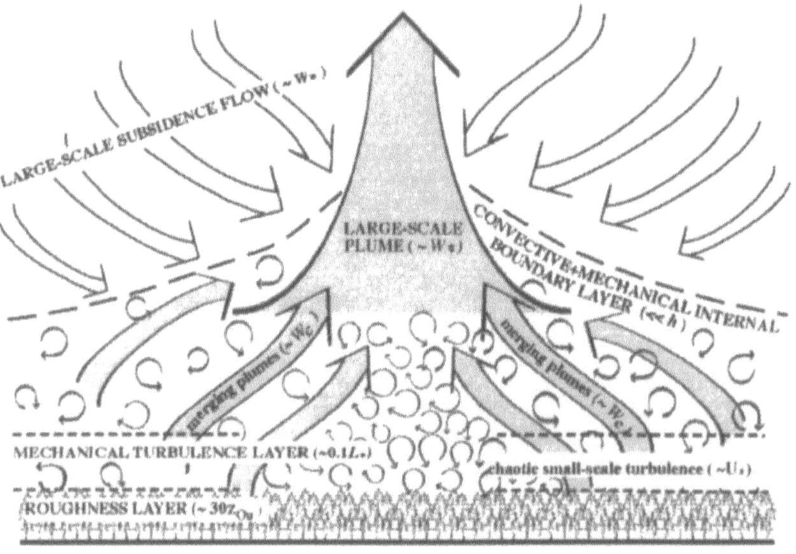

Figure 7. Schematic illustration of the proposed model: merging plumes and chaotic eddies in the internal boundary layer.

In the region of interest, namely, under the downdraught, $R_0 < r < R$, where R_0 is the typical plume radius, the first term on the r.h.s. of Eq. (36) is negative. It can be estimated as $|w_I| \sim W_* h_I / h$. In this case it is small compared to the second term, $C_e (B_s h_I)^{1/3} \sim W_* (h_I / h)^{1/3}$, and it can be neglected as long as $h_I / h \ll 1$. Then, taking a typical value of the horizontal velocity $u_I \propto -W_*$ (which is justified when not too close to the plume axis), Eq. (36) reduces to $dh_I / dr \propto -(h_I / h)^{1/3}$. This yields the radial dependence of the IBL depth,

$$dh_I / h \propto \left[(R - r) / / h\right]^{3/2}. \tag{37}$$

Notice that the radial pressure gradient drops with increasing height, so that the IBL is strongly baroclinic. Away from the surface, radial velocity tends to decrease with increasing height. Then, taking into account the strong unstable stratification in the near-surface flow, the radial velocity maximum can occur at a height proportional to the Monin-Obukhov length, L_*, Eq. (31), rather than at the IBL upper boundary, Eq. (37).

Indeed, by neglecting baroclinicity close to the surface, the vertical gradient of the radial velocity modulus, $\partial u / \partial z$, is conventionally approximated as

$$\frac{\partial u}{\partial z} = \begin{cases} U_* / k_u z & \text{at} \quad z_{0u} << z \le C_* L_* \\ C_U U_*^2 / B_s^{1/3} z^{4/3} & \text{at} \quad z \ge C_* L_* , \end{cases} \tag{38}$$

where C_*=0.1 and C_U=1.7 are dimensionless empirical constants [26].

Eq. (38) implies that strong vertical increments occur at the height intervals $z < C_* L_*$, whereas at heights $z > C_* L_*$ the velocity, $u(z)$, gradually levels off. Hence, in the actual IBL, $u(z)$ can, due to baroclinicity, decrease already at heights somewhat exceeding the level $z = C_* L_*$. This highlights the fact that the velocity maximum height, $\propto L_*$, can hardly be used as the IBL depth scale.

At the height h_I, a typical value of the radial velocity modulus can obviously be taken proportional to the Deardorff velocity scale, W_*, whereas potential temperature and specific humidity can be identified with their mean values, θ_a and q_a, typical of the CBL interior,

$$u(h_I) = C_u W_*, \quad \theta(h_I) = \theta_a, \quad q(h_I) = q_a. \tag{39}$$

Here, the dimensionless coefficient C_u depends on the radial coordinate, r/h, and it turns into a constant after averaging in the horizontal over the cell area.

We extend the conventional parameterization of the mean vertical profiles in the convective surface layer [identical to Eq. (38) for the wind velocity gradient] to the entire internal boundary layer in question, $z_{0u} << z \le h_I$, *viz.*,

$$\frac{\partial u}{\partial z} = \begin{cases} \dfrac{U_*}{k_u z} & \text{at} \quad z \le C_* L_* \\ \dfrac{C_U U_*^2}{B_s^{1/3} z^{4/3}} \left[1 - \xi^{4/3} f_u(\xi)\right] & \text{at} \quad C_* L_* \le z \le h_I , \end{cases} \tag{40}$$

$$\frac{\partial \theta}{\partial z} = \begin{cases} \dfrac{Q_s}{k_T U_* z} & \text{at} \quad z \le C_* L_* \\ -\dfrac{C_T Q_s}{B_s^{1/3} z^{4/3}} \left[1 - \xi^{4/3} f_T(\xi)\right] & \text{at} \quad C_* L_* \le z \le h_I , \end{cases} \tag{41}$$

$$\frac{\partial q}{\partial z} = \begin{cases} \dfrac{E_s}{k_q U_* z} & \text{at} \quad z \leq C_* L_* \\ -\dfrac{C_q E_s}{B_s^{1/3} z^{4/3}} \left[1 - \xi^{4/3} f_q(\xi)\right] & \text{at} \quad C_* L_* \leq z \leq h_I , \end{cases} \tag{42}$$

where k_u=0.38 and $k_T = k_q$=0.42 are the von Kármán constants, and $f_{\{u,T,q\}}$ are correction functions of $\xi \equiv z / h_I$ with regular behaviour at $\xi \to 0$ which account for the aggregate effects of the IBL heterogeneity and baroclinicity.

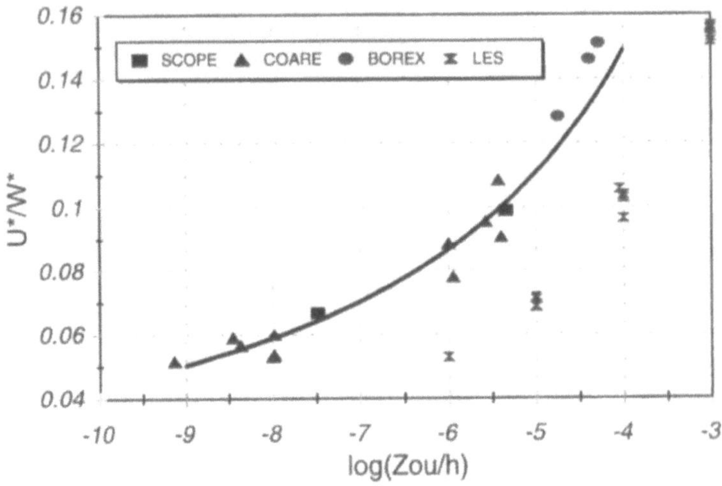

Figure 8. Dimensionless minimum friction velocity, U_* / W_*, versus the roughness length to the CBL depth ratio, z_{0u}/h. Theoretical curve, plotted after Eq. (43), agrees well with data from field experiments SCOPE, COARE and BOREX. Results from LES underestimate U_* / W_*.

Only reference values of the empirical dimensionless constants in Eqs (40)-(42) are available, *viz.*, C_*=0.1, C_U=1.7 and $C_T = C_q = 1$, from measurements in the regular horizontally homogeneous atmospheric surface layer (*e.g.*, Kader and Yaglom[26]).

Eqs. (40)-(42) are integrated in the vertical over the layer $0 < z \leq h_I$, which is why the specific form of the functions $f_{\{u,T,q\}}$ is not needed. By employing the roughness-length parameterization of the flow-surface interaction and holding the upper boundary conditions, Eq. (39), this integration yields local, *r*-dependent resistance and heat/mass transfer laws.

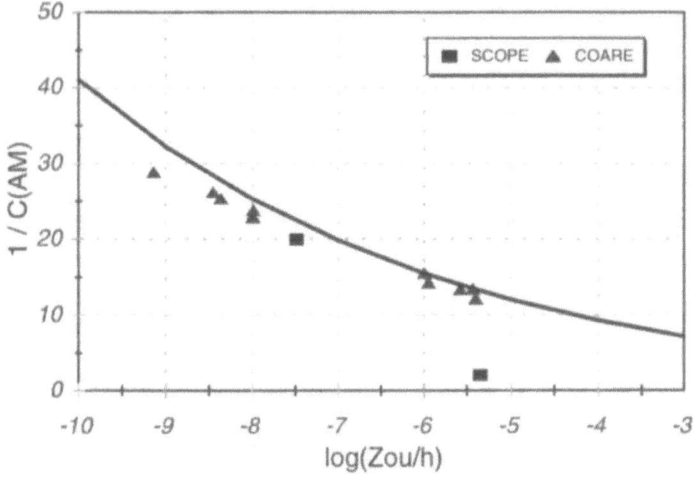

Figure 9. Reciprocal of the aerodynamic heat transfer coefficient, C_{AH}^{-1}, versus z_{0u}/h. The curve, after Eqs. (44) and (43), agrees well with field data. LES underestimate C_{AH}.

Then, by averaging in the horizontal over the convection-cell area, the resulting expressions of the minimum friction velocity, and the heat and mass transfer read

$$\ln\frac{h}{z_{0u}} = C_{u0}\frac{W_*}{U_*} + 3\ln\frac{W_*}{U_*} - C_{u1} + C_{u2}\left(\frac{W_*}{U_*}\right)^{-1}, \tag{43}$$

$$\frac{1}{C_{AH}} \equiv \frac{\Delta\theta U_*}{Q_s} - k_T\ln\left(\frac{z_{0u}}{z_{0T}}\right) = C_{T0}\frac{W_*}{U_*} - C_{T1} + C_{T2}\left(\frac{W_*}{U_*}\right)^{-1}, \tag{44}$$

$$\frac{1}{C_{AM}} \equiv \frac{\Delta q U_*}{E_s} - k_q\ln\left(\frac{z_{0u}}{z_{0q}}\right) = C_{q0}\frac{W_*}{U_*} - C_{q1} + C_{q2}\left(\frac{W_*}{U_*}\right)^{-1}. \tag{45}$$

Here, $C_{\{u0,u1,u2\}}$, $C_{\{T0,T1,T2\}}$ and $C_{\{q0,q1,q2\}}$ are dimensionless constants that are expressed through the unknown coefficient omitted on the r.h.s. of Eq. (37), through the area averaged value of the coefficient C_u in Eq. (39) and also through the unknown constants C_*, C_U, C_T and C_q in Eqs. (40)-(42) appropriate for the IBL. Actually they are to be determined empirically.

Eq. (43) presents the dimensionless friction velocity, U_* / W_*, as a function of the dimensionless roughness length, z_{0u} / h. Eqs. (44) and (45) present the bulk heat and mass transfer coefficients,

$$C_H \equiv Q_s / \Delta\theta U_*, \qquad C_M \equiv E_s / \Delta q U_*, \tag{46}$$

as functions of U_* / W_* (and hence of z_{0u} / h) and also of the roughness length ratios, z_{0u} / z_{0T} or z_{0u} / z_{0q}, respectively.

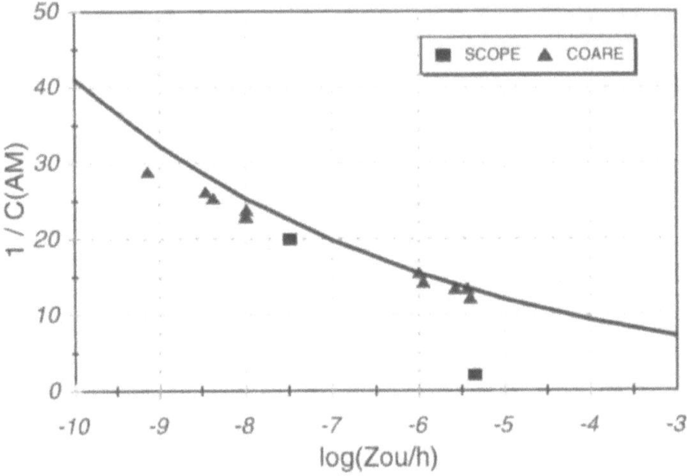

Figure 10. Same as for Fig. 9, but for reciprocal of the mass transfer coefficient, C_{AM}^{-1}. The curve is plotted after Eqs. (45) and (43).

Over natural rough surfaces, the momentum is very efficiently transferred to the roughness elements by the pressure forces, which are orders of magnitude higher than the molecular-viscosity forces. On the contrary, for scalars, such as heat and water vapour, the flow-surface interaction is always controlled by the molecular conductivity and diffusivity, respectively. As a result, the roughness length for wind, z_{0u}, is usually several orders of magnitude higher than the roughness lengths for scalars, z_{0T} and z_{0q}. Moreover, as z_{0u} is practically independent of viscosity, it can be considered as a comparatively invariable, basically geometrical feature of the surface. Specification of z_{0T} and z_{0q} represents a separate challenging task resting beyond the scope of the present paper.

Here, we restrict our consideration to the "aerodynamic coefficients",

$$C_{AH} \equiv Q_s / \Delta\theta_A U_*, \qquad C_{AM} \equiv E_s / \Delta q_A U_*, \tag{47}$$

rather than bulk coefficients, Eq. (46). Here, $\Delta\theta_A$ and Δq_A are the "aerodynamic increments",

$$\Delta\theta_A \equiv \theta_{As} - \theta_a = \left(\theta_s - \frac{Q_s}{k_T U_*} \ln \frac{z_{0u}}{z_{0T}}\right) - \theta_a,$$
$$\Delta q_A \equiv q_{As} - q_a = \left(q_s - \frac{E_s}{k_q U_*} \ln \frac{z_{0u}}{z_{0q}}\right) - q_a, \tag{48}$$

based on the so-called aerodynamic surface temperature and humidity, θ_{As} and q_{As}, *i.e.* the potential temperature and specific humidity extrapolated logarithmically from the logarithmic boundary layer downwards to the level $z = z_{0u}$ (*e.g.*, Mahrt *et al.* [34]). Eqs. (44) and (45) present aerodynamic heat and mass transfer coefficients, C_{AH} and C_{AM}, as universal functions of U_* / W_* (or of z_{0u} / h, which is the same).

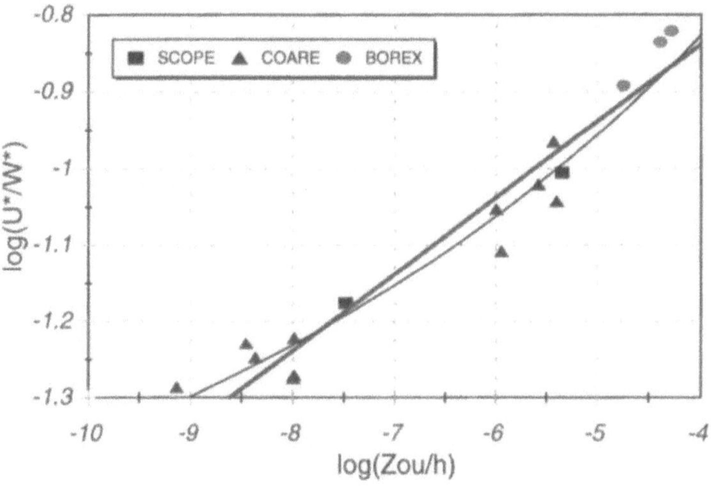

Figure 11. Dimensionless minimum friction velocity, U_* / W_*, versus z_{0u}/h. The curve and the heavy line are plotted after Eqs. (43) and (51), respectively.

The dependencies U_* / W_*, C_{AH}, and C_{AM} on z_{0u} / h given by Eqs. (43)-(45) are shown in Figs. 8-10 together with data from measurements in atmospheric CBLs.

The dimensionless constants deduced from the best fit to the atmospheric data are

$$C_{u0} = 0.6, \quad C_{u1} = 1.2, \quad C_{u2} = 8,$$
$$C_{\theta 0} = 1.5, \quad C_{\theta 1} = 1.7, \quad C_{\theta 2} = 2, \qquad (49)$$
$$C_{q0} = 1.5, \quad C_{q1} = 1.9, \quad C_{q2} = 1.5.$$

The LES data shown in Figs. 8 and 9 diverge drastically from atmospheric data and strongly underestimate the minimum friction velocity and the heat transfer coefficient.

3.6. FLUX-CALCULATION TECHNIQUE

As is obvious from the discussion of the roughness length concept in Section 3.2, the resistance and heat/mass transfer laws, Eqs. (43)-(45) are physically sound when the roughness length is not too high, $z_{0u} / h << 0.1(U_* / W_*)^3 \sim 10^{-4}$.

Alternatively, over very smooth surfaces the effective roughness length can not be smaller than a virtual viscous roughness length, $z_{0u} \sim 0.1\,\nu / U_* \sim \nu / W_*$, which is typically of order 10^{-4}cm (taking $\nu \sim 0.1\text{cm}^2/\text{s}$ and $W_* \sim 1\text{m/s}$). Then taking into account the typical atmospheric CBL depth, $h \lesssim 3\text{km}$, the ratio z_{0u} / h can not be smaller than 10^{-10}.

The problem in question is therefore of interest in the interval

$$10^{-10} \le z_{0u} / h \le 10^{-5}. \qquad (50)$$

Here, referring to Fig. 11, the dependence of U_* / W_* on z_{0u} / h is approximated with reasonably accuracy as

$$U_* / W_* \approx 0.36(z_{0u} / h)^{0.1}. \qquad (51)$$

Referring to Figs. 12 and 13, the aerodynamic heat and mass transfer coefficients, C_{AH} and C_{AM}, are nearly perfectly approximated by reduced versions of Eqs (44) and (45), implying linear dependencies on W_* / U_*,

$$C_{AH}^{-1} \approx 1.5(W_* / U_*) - 1.5 \approx 4.4(z_{0u} / h)^{-0.1} - 1.5, \qquad (52)$$

$$C_{AM}^{-1} \approx 1.5(W_* / U_*) - 1.7 \approx 4.4(z_{0u} / h)^{-0.1} - 1.7. \qquad (53)$$

Hence U_*, C_{AH}, and C_{AM} are immediately determined from the CBL governing parameters W_*, h, and z_{0u}. The flux calculation technique based on Eqs. (51)-(53) can be easily implemented in climate and meso-scale atmospheric models. As seen from Figs. 11-13, these equations are valid up to $z_{0u} / h \sim 10^{-4}$.

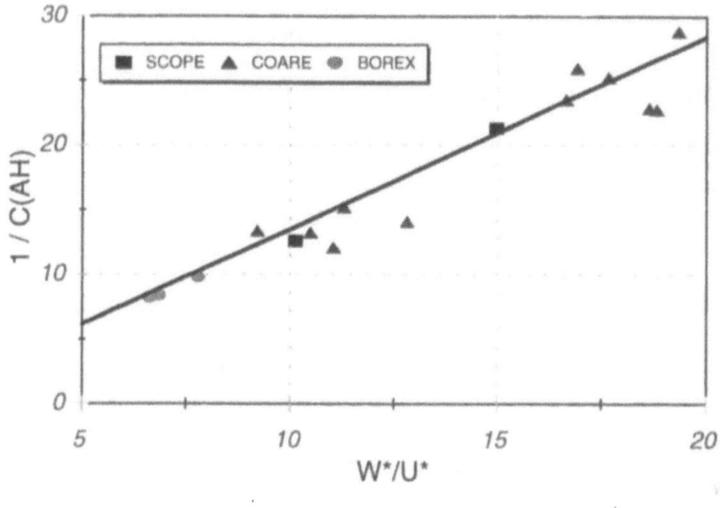

Figure12. Reciprocal of the aerodynamic heat transfer coefficient, C_{AH}^{-1}, versus W_* / U_*. The line is plotted after Eq. (52).

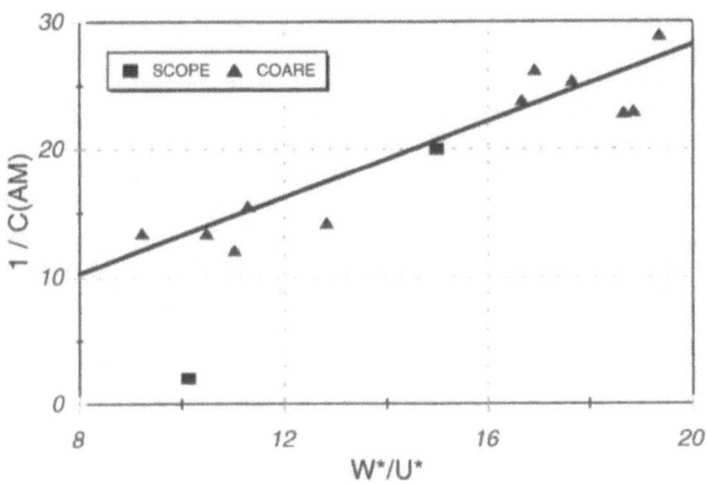

Figure 13. Reciprocal of the aerodynamic mass transfer coefficient, C_{AM}^{-1}, versus W_* / U_*. The line is plotted after Eq. (53).

4. Conclusions

A theoretical model for the minimum friction velocity and non-local heat and mass transfer in the atmospheric surface layer is developed for the calm convection regimes over the water surface and not very rough land surfaces. The model is validated using data from field measurements.

Earlier theoretical models are shown to be not sufficiently advanced. It is also shown that recent large-eddy simulation studies, served as "empirical" basis for earlier models, are hardly applicable to numerical simulation of the surface roughness effects.

The proposed model is valid when the roughness length to the CBL depth ratio, z_{0u} / h, is less than 10^{-4}. A theory of the purely convective atmospheric surface layer over very rough surfaces is a subject of further investigation.

Another challenging problem, that plays a key role in practical calculation of the heat and mass transfer over natural surfaces, is proper determination of the roughness lengths for scalars.

Acknowledgements

The authors are grateful to Ernst Augstein, Chris Fairall, Søren Larsen, Donald Lenschow, Larry Mahrt, Erich Plate, Ulrich Schumann, and Alastair Williams for their comments on earlier drafts of this paper. D. Lenschow called our attention to the fact that LES is unable to resolve the interaction of the near-surface flow with the roughness elements. S. Larsen and C. Fairall provided us with experimental data.

This work has been carried out when one of the authors (SZ) was a visiting scientist at the Alfred Wegener Institute for Polar and Marine Research, Bremerhaven, Max Planck Institute for Meteorology, Hamburg, and GKSS Research Centre, Geesthacht.

An important contribution to this paper was made due to discussions between the two authors (SZ and JH) during the visit of SZ to the Department of Applied Mathematics and Theoretical Physics, University of Cambridge, 10-16 December 1995.

This research was partially supported from (i) EU Project SFINCS - EC Contract ENV4-CT97-0573, (ii) Project INTAS-94-2255, (iii) Institute for Remote Sensing Applications, JRC, Ispra, in the scope of Contract No. 10934-95-04 F1EI ISP, and (iv) National Observatory of Athens, in the scope of Contract No. JOR3-CT95-0008, JOULE Program of DG-XII of the European Commissions.

References

1. Beljaars, A.C.M. (1995) The parameterization of surface fluxes in large scale models under free convection, *Quart. J. Roy. Meteorol. Soc.* **121**, 225-270.
2. Beljaars, A.C.M., and Holtslag, A.A.M. (1991) Flux parameterization over land surfaces for atmospheric models, *J. Appl. Meteorol.* **30**, 327-341.
3. Benoit, R. (1976) A Comprehensive Parameterization of the Atmospheric Boundary Layer for General Circulation Models, NCAR Cooperative Thesis No. 39, NCAR library. P.O. Box 3000, Boulder, CO 80307, NTIS: N77-25757, PB-264-540/6, 298 pp.
4. Bradshaw, P. (1978) Comments on "Horizontal velocity spectra in an unstable surface layer", *J. Atmos. Sci.* **35**, 1768-1769.
5. Businger, J.A. (1973a) A note on free convection, *Boundary-Layer Meteorol.* **4**, 323-326.

6. Businger, J.A. (1973b) Turbulent transfer in the atmospheric surface layer, in D.A. Haugen (ed.), *Workshop in Micrometeorology*, AMS, Boston, pp. 67-100.
7. Castaing, B., Gunaratne, G., Heslot, F., Kadanoff, L., Libchauber, A., et al. (1989) Scaling of hard thermal turbulence in Rayleigh-Benard convection, *J. Fluid Mech.* **204**, 1-30.
8. Caughey, S.J. (1982) Observed Characteristics of the atmospheric boundary layer, in F.T.M. Nieuwstadt and H. Van Dop (eds.), *Atmospheric Turbulence and Air Pollution Modelling*, D. Reidel, Dordrecht, pp. 107-158.
9. Deardorff, J.W. (1970) Convective velocity and temperature scales for the unstable planetary boundary layer, *J. Atmos. Sci.* **27**, 1211-1213.
10. Deardorff, J.W. (1972a) Numerical investigation of neutral and unstable planetary boundary layers, *J. Appl. Meteorol.* **11**, 91-115.
11. Deardorff, J.W. (1972b) Parameterization of planetary boundary layer for use in general circulation models, *Mon. Wea. Rev.* **2**, 93-106.
12. Deardorff, J.W., Willis, G.E., and Lilly, D.K. (1969) Laboratory investigation of non-steady penetrative convection, *J. Fluid Mech.* **35**, 7-31.
13. Fairall, C. W., Bradley, E.F., Rogers, D.P., Edson, J.B., and Young, G.S. (1996a) Air-sea flux parameterization in TOGA COARE, *J. Geophys. Res.* **101**(C2), 3747-3764.
14. Fairall, C. W., Grachev, A.A., Bedard, A.J., and Nishiyama, R.T. (1996b) Wind, Wave, Stress, and Surface Roughness Relationships from Turbulence Measurements Made on R/P FLIP in the SCOPE Experiment. NOAA Technical Memorandum, ERL ETL 268, Boulder, Colorado, USA, April 1996, 37 pp. (Available from the National Technical Information Service, 5285 Port Royal Rd., Springfield, VA 22161)
15. Fairall, C. W., White, A.B., Edson, J.B., and Hare, J.E. (1997) Integrated shipboard measurements of the marine boundary layer, to appear in *J. Ocean. Atmos. Technol.*
16. Fairall, C. W., and Grachev, A.A. (1997) On the sea-surface roughness length under free convection conditions, in *Proc. 8th AMS Conference on Air-Sea Interaction and Conference on the Global Ocean-Atmosphere-Land System (GOALS), 28 January- 2 February 1996, Atlanta, Georgia,* J134-J138. [See also Fairall, C. W., and Grachev, A.A. (1996) Air-sea fluxes and sea surface roughness length in free convection limit, to appear in *J. Atmos. Sci.*]
17. Golitsyn, G.S. (1979) Simple theoretical and experimental study of convection with some geophysical applications and analogies, *J. Fluid Mech.* **95**, 567-608.
18. Golitsyn, G.S., and Grachev, A.A. (1986) Free convection of multi-component media and parameterization of air-sea interaction at light winds, *Ocean-Air Interactions* **1**, 57-78.
19. Grachev, A.A. (1993) Free convection near the air-sea interface, in *Air-Sea Interface Symposium (Marseille, France, June 24-30, 1993), Proceedings.*
20. Grachev, A.A., Fairall, C.W., and Zilitinkevich, S.S. (1997) Surface-layer scaling for the convection-induced stress regime, *Boundary-Layer Meteorol.* **83**, 423-439.
21. Hibberd, M.F., and Sawford, B.L. (1994) A saline laboratory model of the planetary convective boundary layer, *Boundary-Layer Meteorol.* **67**, 229-250.
22. Högström, U. (1990) Analysis of turbulence structure in the surface layer with a modified smilarity formulation for near neutral conditions, *J. Atmos. Sci.* **47**, 1949-1972.
23. Howard, L.N. (1990) Limits of the transport of heat and momentum by turbulent convection with large-scale flow, *Stud. Appl. Math.* **83**, 273-285.
24. Hunt, J.C.R. (1984) Turbulence structure in thermal convection and shear-free boundary layers, *J. Fluid Mech.* **138**, 161-184.
25. Hunt, J.C.R., Kaimal, J.C., and Gaynor, J.I. (1988) Eddy structure in the convective boundary layer - new measurements and new concepts, *Quart. J. Roy. Meteorol. Soc.* **114**, 837-858.
26. Kader, B.A., and Yaglom, A.M. (1990) Mean fields and fluctuation moments in unstably stratified turbulent boundary layers, *J. Fluid Mech.* **212**, 637-662.
27. Kraichnan, R.H. (1962) Turbulent thermal convection at arbitrary Prandtl number, *Phys. Fluids* **5**, 1374-1389.
28. Kropfli, R.A., and Clifford, S.F. (1994) The San Clemente Ocean Probing Experiment: A study of air-sea interactions with remote and in situ sensors, in *Proc. IGARSS'94*, IV, 2470-2409.
29. Lenschow, D.H., and Stephens, P.L. (1980) The role of thermals in the convective boundary layer, *Boundary-Layer Meteorol.* **19**, 509-532.
30. Lenschow, D.H., Wyngaard, J.C., and Pennell, W.T. (1980) Mean-field and second-moment budgets in a baroclinic, convective boundary layer, *J. Atmos. Sci.*, **37**, 1313-1326.
31. Liu, W.T., Katsaros, K.B., and Businger, J.A. (1979) Bulk parameterization of air-sea exchange of heat and water vapour including the molecular constants at the interface, *J. Atmos. Sci.* **36**, 1722-1735.

32. Louis, J.F. (1979) A parametric model of vertical eddy fluxes in the atmosphere, *Boundary-Layer Meteorol.* **17**, 187-202.
33. Mahrt, L. (1991) Eddy asymmetry in the sheared heated boundary layer, *J. Atmos. Sci.* **48**, 472-492.
34. Mahrt, L., Sun, J., Vickers, D., MacPherson, J.I., Pederson, J.R., and Desjardins, R.L. (1994) Observation of fluxes and inland breezes over a heterogeneous surface, *J. Atmos. Sci.* **51**, 2484-2499.
35. Malkus, W.V.R. (1954) The heat transport and spectrum of thermal turbulence, *Proc. Roy. Soc. London* **A225**, 196-212.
36. Mason, P.J. (1988) The formation of areally averaged roughness lengths, *Quart. J. Roy. Meteorol. Soc.* **114**, 399-420.
37. Mellor, G.L., and Yamada, T. (1974) A hierarchy of turbulence closure models for planetary boundary layers, *J. Atmos. Sci.* **31**, 1791-1806.
38. Mikkelsen, T., Jorgensen, H.E., Lofstrom, P., and Lyck, E. (1996) Borex-95: Atmospheric Dispersion Experiment on Concentration Fluctuations, Data Report RISØ-R-927(EN).
39. Miller, M.J. (1992) The sensitivity of the ECMWF Model to the parameterization of evaporation from the tropical oceans, *J. Climate* **5**, 418-434.
40. Moeng, C.-H., and Wyngaard, J.C. (1984) Statistics of conservative scalars in the convective boundary layer, *J. Atmos. Sci.* **41**, 3161-3169.
41. Moeng, C.-H., and Wyngaard, J.C. (1989) Evaluation of turbulent transport and dissipation closures in second-order modelling, *J. Atmos. Sci.* **46**, 2311-2330.
42. Monin A.S., and Obukhov, A.M. (1954) Basic laws of turbulence mixing in the surface layer of the atmosphere, *Trudy Geofiz. Inst. AN SSSR* **24** (151), 163-187.
43. Monin, A.S., and Yaglom, A.M. (1971) *Statistical Fluid Mechanics*. Vol. 1, MIT Press, Cambridge MA, 769 pp.
44. Nieuwstadt, F.T.M., Mason, P.J., Moeng, C.-H., and Schumann, U. (1993) Large-eddy simulation of the convective boundary layer: a comparison of four computer codes, in F. Durst, R. Friedrich, B.E. Launder, F.W. Schmidt, U. Schumann and J.H. Whitelaw (eds.), *Turbulent Shear Flows 8*, Springer-Verlag, Berlin, 343-367.
45. Obukhov, A.M. (1946) Turbulence in thermally heterogeneous atmosphere, *Trudy Inst. Teoret. Geofiz. AN SSSR* **1**, 95-115.
46. Obukhov, A.M. (1960) Structure of temperature and velocity fields under conditions of free convection, *Izvestiya AN SSSR, Ser. Geofiz.*, No. 9, 1392-1396.
47. Panofsky, H.A., Tennekes, H., Lenschow D.H., and Wyngaard, J.C. (1977) The characteristics of turbulent velocity components in the surface layer under convective conditions, *Boundary-Layer Meteorol.* **11**, 355-361.
48. Plate, E.J. (1971) *Aerodynamic Characteristics of Atmospheric Boundary Layer*, U.S. Atomic Energy Commission, 190 pp.
49. Prandtl, L (1932) Meteorologische Anvendungen der Strömungslehre, *Beitr. Phys. Fr. Atmos.* **19**, 188-202.
50. Priestley, C.H.B. (1954) Convection from a large horizontal surface, *Austr. J. Phys.* **7**, 176-202.
51. Schmidt, H., and Schumann, U. (1989) Coherent structure of the convective boundary layer derived from large-eddy simulations, *J. Fluid Mech.* **200**, 511-562.
52. Schumann, U. (1988) Minimum friction velocity and heat transfer in the rough surface layer of a convective boundary layer, *Boundary-Layer Meteorol.* **44**, 311-326.
53. Siggia, E.D. (1994) High Rayleigh number convection, *Ann Rev. Fluid Mech.* **26**, 137-168.
54. Sorbjan, Z. (1989) *Structure of the Atmospheric Boundary Layer*, Prentice Hall, Englewood Cliffs NJ, 317 pp.
55. Stull, R.B. (1994) A convective transport theory for surface fluxes, *J. Atmos. Sci.* **51**, 3-22.
56. Sykes, R.I., and Henn, D.S. (1989) Large-eddy simulation of turbulent sheared convection, *J. Atmos. Sci.* **46**, 1106-1118.
57. Sykes, R.I., Henn, D.S., and Lewellen, W.S. (1993) Surface-layer description under free-convection conditions, *Quart. J. Roy. Meteorol. Soc.* **119**, 409-421.
58. Townsend, A.A. (1961) Equilibrium layers and wall turbulence, *J. Fluid. Mech.* **11**, 97-120.
59. Webster, P.J., and Lukas, R. (1990) TOGA COARE: The Coupled Ocean-Atmosphere Response Experiment, *Bull. Amer. Meteorol. Soc.* **73**, 1377-1416.
60. Wilczak, J.M., and Businger, Joost A. (1984) Large-scale eddies in the unstably startified atmospheric surface layer. Part II: Turbulent pressure fluctuations and budgets of heat flux, stress and turbulent kinetic energy, *J. Atmos. Sci.* **41**, 3551-3567.

61. Williams, A.G., and Hacker, J.M. (1992) The composite shape and structure of coherent eddies in the convective boundary layer, *Boundary-Layer Meteorol.* **61**, 213-245.
62. Williams, A.G., and Hacker, J.M. (1993) Interaction between coherent eddies in the lower convective boundary layer, *Boundary-Layer Meteorol.* **64**, 55-74.
63. Williams, A.G., Kraus, H., and Hacker, J.M. (1996) Transport processes in the Tropical Warm Pool boundary layer. Part 1: Spectral composition of fluxes, *J. Atmos. Sci.* **53**, 1187-1202.
64. Williams, A.G., Kraus, H., and Hacker, J.M. (1997) Transport processes in the Tropical Warm Pool boundary layer. Part 2: Vertical structure and variability, submitted to *J. Atmos. Sci.*
65. Wyngaard, J.C. (1983) Lectures on the planetary boundary layer, in D.K. Lilly and T. Gal-Chen (eds.), *Mesoscale Meteorology - Theories, Observations and Models*, Reidel, Dordrecht, 603-650.
66. Wyngaard, J.C. (1987) A physical mechanism for the asymmetry in top-down and bottom-up diffusion, *J. Atmos. Sci.* **44**, 1083-1087.
67. Wyngaard, J.C., and Cote, O.R. (1974) The evolution of a convective planetary boundary layer - a higher-order-closure model study, *Boundary-Layer Meteorol.* **7**, 289-308.
68. Yaglom, A.M. (1994) Fluctuation spectra and variances in convective turbulent boundary layers: a reevaluation of old models, *Phys. Fluids* **6**, Pt. 2, 962-972.
69. Zilitinkevich, S.S. (1991) *Turbulent Penetrative Convection*, Avebury Technical, Aldershot, 179 pp.
70. Zilitinkevich, S.S. (1994) A generalised scaling for convective shear flows, *Boundary-Layer Meteorol.* **70**, 51-78.
71. Zilitinkevich, S.S. (1995) Non-local turbulent transport: pollution dispersion aspects of coherent structure of convective flows, in H. Power, N. Moussiopoulos and C.A. Brebbia (eds), *Air Pollution III - Volume 1. Air Pollution Theory and Simulation*, Computational Mechanics Publications, Southampton Boston, 53-60.
72. Zilitinkevich, S. (1997) Heat/Mass Transfer in the Convective Surface Layer: Towards Improved Parameterization of Surface Fluxes in Climate Models, Alfred-Wegener-Institut für Polar- und Meeresforschung. Berichte aus dem Fachbereich Physik, Report 76, 34 pp.

CONVECTIVE TRANSPORT THEORY
AND THE RADIX LAYER

R. B. STULL
Atmospheric Science Programme, Department of Geography
The University of British Columbia, 1984 West Mall
Vancouver, BC, V6T 1Z2 Canada

Abstract

Statically unstable atmospheric boundary layers contain buoyant thermals with undiluted cores that transport air vertically across large distances. The rate of convective transport defines a buoyancy velocity, which is used to parameterize surface heat, moisture, and momentum fluxes (drag). Vertical profiles of wind and potential temperature are found for the radix layer, the bottom fifth of the boundary layer, which work better than classical surface-layer similarity theories.

1. Introduction to Nonlocal Mixing by Convection

1.1. THE NATURE OF COHERENT THERMAL STRUCTURES

Free convection occurs in the atmospheric boundary layer (BL) during conditions of light to calm winds when the surface of the earth is warmer than the overlying air. This can occur during fair weather when the land surface is heated by the sun during day, or any time day or night when cooler air is advected over a warmer land or sea surface. The result of this forcing is a statically unstable atmosphere [7] of warmer air underlying cooler air, in the potential temperature sense. Nature attempts to undo this instability by creating overturning in the BL, where turbulence in the form of thermal circulations moves warm air up and cold air down. Continued destabilization of the air by the external forcings causes continuous turbulence, resulting in a well-mixed layer (ML).

The thermals in this mixed layer (ML) have aspect ratios of order one, with diameters and depths roughly equal to the depth z_i of the ML, of order 1 km (Figure 1).

E.J. Plate et al. (eds.), Buoyant Convection in Geophysical Flows, 115–128.

Turbulent velocities are of order of the Deardorff velocity ($w_* = [(g/T_v)\cdot F_H \cdot z_i]^{1/3}$), typically about 1 m/s, where g is gravitational acceleration, T_v is virtual absolute temperature, and F_H is kinematic heat flux at the surface. Turnover time scales are of order of the Deardorff time ($t_* = z_i / w_*$), roughly 15 minutes.

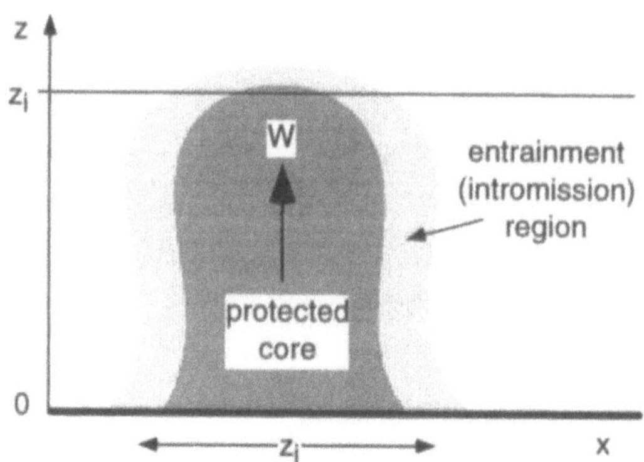

Figure 1. Idealization of a convective thermal [9], with aspect ratio of one and a protected core where surface-layer air is carried undiluted to the top of the ML.

The thermal diameters are sufficiently large that there is a very large central core that is protected from small-eddy lateral entrainment [2]. Within this core, large (1 km diameter) parcels of air from near the surface are moved to the top of the ML by the organized updraft of this thermal coherent structure. The fact that measured cumulus cloud base altitudes are equal to the theoretical lifting condensation level (LCL) of air near the surface is proof that surface-layer (SL) air can reach the ML top with no dilution [11].

1.2. NONLOCAL MIXING VS. TRADITIONAL SHEAR FLOWS

The vertical transport by thermals is not a small-eddy diffusion process, but a large-eddy vertical-advection process [9]. Because these coherent circulations can move air across the whole depth of the ML, the effect is said to be "nonlocal". It is nonlocal in the sense that air at one height in the ML can be directly affected by air at some non-neighboring height, via coherent-eddy advection between these two levels. In the real atmosphere, there is a spectrum of sizes of large-coherent eddies that cause direct transport without dilution between every possible pair of heights within the ML. One would not expect that a small-eddy diffusion theory, such as K-theory, should work well in the interior of the ML where large-eddies dominate.

Also, the nature of feedback between mean flow and turbulence forcing is different for convective turbulence than for shear-driven turbulence, as shown by Stull [10]. For shear flows near a solid boundary, there is a feedback as follows: turbulence transports momentum toward the boundary; momentum transport affects the wind-profile shape; and shear in the wind profile generates turbulence (Figure 2).

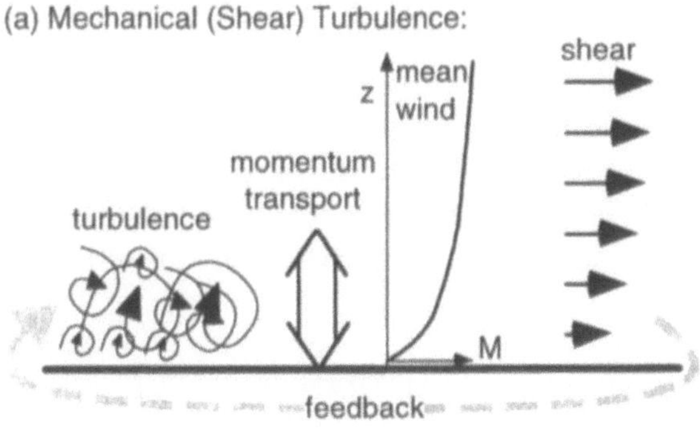

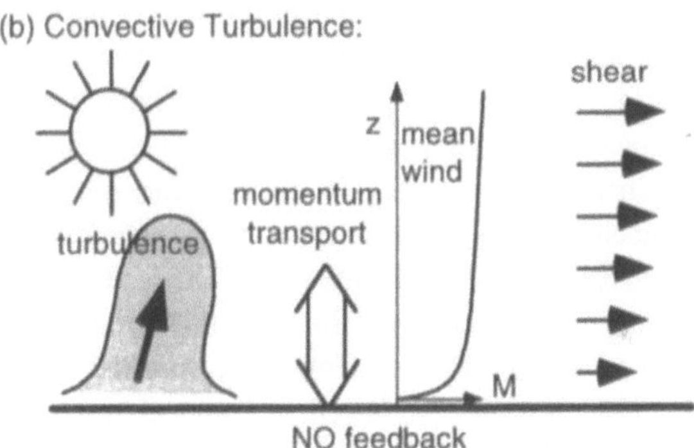

Figure 2. (a) Illustration of the mechanical-turbulence feedback that is utilized in the formulation of Monin-Obukhov similarity theory. (b) No such feedback occurs when turbulence is thermally forced.

However, for convective thermals such a feedback does not exist: turbulence in the form of thermals transport momentum; momentum transport alters the wind profile shape; but shear in the wind profile is not the major mechanism for generating turbulence. Instead it is solar heating, or the surface-to-air temperature difference, that drives the turbulence. Thus, one would not expect that a small-eddy surface-layer theory such as Monin-Obukhov (MO) similarity theory should work well to describe vertical profiles of mean wind and potential temperature within the bottom of a convective ML.

At the bottom interface of the convective ML, we usually observe significant surface heat fluxes on sunny calm days. This surface heat flux can happen even with no mean wind, because the large coherent thermal circulations are removing heated air from just above the surface, and replacing it with cooler air brought down from the mid ML. Traditional surface-drag or bulk-transfer theories for surface flux (such as $F_H = C_H \cdot M \cdot \Delta\theta$, where C_H is a bulk heat transfer coefficient, M is mean horizontal wind speed at 10 m elevation, and $\Delta\theta$ is the potential temperature difference between the surface skin and the air at height z = 10 m) fail for this situation because they require a nonzero mean wind.

The above arguments suggest that many of the traditional wall-shear-flow turbulence theories are anticipated to have difficulty for the convective ML. The theories in question are: K-theory in the ML interior, MO similarity theory in the bottom fifth of the ML, and bulk-transfer or drag laws at the bottom interface of the ML. Alternative theories are needed that properly account for the nonlocal nature of convective circulations.

1.3. IMPLICATIONS FOR VERTICAL TRANSPORT THEORIES

Because the vertical transport mechanism in thermals is known to be advective-like, this allows for improved parameterizations of turbulent processes. Starting from the bottom of the ML and working up, the next two sections describe new theories for transport: across the bottom interface (section 2 on convective transport theory), and within the bottom fifth of the ML (section 3 on the radix layer). Transport within the interior of the ML is discussed in the chapter on transilient turbulence theory in this volume.

Each of these parameterizations describe a form of low-order turbulence closure. Convective-transport theory [8] utilizes the mean conditions at the surface skin and in the ML interior to estimate vertical turbulent fluxes at the bottom interface. Radix-layer theory [5] relates mean wind and potential temperature profiles to surface fluxes and forcings. Transilient turbulence theory describes nonlocal mixing between various heights within the ML, from which turbulent fluxes can be inferred.

These three theories are related to each other by the philosophy that nonlocal transport works differently than diffusive transport. By utilizing the known behavior of coherent thermal structures, we can design nonlocal theories specifically for the convective ML.

2. **Convective Transport Theory**

2.1. TRANSPORT VELOCITY AND FLUX ACROSS AN INTERFACE

Turbulence cannot cause transport across the surface of the earth, because turbulent air parcels cannot penetrate into the ground or ocean during normal conditions. Transport between the surface and the bottom few millimeters of air is via molecular processes of conduction, diffusion, and viscosity. In addition, for heat there is radiative transfer. This thin microlayer of air is quickly modified to take on characteristics of the underlying surface; namely, it attempts to approach equilibrium in temperature, moisture, and momentum. If this equilibrium could be reached, then any further molecular transport between the surface and the air would cease.

However, the microlayer does not reach equilibrium with the surface because turbulent motions continuously remove air from the modified microlayer and replace it with less-modified air from higher in the ML. As it turns out, the weak link in this chain of processes is not the molecular transport, but the turbulent transport that recirculates air into the microlayer. Namely, the net flux across the surface is controlled by turbulence. Hence, to parameterize the surface flux, we must parameterize the ability of turbulence to transport air vertically from the microlayer.

The transport velocity w_T is a measure of the ability of turbulence to cause vertical transport away from the surface [6]. For an arbitrary quantity ψ, which could represent potential temperature, water vapor mixing ratio, horizontal wind components, or pollutant concentration, the turbulent transport across an interface is well parameterized by

$$F_{\psi} = \overline{w' \psi'}_s = w_T \cdot (\psi_{skin} - \psi_{air}) \tag{1}$$

where $F_y = w' \psi'_s$ is the vertical kinematic heat flux of ψ at the surface, ψ_{skin} is the value at the very top of the surface, and ψ_{air} is the value higher in the BL. Either greater difference in ψ between the surface skin and the air, or greater vertical transport velocity, can cause a greater vertical flux.

The crux of this parameterization problem is in describing w_T. For turbulence generated by wind shear near the bottom boundary, one might expect that vertical turbulent motions might scale to the shear, because it is the shear that generates the turbulence. Because wind speed is zero at the ground, the shear is proportional to the mean wind speed M just above the ground, such as at z = 10 m. This argument underlies traditional drag or bulk-transport laws, and results in the following parameterization for vertical transport velocity: $w_T = C_{\psi} \cdot M_{10}$, where C_{ψ} is the empirical bulk transport coefficient for variable ψ . For momentum, this coefficient is called the drag coefficient. When used in (1), the subscript "air" corresponds to a height of z = 10 m. Typical values of the drag coefficient are 0.001 to 0.02 .

2.2. MIXED-LAYER DRAG

Based on discussion in the previous subsection, we would anticipate that the traditional drag-law approach might be inappropriate for free convection, where winds are near calm but where vigorous turbulence nonetheless exists in the form of convective thermals. The transport velocity for free convection should be proportional to the actual vertical velocity within individual thermals [8], which is proportional to the Deardorff velocity. Thus, a logical parameterization would be $w_T = C_{*\psi} \cdot w_*$, where $C_{*\psi}$ is an empirical "mixed-layer transport coefficient". For this case, "air" in (1) corresponds to an altitude in the middle of the ML. Observations suggest that typical values of the mixed-layer transport coefficients are $C_{*M} = 0.023$ for momentum, and $C_{*H} = 0.0063$ for heat.

Using heat as an example, this mixed-layer transport parameterization is $F_H = \overline{w'\theta'}_s = C_{*H} \cdot w_* \cdot (\theta_{skin} - \theta_{zi/2})$, where $F_H = \overline{w'\theta'}_s$ is the surface kinematic heat flux. But w_* contains F_H in its definition. Thus the ML transport formulation is implicit, with F_H on both sides of the equation. However, with a little algebra, the equation can be rearranged into explicit form, described next.

2.3. CONVECTIVE TRANSPORT

An explicit parameterization for surface fluxes during free convection is [8]:

$$F_\psi = \overline{w'\psi'}_s = b_\psi \cdot w_B \cdot (\psi_{skin} - \psi_{UL}) \tag{2}$$

where b_ψ is an empirical coefficient, w_B is a new buoyancy velocity defined in the next subsection, and UL denotes a height within the uniform-layer portion of the ML for which a value at height $z_i/2$ would be representative (see the radix-layer section for details). Typical values of the empirical coefficient are $b_H = 0.0005$ for heat, and $b_M = 0.0018$ for momentum. This approach is called convective transport theory (CTT).

2.4. NEW SCALING VARIABLES AND IMPLICATIONS

A new velocity scale that comes naturally from the previous analysis is called the buoyancy velocity:

$$w_B = \left[(g/T_v) \cdot z_i \cdot (\theta_{v\ skin} - \theta_{v\ UL})\right] \tag{3}$$

where θ_v is the virtual potential temperature defined as $\theta_v = \theta \cdot (1+0.61 \cdot r)$, and r is the water-vapor mixing ratio in unsaturated air. The buoyancy velocity is linearly related to the Deardorff velocity by $w_B = 12.6\ w_*$. The main advantage of w_B is that it depends only on mean variables such as temperature and humidity, and not on turbulent quantities such as the heat or moisture flux.

A convective Richardson number can be defined as:

$$R_* = [w_B / M_{UL}]^2 \tag{4}$$

where M_{UL} is the wind speed in the mid mixed layer (i.e., in the uniform layer). Free convection scaling is appropriate whenever $R_* \geq 3$.

One implication of CTT is that the empirical coefficient should be independent of the surface roughness. The reason is that vertical transport is caused by thermal updrafts, which are independent of roughness. This implication differs from traditional surface-layer drag theories, which depend on surface roughness via the log-wind equation. Whether CTT is indeed independent of roughness is still being tested by other investigators for various data sets.

Another implication is that CTT could be used to help explain the strong heat fluxes over the western tropical Pacific Ocean [3]. In that region winds are often very light, causing traditional bulk-transfer laws to significantly underpredict the heat flux. However CTT gives a better estimate of heat fluxes over that region during non-stormy situations. During storms, other mechanisms contribute to the surface fluxes there, such as mesoscale patches of cold storm-outflow air flowing over the warmer sea surface.

It might be possible to use CTT to estimate surface fluxes remotely from satellite, during fair weather. For heat flux, CTT needs surface skin temperature and mid-mixed-layer temperature, which are more-easily measured by satellite than are surface-layer values. Also, surface-based radiometric sensors such as the Atmospheric Emitted Radiation Interferometer (AERI) could utilize CTT to estimate both heat and moisture fluxes remotely.

3. **Radix Layer**

3.1. COMPONENTS OF THE MIXED LAYER

Figure 3 shows the structure of the convective ML, as defined by Santoso and Stull [5]. In the interior is the uniform layer (UL) where wind and potential temperature variables are so well mixed by turbulence that they are virtually uniform with height. At the very bottom is the classical surface layer (SL), where fluxes vary by less than 10% of their magnitude, and where small-eddy theories such as Monin-Obukhov similarity apply. The top of the SL is below the bottom of the UL. Between these two layers, as well as within most of the SL, small-diameter SL convective plumes merge together to form the roots of convective thermals during conditions of free convection.

This layer is called the radix layer, from the Latin word for “root”. Thus, the radix layer (RxL) is a subdomain of the ML, and the surface layer is a subdomain of the RxL.

There is a superposition of quasi-independent processes that create the UL and the RxL. During free convection, large thermal structures (Figure 4) transport information (slower winds, warmer temperature, higher humidities) from near the surface to a wide range of depths within the UL. Thus, the UL “feels” the effects of the surface directly via these coherent structures, resulting in winds that are subgeostrophic, potential temperatures that are warming, and humidities that are increasing.

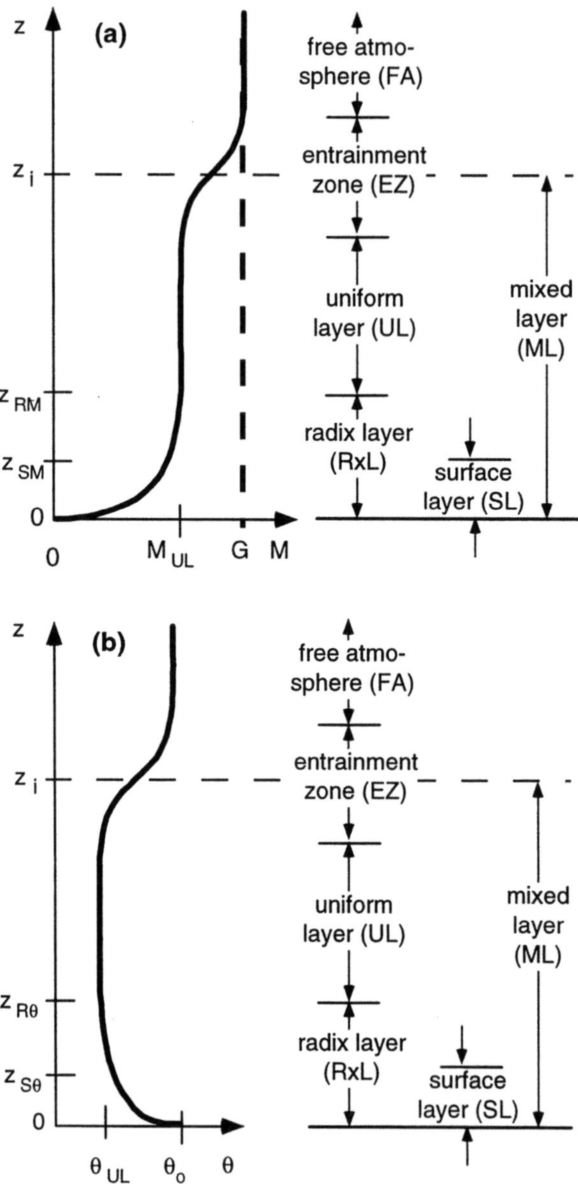

Figure 3. Structure of the mixed layer for (a) wind speed M and (b) potential temperature θ [5].

Superimposed on this deep layer of large-eddy transport is a shallower layer of small-eddy transport. This shallow layer, the RxL, feels all of the friction and heating of the ML, plus additional friction and heating associated with small eddies. Closer to the ground are greater numbers of more-intense small eddies, causing ever increasing curvature in the wind and temperature profiles as the surface is approached from above.

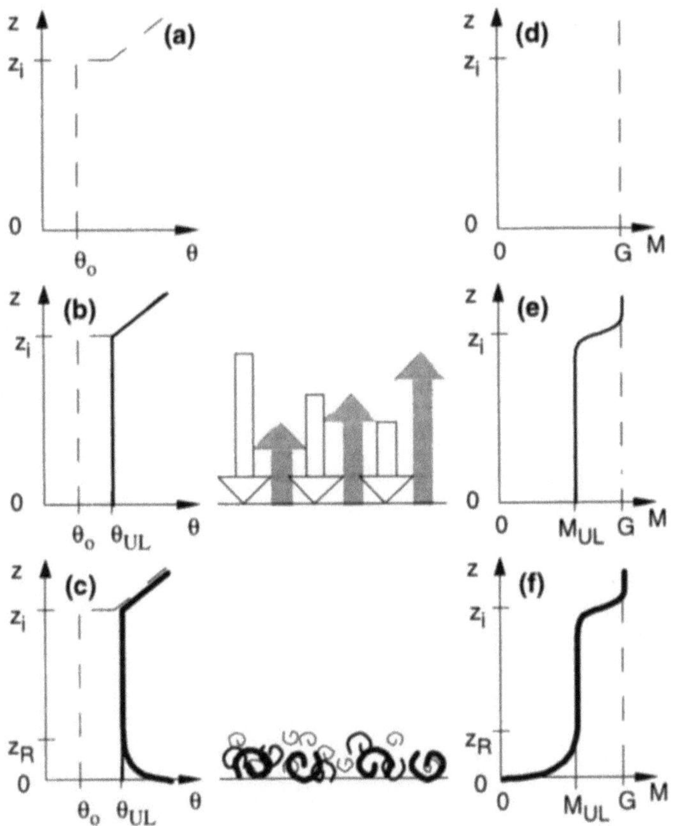

Figure 4. Superposition of large-eddy and small eddies processes in the ML, and their effect on the profiles of mean wind speed M and potential temperature θ. (a) and (d) represent background or initial states. (b) and (e) show how large-eddy transport between the surface and the whole ML modifies the initial states to create the uniform-layer state. (c) and (f) start with the previous figures and superimpose the effects of small eddies in only the radix layer. The result is a radix layer profile that becomes tangent to the uniform layer at a finite height, as is observed [5].

Based on this interpretation of the physics, we would anticipate that the (wind, potential temperature) profiles should start from their near surface values of (0, θ_o) and then (increase, decrease) with height, with smoothly reducing slope until they become exactly tangent to the UL (wind speed, potential temperature) at the top of the RxL. To do this, any profile relationship for the RxL must incorporate the RxL depths, and the characteristics of the UL. Traditional MO similarity relationships fail in the upper half of the RxL because they utilize only surface scales, and do not incorporate information about the UL.

3.2. REQUIREMENTS FOR OBSERVING THE RADIX LAYER

Time or space-averaged mean wind, temperature, and humidity profiles are needed to observe the RxL. These are similar to the requirements for measuring statistically robust profiles in the surface layer. The main difficulty, however, is that often the top of the RxL is above the top of most instrumented masts or towers. Hence, tethered balloons with sensors at many heights along the tether, aircraft flying slant ascent/descent soundings, or remote sensors with sufficient precision are needed to make these measurements.

Unfortunately, cost constraints force many field experiments to utilize different observing platforms to make measurements at different ranges of heights, resulting in inconsistent averaging periods. The most frequent example is a surface-layer mast making time averages of roughly 30 minutes over heights of roughly 30 m, with rawinsonde balloons launched every few hours to get snapshot observations at higher altitudes. As a result, if one plots both the mast data and the rawinsonde data on the same graph, there are often differences or discontinuities where these two systems overlap. Often these discontinuities are near the top of the RxL, making it impossible to identify the actual top of the RxL from the sounding.

This is a plea to researchers to utilize consistent averaging when they perform field experiments. In the example that follows, data from the 1973 Minnesota field experiment [4] was utilized, where instruments at various heights on the tether of a tethered balloon collected observations that were consistently time averaged. A new field experiment, Boundary Layer Experiment 1996 (BLX96), was performed using instrumented aircraft making slant soundings [1, 12] to gather spatially-averaged data in the

3.3. PROFILE RELATIONSHIPS FOR POTENTIAL TEMPERATURE AND WIND

As in any similarity theory, the first step is to anticipate what variables might be relevant to the flow, and then form dimensionless groups from them [6]. The next step is use field experiment data to plot dimensionless groups against each other. Any relationships that become apparent in these graphs can be fit with empirical curves.

We anticipate that for wind, the relevant dimensions groups are M/M_{UL} and z/z_{RM}, where M_{UL} is wind speed in the uniform layer and z_{RM} is the radix layer depth for momentum. Similarly for potential temperature, we anticipate the dimensionless groups to be $(\theta\text{-}\theta_{UL})/(\theta_o\text{-}\theta_{UL})$ and z/z_{R_θ}, where θ_{UL} is the potential temperature in the uniform layer, and z_{R_θ} is the radix layer depth for temperature. Also, since the same turbulent eddies are mixing both wind and temperature, we expect the empirical relationships to have the same functional form.

Based on the 1973 Minnesota data, we find the following empirical profile equations for the RxL:

$$\frac{M}{M_{UL}} = F\left(\frac{z}{z_R}\right) \tag{5}$$

$$\frac{\theta - \theta_{UL}}{\theta_o - \theta_{UL}} = 1 - F\left(\frac{z}{z_R}\right) \tag{6}$$

$$F = \left(\frac{z}{z_R}\right)^A \cdot \exp\left[A \cdot \left(1 - \frac{z}{z_R}\right)\right] \quad \text{for } z \leq z_R \tag{7}$$

with $F = 1$ for $z > z_R$. $A = 0.10 \pm 0.01$ is an empirical constant. Also, the radix layer depth z_R would be different for the two variables, and would also vary from day to day and time to time, analogous to the way that the Obukhov length varies in SL theory. For the Minnesota data, typical RxL depths were of the order of 400 ± 300 m for momentum, and 50 ± 30 for potential temperature.

Figure 5 shows how the data from all 11 runs of the Minnesota experiment collapse to the same similarity curves when plotted in dimensionless form using (5) and (6). The empirical curve (7) is also plotted as the solid line in those figures. For both curves, the correlation coefficient between empirical curve and data is $r = 0.99$. The similarity curves above are valid within the range $0.5 \leq z \leq 500$ m, where the top limit should be set to half the ML depth z_i. Below 0.5 m, the profiles approach zero wind speed and zero temperature difference at a height of $z = 0$, rather than at the height of the roughness length z_o as designed into traditional MO similarity theory. Thus, for investigators wanting profiles below 0.5 m, we recommend that MO theory be used instead of radix-layer theory.

Finally, research is continuing to determine the behavior of the radix layer depth. We anticipate that it will depend on surface roughness, and on the buoyancy velocity. If such a relationship is found, the result will be a new flux-profile relationship covering the radix-layer depth and the uniform layer.

3.4. IMPLICATIONS FOR PROFILE RELATIONSHIPS

Most end users of wind and temperature profiles do not want gradients. They want the mean variables for applications such as wind loads on buildings and bridges, pollutant dispersion, power generation by wind turbines, and temperatures for chemical reaction rates, photosynthesis, and radiation calculations. Our relationships above are in just such form.

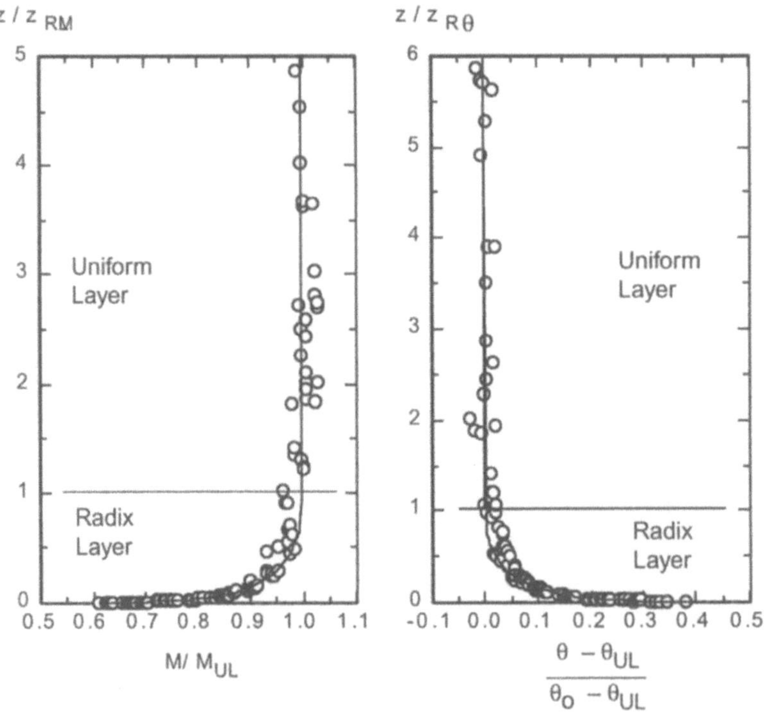

Figure 5. Validation of the radix-layer profile equations (5) - (7) against field data from the 1973 Minnesota experiment [5].

Many errors are hidden if gradients of wind and potential temperature are studied instead of the mean values of wind and temperature. Traditionally, flux-profile relationships are presented using dimensionless wind shear $F_M = (\partial M/\partial z)\cdot k\cdot z/u_*$ and temperature gradients $F_H = (\partial \theta/\partial z)\cdot k\cdot z/\theta_*$, where u_* is the friction velocity, θ_* is the convective temperature scale, and $k = 0.4$ is von Kármán's constant. It is easy to form a similar dimensionless shear variable in the radix layer, such as $(\partial M/\partial z)\cdot z_{RM}/M_{UL}$ using (5) and (7). When all of these gradients of profile relationships are plotted against gradients of observed data, the fits are often quite good. It is difficult to discriminate between different similarity theories when plotted in gradient form.

However, when the traditional SL theories are integrated with height to get mean winds and temperatures, the errors become apparent and often significant [5]. Such errors are not present in RxL theory. We recommend that future validation of wind and temperature profiles against observations be done using the pure variables, rather than their gradients.

4. Conclusions

As the nonlocal nature of convection becomes recognized and better understood, investigators can utilize this knowledge to develop better parameterizations for turbulence processes. In this paper, two such parameterizations were presented: convective transport theory for flux at the surface of the earth, and radix layer theory for the vertical profiles of mean variables in the bottom fifth of the ML. A third theory, transilient turbulence theory, describes nonlocal transport within the interior of the ML, and is presented in a separate chapter of this volume.

Convective transport theory gives surface fluxes as a function of the difference of mean quantities between the surface skin and the middle of the ML. This has the advantage of bypassing the traditional surface layer, and is expected to be independent of surface roughness length. These features open the possibility of utilizing remotely-sensed measurements from satellites to estimate surface fluxes. A new buoyancy velocity scale and convective Richardson number were also defined.

Radix layer theory gives profiles of mean wind speed and potential temperature in the bottom fifth of the ML. It is superior to surface layer theories just below the uniform layer, and is similar in performance within the bottom half of the radix layer. Only below about 0.5 m does classical surface-layer theory perform better during free convection.

References

1. Berg, L.K., Stull, R.B., Santoso, E. and Hacker, J.P. (1997) *BLX96 Airborne Scientists Flight Log*, BLRT-97-1, Dept. of Geography, Univ. of British Columbia, Vancouver, Canada.
2. Crum, T.D., Stull, R.B. and Eloranta, E.W. (1987) Coincident lidar and aircraft observations of entrainment into thermals and mixed layers, *J. Climate and Applied Meteorology* **26**, 774-788.
3. Greischar, L. and Stull, R. (1998) Convective transport theory for surface fluxes, tested over the Western Pacific warm pool, *J. Atmospheric Sciences* **55** (in press).
4. Izumi, Y. and Caughey, J.S. (1976) *Minnesota 1973 Atmospheric Boundary Layer Experiment Data Report*, AFCRL-TR-76-0038, U.S. Air Force Geophysics Lab, Massachusetts.

5. Santoso, E. and Stull, R. (1998) Wind and temperature profiles in the radix layer, the bottom fifth of the convective boundary layer, *J. Applied Meteorology* **37**, (in press).

6. Stull, R.B. (1988) *An Introduction to Boundary Layer Meteorology*, Kluwer Academic Publishers, Dordrecht.

7. Stull, R.B. (1991) Static stability — an update, *Bulletin of the American Meteorological Society* **72**, 1521-1529.

8. Stull, R.B. (1994) A convective transport theory for surface fluxes, *J. Atmospheric Sciences* **51**, 3-22.

9. Stull, R.B. (1995) *Meteorology Today for Scientists and Engineers*, Wadsworth (West) ITP, Belmont, California.

10. Stull, R.B. (1997) Reply (to Comments on "A convective transport theory for surface fluxes."), *J. Atmospheric Sciences* **54**, 579.

11. Stull, R. and Eloranta, E. (1985) A case study of the accuracy of routine fair-weather cloud-base reports, *National Weather Digest* **10**, 19-24.

12. Stull, R.B., Santoso, E., Berg, L., and Hacker, J. (1997) Boundary layer experiment 1996 (BLX96), *Bulletin of the American Meteorological Society* **78**, 1149-1158.

CONVECTION IN VISCOUS AND ROTATING FLUIDS FROM THE VIEWPOINT OF THE FORCED FLOW THEORY

G. S. GOLITSYN
*A. M. Oboukhov Institute of Atmospheric Physics,
Russian Academy of Sciences,
Pyzhevsky 3, 109017 Moscow, Russia*

Abstract

Many geophysical flows, including all types of convection, and some other phenomena can be described in terms of their forcing and certain characteristic time scales. In this paper, a principle of the fastest response is proposed. It states that quadratic characteristics of the physical fields, *i.e.* variances and other second-order moments, can be estimated as the product of their generation rate and a minimum time scale inherent in the system considered. From this standpoint, a review is given of turbulence in three and two dimensions, convection in viscous and rotating fluids, and circulations of the planetary atmospheres. The principle is applicable to a number of forced flows of classical hydrodynamics, among which are the Stokes flow of viscous fluid around a body falling under gravity, and the pipe flow caused by an imposed pressure gradient. In these cases, the principle reduces to the statement that the characteristic velocity of the flow is the applied force times the minimum time scale typical of the flow.

Results of the application of the proposed principle are compared with the predictions of the similarity theory. Neither of the two enables dimensionless coefficients to be evaluated. The principle is more efficient in some cases, however, as it combines a number of external parameters into a single time scale. This is the case, for example, in viscous turbulent flows where the proposed principle immediately yields the viscosity dependence of the flow velocity while the similarity theory leaves the dependence on the Reynolds number undetermined.

In specific cases both forcing and energy are known and the interval between events is to be determined. Earthquakes and cosmic rays provide

E.J. Plate et al. (eds.), Buoyant Convection in Geophysical Flows, 129–155.

good examples that have been thoroughly studied (for cosmic rays, the density flux, *i.e.* the number of particles passing across a unit area in a unit time, is of interest). It is shown that the proposed principle can be successfully used in these cases as well. It predicts the frequency-intensity spectra for the earthquakes and cosmic rays in a good agreement with observations.

1. Introduction

In a broad sense, by convection is meant a fluid flow with inhomogeneous distribution of density in the field of external body force, such as gravity and centrifugal force, the latter force is encountered in some technical applications. If the density inhomogeneity is maintained by an external source of energy (*e.g.* by solar radiation, as is the case in the atmospheric boundary layer), then a statistically stationary (statistically steady) regime of motion sets in, whose velocities and the transport and dispersion properties are of great importance for understanding the host of natural and technical problems. From this general point of view, the atmospheric general circulation can be thought of as a kind of convection arising due to the spherical shape of our planet that leads to the non-uniform heating of the surface and the atmosphere of the planet by the sun. Rotation as well as stratification bring new aspects into the picture of convection.

Forcing of a flow can stem from instabilities of motions at certain scales, leading to the energy propagation to other scales due to the inertial non-linear forces. The energy propagates to either smaller scales where it is dissipated by viscosity, as is the case in three-dimensional (3D) turbulence, or to larger scales, as in two-dimensional (2D) turbulence. The latter effect was called "negative viscosity" until a proper understanding of the forced 2D flows had been achieved (see Starr [60]). At a constant rate of energy pumping into a dissipative system a statistically steady flow regime is realised. This was shown by Kolmogorov [41,42] and his graduate student Oboukhov [50] for 3D turbulent flows, by Batchelor [4] and Kraichnan [43] for 2D flows, and by Charney [12] for quasi-geostrophic flows. Other variables in turbulent flows, *e.g.* passive and active scalars, exhibit similar behaviour to the velocity field. This fact was explained by Oboukhov [52] and Corrsin [14] (see also Monin and Yaglom [48]).

Flows under applied body force like gravity or imposed pressure gradient also belong to this family of phenomena. The rate of energy supply to such flows is the work done by the force f_a over a distance r in a unit time. To put it differently, as Russian kids are taught at school, it is the product of the force and the velocity. The energy can either dissipate in the flow directly, as in the laminar Stokes flow around a sphere, or the secondary flows are formed around a body moving through the fluid in a non-linear regime where inertial forces play an important role.

There are other physical systems that involve a more or less steady energy input but release energy in a different way than in hydrodynamics. As was shown by the present author, earthquakes, EQs [34], and cosmic rays, CRs [35], are examples of such systems. These systems dissipate their energy in "events", the EQs, or in "objects", the CR particles. Their volume energy, the stress drop in the EQs and the pressure in the CRs, are known (*i.e.* these quantities can be, and were, measured). This is a different type of the processing of energy pumped into the system as compared to the turbulence-type systems where the energy is cascaded by the flow through the spatial scales. Of course, the volume energy of the EQs and CRs is eventually dissipated through a variety of elementary processes but that stage apparently does not determine their frequency-intensity spectra.

All the above flows and phenomena can be considered exactly in the same way as some laminar flows, or, with some approximations, statistically, using dimensional analysis and methods of the similarity theory [10,9,59,2, and many other textbooks]. In the Russian scientific circles after World War II there were two popular sayings. The first was "the similarity theory is a similarity (likeness) of a theory". It was reiterated by skeptics. The second was "the similarity theory (in every particular case) is closer to the art than to the science". It was very popular among admirers. The author, who considered a number of flows and phenomena using the similarity theory, should confess that every time massive effort was made and new reasoning was invented before arriving at the final result. Some common features of turbulence, convection (without rotation) and atmospheric circulation were summarised some time ago [30] and formulated as the rule of the fastest response, stating that the kinetic energy of the system is the energy generation rate times a minimum time scale inherent in the system. A number of different physical systems examined ever since (and possibly many others) obey this rule. Now we dare to generalise it and propose that the rule be regarded as a principle.

2. The Principle of the Fastest Response

The principle can be stated as follows: *the second-order moments of physical fields, such as kinetic energy, velocity variances and scalar variances, are the products of their generation rate and a minimum time scale inherent in the physical system considered.*

The generation rate of the second-order moments reflects the flow forcing. We have already mentioned a number of forcings. These are the solar energy for atmospheric (and oceanic) flows, the flow instabilities at certain scales, geothermal heat flux for mantle convection and, ultimately, for earthquakes, and the time rate of work by applied forces such as gravity. We can add here the rate of energy generation by the supernova star explosions, producing energy for cosmic rays, internal galactic magnetic field, and many other things. We can also mention here the rate of generation of the passive scalar inhomogeneities which in our atmosphere and ocean again have the sun as an ultimate source of energy.

Time scales can be internal, such as dynamic time of inertial forces $\tau_d = r/u$, r and u being characteristic length and velocity scales, respectively, and external, such as the rotation period $\tau_\Omega = (2\Omega)^{-1}$ (a reciprocal of the Coriolis parameter), Ω being the angular velocity of rotation. They can be related to the physical properties of the system considered, such as viscous time $\tau_\nu = r^2/\nu$, ν being the kinematic viscosity, and the time of relaxation to the local hydrodynamic equilibrium $\tau_c = r/c_e$, c_e being the velocity of sound (see Landau and Lifshits [45]). In other words, τ_c is the time of relaxation of density and pressure disturbances on global scale to the geostrophic balance [53]. Apparently other time scales could be present. In the hydrodynamic cases, these time scales formally appear when one makes the equations dimensionless.

For other physical systems (of non-fluid type), we know characteristics of energy and its generation rate and seek the intervals between the occurrence of events (or the appearance of objects with different energies). Again the general principle works, and we seek the distribution of intervals by imposing a requirement that the interval should be a minimum (if one were to decide between several possible intervals). Here an assumption should be made that the events are independent of each other.

Comparison of various time scales yields the similarity criteria. For instance, if $\tau_d \ll \tau_\nu$, then $\tau_\nu/\tau_d = ru/\nu = \mathrm{Re} \gg 1$, *i.e.* the Reynolds number is large and the effects of viscosity can be neglected. Similarly, $\tau_c \ll \tau_d$ means a small Mach number, $\mathrm{Ma} = u/c \ll 1$, $\tau_\Omega \ll \tau_d$ means a small Rossby number, $\mathrm{Ro} = u/\Omega r \ll 1$, and $\tau_\Omega \ll \tau_\nu$ means a large Taylor number, $\mathrm{Ta} = 4\Omega^2 r^4/\nu^2 \gg 1$, and a small Ekman number, $\mathrm{Ek} = \mathrm{Ta}^{-1/2} = \nu/2\Omega r^2 \ll 1$. The last relation can be conveniently recast in terms of the Coriolis parameter $f_c = 2\Omega \sin\phi$, ϕ being the latitude, to characterise the Earth conditions where typically $\mathrm{Ek} = \nu/f_c r^2 \ll 1$.

In a sense the principle of the fastest response is equivalent to a set of similarity criteria that offer a means of neglecting certain governing parameters of the physical system in question. In other words, this neglect is a similarity hypothesis with respect to the governing parameters. A. N. Kolmogorov, for example, formulated his second hypothesis in the famous 1941 paper [41] as the possibility to neglect the kinematic viscosity for the inertial range of the 3D turbulence as soon as $\mathrm{Re} \gg 1$. Considering other particular examples in what follows, we will return to the problem of correspondence between our principle and the similarity theory. Now we note that our principle decreases the number of governing parameters by combining some of them into a single time scale. In some cases it allows one to get direct results while the usual similarity-theory procedure yields only the functional dependence on dimensionless combinations of governing parameters. In the terminology of Barenblatt [2], this means that in some cases our principle can decipher the self-similarity of the second kind, when the dependence upon a certain similarity criterion (dimensionless combination of governing parameters) does not vanish no matter how small or large its value may be [see Eq. (7) and the subsequent text]. Of course, there are many situations (a good few of those are described by Barenblatt [2]) when special efforts are required to elucidate the self-similarity of the second kind. Below we present a number of examples of the successful application of our principle. In some classical cases, as in 3D and 2D turbulence, one may consider the principle as a short-cut to illustrate the results, at least for educational purposes.

3. The Kolmogorov-Oboukhov Theory of Locally Homogeneous and Isotropic Turbulence

The theory has been formulated in a series of papers by Kolmogorov [41,42] and Oboukhov [50]. A. N. Kolmogorov introduced the structure functions for the velocity field as the mean square velocity difference between the two points r apart, where $r = |\mathbf{r}|$ is the magnitude of the position vector $\mathbf{r}$. Widely used is the structure function for the velocity component aligned with the vector $\mathbf{r}$,

$$D_u(|\mathbf{r}|) = \left\langle [u_i(\mathbf{x}+\mathbf{r}) - u_i(\mathbf{x})]^2 \right\rangle = \langle v_i^2(\mathbf{r}) \rangle \,, \tag{1}$$

where the angle brackets denote the ensemble mean.

The rate of generation of the turbulence kinetic energy per a unit mass (TKE) can be estimated as

$$\epsilon = (\Delta U)^3 L_0^{-1}, \tag{2}$$

where ΔU and L_0 are the scale of velocity difference and the external length scale, respectively, characteristic of the large scale flow. In a statistically steady state, the rate of TKE generation equals the rate of its dissipation, ϵ, described according to the Stokes formula, as

$$\epsilon = \frac{\nu}{2} \left\langle \left(\frac{\partial v_i}{\partial x_j} + \frac{\partial v_j}{\partial x_i} \right)^2 \right\rangle . \tag{3}$$

The TKE dissipation, ϵ is considered to be an external parameter whose value is known. Using ϵ and ν as the key governing parameters, A. N. Kolmogorov suggested the following microscales for the length, time and velocity:

$$l_k = \nu^{3/4}\epsilon^{-1/4}, \quad \tau_k = \nu^{1/2}\epsilon^{-1/2}, \quad v_k = (\nu\epsilon)^{1/4}. \tag{4}$$

In the range of time scales where the viscous time strongly exceeds dynamic time, $\tau_\nu \gg \tau_d$ and therefore Re $\gg 1$, our principle yields the result,

$$\left\langle v_i^2(\mathbf{r})\right\rangle \approx \epsilon\tau_d = \epsilon r \left\langle v_i^2(\mathbf{r})\right\rangle^{-1/2}, \quad \text{or } \left\langle v_i^2(\mathbf{r})\right\rangle = C_v(\epsilon r)^{2/3}, \tag{5}$$

where the estimate $C_v = 2$ derived from measurements is accurate to 10–20% [61].

In the viscous subrange, where $r \leq l_k = \nu^{3/4}\epsilon^{-1/4}$, the shortest time is $\tau_\nu = r^2\nu^{-1}$. Using our principle, we obtain

$$\left\langle v_l^2(\mathbf{r})\right\rangle = C_\nu \epsilon r^2 \nu^{-1}, \tag{6}$$

where C_ν is a dimensionless constant.

The formula (6) can be derived with an exact value of the dimensionless coefficient C_ν. Expanding $\left\langle v_i^2(\mathbf{r})\right\rangle$ into a Taylor series and retaining only linear terms in $\mathbf{r}$, then applying the Einstein summation convention on repeated indices in Eq. (3), one finds $C_\nu = 1/15$ for isotropic fields. Equation (6) provides a good example of how our principle works. Ordinary dimensional analysis does not allow one to obtain Eq. (6). Indeed, out of the three governing parameters of the problem, ϵ, ν and r, we can form a dimensionless similarity criterion, the Reynolds number Re $= r\epsilon^{1/4}\nu^{-3/4}$. Then, the general expression for $\left\langle v_i^2(\mathbf{r})\right\rangle$ at $r \leq l_k$ takes the form

$$\left\langle v_l^2(\mathbf{r})\right\rangle = (\nu\epsilon)^{1/2} f_v\left(r\epsilon^{1/4}\nu^{-3/4}\right), \tag{7}$$

where f_v is a function of the Reynolds number Re. In order to obtain Eq. (6), we should assume that $f_v = C_\nu \mathrm{Re}^2$ at small Reynolds numbers. Our principle allows one to escape all this trouble. However, the numerical value of dimensionless coefficient C_ν cannot be determined.

With a knowledge of the structure function, one can introduce the spectral density $E(k)$ (see Monin and Yaglom [48] for details):

$$\frac{1}{2}\left\langle v_l^2(\mathbf{r})\right\rangle = \int_0^\infty f_v(ikr)E(k)dk, \tag{8}$$

where k is the wave number, i is unit imaginary number, and $f_v(\mathrm{i}kr)$ is a kernel of the Fourier-type transform. As seen from Eq. (8), the dimension of $E(k)$ is $\mathrm{L}^3\mathrm{T}^{-2}$, where L and T are dimensions of length and time, respectively. From Eq. (5) we obtain

$$E(k) = C'\epsilon^{2/3}k^{-5/3}, \tag{9}$$

where C' is a dimensionless constant. Note that the exponents in Eqs. (5b) and (9) are related as

$$m + n = -1, \tag{10}$$

where in the case in question $m = 2/3$ and $n = -5/3$.

Again from Eq. (5) we see that the dynamic inertial time is

$$\tau_d = ru^{-1} = r^{2/3}\epsilon^{-1/3}. \tag{11}$$

Oboukhov [52] introduced an average molecular destruction rate of a passive scalar θ in a turbulent flow,

$$N = -\frac{1}{2}\frac{d\langle\theta^2\rangle}{dt} = k_d\left\langle\left(\frac{\partial\theta}{\partial x_i}\right)^2\right\rangle, \tag{12}$$

where k_d is the molecular diffusivity for the scalar θ. Equation (12) is similar to the formula for the TKE dissipation rate, Eq. (3). Using the scalar difference characteristic of the large scale flow, $\Delta\Theta$, the following estimate holds:

$$\epsilon = \Delta U(\Delta\Theta)^2 L_0^{-1}, \tag{13}$$

that is similar to the estimate of ϵ given by Eq. (2).

For the structure function of the θ field, our principle gives

$$D_\theta(r) = N\tau_d = C_\theta N\epsilon^{-1/3}r^{2/3}, \tag{14}$$

where C_θ is a dimensionless constant. The dynamic time scale, Eq. (11), has been used to derive the above formula. Equation (14) was obtained by A. M. Oboukhov in 1949.

Corrsin [14] proposed the following formula for the spectral density:

$$E_\theta(k) = N\tau_d = C'_\theta N\epsilon^{-1/3}k^{-5/3}, \tag{15}$$

where C'_θ is one more dimensionless constant. This relation immediately follows from Eq. (14). Dimensionless constants C_θ and C'_θ were estimated at 3 and 0.75, respectively. These estimates are much less accurate than the estimate of C_v in Eq. (5) for the velocity structure function. Although the course of the large scatter in the C_θ and C'_θ estimates is still poorly understood, the "–5/3" slope of the spectra for temperature, humidity, concentration of CO_2, and other scalars, is borne out by numerous observations.

4. Free and Sheared Convection

The closest relative to turbulence is convection. In this section, we consider convection with no rotation. We further simplify the discussion by assuming that the density is inhomogeneous only in the vertical. This is the case in a horizontal fluid layer heated from below or cooled from above (the latter situation is typical of the natural water surfaces). Considering statistically horizontally homogeneous layer with no mean flow, the energy balance equation gives the following expression for the rate of kinetic energy dissipation:

$$\epsilon = \alpha g F(\rho c_p)^{-1}, \tag{16}$$

where α is the thermal expansion coefficient (T^{-1} for the ideal gas, T being the absolute temperature), g is the acceleration due to gravity, F is the heat flux, ρ is the density, and c_p is the specific heat at constant pressure. Equation (16) was first proposed by Prandtl [57].

In the stationary case the r.h.s. of Eq. (16) is equal to the buoyancy flux,

$$B = \rho_0^{-1} g < w'\rho' >, \tag{17}$$

where ρ_0 is the reference density, w is the vertical velocity, primes denote turbulent fluctuations, and the angle brackets denote an ensemble mean. In developing boundary layers, $B \neq \epsilon$ due to entrainment.

Density fluctuations may be caused by the temperature fluctuations or by concentration fluctuations of an admixture, such as salt in the sea water or water vapour in the air. They may occur due to the presence of foreign particles, such as bubbles in water or dust in air. In all these cases, the equation of state should be carefully considered. Examples of the effect of admixtures on the turbulence flow can be found in Barenblatt and Golitsyn [3]. Heavy admixtures (heavy, as compared to the ambient fluid) increase the flow stability because a part of turbulent energy is spent on suspending foreign particles in the flow. Light admixtures decrease the stability by increasing buoyancy, as does the water vapour in the air or the air bubbles in the water.

The most important similarity criteria for the Boussinesq thermal convection are the Rayleigh number,

$$\mathrm{Ra} = \frac{\alpha g \Delta T d^3}{\nu k_T}, \tag{18}$$

where ΔT is the temperature difference across the layer of depth d and k_T is the molecular thermal diffusivity, and the Prandtl number,

$$\mathrm{Pr} = \frac{\nu}{k_T}. \tag{19}$$

The Prandtl number characterises the physical properties of the medium. After the onset of convection the heat (and mass) transfer is intensified. The degree of this intensification is measured by the Nusselt number,

$$\mathrm{Nu} = \frac{Fd}{\rho c_p k_T \Delta T}, \tag{20}$$

which is the ratio of the actual heat flux to the heat flux that would have been observed if there were no motions.

Thus, we have the forcing as given by Eqs. (16) and (17), and the two time scales, the dynamic time scale τ_d and the viscous time scale τ_ν. In the neighbourhood of the heated (cooled) surface the distance z from that surface is the only relevant length scale (convective motions outside of this region "feel" the presence of the other boundary, and the length scale d also comes into play). Then, our principle immediately gives the following scaling estimate for the vertical motions in dynamic regime, when $\tau_d \ll \tau_\nu$:

$$\left\langle [w'(z)]^2 \right\rangle = \epsilon \tau_d = (\epsilon z)^{2/3}. \tag{21}$$

This expression was first obtained by Prandtl [57].

Following Deardorff and Willis [17], we may introduce the bulk velocity scale relevant to the entire fluid layer:

$$w_* = (\epsilon d)^{1/3}. \tag{22}$$

This scale is commonly used nowadays in the studies of convective flows. The length scale $w_* = (B_s d)^{1/3}$, where the surface buoyancy flux B_s is used in place of the TKE dissipation rate ϵ, and d stands for the convective boundary layer depth, was proposed by Deardorff [15,16]. In the meteorological literature, w_* so defined is often referred to as "the Deardorff convective velocity scale".

When the effects of viscousity dominate, we can write down the formulae for the mean velocity,

$$u = C_{c\nu} (\alpha g F / \rho c_p \nu)^{1/2} d, \tag{23}$$

where $C_{c\nu}$ is the dimensionless constant of order 0.1 [29], and for the corresponding viscous time scale,

$$\tau_\nu = d/u = C_{c\nu}^{-1}(\alpha g F/\rho c_p \nu)^{-1/2}. \tag{24}$$

The viscous convective regime plays a very important role in geophysics. The first example, is convection in the mantle. Using a typical value of the geothermal heat flux $F = 0.08$ W·m^{-2}, and the following values of the governing parameters characterising the mantle, $\alpha = 2 \cdot 10^{-5}$ K^{-1}, $g = 10$ m·s^{-2}, $\nu = 2 \cdot 10^{17}$ m^2·s^{-1}, $\rho = 3 \cdot 10^3$ kg·m^{-3}, $c_p = 6 \cdot 10^2$ m^2·s^{-2}·K^{-1} and $d = 3 \cdot 10^6$ m, we obtain the estimate of $u = 2 \cdot 10^{-9}$ m·s^{-1}. This is about 5 cm·year^{-1}, which is precisely the characteristic velocity of the lithospheric plate drift.

The second geophysically important example of viscous convection is the surface cooling of the natural water bodies. It is obvious that the heat loss through the surface which occurs mainly due to evaporation should not depend on the total water depth. This is a good first approximation that implies that the depth scale should drop out from the heat transfer law. Then, the classical "1/3" power law, relating the Nusselt and Rayleigh numbers, holds:

$$\text{Nu} = C_{T\nu}\text{Ra}^{1/3}, \tag{25}$$

where $C_{T\nu} \approx 0.06$ is the dimensionless coefficient. The temperature difference (that in this regime should occur over a thin layer adjacent to the cooled surface) is given by

$$\Delta T = k_T^{1/2}(F/C_{T\nu}\rho c_p)^{3/4}(\nu/\alpha g)^{1/4}. \tag{26}$$

For typical oceanic conditions, $F = 100$ W·m^{-2}, $k_T = 10^{-7}$ m^2·s^{-1}, $\nu = 10^{-6}$ m^2·s^{-1}, $\rho = 10^3$ kg·m^{-3}, $c_p = 4.2 \cdot 10^3$ m^2·s^{-2}·K^{-1}, $\alpha = 2 \cdot 10^{-4}$ K^{-1}, and $g = 10$ m·s^{-2}, we have $\Delta T \approx 0.8$ K. This temperature difference occurs over a layer whose thickness is given by

$$\delta = \rho c_p k_T \Delta T F^{-1}, \tag{27}$$

which in our case is about 3 mm. The layer is indeed thin. This cold skin plays an important role in several applications, the heat and mass exchange between the ocean and the atmosphere among them. A useful recipe for parameterisation of the fluxes of sensible and latent heat using a generalisation of Eq. (25) to the case of a multi-component medium is proposed by Golitsyn and Grachev [38]. That parameterisation applies to calm weather conditions, at wind speeds of up to 3 m·s^{-1} (see also Godfrey and Beljaars [25] and Beljaars [5]).

In windy conditions, the surface friction velocity u_* should be considered as an additional governing parameter. It is a measure of the momentum flux from the atmosphere to the underlying surface, and is ultimately a measure of the strength of the near-surface wind. If the buoyancy flux is non-zero, one can specify the length scale. The application of our principle (neglecting the insignificant molecular effects) yields the relation $u_*^2 = B_s L_O u_*^{-1}$, and thus the length scale (the subscript "s" denotes surface quantities)

$$L_O = u_*^3 B_s^{-1} = \frac{u_*^3}{(g/T)(F_s/\rho c_p)}, \tag{28}$$

which is, of course, the Monin-Oboukhov length. It was first introduced by A. M. Oboukhov in his paper published in Russian in 1946 and translated into English in 1971 [51,54]. Thus the principle may help to easily derive rather complicated expressions that are otherwise difficult to memorise.

5. Convection in Rotating Fluids

Rotation usually increases the stability of the flow. It is also the case for convective flows. Chandrasekhar [11] was the first to study the stability of the rotating horizontal fluid layer in convective conditions. The similarity criterion, that appears to characterise the effect of rotation on the onset of convection, is the Taylor number,

$$\mathrm{Ta} = \frac{4\Omega^2 d^4}{\nu^2}, \tag{29}$$

which is the squared ratio of the Coriolis force to the viscous force. The stability curve, showing the critical Rayleigh number, Ra_{cr}, as function of the Taylor number, Ta, was found. It shows that Ra_{cr} increases only slightly at small and moderate Taylor numbers. At $\mathrm{Ta} > 10^6$ the critical Rayleigh number increases as $\mathrm{Ta}^{2/3}$.

Using the similarity arguments, Golitsyn [31,32] introduced the scale of velocity:

$$u_r = C_r B_s^{1/2} (2\Omega)^{-1/2}, \tag{30}$$

where C_r is a dimensionless constant. Golitsyn [32] crudely tested Eq. (30) against data from experiments with carbonated water performed over a limited range of rotation rates, namely, at 33, 45 and 78 revolutions per minute. Applying Eq. (30) to scale the vertical velocity, he obtained $C_r = 2.3 \pm 0.4$. In the early 80s, an extensive experimental programme was started at the Institute of Atmospheric Physics, Moscow. We had in mind a variety of applications. Results from a large number of studies are summarised in the book by Boubnov and Golitsyn [8]. In the early 90s, several groups began to study rotating convection, with the primary focus on deep convection in the ocean, as was proposed by Golitsyn [31].

Rotating convection serves to illustrate the successful performance of the principle of the fastest response. In case of rapid rotation, *i.e.* when the rotational time scale is the shortest of all time scales inherent in the system, our principle immediately gives

$$u_r^2 = C_r^2 B_s^2 f_c^{-1}, \tag{31}$$

where f_c is the Coriolis parameter. The estimate of $C_r = 1.6 \pm 0.1$ was obtained by Boubnov and Golitsyn [6,7,8], and $C_r = 1.7 \pm 0.2$ was obtained by Fernando *et al.* [18]. Golitsyn [33] presented a diagram of convective regimes for various rotating geophysical flows, and for planets and other astrophysical objects.

Apart from deep ocean convection, there are two more areas of geophysical application of the above results. These are (i) convection in the Earth liquid core [31], and (ii) winds in tropical cyclones [36]. For the Earth liquid core, Eq. (30) gives $u_r \approx 5$ km·year^{-1}, taking characteristic values of the geothermal flux and of the other governing parameters. For the tropical cyclones, we adopt typical values $F_s = 1000$ W·m^{-2}, that corresponds to $B_s = 3 \cdot 10^{-2}$ m^2·s^{-3}, and $f_c = 4 \cdot 10^{-5}$ s^{-1}, that corresponds to 15° latitude. This gives $u_r \approx 30$ m·s^{-1} (with $C_r = 1$), in a good agreement with observations.

6. Two-Dimensional Turbulence

The theory of 2D turbulence was developed by Batchelor [4] and, independently, by Kraichnan [43]. I deliberately put Batchelor first because all major results were presented during his lecture given in Moscow in February 1964, which I was translating for the audience into Russian. Already at that time, there where some indications and even some experimental evidence that spectra of large scale atmospheric motions fall off much steeper than is predicted by the "–5/3" law [55,62], with the exponent of about -2.8 ± 0.2. So the time for the theory to appear was quite ripe (note that it was not until 25 years later that the Kolmogorov-Oboukhov theory received experimental support, see Monin and Yaglom [48] and Yaglom [61]).

In the 2D hydrodynamics, two integrals of motion are conserved in the inviscid limit, namely, energy and enstrophy. The latter quantity is the average vorticity variance, $\langle \omega^2 \rangle$. In the 2D case, ω is a scalar as only the vertical component of curl(**v**) is present. Leith [46] proposed the term "enstrophy" for the quantity $\langle \omega^2 \rangle /2$. The term is derived from the the Greek word $\sigma\tau\rho\omega\phi\eta$ that means rotation (compare geostrophy). The enstrophy is conserved in every fluid particle. If we include viscosity ν, then the vorticity will dissipate. In the Earth's atmosphere, we have permanent generation of vorticity at large scales, mostly due to baroclinic instability of the general circulation at a scale of several thousand kilometres. As in the 3D case, we may assume that the statistically steady balance holds between the generation of vorticity and its dissipation with the rate η given by

$$\eta = \frac{1}{2}\frac{d\left\langle \omega^2 \right\rangle}{dt}. \tag{32}$$

Using η and ν as the key governing parameters, the following scales of length, velocity, and time may be developed:

$$l_\omega = \nu^{1/2}\eta^{-1/6}, \quad v_\omega = \nu^{1/2}\eta^{1/6}, \quad \tau_\omega = \eta^{-1/3}, \tag{33}$$

that are similar to the Kolmogorov microscales, Eq. (4).

We now have all prerequisites to apply our principle. We start with the vorticity structure function,

$$D_\omega(r) = \left\langle [\omega(\mathbf{x}+\mathbf{r}) - \omega(\mathbf{x})]^2 \right\rangle \approx \eta\tau_\omega \approx \eta\omega^{-1} \approx \eta^{2/3} = const(r). \tag{34}$$

The vorticity dissipation rate, η, is constant throughout the inertial range for vorticity. The vorticity spectrum $E_\omega(k)$, whose dimension is LT^{-2}, has the form [compare Eq. (10) relating the exponents in the expressions for the structure function and for the spectral density]

$$E_\omega(k) \approx \eta^{2/3}k^{-1}. \tag{35}$$

Similar spectra can be proposed for passive scalars (see *e.g.* Mirabel and Monin [47] where some comparison with observational data is also presented).

The velocity spectral density $E_u(k)$ is related to the vorticity spectral density $E_\omega(k)$ in a simple way (*e.g.* Monin and Yaglom [48]),

$$E_u(k) = k^{-2}E_\omega(k). \tag{36}$$

Combining Eqs. (35) and (36), we obtain

$$E_u(k) = \eta^{2/3}k^{-3}. \tag{37}$$

Charney [12] developed similar concepts for geostrophic turbulence where potential vorticity Ω_p is conserved, that is

$$\frac{d\Omega_p}{dt} = 0, \quad \Omega_p = \rho^{-1}(\omega + 2\Omega)\nabla s. \tag{38}$$

Here, Ω is the planetary vorticity ans s is the entropy. In the limit of small Rossby number, Ro $\ll 1$, the flow is geostrophically balanced, and relations similar to (33)–(37) are obtained for the potential vorticity.

The rate of the enstrophy generation (dissipation) in the Earth's atmosphere can be estimated at $\eta \sim 10^{-15}$ s^{-3}. A vortex with $r \sim 10^3$ km $= 10^6$ m and velocity $v = 10$ m·s^{-1} has $\omega \sim v/r \sim 10^{-5}$ s^{-1}, which gives an internal time scale $\omega^{-1} \sim \eta^{-1/3} = 10^5$ s, on the order of one day.

Rather elaborate arguments involving the Fjortoft [19] theorem show that in the 2D case the energy of the flow tends to propagate towards the large scales, and the enstrophy, towards the small scales ([44], to my taste, the most lucid description of the whole picture is given by Novikov [49]). Kraichnan [44] showed that for the spatial scales larger than the scale of vorticity generation the energy propagates to even larger scales and eventually feeds the mean flow (zonal circulation of the planetary atmospheres). This is a physical explanation of the phenomenon called "negative viscosity" [60]. This inverse energy cascade with the rate ϵ gives the spectrum $\epsilon^{2/3}k^{-5/3}$. Such a spectrum was observed in some numerical simulations of 2D turbulence. In the Earth's atmosphere, however, there is simply no room for the clear-cut $k^{-5/3}$ spectrum at very large scales, because the wavenumber of the most unstable baroclinic waves ranges from 6 to 8.

Examples of the velocity and temperature spectra corresponding to spatial scales from a few kilometres to several thousand kilometres were published by Gage and Nastrom [20] and were discussed in more detail by Gifford [22]. The data were collected over the period from 1979 to 1981 during thousands of commercial flights in the tropopause, at a height of 10 to 12 km. Spectra of the horizontal velocity components and the temperature show the "–5/3" slope for the spatial scales of up to 300 km. The slope then changes to "–3" for the spatial scales ranging from 500 to 4000 km. In the discussion of the observed spectra, Gifford [22] noticed that the velocity variance σ_v, measured by the area below the spectral curve, varies with latitude as the reciprocal of the Coriolis parameter f_c and is well estimated as ϵf_c^{-1}, where ϵ is the rate of dissipation (generation) of kinetic energy of the large scale motions. Three pints are worthy of mention here, namely, (i) a very strong confirmation of the 2D turbulence spectra of Batchelor and Kraichnan, although I think it is better to speak of the geostrophic turbulence spectra of Charney [12], (ii) the velocity variances, σ_v, do obey our scaling Eq. (31), derived from the principle of the fastest response, and (iii) the validity of the "–5/3" law extends over much larger spatial scales than A. N. Kolmogorov envisaged. The question why theories sometimes work much beyond the limits of applicability set by their authors is a puzzling question. This question should be laid to rest individually in each particular case. In our case, we could expect that the Kolmogorov theory might apply to the heights of measurements and to the spatial scales a few times larger than in a more or less isotropic flow [1] but why the theory applies to the spatial scales of up to several hundred kilometres and to heights of 10 to 12 km remains a puzzle.

7. General Circulation of Planetary Atmospheres

The theory of general circulation was developed by the present author [26,27,28]. Using the planetary radius r, the effective temperature of the outgoing thermal radiation $T_e = (q/\sigma)^{1/4}$ [$q = q_0(1 - A)/4$ is the heat flux, q_0 is the solar constant for the planet considered, A its albedo, and $\sigma = 5.67 \cdot 10^{-9}$ W·m^{-2}K^{-4} is the Stefan-Boltzmann constant], and sound velocity $c_e = (T_e R c_p/c_v)^{1/2}$ (R is the gas constant, and c_v is the specific heat at constant volume), as the characteristic length, temperature and velocity scales, respectively, the hydrodynamic equations yield three dimensionless similarity criteria,

$$\Pi_\Omega = \frac{\Omega r}{c_e}, \tag{39}$$

$$\Pi_g = \frac{RT_e}{gr} = \frac{H}{r}, \tag{40}$$

$$\Pi_M = \frac{q\tau_c}{Mc_pT_e}, \tag{41}$$

where $\tau_c = r/c_e$ is the time scale and M is the mass of the atmospheric column.

The first criterion, Eq. (39), is the rotational Mach number that is much smaller than one for Venus, Pluto and Titan, is of order one for Mars and Earth, and greatly exceeds one for large planets. The second criterion, Eq. (40), stems from the gravity term in the momentum equations, and the third criterion, Eq. (41), stems from the energy equation. Both Π_g and Π_M are small as compared to one for all planets. If we also disregard Π_Ω for the moment, *i.e.* neglect rotational effects, then the following quantity with the dimension of energy can be constructed from the governing parameters,

$$E = 2\pi C_E \sigma^{1/8} c_p^{-1/2} q^{7/8} r^3, \quad 2\pi C_E = const. \tag{42}$$

Since E in the above relation does not depend on mass, it is argued that it should represent the total kinetic energy of the atmospheric circulation. The mean wind velocity can be estimated from Eq. (42). Taking $2\pi C_E = 1$, we get $U \approx 12$ m·s^{-1} for the Earth. For Venus $U \sim 1$ m·s^{-1} and for Mars $U \sim 20$ m·s^{-1}.

Mean temperature difference ΔT along the planetary surface (this difference is the cause of the general circulation) can be estimated from the energy equation in its stationary form,

$$U\Delta T/r = \Pi_M c_e T_e/r.. \tag{43}$$

Equation (43) gives a temperature difference of 50 K for the Earth (typical equator-to-pole temperature difference) and 1 K for Venus. When these results were presented at the International Astronomical Union Symposium No. 38 held at Marfa, Texas, in October 1969, the estimate of $\Delta T = 1$ K was confirmed on the spot by the US radioastronomers, who said that they did not see any appreciable temperature difference over the surface of Venus. Within the limits of observational error that was about 10 K, a small difference of 1 K could not be detected. Later on, the wind velocity on Venus was found to be of order one meter per second, at least for the lower part of the atmosphere containing half of its mass.

A rather complicated Eq. (42) can be recast in concise form,

$$E = Q\tau_c, \tag{44}$$

where $Q = 4\pi r^2 q$ is the total input of the solar energy to the planet and $\tau_c = r/c_e$ is the time scale of relaxation of the pressure perturbations [53]. Equation (44) illustrates our principle in its clearest form. Comparison of τ_c with dynamic time scale $\tau_d = r/u$ shows that our principle requires that the Mach number be small, $\text{Ma} = u/c_e \ll 1$.

We should note that the principle does not directly apply to the giant planets, nor does it apply to the 4-day circulation of the rather thin upper part of the Venus atmosphere [28,30] where well organised structures of the general circulation are observed. In these situations, inverse energy cascade of the 2D or quasi-geostrophic turbulence may apparently result in larger velocities than could be obtained using our simple principle, Eqs. (42) and (44). The time scale that should be used in Eqs. (44) in these cases is still to be found.

8. Examples of Forced Flows in Classical Hydrodynamics

In the previous sections, we have shown how the principle of the fastest response can be successfully applied to determine the second-order moments of physical fields to within a numerical constant (that is most often of order unity). We may try to verify how the principle works in some classical hydrodynamic situations, where exact solutions for the forced flows are known. As stated above, the forcing is the product of the force, f, and the velocity, v. Then, for the velocity field we have an estimate

$$v^2 \approx m^{-1} f v \tau, \tag{45}$$

where m is the mass (see *e.g.* Landau and Lifshits [45] §11). Using Eq. (45), we find

$$v \approx a\tau, \quad a = f/m. \tag{46}$$

This formula can be written for any variable A as follows:

$$A \approx \frac{dA}{dt}\tau. \tag{47}$$

About 150 years ago G. Stokes obtained his famous formula for the drag on a sphere moving through viscous fluids. As it follows from that formula, a small particle falling under gravity has the velocity given by

$$v = \frac{2}{9} g r^2 \nu^{-1} = \frac{2}{9} g \tau_\nu, \tag{48}$$

where r is the particle radius. The particle density is assumed to be much larger than the air density. Our principle immediately gives Eq. (48) with an unspecified numerical coefficient (which is known to be 2/9 from the exact solution).

Using Eq. (46) with $\tau \sim r/v$, we find that at large Reynolds number a body of size r would experience a drag proportional to $\rho v^2 r^2$. The drag coefficient should be determined either experimentally, or computationally. The procedure is described in [59] and in many other textbooks of hydrodynamics.

One more example of a classical forced flow is the Poiseille flow in a pipe of radius r caused by an imposed pressure gradient, $\Delta p/l$, l being the pipe length. For the viscous flow (low Reynolds number), the modified form of our principle given by Eq. (46) yields the result,

$$u = C_p \Delta p (\rho l)^{-1} r^2 \nu^{-1}. \tag{49}$$

The mass flux through the pipe is given by

$$Q = \rho S v = C_p \Delta p (\nu l)^{-1} r^4, \tag{50}$$

where S is the area of the pipe cross-section. For a circular pipe with $S = \pi r^2$, we get $C_p = \pi/8$ (Landau and Lifshits [45] §17).

Similarly, for a pipe with a large diameter, where Re $\gg 1$, we get

$$Q = (\rho \Delta p)^{1/2} l^{-1/2} r^{5/2}, \quad v = (r \Delta p)^{1/2} (\rho l)^{-1/2}. \tag{51}$$

Again the drag coefficient should be determined from additional considerations. Data from laboratory experiments indicate that the drag coefficient is a very weak function of the Reynolds number and that the dependence of the mass flux Q on the pipe radius r in Eq. (51) follows $r^{2.64}$ power low rather than $r^{2.5}$ power law (see Sedov [59]). This is a typical case of the self-similarity of the second kind [2].

Now we consider one more problem which is concerned with everyday life, namely, the problem of room ventilation. The author solved it in 1995 when he moved to his new summer house. The simplest statement of the problem is the following. Suppose we have a room of volume V and temperature T_1 and a window of area S. The temperature outside the room is T_0 and the temperature difference $T_1 - T_0 = \Delta T$ is positive. How will the room temperature change with time if we open the window? Apparently, the forcing here is given by the reduced gravity for a unit mass, $g'(t) = g\Delta T/T_0$. It changes with time because the room is cooling if the window is open. The characteristic velocity of the gravity flow is $(g'h)^{1/2} \sim \left(g'S^{1/2}\right)^{1/2}$, where $h \sim S^{1/2}$ is the characteristic vertical scale of the flow. The characteristic time of the flow is the pendulum time given by $\left(S^{1/2}/g'\right)^{1/2}$. Therefore, Eq. (46) that is the specification of our principle, can be written as

$$v \approx \left(g'S^{1/2}\right)^{1/2} = (g\Delta T/T_0)^{1/2}\, S^{1/4}. \tag{52}$$

The total heat flux through the window is

$$F = C_1 \rho c_p \Delta T S v = C_1 \rho c_p \Delta T \,(g\Delta T/T_0)^{1/2}\, S^{5/4}, \tag{53}$$

where C_1 is a numerical coefficient weakly dependent on the window shape and on the distance of the window from the floor. The excess of the enthalpy in the room (*i.e.* the excess of its heat content) is $I = \rho c_p V \Delta T$. The time rate of change of this quantity is governed by the following equation:

$$\frac{dI}{dt} = -F. \tag{54}$$

Solving this equation is a straightforward exercise (note that we disregard changes in density ρ in Eq. (53), *i.e.* we use the Boussinesq approximation). Using the initial condition $\Delta T(t = 0) = \Delta T_0$, we get

$$\Delta T(t) = \Delta T_0 (1 + bt)^{-2}, \quad b = 2C_1 (g\Delta T_0/T_0)^{1/2} S^{5/4} V^{-1}. \tag{55}$$

According to Eq. (55), the time $t_{1/2}$ of halving of the initial temperature difference ΔT_0 is $b^{-1}(2^{1/2} - 1)$. With $2C_1 = 1$, $\Delta T_0 = 3$ K, $T_0 = 283$ K (10°C), $V = 100$ m^3 and $S = 0.1$ m^2, we obtain $t_{1/2} = 1251$ s ≈ 20 min.

9. Earthquakes

Earthquakes (EQs) occur due to the mantle convection that moves lithospheric plates inhomogeneously thus causing stresses, especially at the plate boundaries. When the stress reaches a critical value the brittle crust breaks up forming a fault which may vary greatly in size. The intensity of the EQ is characterised by its seismic moment,

$$M_s = \mu S u, \tag{56}$$

where μ is the shear modulus of the crust rocks, S is the area of the fault, and u is the mean displacement along the fault. Numerous studies have shown that the stress relieved by the EQ does not vary much. The majority of EQs relieve the stress $\Delta\sigma$ of the order of a few tens of bars [my own histogram for about 150 EQs that are fully described by Purcaru and Berkheimer [58] shows the median of $\Delta\sigma = 44$ bar $= 4.4 \cdot 10^6$ Pa). The two governing parameters, namely, M_s, whose dimension is that of work or energy, and $\Delta\sigma$, which is the elastic energy per a unit volume, form the length scale

$$L = (M_s/\Delta\sigma)^{1/3}. \tag{57}$$

Of major practical importance is the frequency of EQs of various magnitude over a certain area. It dictates the strategy of construction in the seismically hazardous zones. An empirical rule known as the Gutenberg-Richter law serves the purpose. In modern terms, it is usually written as (see *e.g.* Kasahara [40] and Kagan [39])

$$\lg N(\geq M_s) = A - b \lg M_s, \tag{58}$$

where N is a number of EQs with the seismic moment equal of greater than M_s over a certain period of time, and A and b are dimensionless coefficients that depend on the period and the area. Mantle convection is the global geodynamic process. The global EQ catalogues do exist. The driving force for mantle convection is the geothermal heat flux whose global value is $F = 4 \cdot 10^{13}$ W. The coefficient b is found to range from 0.65 to 0.7. For the very strong EQs, b exceeds 1. For the majority of EQs near mid-oceanic ridges, $b \approx 1.05$ [56]. The young crust near the ridges has the thickness, h, of order 5 km, while at the continental margins $h = 25 - 30$ km or more.

The Gutenberg-Richter law specifies the mean interval, $T(\geq M_s) = N^{-1}(\geq M_s)$, between the EQs of a given magnitude characterised by the seismic moment M_s. The seismologists, preoccupied by regional events, do not seem to have thought about the quantification of the forcing. The present author proposed that the geothermal heat flux F be utilised as the forcing for the EQs [34]. As different from all cases discussed above (where the characteristic time was used to determine the other physical quantities), it is the characteristic time itself, *i.e.* $T(\geq M_s)$ as dependent on the seismic moment M_s, that is of prime interest for seismology. Using F as the forcing and invoking our principle, we may rewrite the Gutenberg-Richter law in the following form that does not contain logarithms:

$$M_s = C_T F T(\geq M_s), \tag{59}$$

where the dimensionless coefficient C_T is generally a function of dimensionless parameter

$$\Pi = L/h = (M_s/\Delta\sigma)^{1/3} h^{-1}. \tag{60}$$

The linearised equations of the elastic theory possess various wave-type solutions. However, all these waves have velocities u_w of order several kilometres per second. Therefore, the time scales given by h/u_w are on the order of a few seconds. Obviously, these time scales cannot be related to the frequency of EQs. For the global catalogues, Okal and Romanowitz [56] and Kagan [39] found that the value of b changes from about 2/3 to about 1 at $M_s \approx 1.6 \cdot 10^{20}$ N·m. With $\Delta\sigma = 1.6 \cdot 10^6$ N·m^{-2} Eq. (57) gives the length scale $L = 33$ km, which is just about the mean thickness of the continental crust. Thus, the slope of the curve in the Gutenberg-Richter law changes at $\Pi \approx 1$. For $\Pi \geq 1$, the exponent $b \approx 1$ and therefore the coefficient $C_T(\Pi)$ should be a constant. Its value is found to be close to 0.4, using data from the global catalogues [34]. For the majority of EQs, the seismic moment $M_s \ll 10^{20}$ N·m, so that $\Pi \ll 1$. The coefficient $C_T(\Pi)$ should, therefore, decrease as $\Pi \to 0$. The simplest way to find the dependence of C_T on Π at $\Pi \ll 1$ is to expand $C_T(\Pi)$ into a Taylor series and to retain only the first linear term in the series, $C_T(\Pi) = C_{T1}\Pi = C_{T1}(M_s/\Delta\sigma)^{1/3}h^{-1}$, where C_{T1} is a dimensionless constant. Then, Eq. (59) gives the power law with $b = 2/3$,

$$T(\geq M_s) = M_s^{2/3}(\Delta\sigma)^{1/3} h (C_{T1}F)^{-1}, \tag{61}$$

where the coefficient C_{T1} was estimated at 0.4 [34]. It was also found in *op. cit.* that the length scale L given by Eq. (57) is an appropriate measure of the fault length, the fault area, and the mean displacement along the fault.

As is obvious from the forgoing, once the forcing is specified, the application of our principle immediately yields the Gutenberg-Richter law. The Taylor expansion procedure may be thought of as a way to elucidate the self-similarity of the second kind in Eq. (59). There is fresh evidence that the dependence $T(\geq M_s) \propto M_s^{2/3}$ extends to the values of $M_s = 10^9$ N·m (see Geller *et al.* [21]). This means that Eq. (60) holds for the values of the seismic moment M_s ranging over eleven orders of magnitude.

It was most exciting for the author to learn that the number-intensity distribution of the "starquakes" occurring at some neutron stars also obeys the Gutenberg-Richter law [13]. This is a dramatic illustration of the unity of laws of the Mother Nature.

10. Cosmic Rays

Cosmic rays, CRs, are for the most part protons due to the abundance of hydrogen. The CR flux density $N(\geq E)$, *i.e.* the number of CR particles per unit area in a unit time, is determined as function of their energy E. The cumulative spectrum $I(\geq E)$ for the range of energies from 10^{10} to $3 \cdot 10^{15}$ eV (1 eV$= 1.6 \cdot 10^{-19}$ J) has a very distinct shape, $I \propto E^{-n}$, with the exponent n approximately equal to -1.7 [23]. The CRs result from complicated processes during and after supernova explosions. These explosions also generate magnetic fields. A very interesting and important result of these processes is that the volume density of the CR energy w is on the average equal to the magnetic field energy $H^2/8\pi$. Both energies are of order 0.5 eV$\cdot$cm^{-3}. The condition of the equilibrium is [24]

$$w = H^2(8\pi)^{-1}. \tag{62}$$

We know the rate of generation of CRs, G, to be of order $3 \cdot 10^{39}$ W $= 5 \cdot 10^{52}$ eV$\cdot$s^{-1}. The total energy of CRs in our Galaxy (not counting its halo) is $W = 10^{48}$ J $= 10^{67}$ eV. The thickness of the Galactic disk is $h \approx 200$ pc $= 6 \cdot 10^{18}$ m, and its radius is $R \approx 15$ kpc $= 5 \cdot 10^{20}$ m. Given G and W, we can estimate the CR life time as

$$t_{CR} = WG^{-1} \approx 10^{14}\,\mathrm{s} = 10^7\,\mathrm{years}, \tag{63}$$

which is 3 to 4 orders of magnitude larger than the time it takes for an ultrarelativistic particle ($E > 10^9$ eV $= 1$ GeV) to cross the Galaxy in any direction. It is due to the interactions with magnetic field that CRs live that long and their flux is isotropic in space. The analogy with the earthquakes is useful here. As in the case of EQs, we know the forcing, *i.e.* the rate of the CR generation, G. After a number of transformations, CRs acquire a more or less constant volume energy density, w (analogously, the stress relieved by EQs is also more or less constant, see section 9). In both cases we have the statistics of events, *i.e.* the number of EQs over a certain period, and a number of particles crossing the measurement area over a certain period. The Gutenberg-Richter law deals with the frequency of events (EQs). In the case of CRs, we have the flux of events (particles) across a certain area.

With this difference in mind, we can otherwise proceed as in the previous section.

The cumulative spectrum $I(\geq E)$ has the dimension $\mathrm{L^{-2}T^{-1}}$, so we need a scale of length. By analogy with the EQs, we propose

$$L = (E/w)^{1/3}. \tag{64}$$

Invoking our principle, the time between events (particles of the energy E) can be estimated as $\left[L^2 I(\geq E)\right]^{-1}$ With the only time we have here, our principle gives

$$E = G\left[L^2 I(\geq E)\right]^{-1}. \tag{65}$$

Using Eqs. (64) and (65), we immediately obtain

$$I(\geq E) = C_{CR} G w^{2/3} E^{-5/3}. \tag{66}$$

The exponent "–5/3" differs from the empirical value 1.7 by only 1/30! The derivation of the CR spectrum from the similarity arguments, as well as some other results, can be found in [35].

The physical meaning of the length scale L can be illustrated by considering the molecular chaos. The mean energy of the atom is $\mathcal{E} = \frac{2}{3}kT$, k being the Boltzmann constant, and the volume energy density of chaos, *i.e.* the pressure, is $p = nkT$, n being the volume density of the number of atoms (molecules). By complete analogy with Eq. (64), we can form the length scale

$$l = (\mathcal{E}/p)^{1/3} = (3/2n)^{1/3} \approx n^{-1/3}, \tag{67}$$

which is simply the mean distance between the atoms. The value of l in normal conditions is an order of magnitude smaller than the mean free path given by

$$\lambda = (n\sigma)^{-1}, \tag{68}$$

where σ is the scattering cross-section, usually of order 10^{-15} cm^2.

We can introduce the mean free-path volume

$$V_\lambda = l^2\lambda \approx \sigma^{-1} n^{-5/3} \tag{69}$$

as a part of space related to the mean free path. This formula and the concept behind it are so elementary that they could be found in the XIX century. The author was unable to find Eq. (69) in textbooks, however. The CR particles do not interact with each other. We therefore cannot define mean free path for them (the Larmor radius could serve the purpose). The length scale given by Eq. (64) may be thought of as the mean distance between the particles which collectively have the pressure $w \approx H^2/8\pi$.

11. Discussion and conclusions

A principle of the fastest response is proposed. It states that quadratic characteristics of physical fields (or events), such as kinetic energy, velocity variances and scalar variances, can be found if we know the forcing, *i.e.* their rate of generation, and the characteristic time scales inherent in the physical system in question. The minimum time scale should be chosen. Comparison of the times scales characteristic of hydrodynamic systems yields the well-known similarity criteria and the estimates of their values. This allows one to judge what kind of the flow regime is realised. Then, using the known rate of generation and the chosen time scale, we at once obtain the quadratic characteristics of the system. The principle proves to work well for various flows, such as turbulence in three and two dimensions, convection in viscous and turbulent regimes with and without rotation, and general circulations of the atmospheres of the planets whose rotation rate is not too high. At the rapidly rotating planets, the inverse cascade (towards larger spatial scales) of the geostrophic turbulence significantly increases the relaxation time by organising the circulation flow into orderly (coherent) large-scale structures. This is not accounted for by our principle which underestimates the flow kinetic energy (see Golitsyn [28] §11).

The principle is in many respects equivalent to the habitual similarity theory. The latter was, in fact, originally used to obtain all the results described in the present paper. Neither the similarity theory, nor the principle of the fastest response, provides a way to evaluate dimensionless coefficients. This remains to be done through observations, numerical simulations, or exact solutions whenever possible.

The principle also applies to classical forced flows in hydrodynamics, such as the Stokes flow around a sphere falling under gravity, the pipe flow caused by an imposed pressure gradient, and the gravity flow in a density-inhomogeneous system. In these cases, we are able to determine characteristic scales of velocity and time, but, again, numerical coefficients must be evaluated from additional arguments.

Occasionally, the principle of the fastest response appears to be more efficient than the similarity theory, for instance, in viscous convection or turbulence in the Kolmogorov viscous subrange. Using the principle, a number of external parameters are combined into a single time scale, $\tau_\nu = d^2\nu^{-1}$, while in the straightforward similarity approach the length scale d and the viscosity ν are considered separately. This leads to the dependence of the sought-for quantities on the dimensionless Reynolds number [see the discussion of Eq. (7)]. Here our principle resolves the problem which G. I. Barenblatt calls the self-similarity of the second kind.

An "inverted" form of the principle may be used when the forcing and the quadratic characteristics of the system (*e.g.*, the energy of events or objects) are known, and the time sequence of events (*e.g.*, earthquakes of a certain magnitude, or cosmic ray particles of a certain energy) is of interest. In these cases, we use constant forcing (as a cause for all events) to consider a spectrum of times characteristic of each intensity. Such arguments have made it possible to use the principle of the fastest response to obtain the frequency-intensity characteristics of both earthquakes and cosmic rays, assuming the events or objects of various intensities are independent of each other.

One may expect that the principle works in a variety of other systems. At the time being, it simply states that a certain quantity A can be estimated as its generation rate dA/dt times a time scale τ, which should be a minimum time if there are several time scales characteristic of the system. In this form the principle is almost trivial. At the same time, there is reason to hope that the basic physics behind the principle will be elucidate in the (not very distant) future, that will allow one to more clearly see its potential and limitations. A brief discussion of the material presented here is given in Golitsyn [37].

Acknowledgements

I wish to very much thank Professor Erich Plate and Dr. Evgeni Fedorovich for inviting me to the NATO ASI on Buoyant Convection in Geophysical Flows where I presented two lectures on convection in viscous and rotating fluids. This invitation served as an impetus to look at the forced flows from a broader perspective, and to apply the old (1980) rule of the fastest response to convection with rotation and to some classical hydrodynamical problems. It also made me think how my work on earthquakes and cosmic rays, based on the similarity theory arguments, may be considered from the viewpoint of the above rule. As a result, it is proposed that the rule of the fastest response be regarded as a principle.

References

1. Alexeyev, V. G. and Yaglom, A. M. (1967) Examples of comparison of one- and three-dimensional velocity and temperature spectra, *Izv. Atmos. and Oceanic Phys.* **3**, 897–901.
2. Barenblatt, G. I. (1978) *Similitude, Self-Similarity and Intermediate Asymptotics*, Hydromet. Publ. House, Leningrad.

3. Barenblatt, G. I. and Golitsyn. G. S. (1974) Local structure of mature dust storms, *J. Atmos. Sci.* **31**, 1917–1933.
4. Batchelor, G. K. (1969) Computation of the energy spectrum in homogeneous two-dimensional turbulence, *Phys. Fluids, Suppl. II* **12**, 233–139.
5. Beljaars, A. C. M. (1995) The parameterization of surface fluxes in large-scale models under free convection, *Quart. J. Roy. Meteorol. Soc.* **121**, 255–270.
6. Boubnov, B. M. and Golitsyn, G. S. (1986) Experimental study of convective structures in rotating fluids, *J. Fluid Mech.* **167**, 503–531.
7. Boubnov, B. M. and Golitsyn, G. S. (1990) Temperature and velocity field regimes of convective motions in a rotating plane fluid layer, *J. Fluid Mech.* **219**, 215–239.
8. Boubnov, B. M. and Golitsyn, G. S. (1995) *Convection in Rotating Fluids*, Kluwer Acad. Publ., Dordrecht, 232 pp.
9. Bridgman, O. W. (1931) *Dimensional Analysis*, Yale Univ. Press.
10. Buckingham, E. (1914) On physically similar systems, *Phys. Rev.* **4**, 354–376.
11. Chandrasekhar, S. (1961) *Hydrodynamic and Hydromagnetic Stability*, Claredon, Oxford Univ. Press.
12. Charney, J. (1971) Geostrophic turbulence. *J. Atmos. Sci.* **28**, 1087–1095.
13. Cheng, B., Epstein, R. I., Guyer, R. A., and Young, A. C. (1996) Earthquake-like behavior of soft γ-ray repeaters, *Nature* **382**, 518–520.
14. Corrsin, S. (1951) On the spectrum of isotropic temperature fluctuations in an isotropic turbulence, *J. Appl. Phys.* **22**, 417–423.
15. Deardorff, J. W. (1970a) Preliminary results from numerical integrations of the unstable planetary boundary layer, *J. Atmos. Sci.* **27**, 1209–1211.
16. Deardorff, J. W. (1970b) Convective velocity and temperature scales for the unstable planetary boundary layer and for Rayleigh convection, *J. Atmos. Sci.* **27**, 1211–1213.
17. Deardorff, J. W. and Willis, G. E. (1967) Investigations of turbulent thermal convection between horizontal plates, *J. Fluid Mech.* **28**, 675–704.
18. Fernando, J. H. S., Boyer, D. L., and Chen, R. (1991) Effects of rotation on convective turbulence, *J. Fluid Mech.* **228**, 513–547.
19. Fjortoft, R. (1953) On the changes in the spectral distribution of kinetic energy for two-dimensional nondivergent flow, *Tellus* **5**, 225–230.
20. Gage, K. S. and Nastrom, C. D. (1986) Theoretical interpretation of atmospheric wave number spectra of wind and temperature observed by commercial aircraft during GASP. *J. Atmos. Sci.* **43**, 729–740.
21. Geller, R. J., Jackson, D. D., Kagan, Y. Y., and Mulargia, F. (1997) Earthquakes can not be predicted, *Science* **275**, 1616–1617.
22. Gifford, F. J. (1986) A similarity theory of the tropospheric turbulence energy spectrum. *J. Atmos. Sci.* **45**, 1370–1379.
23. Ginzburg, V. L. (Ed.) (1990) *Astrophysics of Cosmic Rays*, Nauka Publ. House, Moscow. (Engl. transl. by Elsevier, Dordrecht.)
24. Ginzburg. V. L. and Syrovatsky, S. I. (1964) *The Origin of Cosmic Rays*, MacMillan, N. Y.
25. Godfrey, J. S. and Beljaars, A. C. M. (1991) On the turbulent fluxes of buoyancy, heat, and moisture at the air-sea interface at low wind speeds. *J. Geophys. Res.* **96**, 22043–22048.
26. Golitsyn, G. S. (1970) A similarity approach to the general circulation of planetary atmospheres, *Icarus* **13**, 1–24.
27. Golitsyn, G. S. (1971) Derivation of the similarity theory relationships for general circulation of planetary atmospheres by analysis of the hydrodynamics equation system, *Izv. Atmos. and Oceanic Phys.* **9**, 974–978.
28. Golitsyn, G. S. (1973) *Introduction into Dynamics of Planetary Atmospheres*, Hydromet. Publ. House, Leningrad, 104 pp. (NASA trans. No. TTF-15, 627, Dec. 1974.)
29. Golitsyn, G. S. (1979) Simple theoretical and experimental study of convection with some geophysical applications and analogies, *J. Fluid Mech.* **95**, 567–608.

30. Golitsyn, G. S. (1980a) *Studies of convection with geophysical applications and analogies*, Hydromet. Publ. House, Leningrad, 56 pp. (in Russian).
31. Golitsyn, G. S. (1980b) Geostrophic convection, *Trans. (Doklady) USSR Acad. Sci.* **251**, 1356–1360.
32. Golitsyn, G. S. (1981) Structures of convection at rapid rotation, *Trans. (Doklady) USSR Acad. Sci.* **261**, 317–320.
33. Golitsyn, G. S. (1991) Convective regimes on various rotating geophysical and astrophysical objects, *Izv. Atmos. and Oceanic Phys.* **27**, 13–21.
34. Golitsyn, G. S. (1996) Earthquakes from the similarity theory point of view, *Trans. (Doklady) Russ. Acad. Sci., Earth Sci. Sect.* **346**, 536–539.
35. Golitsyn, G. S. (1997a) The spectrum of cosmic rays from the point of view of similarity theory, *Astronomy Letts.* **23**, 127–132.
36. Golitsyn, G. S. (1997b) Statistics and energetics of tropical cyclones, *Trans. (Doklady) Russ. Acad. Sci., Earth Sci. Sect.* **354**, 535–538.
37. Golitsyn, G. S. (1997c) Principle of the fastest response in hydrodynamics, geophysics, and astrophysics, *Trans. (Doklady) Russ. Acad. Sci., Earth Sci. Sect.* **356**, 321–324.
38. Golitsyn, G. S. and Grachev, A. A. (1986) Free convection of multicomponent media and parameterization of air-sea interaction at light winds, *Ocean-Air Interaction* **1**, 57–78.
39. Kagan, Ya. Ya. (1994) Observational evidence for earthquakes as a non-linear dynamic process, *Physica D* **77**, 160–192.
40. Kasahara, K. (1981) *Mechanics of Earthquakes*, Cambridge Univ. Press, 250 pp.
41. Kolmogorov, A. N. (1941a) Local structure of turbulence in incompressible fluid at very large Reynolds numbers, *Trans. (Doklady) USSR Acad. Sci.* **30**, 299–309.
42. Kolmogorov, A. N. (1941b) Dissipation of energy at locally isotropic turbulence, *Ibid.* **32**, 19–21.
43. Kraichnan, R. H. (1967) Inertial ranges in two-dimensional turbulence, *Phys. Fluids* **10**, 1417–1423.
44. Kraichnan, R. H. (1971) Inertial-range transfer in two and three-dimensional turbulence. *J. Fluid Mech.* **47**, 525–536.
45. Landau, L. D. and Lifshits, E. M. (1957) *Mechanics of Continuous Media*, Pergamon Press.
46. Leith, C. E. (1968) Diffusion approximation for two-dimensional turbulence, *Phys. Fluids* **11**, 671–672.
47. Mirabel, A. P. and Monin, A. S. (1983) On statistical regularities of admixture mixing by two-dimensional turbulence, *Izv. Atmos. and Oceanic Phys.* **19**, 902–912.
48. Monin, A. S. and Yaglom, A. M. (1971) *Statistical Hydromechanics*, MIT Press, Vol. I. (Vol. II, *Ibid.* 1975.)
49. Novikov, E. A. (1978) Spectral inequalities for 2D-turbulence, *Izv. Atmos. and Oceanic Phys.* **14**, 668–671.
50. Oboukhov, A. M. (1941) On the energy distribution in the spectrum of turbulent flow, *Trans. (Doklady) USSR Acad. Sci.* **32**, 22–24.
51. Oboukhov, A. M. (1946) Turbulence in the temperature-inhomogeneous atmosphere, *Works (Trudy) of the Institute of Theoretical Geophysics* **1**, 95–115.
52. Oboukhov, A. M. (1949a) Structure of the temperature field in turbulent flows, *Izvestia – Ser. Geogr. Geophys.* **13**, 58–69.
53. Oboukhov, A. M. (1949b) To the problem of geostrophic wind, *Ibid.* **13**, 281–306.
54. Oboukhov, A. M. (1971) Turbulence in the atmosphere with a non-uniform temperature, *Boundary-Layer Meteorol.* **2**, 7–29.
55. Ogura, Y. (1952) The theory of turbulent diffusion in the atmosphere, *J. Meteor. Soc. Japan, Ser. II* **30**, 53–58.
56. Okal, E. A. and Romanowitz, B. A. (1994) On the variation of b-values with earthquake size, *Phys. Earth and Planet. Inter.* **87**, 55–76.
57. Prandtl, L. (1932) Meteorologische Anwendungen der Strömungslehre, *Beitr. Phys.*

fr. Atmosph. **Bd. 19**, 189–202.

58. Purcaru, G. and Berkheimer, G. (1982) Quantitative relations for large earthquakes, *Tectonophysics* **84**, 57–128.
59. Sedov, L. I. (1954) Methods of Similarity and Dimension in Mechanics, 3d Edition, GITTL. (Engl. Transl. by Inforesearch, London, 1959.)
60. Starr, V. (1968) *Physics of Negative Viscosity Phenomena*, McGrow-Hill, N. Y.
61. Yaglom, A. M. (1981) Regularities of small-scale turbulence in the atmosphere and ocean, *Izv. Atmos. and Oceanic Phys.* **17**, 1235–1257.
62. Wiin-Nielsen, A. (1968) On the intensity of the general circulation of the atmosphere, *Rev. Geophys.* **6**, 559–579.

MODELING THE OCEANIC MIXED-LAYER AND EFFECTS OF DEEP CONVECTION

R. H. KÄSE
Institut für Meereskunde Kiel
Düsternbrooker Weg 20
24105 Kiel

1. Introduction

Progress in modeling the oceanic circulation has been achieved in the last few years by increasing the speed of computers and by refining modeling techniques. The dynamics of major current systems such as the Gulfstream-North Atlantic Current and their corresponding eddy variability is reasonably well understood [58, 32]. Climate models predict global warming as a result of increasing CO_2 in the atmosphere and forecast El Nino events in the equatorial Pacific [50]. Freshwater imbalances in the deep convection regions of the polar and subpolar regions of the North Atlantic result in alternating multiple equilibrium states of the global thermohaline vertical circulation - the "conveyor belt" [53]. On the other hand, large scale modeling relies heavily on the parametrization of "subgrid" processes. This is especially true for the oceanic boundary layer. Here the modeling suffers from inappropriate information on the fluxes at the air-sea interface. Most coupled models with simplified fluxes do not represent the surface temperature well enough and water mass characteristics drift away from the initial state. Restoring conditions at the sea surface are needed to force the model back to the observations. The fluxes analyzed from runs with restoring conditions show substantial errors. It is evident that progress in the reliability of long-term predictions of climate variations can only be made with a better representation of mixed layer dynamics.

This article is intended to summarize some basic facts about the aqueous boundary layer under convection and describes a few approaches that illustrate the problem. It is not meant as a complete guide to models of the mixed layer (see Nurser [47] for a more thorough treatment). In the first part a classification of the oceanic mixed layer will be given, followed by

E.J. Plate et al. (eds.), Buoyant Convection in Geophysical Flows, 157–183.

chapters on the governing equations, the closure problem and a discussion of the performance of two different approaches. The impact of coherent structures in the mixed layer will then be addressed for the case of surface streaks associated with Langmuir cells. The second part focuses on deep convection, describes results from non-hydrostatic models resolving the energetic scales of convective plumes and discusses the possibilities of parameterizing the processes that determine the vertical and horizontal spreading of water masses formed during convection events.

2. Shallow Convection and Wind Generated Mixed Layers

2.1. HISTORICAL

The ocean is largely topped by a nearly homogeneous layer with very small gradients in properties overlying a sharp pycnocline. The depth of this "mixed layer" ranges from few tens of meters to more than 1000 m in wintertime in the polar and subpolar regions. Attempts to describe the mean mixed layer depth date back to the early 30's of this century. Defant [14] published charts of the temperature and salinity distribution in the Atlantic Ocean based on results of the METEOR expedition and stated that the depth of the "deck-layer" was 33m shallower than the depth of the maximum temperature gradient . A meridional dependence and the effects of winds were accounted for in Rossby and Montgomery [54] who found the empirical relationship that the depth of the mixed layer was 2.38 times the wind speed in meters per second divided by the sine of the latitude . This relation was used widely and maps based on this relation were still published in the fifties (e.g. Lumby [40]). Only after Deardorff [9, 10] and Deardorff *et al.* [12] began experimenting with convection in laboratory tanks and resolved turbulent structures in numerical experiments, did the dynamics of the oceanic mixed layer received more attention. The first widely accepted and applied ocean mixed layer model was that of Kraus and Turner, henceforth denoted as KT [31]. It assumed that the turbulent layer was well mixed and that the variation in heat content could be completely related to the changes in bulk integrals of the potential energy in the mixed layer. The buoyancy conversion term involved can be eliminated by using the turbulent kinetic energy (TKE) equation. The entrainment velocity w_e, defined as the time rate of change of mixed layer depth during deepening is thus arising from a surplus of TKE supply over consumption in the mixed layer. If the sources and sinks of TKE are known, the deepening of the mixed layer can be calculated in parallel with the corresponding buoyancy and momentum equation.

A variety of papers following KT each offer different approaches for parameterizing the individual terms, but most of them are based on the bulk approach of KT [44, 17]. They have in common, that they usually involve several free tunable parameters to fit the observations. This circumvents the missing information about details of the turbulence, its production/destruction in the near surface stress layer and at the base of the mixed layer. Clearly there is also lack of information about the specification of the dissipation. Progress in the seventies was summarized in the proceedings of the 1975 NATO - Advanced Study Institute [30].

2.2. STRUCTURE OF THE OCEANIC BOUNDARY LAYER

Several processes interact in the oceanic boundary layer, some of which are not even well understood as a single process, making it difficult to parameterize the overall effect [68]. Most obvious are surface waves having an impact on turbulence production by breaking . A mean shear flow (the Stokes drift) is generated as the rectified effect of surface wave action. Convergence and divergence zones of the Stokes drift produce wind rows, related to Langmuir circulation patterns [73, 74]. The directly forced wind drift (Ekman current) responds with vigorous inertial oscillations to changes in the strength and direction of the wind forcing [8]. The shear connected with these flows also induces turbulence. At the bottom of the mixed layer Kelvin-Helmholtz or the symmetric instability of internal waves complicate the picture as well as do overshooting thermals penetrating into the thermocline.

Anis and Moum sketch the convective boundary layer in analogy to the atmosphere as being classified by four different regions, the micro layer, the surface layer, the mixed layer and the thermocline [2]. While the microlayer (cool skin) covers just the top few millimeters dominated by molecular processes, the surface layer exhibiting an unstable density gradient can be as thick as 10 to 20 meters overlying the well mixed layer below. Mixed layer turbulent eddies may have the size of the mixed layer thickness (50-150m) and sinking plumes in deep convection areas can reach the bottom ($>$ 1000 m).

2.3. THE MODELING APPROACH

2.3.1. *The Mixed Layer Equations*

Taking into account the strong dominance of vertical fluxes in the mixed layer, one generally uses horizontally averaged linearized equations for horizontal momentum, with temperature and salinity , and the unknown vertical flux divergence on the right hand side. Let $\boldsymbol{U}, w, T, S$ denote the horizontal velocity vector, vertical velocity component, temperature and salinity, respectively. Then, in a Cartesian coordinate frame with the vertical unit vector $\boldsymbol{k}$ pointing upward, the mixed layer equations (c.f. [19]) read

$$U_t = -fk \times U - \langle U'w' \rangle_z, \tag{1}$$

$$T_t = \frac{F_{sol}}{\rho_0 c_p} I_z - \langle T'w' \rangle_z, \tag{2}$$

$$S_t = - \langle S'w' \rangle_z, \tag{3}$$

where f is the local Coriolis parameter, ρ_0 is a reference density, c_p specific heat of seawater [19] and F_{sol} is the solar irradiance absorbed at the sea surface, while $I(z)$ is its vertically penetrating fraction. The turbulent fluxes at the surface are specified by

$$-\rho_0 \langle U'w'(0) \rangle = \tau, \tag{4}$$

$$-\rho_0 c_p \langle T'w'(0) \rangle = F_{nsol} = H + LE + F_{ir}, \tag{5}$$

$$-\rho_0 c_p \langle S'w'(0) \rangle = E - P. \tag{6}$$

Here τ is the surface wind stress. The sum of sensible (H), latent (LE) and net infrared (F_{ir}) heatflux is usually referred to as the nonsolar flux (F_{nsol}). In this notation heatfluxes are positive when ponting downward.

These mixed layer equations are not closed and require a procedure to describe the fluxes on the right hand side in terms of the mean fields or by replacement with higher moments that finally need a closure. A very general higher order closure technique that is used widely nowadays in meteorology and oceanography is given by Mellor and Yamada [41]. This scheme is discussed elsewhere in this book and will not be used here. Instead, the focus will be on two closures as an example of well performing models used recently in oceanography. First, a method to close the equations, which is based on a length scale formulation of the turbulent kinetic energy (TKE) equations [19] and has been used to simulate the annual cycle at a Pacific station (Ocean Weathership P) will be given more attention here. Another scheme [34], based on a universal profile formulation of the fluxes and only recently formulated for the ocean will follow subsequently .

2.3.2. *The TKE Closure of Gaspar, Gregoris and Lefevre (GGL)*

The GGL scheme [19] was specifically designed for easy embedding into numerical circulation models. More complicated models, as the Mellor and Yamada [41] level 2 and 2.5 closure, either require the determination of a master length scale by solving a prognostic equation in addition to the prognostic turbulent kinetic energy equation, or its specification as a function of characteristic turbulent scales. The former is rather cost intensive if applied to a numerical circulation model while the latter has been restricted mainly to the boundary layer and not proven valid in the entire region from surface to bottom. GGL follow a suggestion of Bougeault and Lacarrere who successfully modeled both boundary layer and clear air turbulence [4]. The formulation uses the concept of eddy diffusivity, in which the eddy coefficients are functions of a length scale and the turbulent kinetic energy.

$$-<\boldsymbol{U}'w'>=k_m\boldsymbol{U}_z, \tag{7}$$

$$-<T'w'>=k_hT_z, \tag{8}$$

$$-<S'w'>=k_sS_z, \tag{9}$$

$$k_m=c_kl_k<e>^{1/2}, \tag{10}$$

$$k_s=k_h=\frac{k_m}{Pr}. \tag{11}$$

While it is possible to allow for different diffusion of heat and salt, GGL restrict their model to a turbulent Prandtl number $Pr=1$. GGL also did not include the original countergradient flux correction term of Bougeault and Lacarrere [4] as suggested by Deardorff [11] for the atmosphere and strongly recommended by Stull[65] (see section 2.4.3.) Their argument is simply that the observations of Shay and Gregg [60] do not support stable gradients during convection.

With the assumption of horizontal homogeneity and dominance of vertical fluxes, the tendency equation for turbulent kinetic energy reduces to

$$<e>_t=-(<e'w'>+\frac{<p'w'>}{\rho_0})_z-<\boldsymbol{U}'w'>\cdot\boldsymbol{U}_z+<b'w'>-\varepsilon \tag{12}$$

where primed quantities denote fluctuations and pressure p, buoyancy b and turbulent dissipation ε are used in standard notation. To close (12), GGL apply Kolmogorov scaling for the dissipation and again use eddy diffusivity concepts for the vertical energy flux divergence, assuming that energy and momentum diffusivities are equal ($k_e=k_m$).

$$\varepsilon=\frac{c_\varepsilon<e>^{3/2}}{l_\varepsilon}, \tag{13}$$

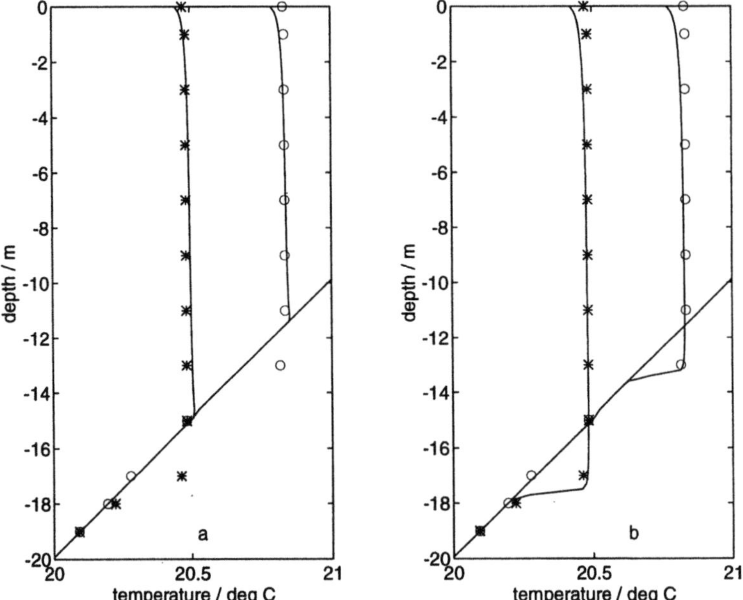

Figure 1. Mixed layer temperature development for $100Wm^{-2}$ cooling at day 3 and 5 using $c_\varepsilon = 0.7$, $c_k = 0.1$ and $k_e/k_t = 1$ (a) and $k_e/k_t = 14, Pr = 0.44$ (b). The symbols correspond to simulations with the KPP-model of 2.3.3

$$-(< e'w' > + \frac{< p'w' >}{\rho_0}) = k_e < e >_z . \tag{14}$$

The most crucial point of eddy diffusivity closures is always the length scale formulation. Here GGL follow Therry and Lacarrere who defined two primary length scales l_u and l_d as the path length of an upward or downward traveling particle that converts its initial TKE to potential energy [67]:

$$l_\varepsilon = (l_u l_d)^{1/2}; l_k = min(l_u, l_d), \tag{15}$$

$$\int_z^{z+l_u} \frac{g}{\rho_0}[\rho(z) - \rho(z')]dz' =< e(z) >, \tag{16}$$

$$\int_z^{z-l_d} \frac{g}{\rho_0}[\rho(z) - \rho(z')]dz' =< e(z) > . \tag{17}$$

In an unstable density gradient l_u and l_d must be replaced by the distance to the boundary. The dissipation and mixing length scales, l_ε and l_k

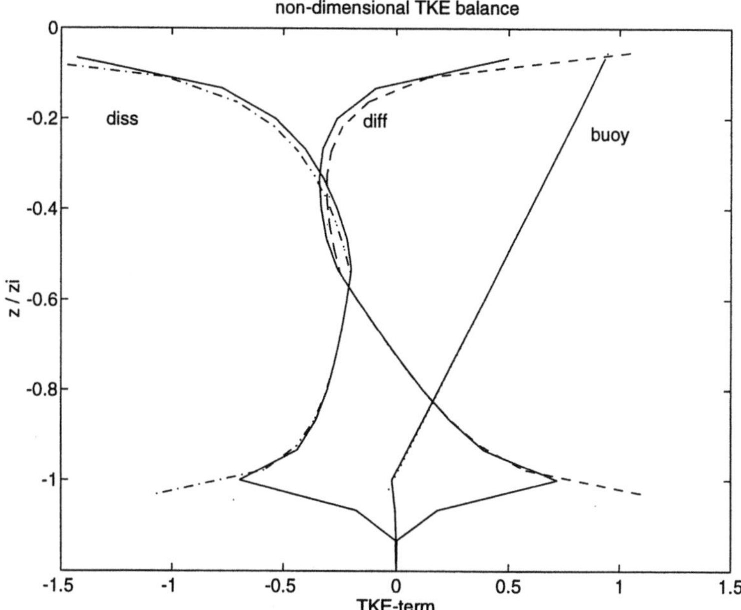

Figure 2. Normalized TKE-balance in GGLmodel with $c_\varepsilon = 0.7$, $c_k = 0.1$. $k_e/k_t = 30$. Dotted, dashed and dash-dotted line: analytical equilibrium solution for piecewise linear length scale. Full lines: numerical solution for 1m vertical resolution. The abbreviations stand for dissipation, diffusion and buoyancy production terms.

just combine the primary scales and do not require additional calibration. In a stratification with constant stability frequency the two primary length scales are equal and represent the scale at which a complete overturn can occur at a given turbulent kinetic energy level. In the convective case,this leads to a linear increase of the length scale away from boundaries or from a strongly stratified region.

To fit the annual cycle at Weathership P GGL used $c_\varepsilon = 0.7$ and $c_k = 0.1$. Their choice of $c_\varepsilon = 0.7$ was motivated by the convection studies of Bougeault and Lacarrere [4], while $c_k = 0.1$ is based on mixing efficiency considerations. This value is equivalent to a mixing efficiency $\gamma = 0.3$, the median estimate of Moum *et al.* [43].

The performance of the GGL scheme for a case of pure convection where a stable temperature gradient of 1 degree per 10 meters is eroded by constant heat loss at 100 Wm^{-2} is evaluated in Figure 1 and compared to the KPP simulation (c.f. 2.3.3) of Large *et al.* [34]. The model tuning of GGL appears to be inappropriate for the convective case. The choice of $c_k = 0.1$ does not allow enough diffusion of turbulent kinetic energy. Instead, most of the production is consumed via dissipation. Equation (12)

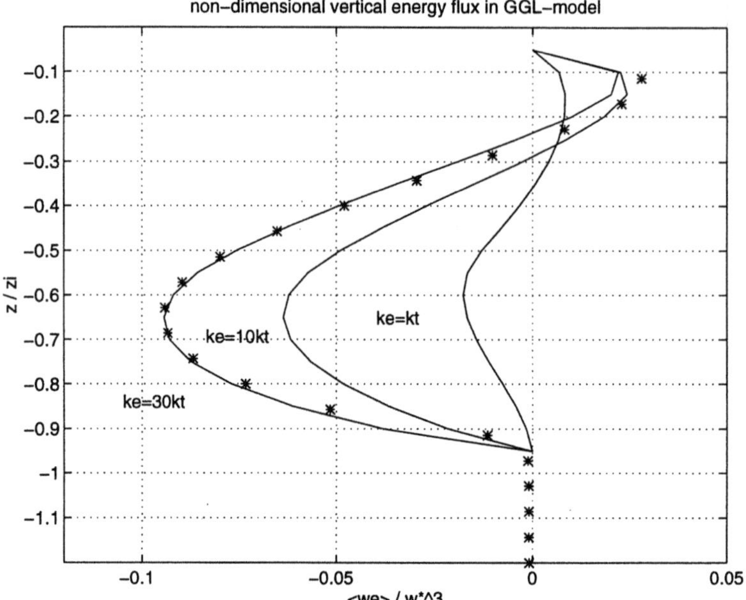

Figure 3. Normalized vertical flux of energy in equilibrium solution of the GGL-model for $c_\varepsilon = 0.7$, $c_k = 0.1$ and $k_e/k_t = 1, 10, 30$. The stars represent the flux at day 5 in a numerical simulation with 1m resolution.

is in equilibrium with the buoyant forcing on a time scale of minutes. In unstable conditions, the length scale in (15)-(17) increases linearly to half of the mixed layer depth and then decreases linearly until the stable region is reached. Since $< w'b' >$ decreases linearly from the surface, the solution of the equilibrium state can be approximated by solving

$$((|\zeta + 0.5| - 0.5)E_\zeta^{3/2})_\zeta - \lambda^2 \frac{E^{3/2}}{(|\zeta + 0.5| - 0.5)} = \frac{\lambda^2}{c_\varepsilon}(1 + \zeta) \tag{18}$$

with $\lambda^2 = \frac{1.5c_\varepsilon}{c_k}$ and $\zeta = \frac{z}{h}$, $E = \frac{e}{(B_0 h)^{2/3}}$. The first term on the left hand side is the diffusion , the second the dissipation term. The solution with vanishing flux at the surface ($\zeta = 0$) and at the mixed layer base ($\zeta = -1$) involves linear combinations of ζ^λ,$\zeta^{-\lambda}$, and ζ with coefficients determined by the boundary and matching conditions at $\zeta = 0, -.5, -1$. For the original GGL parameters $\lambda^2 = 10.5$, indicating a principle balance between dissipation and buoyancy production. The non-dimensional penetration scale is λ^{-1}. Only if this scale is O(1) can turbulent energy be diffused down to the bottom of the mixed layer and be available for penetrative convection. This is seen in Figure 1b where the simulated temperature with an adjusted ratio of $k_e = 14k_t$ is virtually identical with the simulations incorporating better penetration and diffusion into the thermocline as described in 2.3.3.

The TKE-balance of equation (18) is compared to a numerical solution of the full equations (A. Oschlies, personal communication). As in the LES simulations of Moeng and Wyngaard [42] the total energy transport in the mid of the mixed layer represents a sink term (Figure 2). The dependence of the vertical energy flux on the ratio k_e/k_t is shown in Figure 3. The choice for Figure 1b has a non-dimensional flux comparable to atmospheric observations, while that of Figure 1a is much too small and representative of contemporary higher order closure models (c.f. Wyngaard [75]).

2.3.3. *The k-profile Parameterization of Large, McWilliams and Doney*
This formulation (KPP) is also based on meteorological experience. Large *et al.* [34] adopt the Troen and Mahrt [70] atmospheric boundary layer model for the ocean. It is mainly an empirical parametrization that uses similarity scaling in the near surface layer based on the Monin-Obukhov theory. Mixing is assumed to not be complete, allowing for remaining gradients. The total boundary layer depth can be much deeper than the actual mixed layer and consequently there is distinction between the mixed layer depth h_m where the vertical buoyancy gradient exceeds a certain threshold, the entrainment depth h_ε , where the entrainment flux reaches a maximum and the total boundary layer depth h where a bulk Richardson number is subcritical relative to a specified value. The consequence is that this approach allows for possible strong mixing below the well mixed layer.

Near Surface Similarity Scaling. Here KPP relies heavily on the fact that in near wall turbulence the fundamental turbulent parameters are the friction velocity u^* , the scale of scalar property fluctuations S^* and the Monin-Obukhov length scale L:

$$u^{*2} = (< w'u' >^2 + < w'v' >^2)^{1/2}, \tag{19}$$

$$S^* = \frac{-< w's' >}{u^*}, \tag{20}$$

$$L = \frac{u^{*3}}{(KB_f)}, \tag{21}$$

where $K = 0.4$ is von Karman's constant and B_f the surface buoyancy flux. The corresponding universal flux profiles of momentum and scalars as functions of the stability parameter $\zeta = \frac{d}{L}$ (in the atmospheric and ocean bottom boundary layer $d = z$, in the oceanic surface layer $d = -z$) are

$$\Phi_m(\zeta) = \frac{Kd}{u^*} \mid U_z \mid, \tag{22}$$

$$\Phi_s(\zeta) = \frac{Kd}{S^*} S_z. \tag{23}$$

In neutral and stable conditions fluxes of scalars and momentum are the same and > 1, however they differ for unstable situations, where the scalar fluxes are smaller, i.e. $\Phi_s < \Phi_m < 1$. In the surface layer KPP therefore allows for a Prandtl number different from unity. It is not clear whether the proposed form for the universal functions based on meteorological situation is relevant for the ocean at all, since in the near surface region, the rapid absorption of the penetrating solar radiation, the presence of surface waves and Langmuir circulation are not considered [1, 38].

Diffusivity and Nonlocal Transport. Below the surface layer the vertical flux of a general property is treated similar to GGL via eddy diffusivity. However, a countergradient flux term is introduced

$$< w'x' >= -k_x(X_z - \gamma_x) \tag{24}$$

Across the boundary layer depth h , it is assumed that the diffusivities as well as their vertical derivatives are continuous and that the countergradient flux vanishes just inside the boundary layer. The diffusivities k_x are directly proportional to h and a product of a vertical velocity scale w_x and a nondimensional shape function G (a polynomial expansion of the order 3).

$$K_x = h w_x(\frac{z}{h}) G(\frac{z}{h}) \tag{25}$$

The four coefficients in G and the form of the vertical velocity scale are determined by matching with the surface layer and the thermocline. It is found that:

$$w_x = K u^* \Phi^{-1}(\frac{\varepsilon h}{L}); \varepsilon < \frac{d}{h} < 1; \zeta < 0; \tag{26}$$

$$w_x = K u^* \Phi^{-1} \frac{d}{L}; otherwise \tag{27}$$

In the neutral case $\Phi(0) = 1$, the velocity scale is Ku^* and it increases with depth in unstable conditions and decreases in stable conditions. The countergradient flux correction is only nonzero for scalars in the unstable convective case according to Deardorff [11] . For details we refer to Large et. al. [34]. Nurser [47] estimates that the nonlocal transport can be as large as the surface flux in the middle of the boundary layer allowing the gradient flux to be opposite to the total flux. The total flux thus appears to be countergradient.

Boundary Layer Depth. KPP use a bulk Richardson number criterion for the total boundary layer depth, i.e. h is the largest possible value for which the Richardson number

$$Ri_b(d) = \frac{(B_r - B(d))d}{(|\boldsymbol{v}_r - \boldsymbol{v}(d)|^2 + \boldsymbol{v}_t^2)} \tag{28}$$

is still below a predefined value. For the velocity and buoyancy reference ($\boldsymbol{v}_r$ and B_r), KPP avoid numerical resolution dependence by taking averages over the surface layer $0 < \frac{d}{h} < \varepsilon$. This circumvents the shortcoming that a mean shear produced by numerical models on a horizontal grid scale of the internal Rossby radius will usually not become critical and generally results in mixed layers that are too shallow. The choice in Large *et al.* [34] also allows for penetrative convection, because it keeps the turbulent shear in the Richardson number. However the turbulent velocity v_t must be parameterized further. Referring to simulations of penetrating convection, they argue that the ratio of entrainment flux to surface flux ratio should be 0.2, and adjust the proportionality constant in

$$v_t^2 \propto Nhw_s \tag{29}$$

accordingly. This calibration is based on the purely convective case. Their argument is similair to GGL's justification of the inherent length scale as the maximum path of a particle necessary to perform an overturn against the buoyancy at a given kinetic energy. With the inclusion of the turbulent velocity shear Large *et al.* have the possibility of allowing critical Richardson numbers of 0.25, and do not have to increase it to 0.65 for more realistic deepening as was necessary in Price *et al.* [34, 52].

Thermocline Mixing. Here an approach is taken that incorporates the main effects of instability due to resolved shear, internal wave breaking and double diffusion as a sum of three diffusivities $\nu_x = \nu_x^s + \nu_x^w + \nu_x^d$. The dependence of the shear contribution on the Richardson gradient number is justified by measurements in the equatorial region by Peters *et al.* [49]. The internal wave contribution is taken to be constant with $\nu_m^w = 10^{-4}m^2s^{-1}$ and $\nu_s^w = 10^{-5}m^2s^{-1}$. The density ratio dependence for the saltfingering and double diffusive layering case is motivated by studies of Schmitt [56] and Kelly [29]. The resulting maximum values are rather large, while the temperature-salt ratio of 0.7 is a classical value. This parametrization enhances mixing in the thermocline overcoming a deficit in most previously used mixing schemes.

Net Effect. The Large *et al.* KPP scheme appears to be an improvement in the overall behavior of a 1D boundary layer model. It allows penetrative convection, higher diffusion of properties in the thermocline even in wind forcing cases and is capable of maintaining often observed property

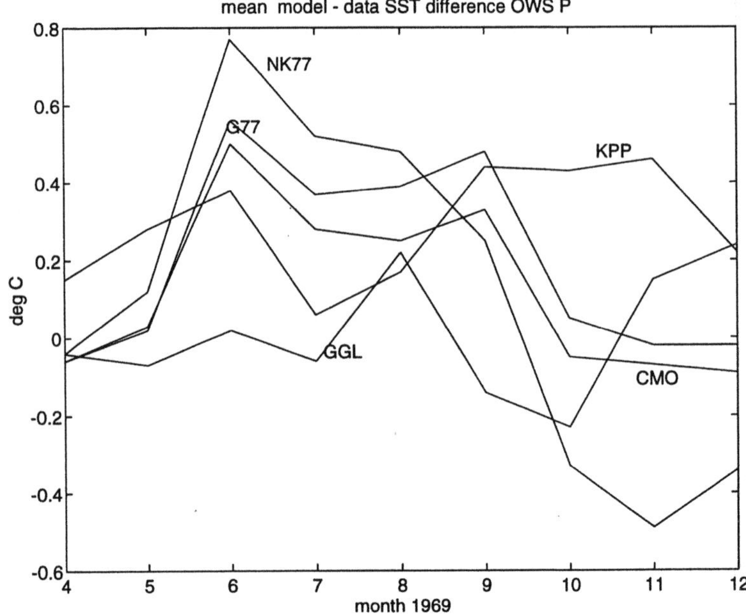

Figure 4. SST differences at OWS-P between different mixed layer models and observations in 1969. Values taken from tables in Large *et al.* [34]. The letter code is as follows: G77: Garwood [17], NK77: Niiler and Kraus [45], CMO: Gaspar [18], GGL and KPP as defined in this section.

gradients in the mixed layer. Figure 4 summarizes the comparison between different approaches in terms of the sea surface temperature (SST) difference between OWS-P and the individual model from Table 6 of [34]. SST is a crucial parameter in coupled ocean - atmosphere models, because it is involved in most meteorological flux parameterizations. Both GGL and KPP perform equally well in the SST comparison. This is somewhat surprising, since the convection case with the choice of GGL's parameters was not performing very well. Large *et al.* argue that GGL needed to use a correction for the heatflux input in a simulation of the "LOTUS" large diurnal cycle and that their model performed better with uncorrected fluxes. The surface heating with uncorrected fluxes would have raised the GGL SST in the upper 5m by as much as 0.5 °C. The clear advantage of KPP lies in the thermocline behavior because of its superior mixing schemes. GGL essentially underestimates the entrainment mixing, indicated by the missing penetrative convection. However, a better choice of the k_e/k_t-ratio can improve this (Figure 1b). Both, GGL and KPP express the fluxes in terms of eddy coefficients. The KPP mixing coefficients compared to the GGL correspondents are rather smooth. In the free convection case, GGL's TKE equation tends rapidly toward a stable normalized solution. In this respect, both methods are congruent, with KPP having the leading edge. This is however, not surprising due to the deliberate shaping of the k-profile. Existing oceanic measurements focussed strongly on the dissipation profiles [39]. They were motivated by atmospheric similarity scaling. It seems that the role of the TKE-transport in the ocean was never adequately addressed. New measuring techniques will be necessary to tackle this problem.

2.4. THE ROLE OF LANGMUIR CELLS

Surface slicks are an ubiquitous feature on the sea surface that appear shortly after increasing wind conditions [74]. The longtime dispute whether these long streaks nearly parallel to the wind are a result of cooling or wind generation seems to be settled in favor of the latter. The principal mechanism first described by Langmuir [33] has been identified in the late 70's to be the interaction between surface waves and wind driven currents [7, 16, 35]. The theoretical framework together with an identifiable, observable process has motivated several groups to use today's computer resources to extend the initial stability analyses [6] via dynamics of the rolls [66] to a full account of the turbulence in the surface layer [61].

2.4.1. *A Simple 2-D Model.*

The Craik-Leibovich concept suggests that wind driven horizontally sheared jets interacting with the vertically sheared Stokes drift generate roll like perturbation in the streamwise vorticity. The scaled 2-D equations involve the streamwise momentum and vorticity according to

$$u_t + vu_y + wu_z = La\nabla^2 u, \tag{30}$$

$$\Omega_t + v\Omega_y + w\Omega_z = La\nabla^2\Omega - u^s_z u_y; La = (\frac{\nu b}{u^*})^{3/2}(\frac{u^*}{u^s_0})^{1/2}. \tag{31}$$

Here $\Omega = \nabla^2\Psi$ is vorticity,Ψ streamfunction, $v = \Psi_z$, $w = -\Psi_y$, and $u^s = 2u^s_0 e^{2bz}$ is the Stokes drift current velocity with the penetration depth b^{-1} . The only controlling parameter in a nonstratified fluid is the Langmuir number La , describing the ratio of viscous damping to forcing by the geometric mean of the shear due to background and the Stokes drift [36].

If stratification were included, a Richardson number based on the background stability frequency and the same forcing would come in as a second parameter. The equations further include the Prandtl number, but its range has not yet been explored in detail. The Langmuir equations are similar to the Rayleigh- Benard equations and have been used to investigate the effects of rolls on a pre- existing stratification.

2.4.2. *Attempts to Parameterize.*

Li and Garrett explore numerically whether Langmuir cells can add to mixed layer deepening [37]. Their modeling shows generation of typical Langmuir cells that stop penetrating into the stratified fluid if the Froude number, $Fr = \frac{w_{dn}}{(h\Delta b)^{1/2}}$ reaches a critical value of 0.9.This Froude number is based on the maximum downwelling cell speed w_{dn} , a characteristic wave speed based on the mixed layer thickness (h) and the buoyancy jump at the base (Δb). Using model results for $\frac{w_{dn}}{u^*}$, they establish a buoyancy jump criterion and show that deepening stops at $h\Delta b = cu^{*2}$ with $c \approx 50$ for fully developed seas. In view of the Price *et al.* Richardson [52] number criterion for wind deepening, Li *et al.* [38] examine the two normalized quantities, $Rb_n = \frac{0.65|\Delta u|^2}{h\Delta b}$ for the wind deepening and $LC_n = 50\frac{u^{*2}}{h\Delta b}$ for Langmuir cell based deepening. They conclude that whenever one of these numbers exceeds 1, mixed layer deepening should occur and they were able to identify such events in data from the "LOTUS" experiment [52]. A possible inclusion of the Langmuir effect into the PWP model is then suggested that modifies the PWP criterion to $(h\Delta b) \leq max(50u^{*2}, 0.65|\Delta u|^2)$.

2.4.3. *Recent Measurements.*

Meanwhile the observational techniques have improved considerably. It is now possible to show the temporal and spatial development of the surface circulation by using sideward projecting sonar beams . Langmuir cells develop instabilities, separate into vortex pairs or merge with neighboring cells [69, 51]. Smith [62], on the other hand , gave evidence that even at wind speeds larger than 8 m/s there was no visible sign of Langmuir cells, while they developed within 15 minutes after the wind increased to 13 m/s. A detailed survey of the Langmuir cell evolution during a storm with a novel Secscan sonar system onboard the stable platform R/P FLIP was done by Smith [63]. Smith obtained movie-like sequences of 2D maps over an area of 250m by 150m, with 5m by 10m spatial resolution. He states that "the maps permit objective assessment of strength, orientation, spacing, and degree of organization of features associated with mixed layer motions (e.g. Langmuir circulation)". First results reveal that previously suggested scaling of the Langmuir cell velocity variance like $u^*u_0^s$ by Plueddemann *et al.* [51], scales closer with u_0^{s2} alone. It appears that the nonlinear dynamic simulation result of vacillation under strong forcing is an observable feature, especially at high wave states [66]. However, the beginning wealth of observations still seems to be adding more pieces to the puzzle.

3. Deep Convection

3.1. LARGE SCALE DEEP CIRCULATION

The source of the global atmospheric and oceanic circulation is the pole-to-equator gradient of the earth's radiative heating. While the atmosphere is driven by the fluxes at its lower boundary and radiative cooling at higher levels, the ocean currents are forced by windstress from the atmosphere, a differential heating gradient, and a freshwater flux resulting as the net effect of precipitation, evaporation, ice formation and river discharge. The direct wind driven component is generally restricted by the mixed layer depth and deeper reaching currents result from horizontal density gradients or energy diffused downward from the surface layer (c.f. Niiler [46]). Compared to the atmosphere, the ocean thermal circulation is much weaker, because heating occurs in the areas where highest temperatures prevail. The cooling, on the other hand, takes place in the high latitudes at low temperatures, thus reducing the possibility of a strong thermal circulation [72]. As an important consequence, the slow ocean thermohaline circulation appears to be a major factor in maintaining the present climate and controlling variations on a variety of scales. The first global picture of a coherent deep ocean circulation was developed by Stommel [64].He showed that the crucial part of the abyssal path consists of North Atlantic Deep Water formed in the polar and subpolar latitudes and flows into the Southern Ocean along the western boundary. The present picture of the deep circulation as a "conveyor belt" is sketched by Gordon and Broecker [21, 5]. Deep water production takes place only in isolated regions and is linked to deep convection events and the phase of horizontal spreading out of this areat that follows. Examples of deep convection sites are the Greenland Sea, Labrador Sea, Weddell Sea, as well as the Gulf of Lions in the Mediterranean Sea. It is important to understand the reasons for the strong localisation and the agents that transport the convectively formed watermass out of the sinking regions.

3.2. CONVECTION WITH ROTATION

3.2.1. *Scaling.*

Laboratory experiments [15, 20] and numerical simulations [24] suggest that a convective Rossby number $Ro = \frac{w^*}{fh}$, where w^* is the convective velocity scale and h the convection layer depth is the controlling parameter for convection with rotation. With Deardorff 's [10] velocity scale $w^* = (Bh)^{1/3}$, where B is the surface buoyancy flux, the Rossby number can be written

$$Ro = (\frac{B}{h^2 f^3})^{1/3} \tag{32}$$

This can also be rewritten in terms of the Rayleigh flux (Ra_f) and Taylor (Ta) number as:

$$Ro = (\frac{Ra_f}{Ta^{3/2}} Pr^2)^{1/3}, \tag{33}$$

with

$$Ra_f = \frac{Bh^4}{\nu\kappa^2}, Ta = \frac{f^2h^4}{\nu^2}, \tag{34}$$

viscosity ν and thermal diffusivity κ. A typical velocity scale during deep convection is 0.01 m/s, so the convective Rossby number is about 1 for a 100 m thick convection layer. It is therefore expected that deeper reaching plumes are constrained by rotation when $Ro < 1$.

3.2.2. *Model Equations.*
Convection studies with rotation mainly use a linear temperature-density relationship and are based on the Rayleigh-Benard equations modified to include the effect of the vertical component of the Coriolis force. A typical scaled formulation is

$$\nabla^2 w_t [\boldsymbol{\nabla} \cdot (\boldsymbol{\omega} + f\boldsymbol{k}) x\boldsymbol{u}]_z + \nabla^2 [\boldsymbol{k} \cdot (\boldsymbol{\omega} + f\boldsymbol{k}) x\boldsymbol{u}] = Pr[\nabla^4 w + Ra_f \nabla_h^2 T], \tag{35}$$

$$T_t + \boldsymbol{u} \cdot \boldsymbol{\nabla} T = \nabla^2 T, \tag{36}$$

$$\omega_t^z + \boldsymbol{k} \cdot [\boldsymbol{\omega} + f\boldsymbol{k} x\boldsymbol{u}] = Pr \nabla^2 \omega^z. \tag{37}$$

Here $\boldsymbol{u}$ and $\boldsymbol{\omega}$ are the velocity and vorticity vector with vertical components w and ω^z, respectively. Julien *et al.* [27, 25] have examined the behavior of rotating convection as a function of the boundary conditions, Prandtl number, and Rayleigh number. The boundary condition for the ocean is (in the absence of wind forcing), a stress free surface with constant cooling via a destabilizing heatflux. The bottom conditions are preferably also stress free. In order to avoid reflection of impinging waves, some authors apply a radiation condition.

3.2.3. *Results of High Resolution Simulations.*
Visualizations by Julien *et al.* [26] reveal the striking difference between plume structures in the non-rotating and strongly rotating flows. Without rotation, active plumes plunge deep into the stratified underlying layer and lift the heavier fluid into the mixed layer under vigorous mixing, thus accounting for the change in the sign of the buoyancy flux. With strong

rotation the width of the cells is significantly reduced and their number is increased. The emitting plumes show cyclonic vorticity, while mixing in the upwelling zones is reduced in turn, decreasing the rate of penetrative convection. It is remarkable that the TKE budget analysis in the Julien *et al.* [26] large eddy simulation (LES) with rotation (their Figure 8b,c) except for the very near top and bottom reveals the principle balance of terms as in our parameterized convection run of 2.2.1 (Figure 2). The rate of mixed layer deepening has almost the same time dependence as that suggested by Deardorff *et al.* [13] for the non-penetrating case, namely

$$h = [h^2(t=0) + 2t\frac{-\kappa T_z|^{surf}}{T_z|_{bot}}]^{1/2}. \tag{38}$$

However, it is slightly deeper in the LES with little difference between both cases. The implication of this behavior for deep convection is that with a limited amount of cooling, represented by the unstable temperature gradient $-T_z|^{surf}$, one really needs a significantly reduced background stratification, (a *preconditioning* of the convection site). The convective deepening in 2.2.1 was 17 m in 5 days with a background temperature gradient of one degree per ten meters. The above rule would predict 15 m for a heat loss of 100 W/m^2, consistent with the findings of Julien *et al.* [26]. Mixing a very thick layer of 2000m at a net heat loss of 500 W/m^2 requires 80 days when the initial temperature difference is one degree. Accordingly, a nearly neutral stratification enables deep layers to cool rapidly. This clearly shows that numerical simulations of convection with a neutral background stratification [24] are not really applicable to realistic conditions and generally predict too large a deepening and thus a larger influence of the rotation.

3.2.4. *Impact of Localized Forcing.*
A thorough description and discussion of the convection in deep water formation areas has been given by Schott el al. [57]. They provide evidence of the presence of plumes smaller than 1 km-scale in a deep convection region which is typically 50 - 100 km in horizontal extent. Another feature predicted by Jones and Marshall [24], eddies of O(5-10km) resulting from the breakup of the rim of the convection area due to baroclinic instability. They seem to be present in timeseries of moored instruments and temperature sections. However, these instability eddies are not really widely surveyed features in convection regimes, but merely are projections of simulation results on the observations. An in situ cause-effect study and monitoring of their translation must still be performed. Jones and Marshall's [24] study favors the generation of instability eddies, because their forcing area has an abrupt drop to zero, creating a rim current within the inertial timescale. On the other hand, long lived coherent structures are evident in many

geophysical flows, and theoretical studies with low order physics such as the quasi-geostrophic potential vorticity conservation (e.g. [22, 23, 28]) are capable of explaining the basic mechanism of the transport characteristics.

3.3. COHERENT STRUCTURES IN DEEP CONVECTION

3.3.1. *Concepts.*

Long lived coherent structures can result from self-organization of turbulent flows [3]. Injection of a turbulent jet into a quiescent fluid causes a dipol like structure at the nose, which moves at a speed given by the non-linear self-interaction of the vortex pair (c.f. [71]). Integration of an initially randomly distributed potential vorticity over time, reveals an inverse cascade from small to larger scales, which finally evolve into a vertically aligned system of two coherent vortices of opposite polarity (J. Weiss, Internet communication). Thus it is likely that localized convection in the minimal model approach can be treated as the interaction of elementary point vortices that cancel each other's vorticity in the convection region. At the perimeter they superimpose to form the geostrophic rim current. In the stable case this is a purely baroclinic feature. In a continuos model, the stability of such a system requires the determination of the imaginary part of the frequency of sinusoidal perturbations, while in the discrete model the alignment behavior of point vortex ensembles is needed. Gryanik [22] has derived equations that describe the interaction of singular vortices with themselves and background currents. Later, Hogg and Stommel [23] introduced the "heton", a baroclinic point vortex and simulated the explosion of a "heton-cloud". The main result of these studies is that hetons can effectively spread properties over large distances. Numerical models of the general circulation do not resolve these small size features. The parameterization of the net effect of the spreading by hetons is a major problem and not yet solved.

3.3.2. *Numerical Simulation of the Post-Convection Phase.*

Even the simplified heton equations need numerical investigation due to the nonlinear structure. It is therefore interesting to perform numerical integration of the original Navier- Stokes equations in hydrostatic approximations, in order to allow for a larger range of validity. The main restriction of the quasi-geostrophic equation results from the restriction to low values of the ratio $\frac{\omega^z}{f}$, restricting the model to vertical vorticity much smaller than the planetary vorticity. The heton models are also formulated as a

Mar 1996 Convective Lens Vorticity z=0, day 14, 256x256 pts dx=0.5km [R.Kaese,IfM Kie l]

Mar 1996 Convective Lens Vorticity z=0, day 26, 256x256 pts dx=0.5km [R.Kaese,IfM Kiel]

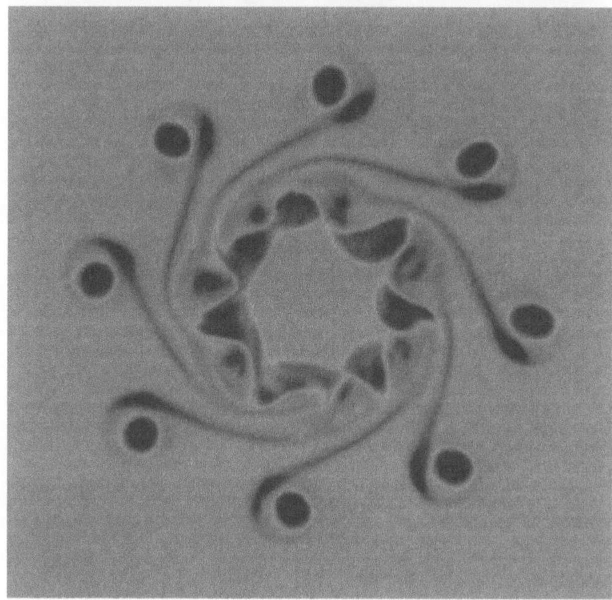

Figure 5. Quasi-geostrophic modeling of hetonic explosion in a convection region: Surface vorticity of the beginning instability (top) and in the spreading phase (bottom).

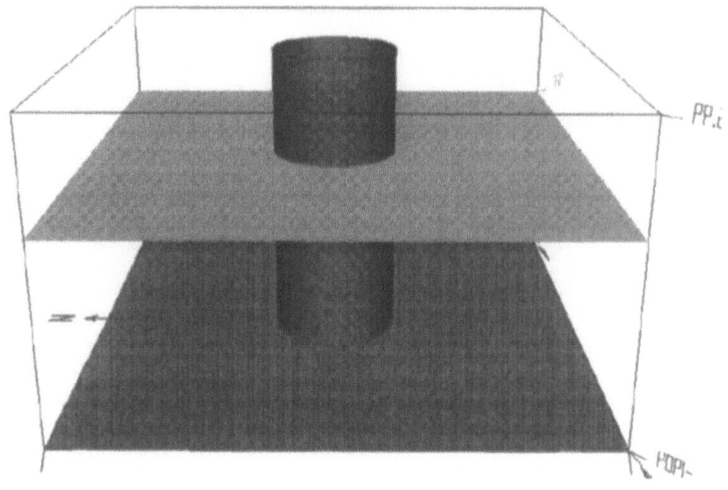

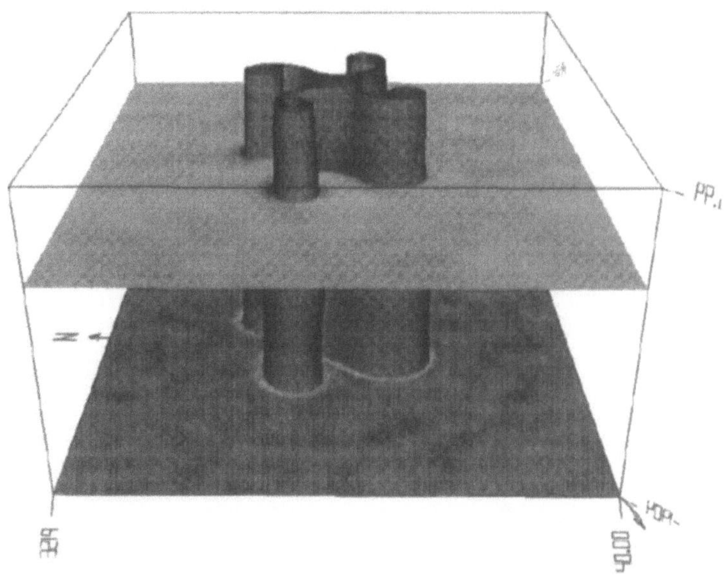

Figure 6. Hetonic explosion of a convection region: Initial state (top) and beginning instability (bottom).

pure baroclinic problem, based mainly on the integral constraint of vanishing vertically integrated horizontal momentum. This is only justified for missing boundary layers and free slip conditions. Integration by the au-

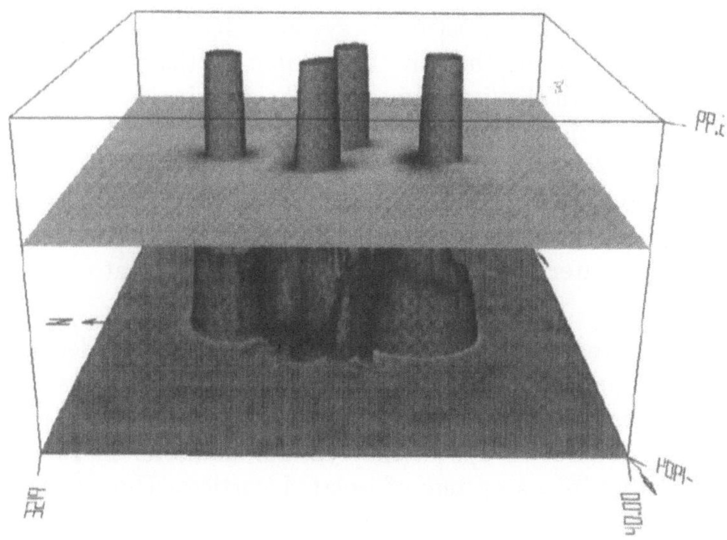

Figure 7. Hetonic explosion of a convection region: Spreading phase.

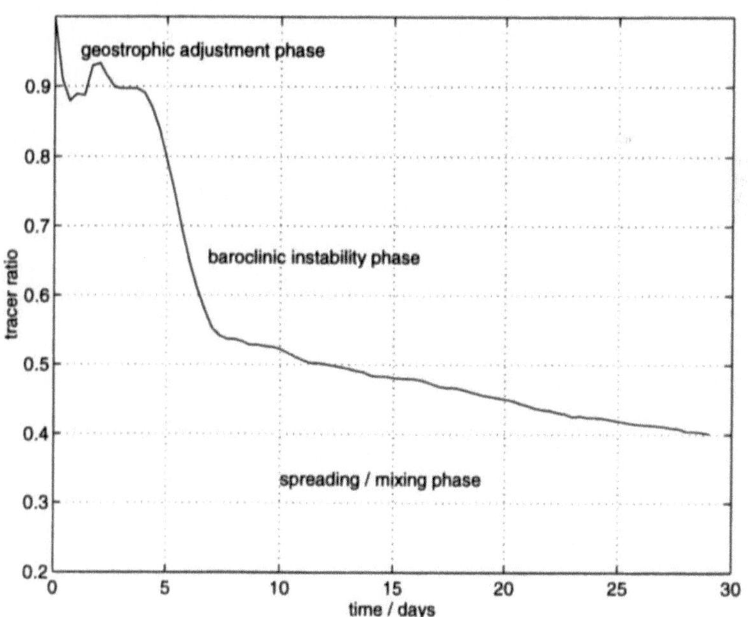

Figure 8. Ratio of fractional tracer content of top half to bottom half.

thor of a multi-level quasi-geostrophic system initialized with homogenous potential vorticity reveals that the barotropic vorticity is definitely non-zero, due to the vertical misalignment of the forming hetons. Essentially, the propagation of the exterior heton cloud is controlled by the barotopic dipoles. Figure 5 displays the near surface potential vorticity field at two stages of development in a high horizontal resolution (256x256x5) simulation. Apart from the aesthetic appeal of the simulation, it clearly shows that the original interior large scale patch breaks up into quasi-discrete vortices.

Due to computer restrictions the primitive equation modeling was performed with a smaller initial patch. The 3D development is shown for an initially cylindrical column of homogeneous, but heavier water, corresponding to a Rossby radius of 1/8 of the cylinder diameter in Figure 6. The cylinder breaks up into a system of isolated vortices. Due to the presence of rotation, these vortices trap heavier water near the surface. The final fate of the trapped water with the same density characteristics as the bottom layer, depends on the consecutive spreading and interaction of these vortices and the mixing processes, which act only on the gridscale in the model. Figure 8 displays the fractional partition of a tracer in the top and bottom half. The spreading history consists of three different phases, the initial cylinder collapse, the baroclinic instability phase, and the slow mixing phase. Since it is likely that the eddies will encounter mixing across their density fronts, due to a variety of unresolved small scale processes, the initial convectively formed water mass will not eventually sink to its neutral density level, and therefore will also contribute to the mixing in the entire water column. This makes it difficult to estimate the rate of the deep water formation from data in the convection region alone without detailed knowledge of the mixing history along the pathways of the vortices. There is a good chance that they reach the ocean boundaries rapidly and interact with topography and the boundary currents. Current observational activities by several countries in the subpolar North Atlantic promise exciting new insight into the modes that control the deep dater production, and the spreading and injection into the global thermohaline circulation.

4. Discussion

The improvement of the mixed layer models in the last decade came about through careful analysis of simulations using LES techniques. Though mainly applied to the atmospheric boundary layer, they lead to important insight into the role of countergradient fluxes and a better understanding of the balance of terms in the TKE equation. In convective deepening, strong fluxes of TKE remove energy out of the central mixed layer, which is then used to erode the thermocline underneath. If the models do not take into account penetrative convection, this leads to an unrealistic increase in dissipation. It is essential to include a proper mixing scheme in the "transition layer" and the upper thermocline. New simulations of penetrative convection with rotation [26] suggest that the entrainment flux is reduced by rotation. The atmospheric value of the ratio of entrainment flux to surface buoyancy flux of 0.2 used by Large *et al.* [34] seems to be an overestimate for deep convective layers that are influenced by rotational effects in this perspective.

The surface layer is still not adequately parameterized because the effect of Langmuir circulation is not included, although a suggestion of Li *et al.* [38] in the form of a modified bulk Richardson number criterion exists. However, all slab mixed layer models generally produce an unrealistic property jump. The energy dissipation in observations of convection near the surface is generally higher than the sum of shear and buoyancy production [2]. The additional source of turbulence is not clearly identified and may not only come from breaking surface waves, but also from the semi-chaotic "Langmuir turbulence" of Skyllingstad and Denbo [61].

The role of shear associated with inertial oscillations at the mixed layer base must be given more attention according to results by d'Asaro *et al.* [8]. The inertial shear has strong horizontal variation caused by mesoscale eddies. This brings up the general question of the validity of a 1-D approach.

New observations such as from the Arabian Sea Monsoon experiment [55] have a high potential for further testing the performance of mixed layer models. The storms affecting the monsoon region are stronger and steadier than those in midlatitude locations. The monsoon-cycle results in two periods of deepening per year. The winter mixed layer is deeper than 100m during NE monsoon , while it is only about 70m deep in the much stronger SW monsoon. The deeper winter mixed layer arises from the strong surface cooling due to the large air-sea temperature difference.

The present questions in deep convection are related to the combined effect of convective cooling and wind generated shear, the role of site selection mechanisms ("preconditioning"), the net downward transport of newly formed deep water, its injection into the general circulation, and the process of restratification. According to Send and Marshall [59] there is no net sinking in a localized convection region, plumes just restructure the water column vertically in a non-penetrative manner. The role of coherent vortices in the horizontal spreading of the deep water needs more clarification. The coming years will probably see a shift of attention from the vertical to the horizontal structure of the mixed layer and attempts to parameterize the corresponding horizontal transport processes and fluxes will be undertaken.

5. Acknowledgement

The author is indebted to Jürgen Willebrand who made his TKE-code available, to Andreas Oschlies for helpful discussions about his TKE-model implementation, as well as to Drs. C. Garrett, M. Gregg, M. Li, G. Nurser, U. Send. J. Smith, and D. Rudnick for supplying material used in the lecture.

References

1. Agrawal, E. C., Terray, E., Donelan, M., Hwang, P., III, A. W., Drennam, W., Kahma, K., and Kitaigorodskii, S. (1992). Enhanced dissipation of kinetic energy beneath surface waves. *Nature*, 259:219–220.
2. Anis, A. and Moum, J. (1992). The superadiabatic surface layer of the ocean during convection. *J. Phys. Oceanogr.*, 22:1221–1227.
3. Boubnov, B. M. and Golitsyn, G. (1986). Experimental study of convective structures in rotating fluids. *J. Fluid. Mech.*, 167:503–531.
4. Bougeault, P. and Lacarrere, P. (1989). Parameterization of orography-induced turbulence meso-beta scale model. *Mon. Weather Rev.*, 117:1872–1890.
5. Broecker, W. S. (1991). The great ocean conveyor. *Oceanography*, 4.
6. Craik, A. D. D. (1977). The generation of langmuir circulation by an instability mechanism. *J. Fluid Mech.*, pages 209–223.
7. Craik, A. D. D. and Leibovich, S. (1976). A rational model for langmuir circulation. *J. Fluid Mech.*, 73:401–426.
8. d'Asaro, E. A., Eriksen, C., Levine, M., Niiler, P., Paulson, C., and van Meurs, P. (1995). Upper-ocean inertial currents forced by a strong storm. part i: Data and comparisons with linear theory. *J. Phys. Oceanogr.*, 25(11):2909–2936.
9. Deardorff, J. W. (1966). The counter gradient heat flux in the atmosphere and in the laboratory. *J. Atmos. Sci.*, 23:503–506.
10. Deardorff, J. W. (1970). A numerical study of three-dimensional channel flow at large reynolds numbers. *J. Fluid Mech.*, 23:453–480.
11. Deardorff, J. W. (1972). Theoretical expressions for the counter-gradient vertical heat flux. *J. Geophys. Res.*, 77:5900–5904.
12. Deardorff, J. W., Willis, G., and Lilly, D. (1969). Laboratory investigation of non-steady penetrative convection. *J. Fluid Mech.*, 35:7–31.
13. Deardorff, J. W., Willis, G., and Stockton, B. (1980). Laboratory studies of the entrainment zone of a convective mixed layer. *J. Fluid Mech.*, 100:41–64.
14. Defant, A. (1936). Schichtung und zirkulation des atlantischen ozeans. *Wiss. Ergebn. Deutsch. Atlant. Exp. Meteor 1925-27*, 6(1):289–411.
15. Fernando, H. J. S., Chen, R.-R., and Boyer, D. (1991). Effects of rotation on convective turbulence. *J. Fluid Mech.*, 228:513–547.
16. Garrett, J. (1976). Generation of langmuir circulation by surface waves- a feedback mechanism. *J. Mar. Res.*, 34:117–130.
17. Garwood, R. W. (1977). An oceanic mixed layer model capable of simulating cyclic states. *J. Phys. Oceanogr.*, 7:455–471.
18. Gaspar, P. (1988). Modeling the seasonal cycle of the upper ocean. *J. Phys. Oceanogr.*, 18:161–180.
19. Gaspar, P., Gregoris, Y., and Lefevre, J.-M. (1990). A simple eddy kinetic energy model for simulations of the oceanic vertical mixing: Tests at station papa and long-term-upper ocean study site. *J. Geophys. Res.*, 95:16,179–16,193.
20. Golitsyn, G. (1998). Convection in viscous and rotating fluids from the viewpoint of the forced flow theory. See this volume.
21. Gordon, A. (1986). Interocean exchange of thermohaline water. *J. Geophys. Res.*,

91:5037–5046.

22. Gryanik, V. (1983). Dynamics of singular geostrophic vortices in a two-level model of the atmosphere (or ocean). *Atmos. Ocean Phys.*, 91(3):171–179.
23. Hogg, N. and Stommel, H. (1985). The heton, an elementary interaction between discrete baroclinic geostrophic vortices and its implication concerning eddy heat-flow. *Proc. Roy. Soc. London*, A(397):1–20.
24. Jones, H. and Marshall, J. (1993). Convection with rotation in a neutral ocean: a study open-ocean deep convection. *J. Phys. Oceanogr.*, 23:1009–1039.
25. Julien K., S. Legg, J. M. and Werne, J. (1996a). Hard turbulence in rotating rayleigh-benard convection. *Phy. Rev. E*, 53:5557.
26. Julien K., S. Legg, J. M. and Werne, J. (1996b). Penetrative convection in rapidly rotating flows: Preliminary results from numerical simulation. *Dynamics of Atmospheres and Oceans*, 24:237–249.
27. Julien K., S. Legg, J. M. and Werne, J. (1996c). Rapidly rotating turbulent rayleigh-benard convection. *J. Fluid Mech.*, 322:243.
28. Käse, R.H. and W. Zenk (1996). Structure of the mediterranean water and meddy characteristics in the northeastern atlantic. In Krauss, W., editor, *The warmwatersphere of the North Atlantic Ocean*, chapter 12, pages 365–395. Bornträger, Berlin and Stuttgart.
29. Kelly, D. (1990). Fluxes through diffusive staircases: a new formulation. *J. Geophys. Res.*, 95:3365–3371.
30. Kraus, E. (1977). *Modelling and Prediction of the Upper Layers of the Ocean.* Pergamon, New York.
31. Kraus, E. and Turner, J. (1967). A one-dimensional model of the seasonal thermocline, ii, the general theory and its consequences. *Tellus*, 19:98–105.
32. Krauss, W., editor (1996). *The warmwatersphere of the North Atlantic Ocean*, page 444. Bornträger, Berlin and Stuttgart.
33. Langmuir, I. (1938). Surface motion of water induced by wind. *Science*, pages 119–123.
34. Large, W., McWilliams, J., and Doney, C. (1994). Ocean vertical mixing: A review and a model with a nonlocal boundary layer parameterization. *Rev. Geophys.*, 32:363–402.
35. Leibovich, S. (1977). Convective instability of stably stratified water in the ocean. *J. Fluid Mech.*, 82:561–581.
36. Leibovich, S. (1983). The form and dynamics of langmuir circulations. *Ann. Rev. Fluid Mech.*, 15:391–427.
37. Li, M. and Garrett, C. (1997). Mixed-layer deepening due to langmuir circulation. *J. Phys. Oceanogr.*, 27:121–132.
38. Li, M., Zahariev, K., and Garrett, C. (1995). Role of langmuir circulation in the deepening of the ocean surface mixed layer. *Science*, 270:1955–1957.
39. Lombardo, C. P. and Gregg, M. (1989). Similarity scaling of viscous and thermal dissipation in a convecting surface boundary layer. *J. Geophys. Res.*, 94:6273–6284.
40. Lumby, J. (1955). The depth of the wind-produced mixed layer in the oceans. *Fish. Invest.*, 2(20):1–12.
41. Mellor, G. and Yamada, T. (1982). Development of a turbulence closure model for geophysical problems. *Rev. Geophys. Space Phys.*, 20:851–875.
42. Moeng, C.-H. and Wyngaard, J. (1989). Evaluation of turbulent transport and dissipation closures in second order modeling. *J. Atmos. Sci.*, 46:2311–2330.
43. Moum, J., Caldwell, D., and Paulson, C. (1989). Mixing in the equatorial surface layer and thermocline. *J. Geophys. Res*, 94:2005–2021.
44. Niiler, P. (1975). Deepening of the wind mixed layer. *J. Mar. Res.*, 33:405–422.
45. Niiler, P. and Kraus, E. (1977). One dimensional models of the upper ocean, in modelling and prediction of the upper layers of the ocean. In Kraus, E., editor, *Modelling and Prediction of the Upper Layers of the Ocean*, pages 143–172. Pergamon, New York.

46. Niiler, P. P. (1992). The ocean circulation. In Trenberth, K., editor, *Climate System Modeling*, pages 117–148. Cambridge University Press, Englewood Cliffs, N.J.
47. Nurser, A. (1996). A review of models and observations of the oceanic mixed layer. Technical Report 14, Southampton Oceanographic Centre.
48. Paulson, C. (1970). Representation of windspeed and temperature profiles in the unstable atmospheric boundary layer. *J. Appl. Meteorol.*, 9:857–861.
49. Peters, H., Gregg, M., and Toole, J. (1988). on the parameterization of equatorial turbulence. *J. Geohys. Res.*, 93:1199–1218.
50. Philander, S. (1990). *El Nino, La Nina and the Southern Oscillation.* Academic Press, San Diego.
51. Plueddemann, A., Smith, J., Farmer, D., Weller, R., Crawford, W., Pinkel, R., Vagle, S., and Gnanadesikan, A. (1996). Structure and variability of langmuir circulation during the surface waves processes program. *J. Geophys. Res.*, 101:3525–3543.
52. Price, J., Weller, R., and Pinkel, R. (1986). Diurnal cycling: observations and models of the upper ocean response in diurnal heating, cooling and wind mixing. *J. Geophys. Res.*, 91:8411–8427.
53. Rahmstorf, S. (1995). Bifurcations of the atlantic thermohaline circulation in response to changes in the hydrological cycle. *Nature*, (378):145–149.
54. Rossby, C. and Montgomery, R. (1935). The layer of frictional influence in wind and ocean current. *Pap. Phys. Oceanogr. Met.*, 3(3):1–101.
55. Rudnick, D., Weller, R., Eriksen, C., Dickey, T., Marra, J., and Langdon, C. (97). Moored instruments weather arabian sea monsoons, yield data. *EOS*, 78(11):117,120–121.
56. Schmitt, R. (1981). Form of the temperature-salinity relationship in the central water: Evidence for double-diffusive mixing. *J. Phys. Oceanogr.*, (11):1015–1026.
57. Schott, F., Visbeck, M., and Send, U. (1994). Open ocean deep convection, mediterranean and greenland seas. In Malanotte-Rizzoli, P. and Robinson, A., editors, *Ocean Processes and Climate Dynamics*, pages 203–225. Kluwer Academic.
58. Semtner, A. (1995). Modeling ocean circulation. *Science*, (269):1379–1385.
59. Send, U. and Marshall, J. (1995). Integral effects of deep convection. *J. Phys. Oceanogr.*, (21):558–574.
60. Shay, T. and Gregg, M. (1986). Convectively driven turbulent mixing in the upper ocean. *J. Phys. Oceanogr.*, (16):1777–1798.
61. Skyllingstad, E. and Denbo, D. (1995). An ocean large-eddy simulation of langmuir circulation and convection in the surface mixed layer. *J. Geophys. Res.*, (100):8501–8522.
62. Smith, J. (1992). Observed growth of langmuir circulation. *J. Geophys. Res.*, 97:95–110.
63. Smith, J. (1997). Evolution of langmuir circulation during a storm. submitted.
64. Stommel, H. (1958). The abyssal circulation. *Deep Sea Res.*, 5:80–82.
65. Stull, R. (1998). Convective transport theory and the radix layer. See this volume.
66. Tandon, A. and Leibovich, S. (1995). Simulations of the three-dimensional langmuir circulation in water of constant density. *J. Geophys. Res.*, (100):22613–22623.
67. Therry, G. and P.Lacarrere (1983). Improving the the eddy kinetic energy model for planetary boundary layer description. *Boundary-Layer Meteorol.*, 19:63–88.
68. Thorpe, S. (1985). Small scale processes in the upper ocean boundary layer. *Nature*, (318):519–522.
69. Thorpe, S. (1992). The breakup of langmuir circulation and the instability of an array of vortices. *J. Phys. Oceanogr.*, (22):350–360.
70. Troen, I. and Mahrt, L. (1986). A simple model of the atmospheric boundary layer: Sensitivity to surface evaporation. *Boundary Layer Meteorol.*, (37):129–148.
71. van Heijst, G. (1993). Self organization in two-dimensional flow. *Nederlands Tijdschrift voor Natuurkunde*, 59(20):321–325.
72. Webster, P. (1994). The role of hydrological processes in ocean-atmosphere interactions. *Rev. Geophys.*, 32(4):427–476.

73. Weller, R., Dean, J., Marra, J., Price, J., Francis, E., and Boardman, D. (1984). Three-dimensional flow in the upper ocean. *Science*, (227):1552–1556.
74. Weller, R. and Price, J. (1988). Langmuir circulation within the oceanic mixed layer. *Deep Sea Res.*, 35:711–747.
75. Wyngaard, J. (1998). Experiment, numerical modeling, numerical simulation, and their roles in the study of convection. See this volume.

OBSERVATIONS OF CLEAR AND CLOUD-CAPPED CONVECTIVE BOUNDARY LAYERS, AND TECHNIQUES FOR PROBING THEM

D. H. LENSCHOW
National Center for Atmospheric Research
P.O.Box 3000, Boulder, CO 80307 USA

1. Introduction

The planetary boundary layer (PBL) is that part of the atmosphere that interacts directly with the Earth's surface on a time scale of a few hours or less. It is the layer which transports the gases that are essential to life, as well as the effluents of our civilization. It is the conduit through which energy from the surface is transferred to the level where clouds are generated. The earth-atmosphere system absorbs about 740 W m^{-2} from the sun averaged over the entire globe, and almost three-quarters of this energy is absorbed at the surface, to be released into the atmosphere through sensible and latent heat flux, and by infrared radiation exchange. Surface heat fluxes generate turbulence which leads to efficient mixing of the lower atmosphere and may lead to the formation of clouds near the top of the turbulently-mixed PBL. Clouds may then modify this process by shading the surface, reflecting and absorbing solar radiation (globally they reflect about 20% of the total incoming radiation), and, if deep enough, by generating their own secondary circulations through the release of latent heat by condensation. In this way, clouds are a mechanism for releasing latent heat emitted from the surface, exchanging PBL properties with the overlying atmosphere, and closing the atmospheric water budget through precipitation.

Over land, these energy exchange processes have a strong diurnal modulation. A schematic of these evolutionary processes and how they relate to the rest of the atmosphere is shown in Fig. 1. Observations are one approach to learning more about the PBL. Summaries of observed characteristics of the PBL are given by e.g. Caughey [2], Stull [32], and Kaimal and Finnigan [10]. As shown schematically in Fig. 2, numerical and laboratory simulations are complementary to observations. Each has its advantages

E.J. Plate et al. (eds.), Buoyant Convection in Geophysical Flows, 185–206.

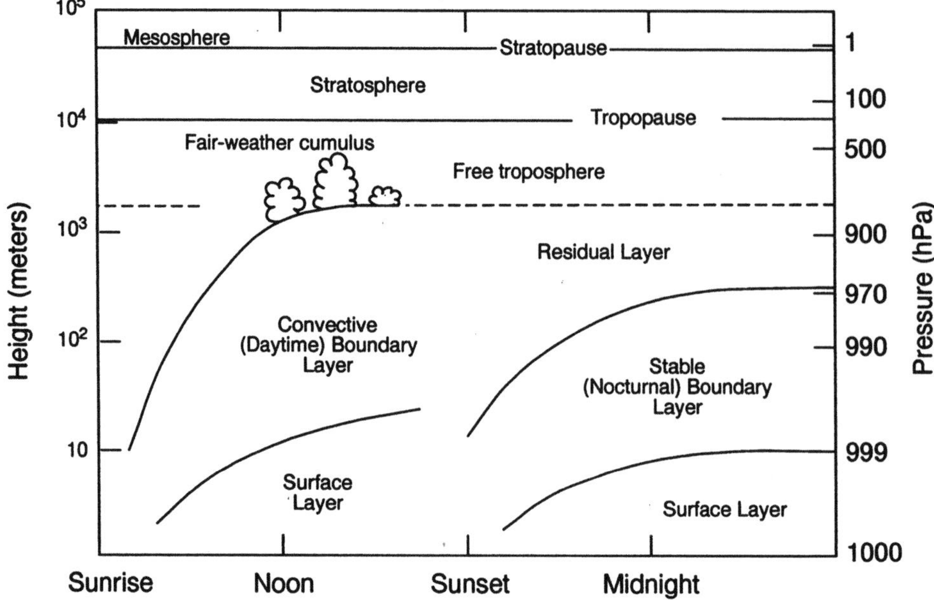

Figure 1. Schematic of the structure of the atmosphere, plotted versus the logarithm of height, showing a typical diurnal cycle over land.

and disadvantages, as you will hear throughout the rest of the lectures. However, the touchstone of PBL research lies solely with observations. It is nature which we are attempting to predict, and ultimately it is observations of nature which provide the basis for testing our models of its behavior.

2. Clear air convective boundary layer

We start with the clear-air convective PBL (CBL) which forms over heated surfaces when the stability parameter $z_i/-\mathcal{L}$, where z_i is the depth of the CBL and $\mathcal{L}$ is the Obukhov length, is large (greater than about 5 or so). We assume horizontal homogeneity of the surface and the turbulence statistics. In this case, vertical gradients and many turbulence statistics in the *surface layer*, which is the lowest 10% or so of the CBL where fluxes can be assumed constant, obey surface-layer similarity, discussed here by Zilitinkevich *et al.* [40]. In the *mixed layer*, which is that part of the CBL which lies above the surface layer and below the *entrainment zone*, efficient mixing by eddies that extend throughout the layer forces the vertical gradients of scalars to be small. Here, vertical gradients and many turbulence statistics are functions of z/z_i, and can be scaled by z_i and the *Deardorff convective*

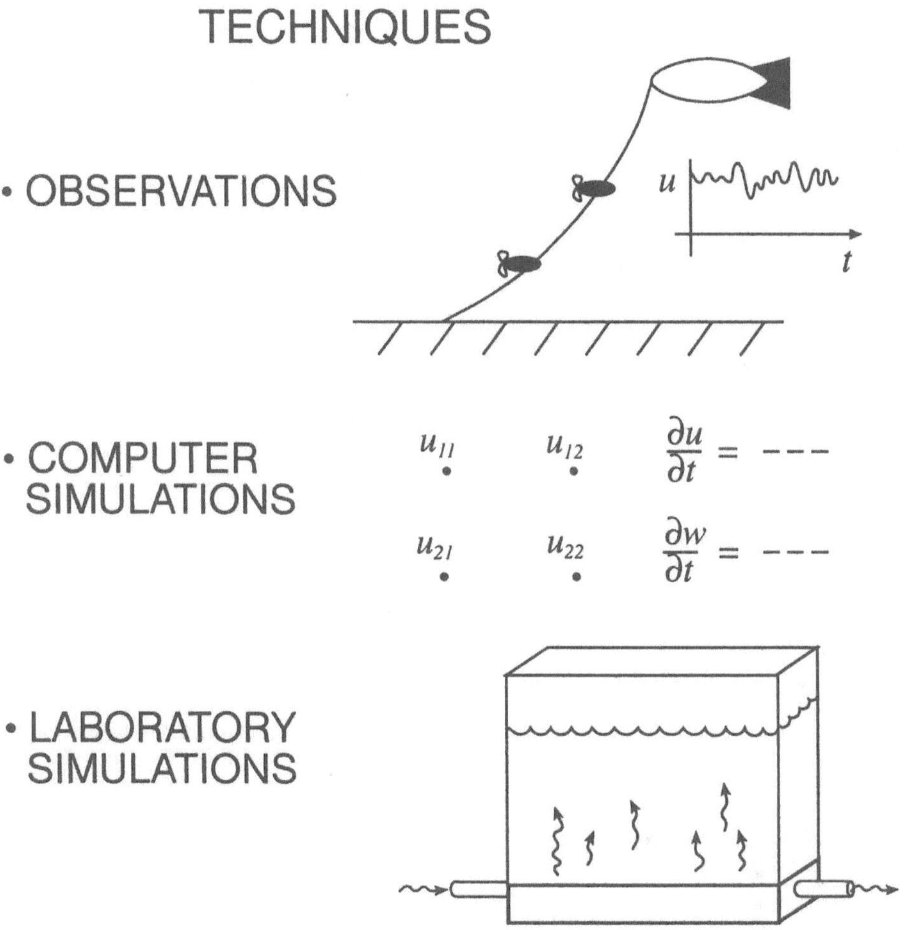

Figure 2. Observations, computer simulations and laboratory simulations are three complementary ways to study the structure and evolution of the convective PBL.

velocity scale,

$$w_* = \left[\frac{g}{\overline{T}}(\overline{wT_v})_0 z_i\right]^{1/3}, \tag{1}$$

where $g/\overline{T}$ is the buoyancy parameter and $(\overline{wT_v})_0$ is the surface virtual temperature flux. For example, the standard deviation of vertical velocity can be written as

$$\sigma_w(z/z_i) = w_* f(z/z_i). \tag{2}$$

The structure of the CBL is illustrated schematically in Fig. 3. From an observational perspective, (2) can be used to estimate the surface buoyancy flux from σ_w. That is, it is possible to estimate the surface buoyancy flux

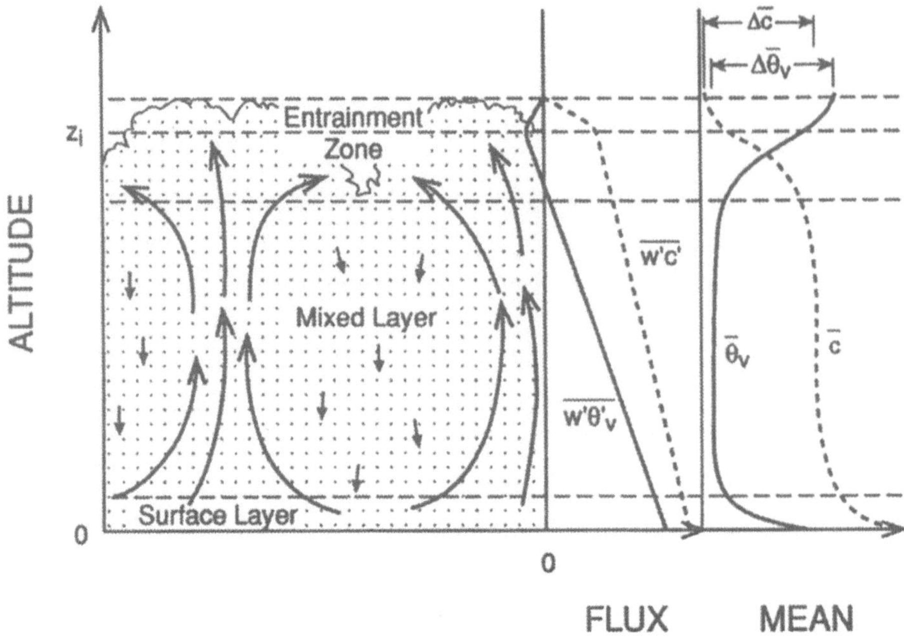

Figure 3. Schematic diagram of the CBL, showing the surface, or constant-flux layer (typically 10 to 100 m depth), the mixed layer (typically 500 to 2000 m depth) and the entrainment zone (typically 50 to 200 m depth). Scalar fluxes (here θ_v is virtual potential temperature and c is a passive scalar) are normally nearly linear, and mean scalar gradients are small throughout the mixed layer; both fluxes and gradients show considerable variation in the entrainment zone, while in the surface layer fluxes are nearly constant but the magnitudes of gradients can be large.

from measurements of z_i and vertical velocity fluctuations well above the surface layer. This is one approach to estimating surface buoyancy flux remotely; for example, vertically pointing Doppler radars and lidars can measure the variance of w, as well as other turbulence statistics, as functions of height.

In the upper part of the CBL, this scaling may not be adequate because of the impact of entrainment. Here the turbulence statistics may also be affected by the flux through the top of the CBL—the entrainment flux. In this case, as shown by Wyngaard and Brost [39], and Moeng and Wyngaard [25] from large-eddy numerical simulations (LES) of the CBL, gradients of a scalar $\mathcal{S} = S + s$, where S is the mean and s the fluctuation of $\mathcal{S}$, can be parameterized by incorporating the separate effects of surface and entrainment fluxes in the form

$$-\frac{w_* z_i}{(\overline{ws})_0}\frac{\partial S_0}{\partial z} = g_0(z/z_i) \tag{3}$$

and

$$-\frac{w_* z_i}{(\overline{ws})_{z_i}} \frac{\partial S_i}{\partial z} = g_i(z/z_i), \tag{4}$$

where $\frac{\partial S_0}{\partial z}$ and $\frac{\partial S_i}{\partial z}$ are the separate contributions to $\frac{\partial S}{\partial z}$ from the surface and entrainment fluxes, respectively. The functions $g_0(z/z_i)$ and $g_i(z/z_i)$ were estimated empirically from LES in the above studies. Integrating these equations over two height differences leads to two equations which can be solved for the surface and entrainment fluxes. Thus, in principle, the two fluxes can be estimated from mean concentration measurements at three levels in the mixed layer. Similar relations for scalar variances in the mixed layer were obtained by [25] and [39]. Davis [4] showed that aircraft observations were consistent with these relations and discussed how more accurate estimates of both the gradient and variance relations could be obtained from observations. Davis *et al.* [5] applied this gradient technique to estimating fluxes of non-methane hydrocarbons from tethered balloon observations, and Kiemle *et al.* [14] have compared remotely-measured water vapor profiles (using a differential absorption lidar looking downward from an aircraft) with the LES results and obtained estimates of entrainment flux remotely.

This mixed-layer scaling approach provides a powerful tool for predicting the behavior of the clear CBL. However, it does not provide explicit formulations for such important variables as the entrainment flux, which is primarily a function of w_* but also depends on the density structure and wind shear at, and just above, the CBL top, and on gravity waves in the overlying stably-stratified free atmosphere. Addressing questions such as entrainment velocity requires going beyond studies of the statistical properties of the CBL to consideration of the underlying turbulent eddy structures. This has led to a renewed interest in studying the coherent structures, such as thermals [22] and longitudinal roll vortices [7] that are embedded in what we observe as PBL turbulence. Coincidentally, we now have improved tools for observing these structures (especially remote sensors such as radars and lidars), as well as the increasing capability of LES to resolve them (e.g. [31]).

However, it is well to keep in mind that our observations of such structures predate "modern" science. For example, observations of thermals and rolls, and under what stability conditions each occur were reported by Woodcock [36] on the basis of the behavior of soaring sea gulls. Similarly, visual observations of the formation of small cumuli at the top of the CBL give us information about e.g. the scale and organization of underlying thermals and longitudinal rolls.

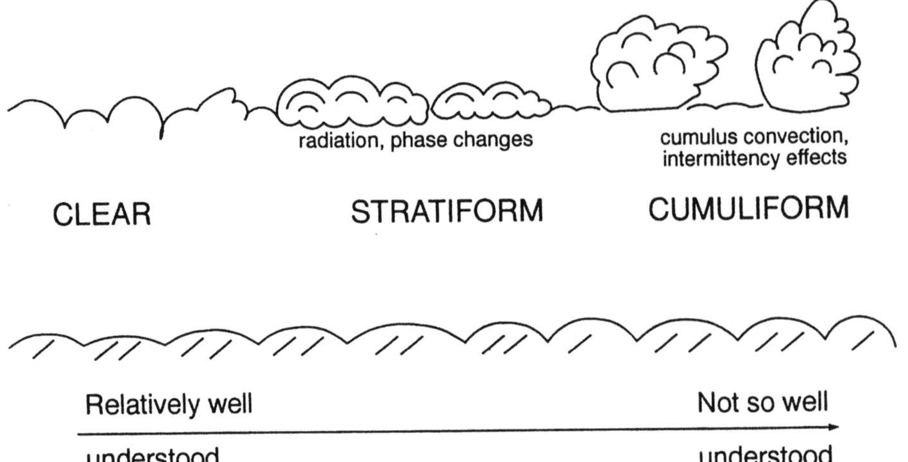

Figure 4. Schematic illustration showing the increasing complexity of the CBL going from clear (driven by surface heating) to stratiform-capped (involves also radiative cooling and phase changes) to penetrating cumuliform clouds (involves also stability and thermodynamic structure of the overlying free atmosphere; highly intermittent).

3. Boundary layers with clouds

Clouds in (or just above) the CBL add considerably to the complexity of its structure. As shown in Fig. 4, the next step in complexity beyond the clear CBL is the stratus-capped CBL. Fig. 5 shows schematically the difference in stability across the capping CBL inversion that determines whether stratiform cloud develops beneath the top of the CBL or whether cumuliform cloud develops at the CBL top and penetrates into the overlying free troposphere. Fig. 6 shows schematically the added complexity that is introduced when stratiform clouds cap the CBL. I will describe some observational techniques and results; C.-H. Moeng [23] describes more fully the physical processes important in the stratus-capped CBL.

Because of the climatic implications of stratiform cloud (a 4% increase in global stratus cover may be sufficient to offset the 2–3 K predicted rise in global temperature due to a doubling of CO_2 [29]), there has recently been considerable interest in studying their structure—especially marine stratocumulus—and several major field programs have been mounted to investigate them. One of the major conclusions of these studies is, not surprisingly, that marine stratus are more complicated than was previously thought. They only occasionally live up to the ideal horizontally uniform cloud decks that were once envisioned and are still modeled. A major complication is the widespread occurrence of "decoupled" stratocumulus decks.

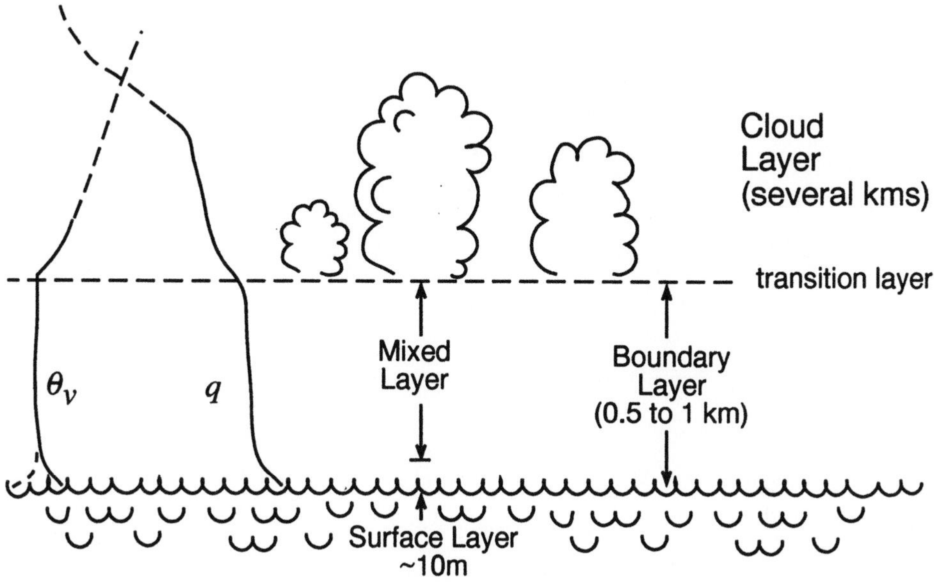

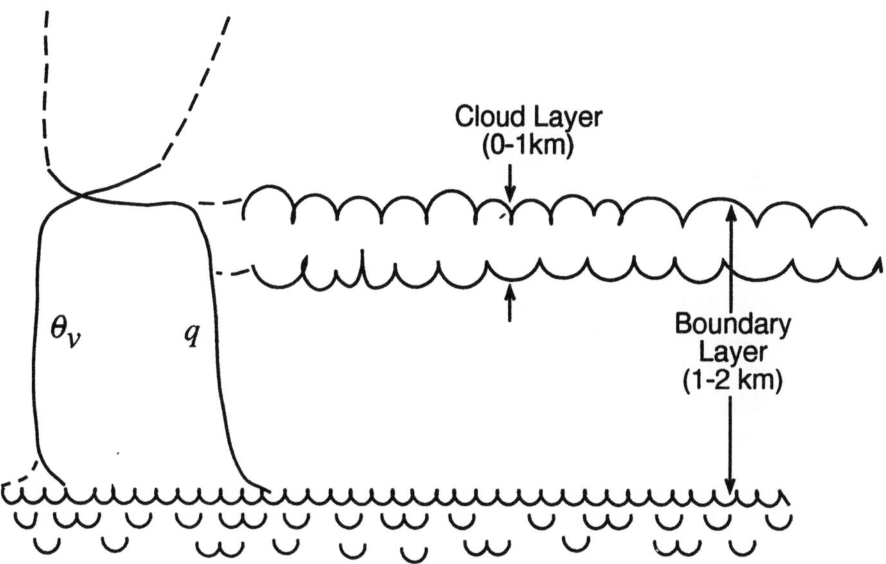

Figure 5. Schematic illustration of the different thermodynamic and cloud structures associated with cumuliform and stratiform-capped marine CBLs.

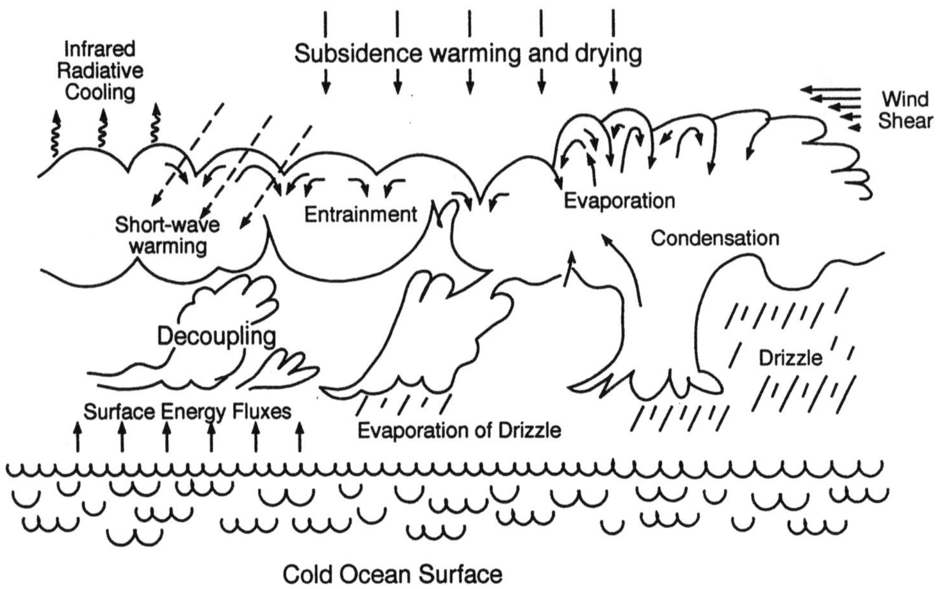

Figure 6. A schematic illustration of the many physical processes involved in the structure and evolution of the stratiform-capped PBL.

That is, the turbulence generated by surface buoyancy flux and wind stress, and by radiative cooling at cloud top is often insufficient to maintain continuous turbulent exchange between the air near the surface and that near the cloud layer. Nicholls [27] was apparently the first to document observations of this structure, and subsequent field programs (e.g. FIRE, ASTEX, and ACE-1) have found that it occurs frequently.

The occurrence of decoupling is determined by z_i, the top and bottom buoyancy fluxes, and sometimes the mean wind speed. Additional processes that can accentuate this decoupling are evaporation of drizzle and short-wave radiative warming. Fig. 7 shows an example of the effect of drizzle on a low-level flight leg from FIRE off the California coast [28]. We see that areas in which the aircraft encountered liquid water have lower temperature and higher humidity, indicating evaporation, and that these regions occur on scales > 10 km. Thus, drizzle is not only a stabilizing process tending to cool the air below cloud, but is also a process that can generate mesoscale variability, and create conditions ripe for the development of cumuli beneath

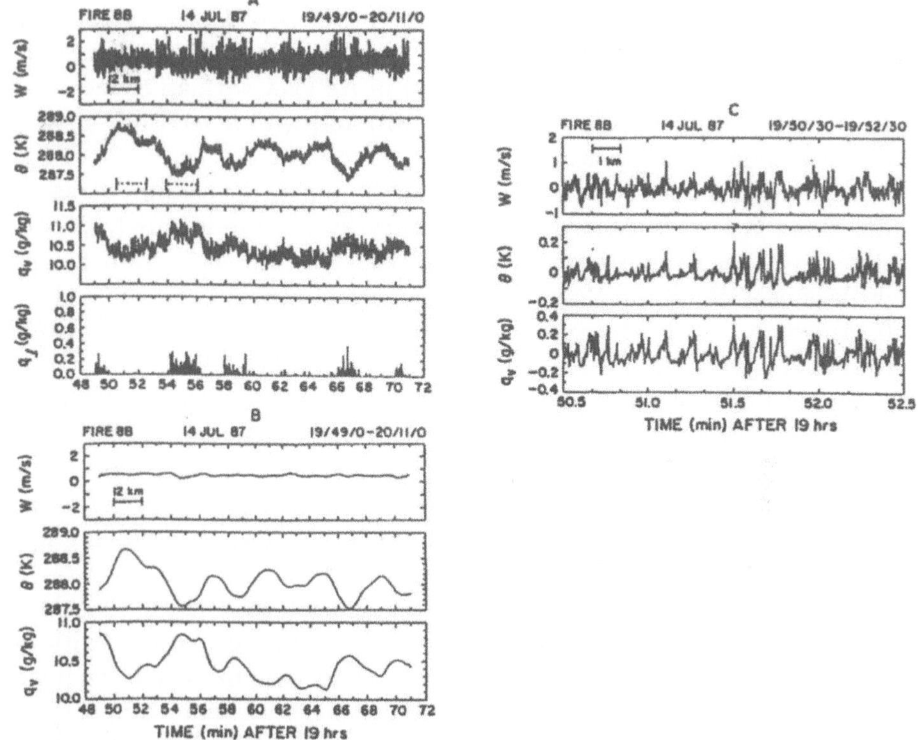

Figure 7. (A) Time series (vertical air velocity, potential temperature, humidity, and liquid water mixing ratio) from an aircraft flight leg near the surface of a stratiform-capped mixed layer off the southern California coast during FIRE. (B) Same time series as in (A) after removing contributions at wavelengths < 10 km with a low-pass filter, showing negative correlation between water vapor and potential temperature resulting from evaporated drizzle. (C) A two-minute segment (19:50:30–19:52:30) of the time series after filtering out wavelengths > 2 km showing positive correlation between vertical velocity, potential temperature and water vapor due to positive surface sensible and latent heat fluxes [28].

the stratiform cloud deck. This, in turn, can further enhance the breakup of the cloud deck into fields of cumulus clouds.

In a sense, the term "decoupling" may be somewhat of a misnomer, since the two layers are still intermittently connected by cumulus convection that is generated near the top of the lower layer (the actual "boundary layer" in a classical sense) by accumulating water vapor from surface evaporation. This convection then replenishes the upper layer water vapor which is depleted by entrainment of drier air from above. The typical result is a fairly solid upper layer of stratocumulus within which convective events can be found that originate from cumulus convection at the top of the lower turbulent layer, and result in a characteristic "mushroom" shaped embedded convective cloud which is made visible by overshooting the surrounding cloud bank and by higher reflectivity due to enhanced liquid water content. These convective structures were observed in ASTEX with an island-based Doppler radar (Fig. 13; Kropfli and Orr [16]) and with a vertically-pointed backscatter lidar from an aircraft [35].

Observations of the turbulent eddies at the top of the capping stratocumulus cloud layer generated by the buoyancy flux due to infrared radiative cooling have been reported by Nicholls [26]. He estimated from aircraft flight legs near cloud top that the average size of elements intersected was $\sim 0.2h$, where h is the distance from cloud top through which radiatively-driven mixing extends (for his cases, $h \sim 400 - 700$ m). The distribution of element sizes intersected by the aircraft was consistent with a cellular pattern of downdrafts $\sim 0.1h - 0.15h$ in width around the periphery of updrafts $\sim 0.5h - 0.75h$ in diameter. The corresponding spectral peak was $\sim 0.6h$, increasing to $\sim 1.4h$ near the bottom end of the radiatively-driven mixed layer. This morphology is consistent with airborne downward-looking lidar observations of the cloud-top structure (e.g. [9]).

When the atmospheric stability is insufficient to contain the clouds within the CBL, we have the development of cumulus convection with the added complexity of larger length scales, cloud dynamical effects in the CBL, transport of CBL air into the free troposphere, intermittent shading, and more intense precipitation events. While interactions of cumulus clouds with the CBL are beyond the scope of this discussion, I mention in passing that cumulus convection is the dominant mechanism for injecting PBL air into the free troposphere. This process, called *cloud venting*, is shown schematically in Fig. 8 and described by e.g. Cotton *et al.* [3]. The CBL air injected into the stably stratified free troposphere can then be spread out horizontally by wind shear into rather thin laminae, as shown in Fig. 9.

As an example, if we assume an initial square cross-section of air injected into the free atmosphere, a constant unidirectional wind shear $\partial U/\partial z$, and conservation of mass (area), then for

$$\tau \frac{\partial U}{\partial z} \gg 1, \tag{5}$$

where τ is the elapsed time since injection,

$$h(\tau) \simeq \frac{h_0}{\tau(\partial U/\partial z)} \quad \text{and} \quad L(\tau) \simeq L_0 \tau \frac{\partial U}{\partial z}. \tag{6}$$

For $\partial U/\partial z = 3$ m s^{-1}/1000 m and $\tau = 10$ hours,

$$\tau \frac{\partial U}{\partial z} \simeq 100. \tag{7}$$

Figure 8. Schematic illustration of how CBL air is injected into the overlying atmosphere by cloud venting, with the result that layers of air containing a mixture of CBL and above-CBL air are formed.

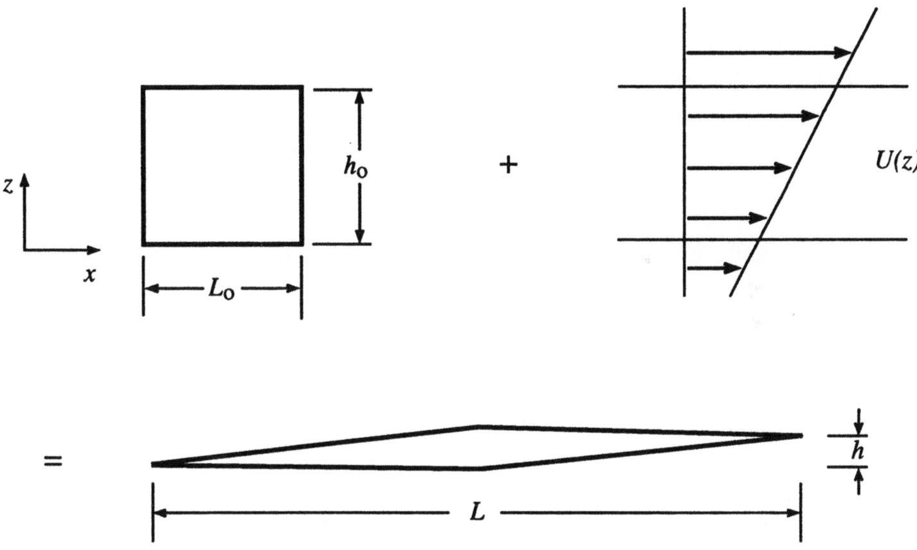

Figure 9. Schematic of how CBL air injected into the overlying atmosphere can be stretched out into extended thin layers by wind shear.

If h_0 and L_0 are 1000 m, after 10 hours $h \simeq 10$ m and $L \simeq 100$ km. Mean scalar profiles in the free troposphere measured by aircraft or balloons often show such a finely layered structure, especially in regions of subsiding flow, when sufficient time has elapsed since convection has occurred to stretch out the injected parcels into thin lamina. Kawa and Pearson [13] (their Fig. 7) give an example of this structure obtained from an aircraft sounding in the subsiding subtropical flow off the California coast in summer.

4. Sampling requirements for boundary-layer measurements

An important consideration for observing the CBL is the sample length necessary for obtaining statistically significant measurements. Because of the inherent randomness of the turbulent processes responsible for vertical transport, we develop and use techniques for studying the statistical characteristics of turbulence that are based on the statistics of random variables. Ideally, we would like to measure the statistical variables over temporal or spatial scales long enough that enough samples are obtained to insure sufficient accuracy. In practice, this may be difficult to achieve because in the time (or distance) required to obtain sufficient samples, there may be systematic changes in the conditions or processes that control the variable. Therefore, it is important to have quantitative measures for estimating the accuracy of statistical measurements as a function of averaging time (or length). A basic assumption in this discussion is that the turbulent field is horizontally homogeneous.

We define the autocovariance function for e.g. w as

$$R_w(\xi) \equiv \overline{w(x)w(x+\xi)}, \tag{8}$$

which is a measure of how well w is correlated with itself as a function of separation distance ξ. The autocovariance is useful in defining the integral scale, which is defined as

$$\ell_w \equiv \frac{1}{\sigma_w^2} \int_0^\infty R_w(\xi) d\xi, \tag{9}$$

where σ_w^2 is the variance of w. Of course, in practice, the limit of integration cannot be infinite; it is sufficient to integrate over a separation distance that extends far enough that the autocovariance shows no correlation and fluctuates about zero.

The integral scale is useful in estimating the random error in turbulence statistics due to the limited length of any real set of measurements in the CBL. We consider here the error variance of a flux measurement $F(L) = \overline{wc}$ over a sample length L which is given by

$$\tilde{\sigma}_f^2(L) \equiv \langle [F(L) - \langle F(L) \rangle]^2 \rangle. \tag{10}$$

The angle braces $\langle\,\rangle$ denote an ensemble average, which is an average over an infinite number of identical realizations of the same experiment. Lenschow and Kristensen [19] showed that

$$\tilde{\sigma}_f^2 \leq 4\sigma_w^2\sigma_c^2 \frac{\min(\ell_w \ell_c)}{L}, \tag{11}$$

where σ_c^2 is the variance of c and $\min(\ell_w \ell_c)$ is the smaller of the integral scales of w and c. Normally, in the ABL, $\ell_w \leq \ell_c$, so that

$$\tilde{\sigma}_f^2 \leq 4\sigma_w^2\sigma_c^2 \frac{\ell_w}{L}. \tag{12}$$

From the spectral results of Kaimal *et al.* [12], obtained from tower measurements in the surface layer, the value of ℓ_w/z, where z is the height above the ground, ranges from ~ 0.3 to ~ 1.0 as the hydrodynamic stability ranges from neutral to very unstable. In the mixed layer above the surface layer, Lenschow and Stankov [21] found from aircraft measurements that

$$\ell_w/z_i \simeq 0.24 \left(\frac{z}{z_i}\right)^{1/2}, \tag{13}$$

where z_i is the depth of the mixed layer. These results are plotted in Fig 10. For practical applications, we consider the relative flux measurement error obtained by dividing the square root of (8) by the absolute value of the flux,

$$\frac{\tilde{\sigma}_f}{|\overline{wc}|} \leq \frac{2}{|r_{wc}|} \left(\frac{\ell_w}{L}\right)^{1/2}, \tag{14}$$

where $r_{wc} \equiv \overline{wc}/(\sigma_w \sigma_c)$ is the correlation coefficient between w and c. For neutral to unstable stratification in the surface layer, $0.40 \leq |r_{wc}| \leq 0.58$ [20], where the upper value is the unstable limit and the lower value is at neutral stability. In the mixed layer r_{wc} can vary considerably, but typically in the lower part of the mixed layer we would expect it to be $\simeq 0.4$. Therefore, the relative error in the surface layer is

$$\frac{\tilde{\sigma}_f}{|\overline{wc}|} \leq 5 \left(\frac{z}{L}\right)^{1/2}, \tag{15}$$

and in the lower part of the mixed layer,

$$\frac{\tilde{\sigma}_f}{|\overline{wc}|} \leq 2.4 \left(\frac{z}{L}\right)^{1/2} \left(\frac{z_i}{z}\right)^{1/4}. \tag{16}$$

This says that, for example, at a measurement height of 10 m in the surface layer, the required averaging length for a relative error of $< 20\%$ is $L \geq 6.2$ km, or at a wind speed of 5 m s^{-1}, the required averaging time is ≥ 20

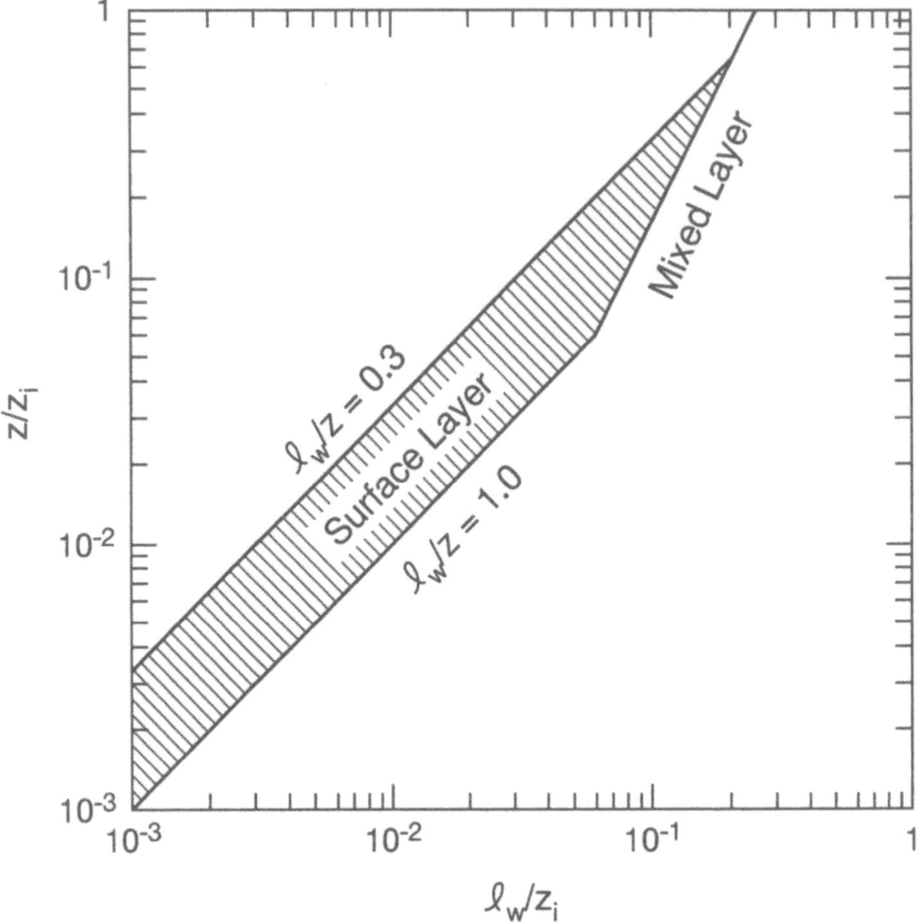

Figure 10. Plot of the normalized integral scale of vertical velocity vs. normalized height through both the surface layer and the mixed layer. In the surface layer, the integral scale is proportional to height, while in the mixed layer it is proportional to the square root of normalized height.

minutes. At a height of 50 m in a 1000 m deep ABL, the mixed-layer measuring length will be $L \geq 32.2$ km. If the airplane is flying at 100 m s^{-1}, the averaging time will be ≥ 5.4 minutes.

As mentioned earlier, these considerations apply if the turbulent field is stationary or horizontally homogeneous. If this is not the case, it may become difficult to define an integral scale; for example, if significant variability exists on scales that are a significant fraction of the record length, the integral scale will be a significant fraction of the length of the time series, and thus cannot be accurately estimated. Generally, the effect of inhomogeneity, even if it occurs on scales a small fraction of the record length (e.g. a patchwork of different crops), is likely to increase the integral scale, and thus longer measurement times would be necessary. This is, however, an area in need of further research.

5. Measurement techniques

A variety of measurement platforms and techniques can be brought to bear on probing the PBL. One reference for PBL measurement technology is Lenschow [17]. A more recent description is available in Kaimal and Finnigan [10], who give a detailed description of direct measurement techniques in the surface layer. For the most part, surface layer measurements are obtained from towers or ships by in situ sensors. Mixed-layer measurements are obtained mostly from aircraft, and from both airborne and ground-based remote sensors such as radars and lidars. Of course, some towers have been used that extend well into the mixed layer (e.g. the 300 m Boulder Atmospheric Observatory [11]), and likewise, some aircraft have been used to probe the upper part of the surface layer (e.g. Dobosy *et al.* [6]). In addition, tethered balloons have also been used to obtain extensive observations of the mixed layer [2].

5.1. DIRECT SENSING

Here we summarize briefly the technology used for making direct observations in the CBL primarily by aircraft. The main advantage of aircraft is their mobility. They can sample typically an order of magnitude faster than fixed-point sensors, and thus obtain more accurate turbulence statistics. Taylor's "frozen turbulence" hypothesis (e.g [34]), for example, is more likely to be satisfied. Furthermore, they can obtain line averages over irregular terrain and carry out extensive survey measurements over remote locations such as ocean, forests, prairies, swamps, tundra or ice caps. On the other hand, their speed leads to some problems. Errors occur which are proportional to the square of the airspeed. This leads to dynamic heating of in situ thermometers and flow distortion effects on sensors induced by flow around the aircraft and the probes themselves ([17], [37], [38]). Finally, aircraft cannot make continuous measurements over long time periods.

Although the mobility of aircraft is a great advantage in making detailed observations throughout the CBL, it also presents a problem—particularly for air motion measurements. The position, velocity, and angular orientation (PVA) of the aircraft need to be continuously measured in an earth-based coordinate system in order to measure air motions. The air velocity is obtained from the vector sum of the aircraft velocity and the velocity of the air relative to the aircraft. These measurements need to be carried out over the frequency range required of the air motion measurements. Inertial navigation systems (INS) have been commonly used to measure the PVA of the aircraft [17], but they have limitations. Since the PV measurements are obtained from integrating accelerometer outputs, they have an inherent drift (typically ~ 0.5 m s^{-1}). This can now be reduced to a few centimeters per second by using measurements from the satellite-based Global Positioning System (GPS).

Air velocity measurements relative to the aircraft are commonly obtained from pressure differences between ports located on the aircraft nose, or on a boom projecting forward of the nose ([1], [17]). The longitudinal component is obtained from the difference between the pressure at a forward-looking port that measures the total dynamic pressure, and a side-facing port that measures the static pressure (which may be located well back on the fuselage), while the transverse components are obtained from ports located at angles of typically 30° to 45° on each side of the longitudinal axis of the aircraft. Distortion of the flow around the aircraft means that probes of this type can normally be calibrated only by flight tests—carrying out aircraft maneuvers that modulate the terms in the air velocity equation, and adjusting the calibrations to remove modulation of the air velocity components in an earth-based coordinate system. This limits the absolute accuracy of air motion measurements to ~ 0.5 m s^{-1}, although velocity fluctuations may be measureable to perhaps an order of magnitude better than this. Recently, Doppler laser systems have been developed that offer the possibility of more accurate absolute accuracy. One implementation of this system uses a conically-scanning laser beam focused at ~ 5 to 10 m, which can be used to resolve the three velocity components far enough away from the aircraft that flow distortion has negligible effect on the measurement. This system has been tested on an NCAR aircraft, but results are not yet available.

One application of more accurate mean air motions is the measure of divergence and vorticity in the CBL. Divergence measurements at several levels through the CBL can then be integrated from the surface to z_i to obtain the mean vertical motion at the CBL top, W_{zi}. If dz_i/dt is also measured, the entrainment velocity can be obtained from

$$w_e = \frac{dz_i}{dt} - W_{zi}. \tag{17}$$

Estimates of dz_i/dt can be obtained from a vertically-pointed lidar measuring backscatter (Typically the CBL contains more particulates in the size range sensed by lidar than the free atmosphere, so the CBL top manifests itself by a large gradient in aerosol backscatter.) or from a series of soundings. The strategy of this measurement technique is described by Lenschow [18]. Some early results of applying this technique are presented by Russell *et al.* [30]. They also present comparisons with estimates of the entrainment velocity obtained by two other techniques. The first is using the relation

$$w_e = -\frac{(\overline{ws})_{zi}}{\Delta S}, \tag{18}$$

where $(\overline{ws})_{zi}$ is the flux of a scalar S at z_i and ΔS is the jump in S across the top of the CBL. The second is evaluating all the terms in the budget of a trace species, except for the entrainment flux, over the course of about 1.5 days during an aircraft Lagrangian experiment. The entrainment rate can then be obtained as the unknown quantity to balance the species budget in the PBL.

Entrainment is an important quantity to measure in both clear and cloudy PBLs since it determines the subsequent evolution of the PBL, and different numerical models give conflicting results [24]. However, it is difficult to measure because it typically is < 1 cm s^{-1}, compared to typical turbulent fluctuations of ~ 1 m s^{-1}.

Temperature and humidity measurements are also of fundamental importance for characterizing the CBL since their fluxes determine the surface virtual temperature flux $(\overline{wT_v})_0$ (used to calculate the Deardorff velocity (1)) by the relation

$$(\overline{wT_v})_0 = (\overline{wT})_0 + 0.61\overline{T}(\overline{wq})_0, \tag{19}$$

where $\overline{wq}$ is the flux of water vapor mixing ratio. Temperature is usually measured with a resistance element of small thermal mass, e.g. fine platinum wire or a thermistor. On an aircraft, dynamic heating of the element needs to be corrected, since full adiabatic heating is about 5 C at 100 m s^{-1} and increases with the square of the airspeed. Mean humidity is usually measured with a chilled-mirror dew-point hygrometer, which is accurate, but not fast enough for eddy flux measurement. Humidity fluctuations are usually obtained from radiation absorption measurements, either in the ultraviolet (Lyman-alpha hygrometer) or the infrared [6]. Other scalars are also of interest as tracers of processes in the CBL. For example, ozone and aerosol have been used as tracers of entrainment through the top of the CBL [13, 30].

5.2. REMOTE SENSING

Remote sensing is becoming increasingly important as a tool for both visualizing CBL structures, as well as obtaining quantitative measurements of turbulence statistics. In addition to measuring statistics also readily obtained by direct measurements, remote sensing offers the opportunity to obtain nearly simultaneous measurements throughout the CBL and thus

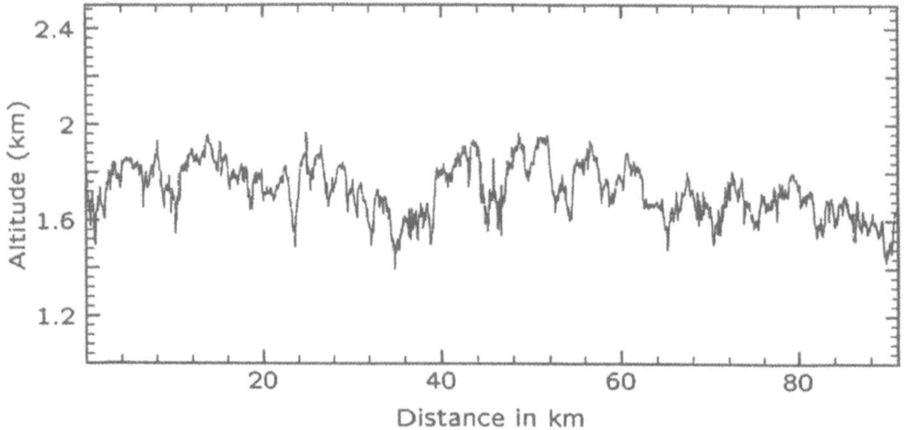

Figure 11. Plot of CBL height measured with a downward-pointing airborne lidar. The height is determined by the maximum gradient in aerosol backscatter. The lidar and observational study (BOREAS-1994) are described by Kiemle *et al.* [14].

study two-point turbulence statistics and coherent structures. Both radars and lidars are now being used to probe the CBL. Radars have the advantage of penetrating clouds, and thus can measure their dynamical and microphysical structure. On the other hand, lidars can generally resolve smaller structures and can measure closer to the surface.

Dr. Kiemle discussed differential absorption lidar measurements of the CBL, so I will not further discuss lidar here, except to show (Fig. 11) an example of a measurement of z_i from a downward looking airborne lidar. This is an example of a simple, but important observation that is difficult to obtain with good accuracy and resolution from direct measurements consisting of repeated penetrations through the CBL top. The resolution is sufficient to make meaningful comparisons with large-eddy simulations in the entrainment zone of the CBL [33]. Doppler lidars are now also available that can provide radial velocity measurements throughout the CBL and above, again using aerosol as the backscatter target.

Another powerful remote sensing technique applied to the CBL is Doppler radar. Fig. 12 shows an example of time series of vertical velocity as a function of height from a ground-based vertically-pointed 8.66 mm wavelength radar [15] which has a vertical resolution of 37.5 m. This cross-section was obtained over the Canadian boreal forest (BOREAS-1994) during daytime using mainly insects in the CBL as backscatter targets. This example also shows that the radar, in contrast to lidar, can delineate structure within the fair-weather cumulus clouds protruding through the top of the CBL. In this figure, we can see motions within the fair-weather cumulus clouds

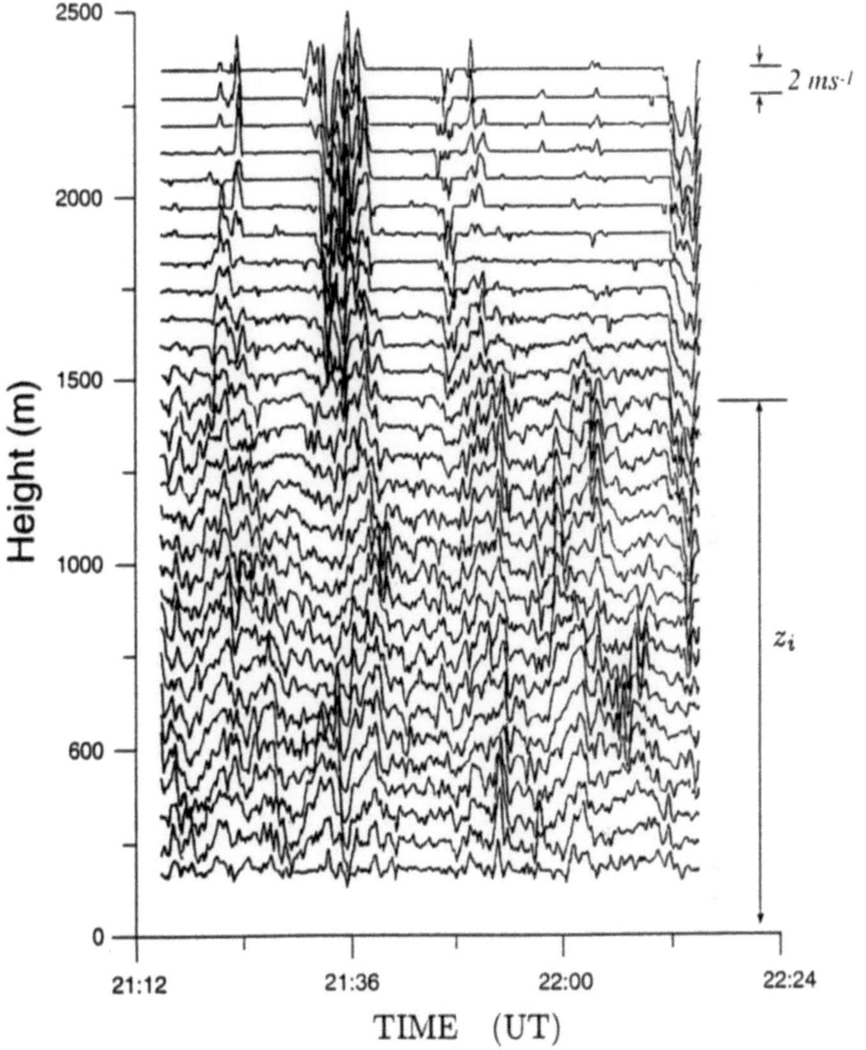

Figure 12. Doppler radar measurements of vertical velocity in the CBL. Regions above 2000 km showing large velocity fluctuations are fair-weather cumulus. Flat areas are regions with insufficient radar backscatter. Location of measurements near Thompson, Manitoba, Canada; local time = UT − 6 hrs.

protruding through the CBL top, and the dynamical effects of the clouds on the underlying CBL.

Finally, Fig. 13 shows a range-height velocity cross-section obtained by scanning the above radar through a convective structure embedded in a marine stratocumulus regime during ASTEX [16]. Inflow is evident at levels below about 1 km, and outflow from about 1.2 km through the top of the stratocumulus at about 1.8 km, with a strong updraft in the middle

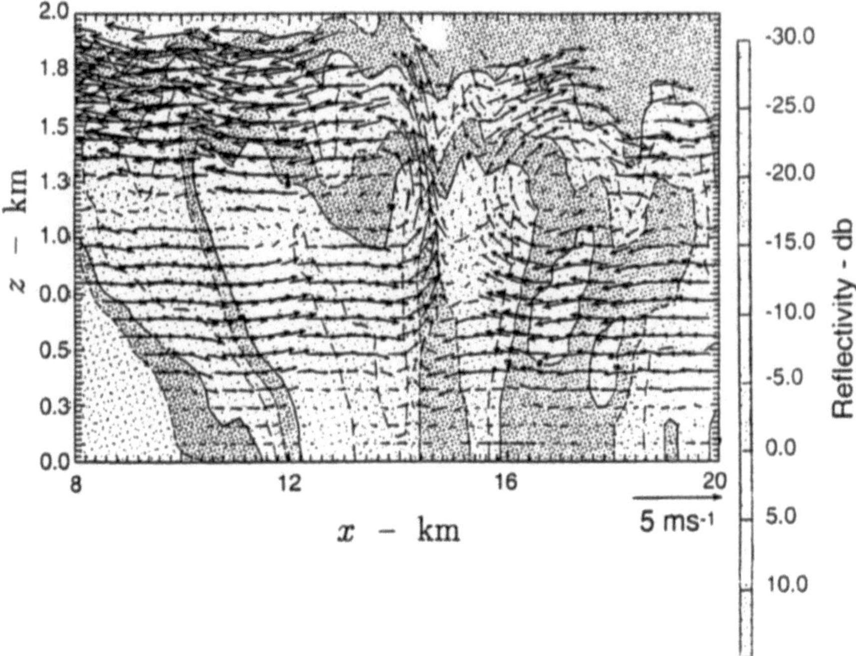

Figure 13. A range-height cross-section showing backscatter and velocity vectors of a "microcell" embedded in a marine stratocumulus regime.

of the structure extending from about 0.5 km height through the top. This example shows the ability of a radar to resolve the entire cross-section of a convective structure in the CBL and thus showing how these different regions relate to the overall structure.

References

1. Brown, E.N., Friehe, C.A., and Lenschow, D.H. (1983) The use of pressure fluctuations on the nose of an aircraft for measuring air motion, *J. Climate Appl. Meteor.* **22**, 171–180.
2. Caughey, S.J. (1982) Observed characteristics of the atmospheric boundary layer, in F.T.M. Nieuwstadt and H. van Dop (eds.), *Atmospheric Turbulence and Air Pollution Modeling*, D. Reidel, pp. 107–158.
3. Cotton, W.R., Alexander, G.D., Hertenstein, R., Walko, R.L., McAnelly, R.L., and Nicholls, M. (1995) Cloud venting – A review and some new global annual extimates, *Earth-Science Reviews* **39**, 169–206.
4. Davis, K.J. (1992) *Surface Fluxes of Trace Gases Derived from Convective-Layer Profiles*, Ph.D. dissertation, University of Colorado. Available as NCAR CT-139, NCAR, Boulder, CO.
5. Davis, K.J., Lenschow, D.H., and Zimmerman, P.R. (1994) Biogenic non-methane hydrocarbon emissions estimated from tethered balloon observations, *J. Geophys. Res* **99**, 25587–25598.
6. Dobosy, R.J., Crawford, T.L., MacPherson, J.I., Desjardins. R.L., Kelly, R.D., Oncley, S.P., and Lenschow, D.H. (1997) Intercomparison among the four flux aircraft

at BOREAS in 1994, *J. Geophys. Res.*, in press.
7. Etling, D. and Brown, R.A. (1993) Roll vortices in the planetary boundary layer: a review, *Boundary-Layer Meteor.* **65**, 215–248.
8. Giez, A., Ehret, G., Schwiesow, R.L., Davis, K.J., and Lenschow, D.H. (1997) Water vapor flux measurements from ground-based vertically-pointed water vapor differential absorption and Doppler lidars, submitted to *J. Atmos. and Ocean. Tech.*
9. Jensen, J.B. and Lenschow, D.H. (1989) Inferences of stratocumulus structure from airborne lidar observations. Preprint Volume, *Symposium on the Role of Clouds in Atmospheric Chemistry and Global Climate*, 30 Jan–3 Feb 1989, American Meteorological Society, Anaheim, CA., 118–123.
10. Kaimal, J.C. and Finnigan, J.J. (1992) *Atmospheric Boundary Layer Flows – Their Structure and Measurement*, Oxford University Press, New York.
11. Kaimal, J.C. and Gaynor, J.E. (1983) The Boulder Atmospheric Observatory, *J. Clim. Appl. Meteor.* **22**, 863–880.
12. Kaimal, J.C., Wyngaard, J¿C., Izumi, Y., and Coté, O.R. (1972) Spectral characteristics of surface-layer turbulence, *Quart. J. Roy. Meteor. Soc.* **98**, 563–589.
13. Kawa, S.R. and Pearson, R. Jr. (1989) An observational study of stratocumulus entrainment and thermodynamics, *J. Atmos. Sci.* **46**, 2649–2661.
14. Kiemle, C., Ehret, G., Giez, A., Davis, K.J., Lenschow, D.H., and Oncley, S.P. (1997) Estimation of boundary-layer humidity fluxes and statistics from airborne DIAL, *J. Geophys. Res.*, in press.
15. Kropfli, R.A., Bartram, B.W., and Matrosov, S.Y. (1990) The upgraded WPL dual-polarization 8-mm wavelength Doppler radar for microphysical and climate research, Preprints, *1990 Conf. on Cloud Physics*, San Francisco, CA, Amer. Meteor. Soc., 341–345.
16. Kropfli, R.A., and Orr, B.W. (1993) Observations of microcells in the marine boundary layer with 8-mm wavelength Doppler radar, Preprints, *26th Int. Conf. on Radar Meteorology*, Norman OK, Amer. Meteor. Soc., 492–494.
17. Lenschow, D.H., (ed.), (1986) *Probing the Atmospheric Boundary Layer*, American Meteorological Society, Boston, MA.
18. Lenschow, D.H. (1996) A proposal for measuring entrainment into the cloud-capped boundary layer, in A.S. Frisch, D.A. Randall and W.H. Schubert (eds.), *Proceedings of the ETL/CSU Cloud-Related Process Modeling and Measurement Workshop*, Boulder, Colorado, 23–25 October 1995, NOAA ETL Report, pp. 29–55.
19. Lenschow, D. H. and Kristensen, L. (1985) Uncorrelated noise in turbulence measurements, *J. Atmos. and Ocean Technology* **2**, 68–81.
20. Lenschow, D.H. and Kristensen, L. (1986) Comments on "Sampling errors in flux measurements of slowly depositing pollutants, *J. Climate Appl. Meteor.* **25**, 1785–1787.
21. Lenschow, D. H. and Stankov, B.B. (1986) Length scales in the convective atmospheric boundary layer, *J. Atmos. Sci.* **43**, 1198–1209.
22. Lenschow, D. H., and Stephens, P.L. (1980) The role of thermals in the convective boundary layer, *Boundary-Layer Meteor.* **19**, 509–532.
23. Moeng, C.-H. (1998) Stratocumulus-topped atmospheric planetary boundary layer. See this volume.
24. Moeng, C.-H., *et al.* (1996) Simulation of a stratocumulus-topped planetary boundary layer: intercomparison among different numerical codes, *Bull. Amer. Meteor. Soc.* **77**, 261–278.
25. Moeng, C.-H. and Wyngaard, J.C. (1989) Evaluation of turbulent and dissipation closures in second-order modeling, *J. Atmos. Sci.* **46**, 2311-2330.
26. Nicholls, S. (1989) The structure of radiatively driven convection in stratocumulus, *Q. J. Roy. Meteorol. Soc.* **115**, 487–511.
27. Nicholls, S., and Leighton, J. (1986) An observational study of the structure of stratiform cloud sheets: Part I. Structure, *Quart. J. Roy. Meteor. Soc.* **112**, 431–460.

28. Paluch, I.R. and Lenschow, D.H. (1991) Stratiform cloud formation in the marine boundary layer, *J. Atmos. Soc.* **48**, 2141–2158.
29. Randall, D. A., Coakley, J.A., Jr., Fairall, C.W., Kropfli, R.A., and Lenschow, D.H. (1984) Outlook for research on subtropical marine stratus and stratocumulus clouds, *Bull. Amer. Meteor. Soc.* **65**, 1290–1301.
30. Russell, L.M., Lenschow, D.H., Laursen, K.K., Krummel, P.B., Siems, S.T., Bandy, A.R., Thornton, D.C., and Bates, T.C. (1997) Bidirectional mixing in an ACE-1 marine PBL overlain by a second turbulent layer, *J. Geophys. Res.* in press.
31. Schmidt, H. and Schumann, U. (1989) Coherent structure of the convective boundary layer from large-eddy simulations, *J. Fluid Mech.* **200**, 511–562.
32. Stull, R.B. (1988) *An Introduction to Boundary Layer Meteorology*, Kluwer Academic Publishers, Dordrecht.
33. Sullivan, P.P., Moeng, C.-H., Stevens, B., Lenschow, D.H., and Mayor, S.D. (1997) Entrainment and structure of the inversion layer capping the convective planetary boundary layer. Submitted to *J. Atmos. Sci.*
34. Tennekes, H. and Lumley, J.L. (1972) *A First Course in Turbulence*, The MIT Press, Cambridge, MA.
35. Wang, Q. and Lenschow, D.H. (1995) An observational study of the role of penetrating cumulus in a marine stratocumulus-topped boundary layer, *J. Atmos. Sci.* **52**, 2778–2787.
36. Woodcock, A.H. (1942) Soaring over the open sea, *Scientific Monthly* **55**, 226–232.
37. Wyngaard, J.C. (1981) The effects of probe-induced flow distortion on atmospheric turbulence measurements, *J. Appl. Meteor.* **20**, 784–794.
38. Wyngaard, J.C. (1988) The effects of probe-induced flow distortion on atmospheric turbulence measurements: extension to scalars *J. Atmos. Sci.*, **45**, 3400–3412.
39. Wyngaard, J.C. and Brost, R.A. (1984) Top-down and bottom-up diffusion of a scalar in the convective boundary layer, *J. Atmos. Sci.*, **41**, 102–112.
40. Zilitinkevich, S.S., Grachev, A., Hunt, J.C.R. (1998) Surface frictional processes and non-local heat/mass transfer in the shear-free convective boundary layer. See this volume.

AIRBORNE WATER VAPOR DIFFERENTIAL ABSORPTION LIDAR STUDIES OF THE CONVECTIVE BOUNDARY LAYER

C. KIEMLE AND G. EHRET

Institut für Physik der Atmosphäre,
Deutsches Zentrum für Luft- und Raumfahrt (DLR),
D-82234 Oberpfaffenhofen, Germany

K.J. DAVIS

Department of Soil, Water, and Climate,
University of Minnesota, St. Paul, MN 55108, USA

AND

D.H. LENSCHOW AND S.P. ONCLEY

National Center for Atmospheric Research (NCAR),
Boulder, CO 80300, USA

Abstract.

We present measurements of aerosol and water vapor using a Differential Absorption Lidar (DIAL) mounted downward pointing on board the NCAR Electra aircraft during the Boreal Ecosystem–Atmosphere Study (BOREAS). Lidar legs were flown in 1.5 – 3.5 km above the boreal forest of Saskatchewan and Manitoba, Canada. DIAL cross sections of aerosol backscatter and water vapor are used to study the size and structure of thermals in the convective boundary layer (CBL), and of entrainment processes at the CBL top. Applying spectral and autocovariance analyses across horizontal DIAL water vapor series gives insight into the turbulent structure of the atmosphere. Vertical fluxes of humidity at the top of the boundary layer can be estimated from DIAL variance water vapor profiles using a set of empirically derived equations. Such measurements are of high climatological interest, since they enable to evaluate and monitor evaporation and biosphere–atmosphere exchange processes over land surfaces.

E.J. Plate et al. (eds.), Buoyant Convection in Geophysical Flows, 207–238.

1. Introduction

Water vapor in the boundary layer plays a critical role in atmospheric dynamics, the hydrological cycle, and the earth radiation budget. Relevant processes include surface evapotranspiration, transport and diffusion through the boundary layer, and cloud formation and dissipation. Measurements of water vapor and water vapor fluxes are of fundamental importance to the understanding of all these processes. Since water vapor is a passive, conserved scalar in the cloud–free convective boundary layer (CBL), measurements of its mean value and higher order moments such as variance and skewness can also be used to study turbulent processes.

Past studies of boundary–layer turbulence have used in-situ data collected by aircraft ([10], [13]), or large eddy simulations (LES, e.g. [14], [16], [18]). Aircraft in-situ measurements of the statistical properties of the water vapor field in the CBL are typically based on high–frequency humidity sensors. Vertical fluxes are measured by combining vertical wind measurements with fast humidity measurements and using the eddy correlation technique. This method provides a direct measurement of the vertical flux. Obtaining surface and entrainment fluxes from an aircraft requires flying at different altitudes within the CBL, and linearly extrapolating the observed fluxes to the surface roughness height z_0, and the CBL top, z_i. One disadvantage of such in-situ measurements is the time required to fly multiple legs at different altitudes, during which time the CBL is assumed to be quasi–steady. Furthermore, high–frequency response sensors have to be used, and measurement of the vertical wind speed is sensitive to misalignment of the sensor, inaccurate removal of aircraft motions from the air velocity measurement, and flow distortion [9]. LES allow sampling of an entire three-dimensional field of simulated turbulence. LES, however, are limited in their ability to simulate mesoscale variability because of their domain size, and the details of entrainment because of limited resolution, though nesting techniques may alleviate the resolution problem.

An airborne water vapor Differential Absorption Lidar (DIAL) offers the advantages of reduced measurement time and increased sampling compared to in-situ measurements, since measurements can be made throughout the boundary layer with only one flight leg. Thus it can, for example, capture mesoscale variability at a resolution similar to LES. In addition, lidars provide other useful information such as CBL height, and an insight into entrainment zone structure. A major disadvantage of currently available remote sensing technology is the relatively high statistical error due to instrumental noise, and poor temporal and spatial resolution compared to in-situ sensors. Furthermore, remote sensing techniques capable of measuring fluxes, such as Doppler lidar for measuring vertical wind, and

DIAL for detecting trace gases, are rather new and complex. Thus far, few lidar–based flux measurements have been reported, and they have all been ground–based.

Water vapor fluxes in the surface layer were obtained with a Raman lidar using Monin–Obukhov similarity theory and dissipation techniques [5]. A water vapor DIAL and a radar–RASS (Radioacoustic Sounding System) were combined to measure water vapor flux profiles in a CBL using eddy-correlation [20]. CBL fluxes of water vapor have been measured with zenith–looking combined DIAL and Doppler lidar, also by means of eddy correlation [6]. Both latter experiments used active remote sensing systems for both scalar (water vapor) and vertical wind measurements. Significant technical difficulties, including maintaining proper inter-sensor alignment on a vibrating platform, would need to be overcome to repeat experiments such as these on an aircraft.

Alternative techniques, requiring only one active airborne remote sensing system, a DIAL, were discussed in [1]. By measuring with a DIAL the mean vertical mixing ratio profile of a conservative scalar (e.g. water vapor or ozone), and/or its variance profile in the CBL, one can use flux–gradient and flux–variance relationships derived from LES by [15] to solve for the scalar surface and entrainment flux. As these flux–gradient and flux–variance relationships apply to the mid–CBL, we will refer to them as the mixed–layer gradient (MLG) and mixed–layer variance (MLV) techniques.

During the Boreal Ecosystem–Atmosphere Study (BOREAS), the DIAL of the DLR (Deutsches Zentrum für Luft- und Raumfahrt) was mounted in the National Center for Atmospheric Research (NCAR) Electra aircraft and operated on all research flights. We discuss the ability of the lidar to resolve detailed mean and turbulence water vapor structure throughout the upper part of the mixed layer and especially in the entrainment zone (EZ) at the top of the CBL. In a case study, we compare the DIAL observations to in-situ observations and LES–based predictions of mixed–layer scaling formulations, and estimate water vapor entrainment using DIAL data with the MLG and MLV techniques. We also present an overview of the various types of boundary–layers observed by the DLR DIAL during BOREAS. The influences of clouds, smoke, winds and lakes on CBL development are illustrated. This work, a major part of which has already been published [8], represents a significant advance in our ability to retrieve micrometeorological data from airborne DIAL.

2. The DLR Airborne Water Vapor DIAL

LIDAR (Light Detection And Ranging) is a laser–based active remote sensing technique capable of measuring the range–resolved distribution of aerosols and optically thin clouds. Typically, a lidar instrument emits short laser pulses into the atmosphere and detects the backscattered radiation. From the time elapsed between the emission of the laser pulse and detection of the backscattered signal, the range between laser transmitter and backscatter medium can be determined with high precision. The CBL is characterized by high aerosol and moisture loading. The aerosol and moisture content is generally much lower in the free troposphere, and a sharp change in the vertical aerosol and humidity profile occurs at the CBL top. As reported in former studies the CBL top and its variability can be accurately measured with high spatial and temporal resolution by means of a downward–looking airborne aerosol lidar system, e.g. [12].

The Differential Absorption Lidar (DIAL) technique is appropriate for remotely sensing both aerosol and water vapor distributions in the CBL. A near infrared DIAL instrument emits two narrow–band pulses into the atmosphere at distinct but neighboring wavelengths usually denoted by on- and off-line. While the atmospheric transmission of the on-line wavelength is influenced by water vapor absorption, the lidar measurement of the non–absorbing off-line wavelength is not. The comparison of the backscatter at these wavelengths yields the water vapor concentration profile as a function of distance from the lidar [19]. From the off-line data aerosol backscatter profiles can be determined.

A near infrared airborne water vapor DIAL instrument has been developed by the Deutsches Zentrum für Luft- und Raumfahrt (DLR) at Oberpfaffenhofen, Germany [3]. This system uses a narrow-band tunable dye laser for sequentially generating the on-line and off-line wavelengths at 724 nm with a frequency of 4.5 Hz. The dye laser is pumped by a frequency doubled Neodym Yttrium Aluminium Granat (Nd:YAG) laser which operates in Q–switch mode providing short pulses with a repetition rate of 9 Hz. The wavelength calibration is performed by use of the water vapor absorption spectrum from a photo-acoustic cell. The system's receiver consists of a 35 cm diameter Cassegrainian–type telescope which collects the photons backscattered from the atmosphere. They are detected with a photo-multiplier. The amplified signal is digitized with a resolution of 12 bits at a sampling rate of 20 MHz for each laser shot. A real-time micro computer controls the data acquisition and stores these data on a removable disk for further processing on ground.

The DLR airborne DIAL system has successfully participated in two previous field campaigns: the European Field Experiment in a Desertification Threatened Area (EFEDA'91) [7] and the Cloud Experiment Oberpfaffenhofen and Transports (CLEOPATRA'92) [11]. The papers focus on the general ability of the DIAL to detect water vapor structures in the entrainment zone and on comparisons with other sensors. During these previous field campaigns the system was installed on the DLR Falcon 20 research aircraft. In order to prepare the DIAL instrument for the BOREAS'94 field experiment it had to be modified to fit into the NCAR Electra research aircraft, a four–engine turboprop aircraft with considerable vibrational noise. Thus, a pre-BOREAS test campaign took place in the Imperial Valley of California in December 1993 to test the performance of the modified system on the new platform [4].

3. Observations

3.1. BOREAS CBL REGIMES

Lidar backscatter images provide vivid two-dimensional illustrations of the character of the planetary boundary layer. Here we present examples of the various regimes encountered during BOREAS.

Figure 1 from 4 August 94 shows the classical CBL–deep, horizontally homogeneous clear–air convection dominated by irregularly–spaced thermals with diffuse areas suggesting engulfment/entrainment between thermals. We will use this day as a case study and examine the statistics of the corresponding water vapor field in detail. Such CBLs were fairly common during BOREAS.

By contrast, Fig.2 from 7 September 94 shows a classic case of regularly–spaced horizontal roll vortices. Moderate winds blowing from the SSE into the plane of the figure led to the moderate instability needed for roll convection. Distinct roll structures were not commonly encountered during BOREAS. We also note that the periodic structure seen in z_i is reflected in a layer of increased backscatter on the right half of the figure about 100 m above z_i, which may be due to wave motions induced in the overlying stably stratified region by the undulations in z_i.

Figure 3 captures the rapid change in the morning CBL. The variability in boundary–layer height, which is likely a combination of spatial and temporal change, was observed on a transect across the Southern Study Area. This is an example of how rapidly the morning CBL can encroach into the nearly neutrally stratified layer which is a residual from the previous day's CBL. We see growth from about 1.0 km to about 2.5 km height over a time of about 15 minutes and a distance of about 100 km.

Convective clouds formed at the CBL top fairly frequently during the BOREAS experiment. The fluctuating sunlight caused by these clouds has been shown to influence the photo-synthetic response of jack pines growing in the boreal forest. These clouds effect significant exchange of air between the CBL and the lower troposphere, especially when they extend more than a few hundred meters above the CBL into the free troposphere. Figure 4 shows such a case where CBL clouds, with tops between 1.5 km (the height

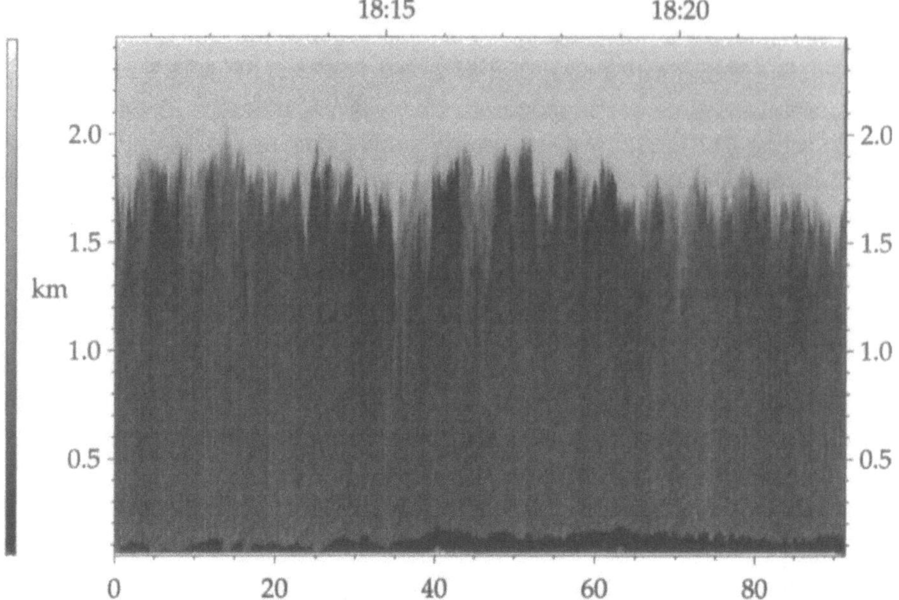

Figure 1. Aerosol backscatter from the lidar looking downward from 2450 m a.s.l., on August 4, 1994. Darker shades are stronger backscatter and the grey scale is linear. Clearly visible is the irregular top of the aerosol loaded CBL. The black curve in the bottom is the ground return. The ordinate values are the altitude a.s.l. in km. The lower abscissa indicates the distance from the leg start point in km, the upper abscissa gives the time in UTC. The local solar time is approximately 6.7 hours behind UTC, so the flight leg is centered around local 11:30 am.

of the CBL) and 3.3 km, can generate mixing of CBL and free tropospheric air. The aerosol–rich CBL air transported up into this convective cloud layer is evident in the figure.

Finally, smoke from forest fires frequently advected over the region, causing persistent haze layers which have a significant impact on the forest radiation budget. This smoky haze and its interaction with the CBL was clearly shown by the lidar. Figures 5 and 6 give examples of this haze advected into the region above the boundary layer and contained within the boundary layer. Figure 5 shows the early morning boundary layer, before convection has been established. A residual well–mixed CBL extends from the ground to about 0.8 km above sea level. From about 0.8 to 3.0 km, several smooth haze layers of varying thickness are evident. These layers were evident to the naked eye. Later in the day, CBL turbulence significantly perturbs these layers, though not to the point of breaking them up. Figure 6 shows a case where the haze layers reflect the convolutions of the CBL top, but the separation between the CBL top and the haze layers is very distinct. The only exception is the point where CBL clouds have penetrated

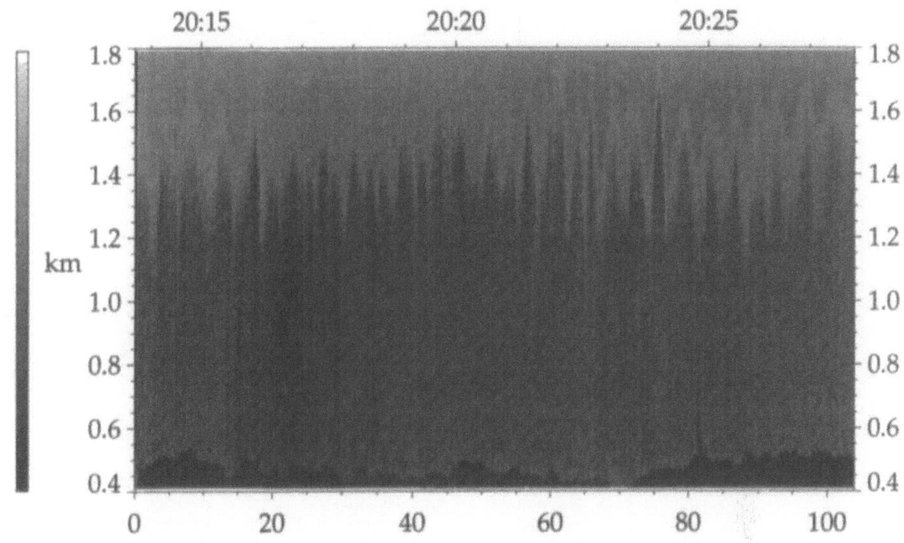

Figure 2. Lidar aerosol backscatter for September, 7, 1994, around local 13:30.

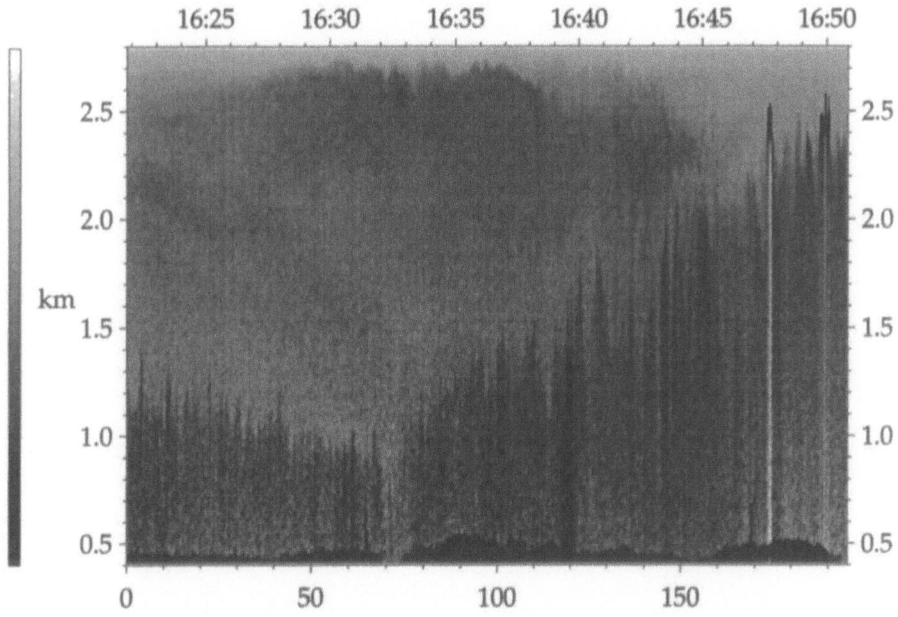

Figure 3. Lidar aerosol backscatter for May, 26, 1994, around local 10:00. In the very right, clouds have been detected, visible through the shadow they produce in the lidar plot. See caption of Fig.1.

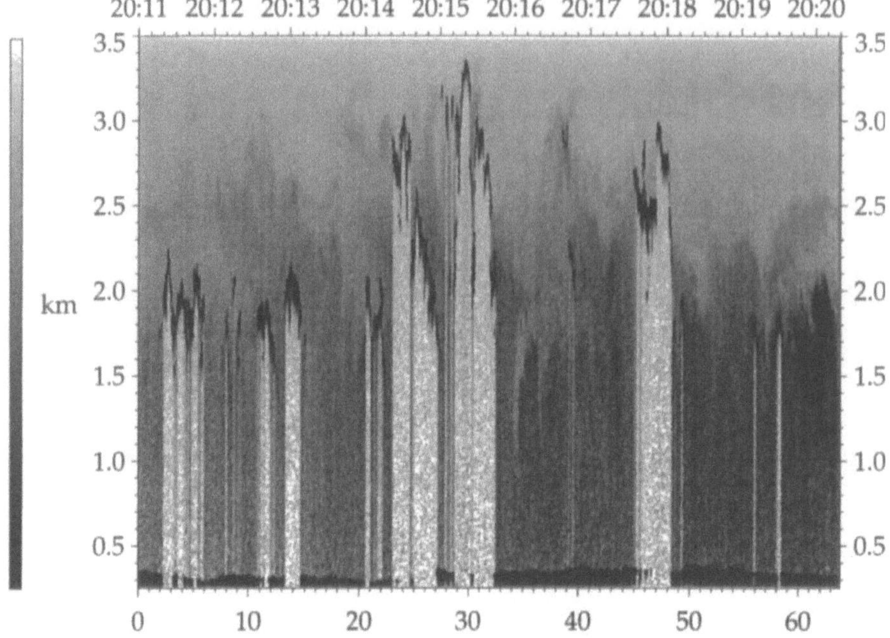

Figure 4. Lidar aerosol backscatter for July, 20, 1994, around local 13:30.

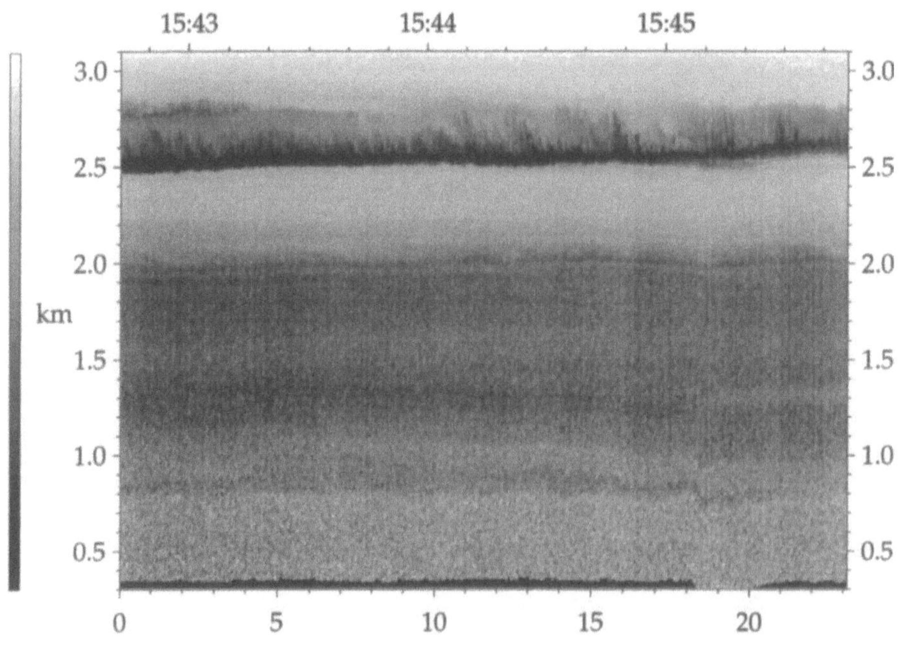

Figure 5. Lidar aerosol backscatter for July, 30, 1994, around local 9:00.

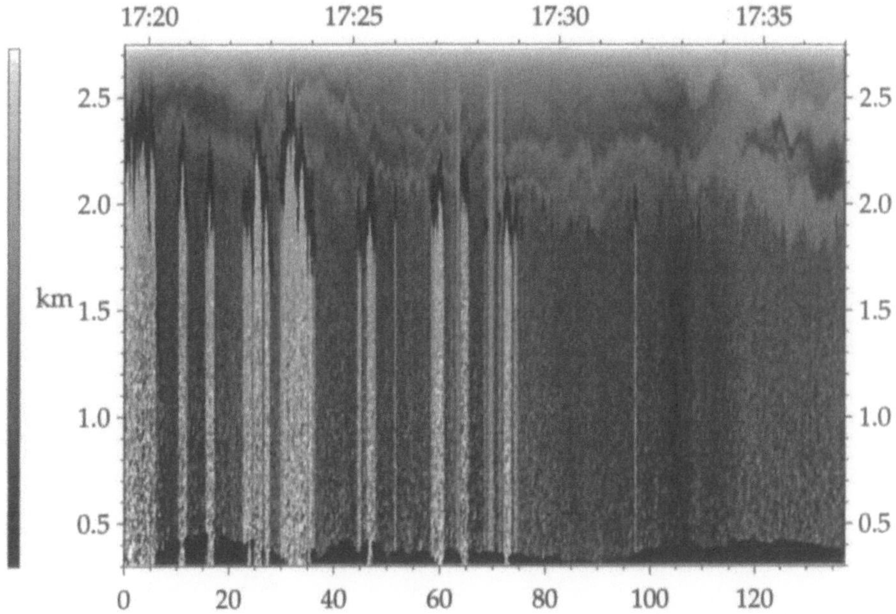

Figure 6. Lidar aerosol backscatter for August, 31, 1994, around local 10:30.

into the overlying haze layer at 17:23:30.

All of these cases save figures 1 and 5 come from the transect across the Southern Study Area. A description of the Electra flight track is found in [17]. It is worth noting that in every one of some 25 lidar runs over Candle Lake spanning late May to mid–September of 1994, none reveal unequivocal evidence of the lake influence in the structure of the CBL. The lake clearly influences flows within the CBL, but it appears that these flows are entirely contained within the CBL and do not modulate the depth of the CBL over the lake or the aerosol backscatter characteristics within the CBL.

3.2. TWO-DIMENSIONAL CBL WATER VAPOR FIELDS

Two-dimensional water vapor fields are shown in figures 7 and 8, corresponding to the cloud- and smoke-free cases of figures 1 and 2. In order to reduce noise in these images, the data have been smoothed vertically over 134 m and horizontally over 2 km. Compared to the aerosol cross sections, the vertical extent has been limited somewhat due to DIAL data retrieval and noise constraints. The remaining relative statistical error due to uncorrelated instrumental noise (cf. equation 2 in the variance profiles section below) is, on average, less than 20 % for the low water vapor mixing ratio values of the free atmosphere above the CBL top, and less than 10 % within the CBL and the EZ. Note that these error estimates represent area

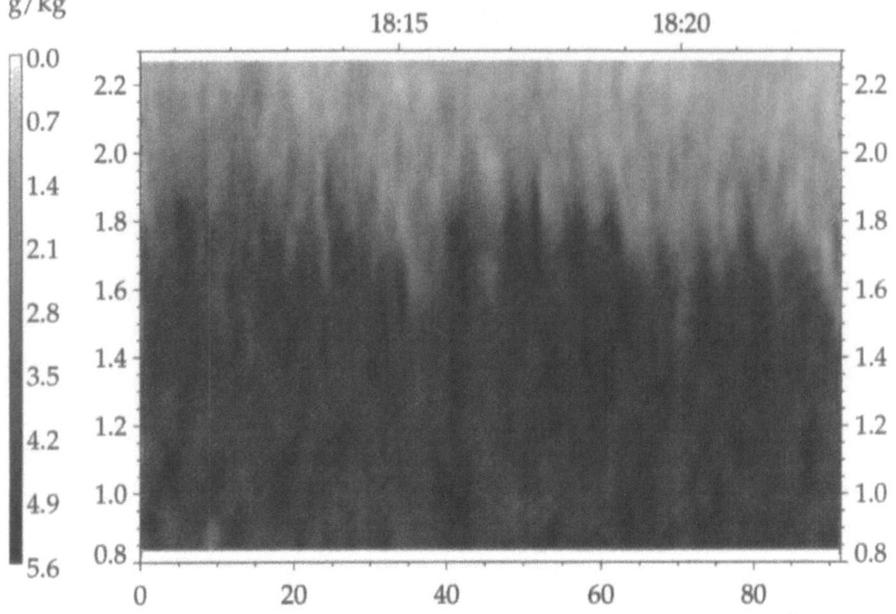

Figure 7. Water vapor mixing ratio on August, 4, 1994, corresponding to Fig.1. Same axes as for the aerosol plot, only, the vertical extent has been limited somewhat due to DIAL data retrieval and noise constraints. The mixing ratio in g/kg is encoded in the color scale. For better visualization, the noisy raw data have been smoothed vertically by 134 m and horizontally by 2 km. The CBL top is well reproduced for features larger than the horizontal smoothing width of 2 km. Rising thermals seen in the aerosol plot correspond to areas of increased humidity in this water vapor cross section.

averages across the whole leg. Locally, the statistical error may be larger or smaller than the average estimate. The smoothing procedure yields a "bubblelike" structure in the plots of varying prominence, the bubbles having the size of the smoothing window. Thus, the plots have to be interpreted cautiously; only features larger than the size of the smoothing window have significance.

In both figures, the EZ is well reproduced for features larger than the horizontal smoothing width of 2 km. Rising thermals seen in the aerosol figures correspond to areas of increased humidity in the water vapor cross sections. An example of such a moist thermal can be found at 18:16 in figures 1 and 7. The thermal shows enhanced moisture through the entire CBL depth shown in the figure. In between these structures, free atmosphere air, drier and less aerosol-laden, is entrained. Entrainment can penetrate very deeply into the CBL, as can be seen in 1 and 7 from features located at 18:15 and 18:20. Here, lower aerosol backscatter and drier regions can be observed down to roughly 900 m above sea level (ASL), corresponding to about 0.5 z_i. These water vapor structures are significant, since their vertical extension

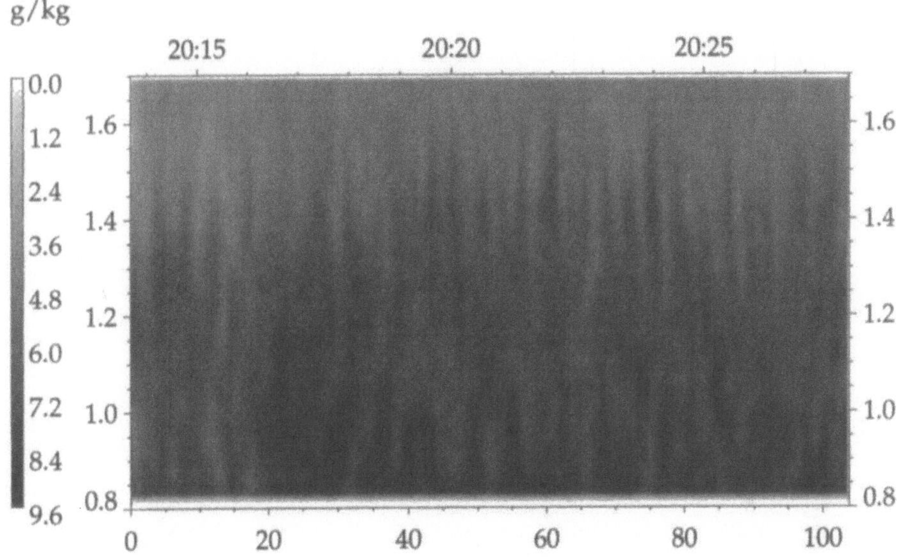

Figure 8. Water vapor mixing ratio on September, 7, 1994, corresponding to Fig.2.

is larger than the vertical smoothing interval. Deep entrainment is not observed between smaller thermal structures, where entrainment and mixing presumably occur at smaller spatial scales, not resolved in the water vapor plots. Such deep entrainment is also not observed in current LES of the CBL (P. Sullivan, personal communication) and may represent mesoscale features which cannot be captured in the limited domain of the LES (typically 5×5×2 km).

3.3. HUMIDITY STATISTICS

While the previous sections presented an overview of the boreal forest CBL as observed with the DIAL, we now focus on deriving humidity statistics from measurements performed on August, 4, 1994 (figures 1 and 7). On that day, data were obtained during a sequence of repeated flight legs at varying altitudes over the Northern Study Area (NSA), a region of forest (primarily jack pine and black spruce), bogs and small lakes in northern Manitoba, Canada. In-situ measurements of water vapor were obtained from two instruments: Accurate mean humidity was obtained from a chilled–mirror dew-point hygrometer. However, its time response is not adequate for measuring turbulence statistics. Turbulent fluctuations of humidity were measured with a Lyman–alpha hygrometer at a rate of 25 s^{-1}. High–frequency wind observations needed for eddy–correlation flux measurements were ob-

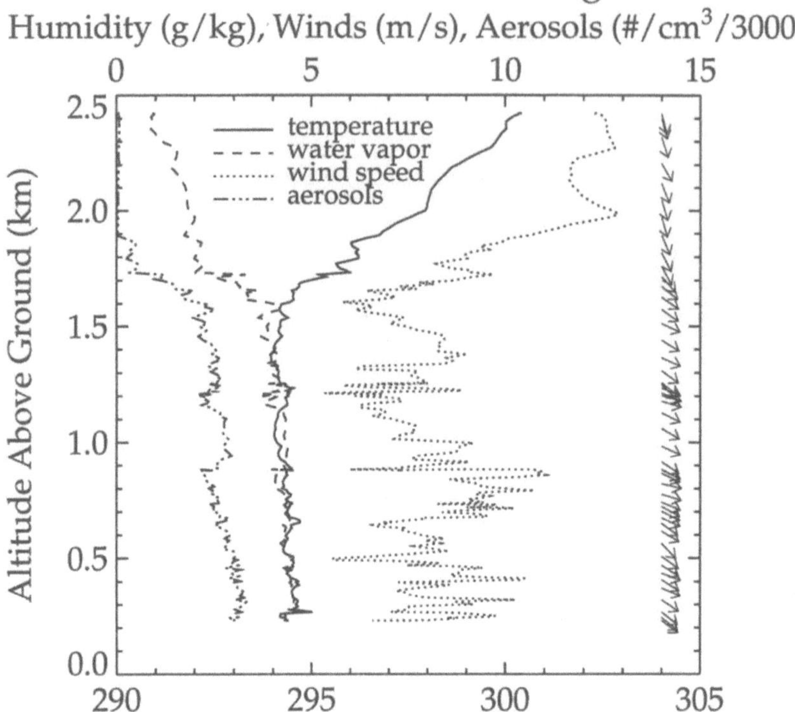

Figure 9. Vertical sounding from the NCAR Electra on 4 August, 1994 over point O in the BOREAS Northern Study Area (see [17] for location of the flight track). Wind arrows show the direction of the wind, where an arrow pointing up describes a wind blowing to the north, and an arrow pointing to the right refers to a wind blowing to the east.

tained with the Electra's differential pressure radome and the airplane motions from an inertial reference unit [9].

Figure 9 shows an in-situ Electra sounding from this day. The CBL winds were around 7 m/s from WNW. Across the inversion the wind speed jumped 5–7 m/s and turned to NNW. This clear, light–shear CBL represents the best combination of DIAL performance and ideal CBL conditions encountered during BOREAS. The following results illustrate what we believe is the state-of-the-art in micro-meteorological analysis of airborne DIAL data.

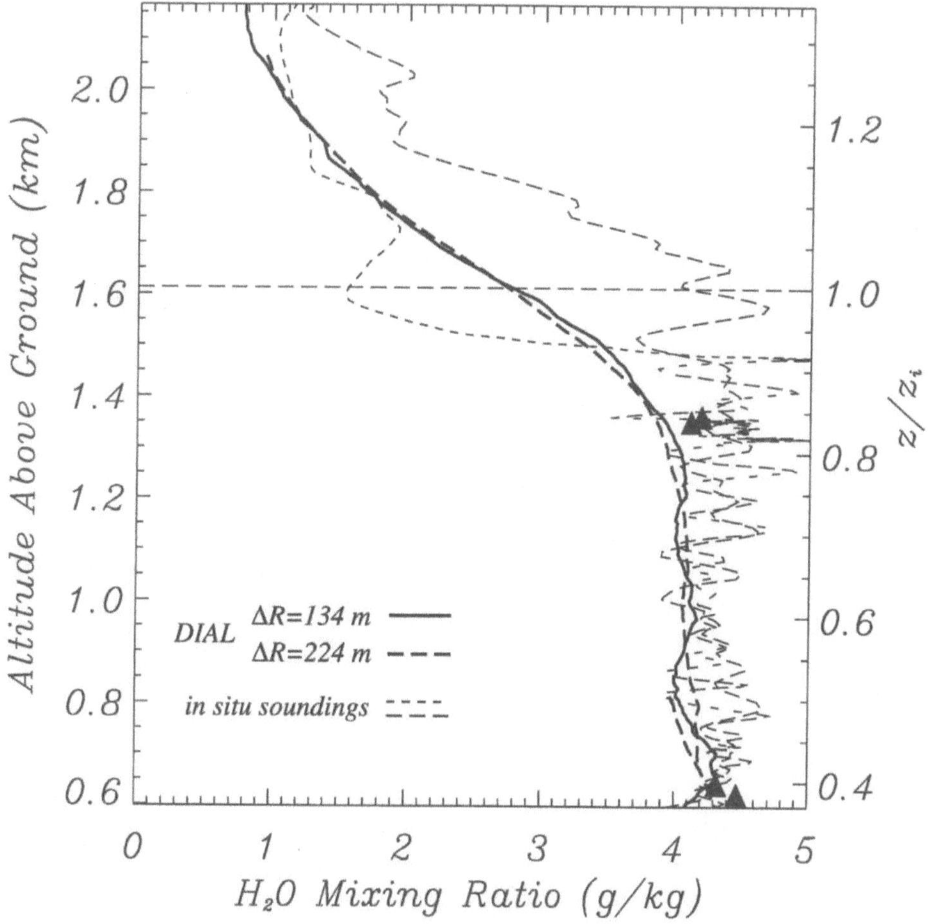

Figure 10. Mean DIAL water vapor mixing ratio profiles (thick lines) for the flight leg of Fig.7, for two different vertical resolutions: ΔR = 134 m (thick solid) and ΔR = 224 m (thick dashed). They are compared to in-situ profiles obtained from aircraft soundings performed just before (short dashed) and after the DIAL measurement (dashed line), and level–leg in-situ mean values (triangles).

3.3.1. *Mean Profiles*

A comparison of mean DIAL water vapor profiles (thick lines) with in-situ profiles obtained from vertical soundings of the Electra aircraft just before and after the flight leg of Fig.7 is shown in Fig.10. The agreement between the DIAL and in-situ profiles within the well-mixed portion of the CBL (below 0.8 z_i) and between the level–leg in-situ mean water vapor values (triangles) and the DIAL is good: below 0.8 z_i, DIAL and in-situ data are found to lie within the range of atmospheric variability (roughly $\pm$0.2 g/kg, which corresponds to $\pm$5 %) shown by the in-situ soundings (thin lines). Since the DIAL profiles represent averages across the whole flight leg, the DIAL entrainment zone profile is much smoother than the in-situ soundings, and shows a gradual CBL to free–troposphere transition. In contrast, the in-situ data is along a slanted line through the undulating CBL top. Figure 1 shows this variability in z_i. We also see good agreement between the beginning and end of the DIAL cross-section, and the in-situ soundings conducted at each end of the cross-section.

To illustrate the effects of DIAL signal processing, Fig.10 shows two mean DIAL profiles from the same flight leg, with different vertical resolutions. The technique of DIAL data retrieval makes it necessary that a so-called vertical range cell ΔR be used for deriving water vapor [3]. The resulting water vapor value represents an average value across that range cell. Thus, ΔR is a measure of the vertical resolution. The solid line is obtained from a resolution ΔR of 134 m, while the dashed line is for ΔR = 224 m. As the figure illustrates, higher ΔR values lead to a smoother mean humidity since small-scale vertical fluctuations are smeared out. In addition, the profile gradient around z_i is decreased. The DIAL system also operated during two legs flown at around 0.8 z_i on this day. The dash-dotted line below $0.6z_i$, which is a mean water vapor profile obtained from one of these flight legs, shows good agreement with the other DIAL profiles and the mean in-situ values at 0.35 z_i.

3.3.2. *Variance Profiles*

Since the horizontally unsmoothed DIAL water vapor data contain significant instrumental noise, it is not possible to compute water vapor variances directly from the standard deviation of a horizontal series taken out of the two-dimensional water vapor field. The true atmospheric variances are mostly overridden by the noise. In the following analysis, we use a 139 m horizontal sample interval consisting of five on- and off-line laser-shots which are averaged before water vapor mixing ratios are computed from the averaged shot-pair, in order to reduce the noise before performing the noise–sensitive and nonlinear DIAL mixing ratio calculation. Since each laser-shot is independent, the noise between shots is uncorrelated. This allows us to use the lagged autocovariance function (ACF) to compute the true atmospheric variance from detrended horizontal DIAL water vapor time series in the presence of random uncorrelated noise. When applying this function to noisy data, both the signal variance and the uncorrelated noise variance are contained in the zero lag value of the ACF. If the sample interval between two adjacent DIAL profiles is much less than the integral scale of the sampled process (water vapor density in the CBL) the true atmospheric variance appears as positive values around the ACF zero lag noise peak. Extrapolation from the small lag ACF values (free from instru-

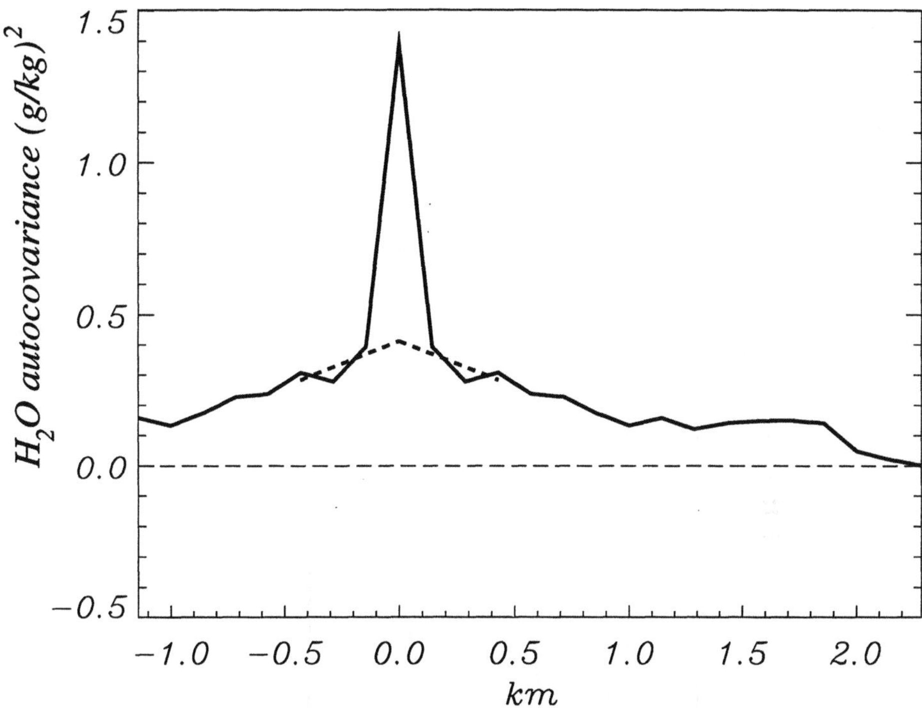

Figure 11. Autocovariance function shape around the zero lag, obtained from a horizontally unsmoothed time series of Fig.7 at an altitude of 1530 m agl., corresponding to 0.95 z_i. The sample interval, hence the spacing between adjacent ACF values, is 139 m. The short dashed lines trace a linear fit through three zero lag neighboring ACF values, and its extrapolation to zero lag, to obtain an estimation of the signal variance (see text).

mental noise) to zero lag provides a good estimate of the real atmospheric variance. As the horizontal integral scale for water vapor fluctuations in the CBL is at least 1000 m [10] and the spacing between DIAL profiles only 139 m, this is certainly the case for the data presented here.

The signal variance retrieval approach described above was realized by extrapolating a linear least squares fit through the three smallest non-zero lagged ACF values on each side of the zero–lag ACF to zero lag. Figure 11 shows an example of the ACF of a horizontal DIAL water vapor time series taken from our selected case (Fig.1) at an altitude of 1530 m above ground (about 0.95 z_i). A vertical resolution of $\Delta R = 134$ m was used to compute the water vapor mixing ratios for both Fig.11 and Fig.12. The short dashed lines trace the linear fit and its extrapolation to zero lag. The noise peak accounts for a variance of 1.4 g^2 kg^{-2}; the estimated signal variance is 0.4 g^2 kg^{-2}. The difference between using this linear extrapolation and simply assuming that the ACF value at the first lag (139 m) equals the signal variance is, for this example, rather small (roughly a 5% error). For other

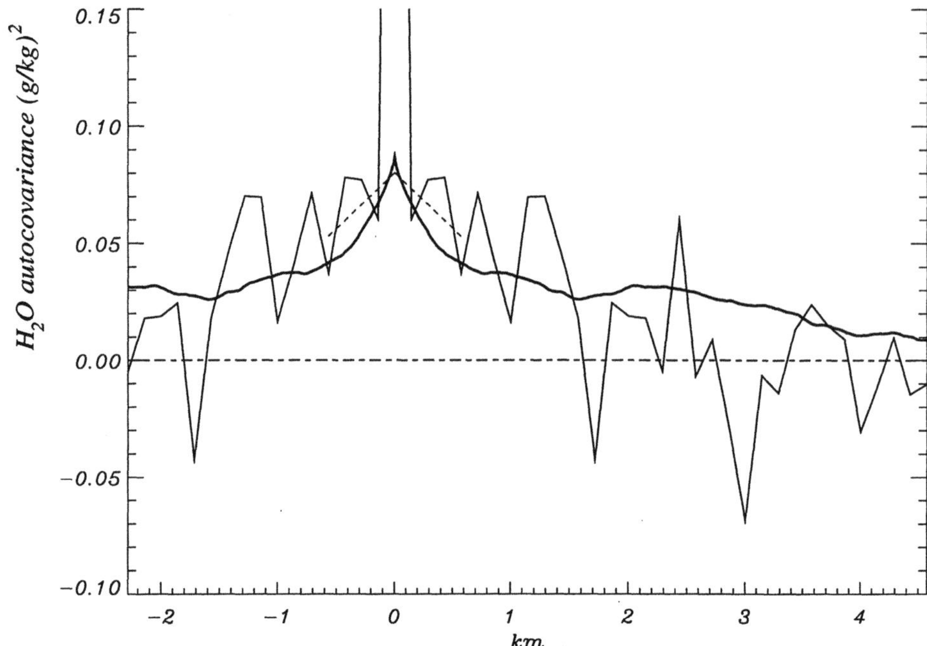

Figure 12. Water vapor autocovariance function obtained from a horizontal cut of Fig.7 (thin solid line) at an altitude of 1340 m agl., corresponding to 0.85 z_i. Same representation as for Fig.11, but the autocovariance function of the fast–response in-situ Lyman–alpha hygrometer on a leg flown 15 minutes later at 0.85 z_i is overplotted (thick solid line).

ACF shapes encountered at different heights, however, the error can reach 20%, as Fig.12 shows. Here we have overplotted the autocovariance function of the in-situ Lyman–alpha hygrometer for the same area and altitude. Despite the fact that the DIAL instrumental noise is considerably higher here at 0.85 z_i (the noise peak is cut off) and affects the reliability of single ACF data points (ACF is more "noisy"), the extrapolation of the linear fit to zero lag (short dashed) nearly coincides with the in-situ value of 0.09 g^2 kg^{-2}. The Lyman–alpha signal can be considered free of uncorrelated noise since the ACF is very smooth and no noise peak at the zero lag ACF data point is apparent.

The following inter-comparisons between several DIAL water vapor variance estimates and between DIAL and in-situ variance estimates will demonstrate that Fig.12 is not a single case, but that the linear extrapolation approach gives good results for an altitude range down to 0.3 z_i. In particular, this approach yields acceptable variance estimates both when large signal variances are found, as is the case within the EZ (Fig.11), and when the ACF shape is shallow and noisy around the noise peak, as is frequently encountered within the well–mixed layer below the EZ, where the water vapor variance is smaller. The linear fit gives a robust and trustworthy estimate even for "noisy" ACFs since it smoothes out the fluctuations (cf. Fig.12). Other attempts to derive the signal variance from the lagged ACF were no more successful than the simple linear extrapolation to zero lag. This is not to say, however, that this is an optimal extrapolation technique. It is likely that it is possible to obtain better estimates of the variance, but we have not yet pursued that aspect.

The method described above makes it possible to separate the uncorrelated noise variance σ_n^2 from the signal variance σ_q^2. If the series is long enough that the turbulence statistics are stable, the statistical error σ_n caused by uncorrelated instrumental noise reads:

$$\sigma_n = \sqrt{ACF_0 - \sigma_q^2}, \tag{1}$$

where ACF_0 is the autocovariance function at zero lag. The relative statistical error $\sigma_{n_{rel}}$ is the ratio of the statistical error to the mean water vapor content $\bar{Q}$. The signal–to–noise ratio is defined as:

$$SNR = \frac{\bar{Q}}{\sigma_n} = \frac{1}{\sigma_{n_{rel}}}. \tag{2}$$

The statistical error and the signal–to–noise ratio obtained this way are useful estimators of the data quality, although they represent only a mean error estimate over the whole series that does not account for local heterogeneities.

Turbulent mixing in the CBL spans spatial scales from the CBL depth z_i down to the molecular dissipation range of a few millimeters. Thus, the variance observed using a given measurement technique is a function of the resolution of the measurement. In our case the observations cover a two-dimensional region, and the observed variance is dependent on both the vertical and the horizontal resolution. In the following examples we find that the scales of significant variance are very dependent on the location of the measurement with respect to the boundaries of the CBL.

Figure 13 shows the result of the method described above, applied to DIAL data from the flight leg shown in Fig.7 for various vertical resolutions ΔR. A similar plot is obtained when varying the horizontal resolution. In the EZ, roughly situated at $(1 \pm 0.15) z_i$, the resulting signal variance is very much a function of the resolution. At z_i, the variance roughly decreases linearly with increasing ΔR. In the well-mixed part of the CBL below about 0.75 z_i, however, the variance is no longer a function of the vertical or horizontal (not shown here) resolution. This is because in the EZ, entrainment processes and the nearby stable inversion produce small spatial scale

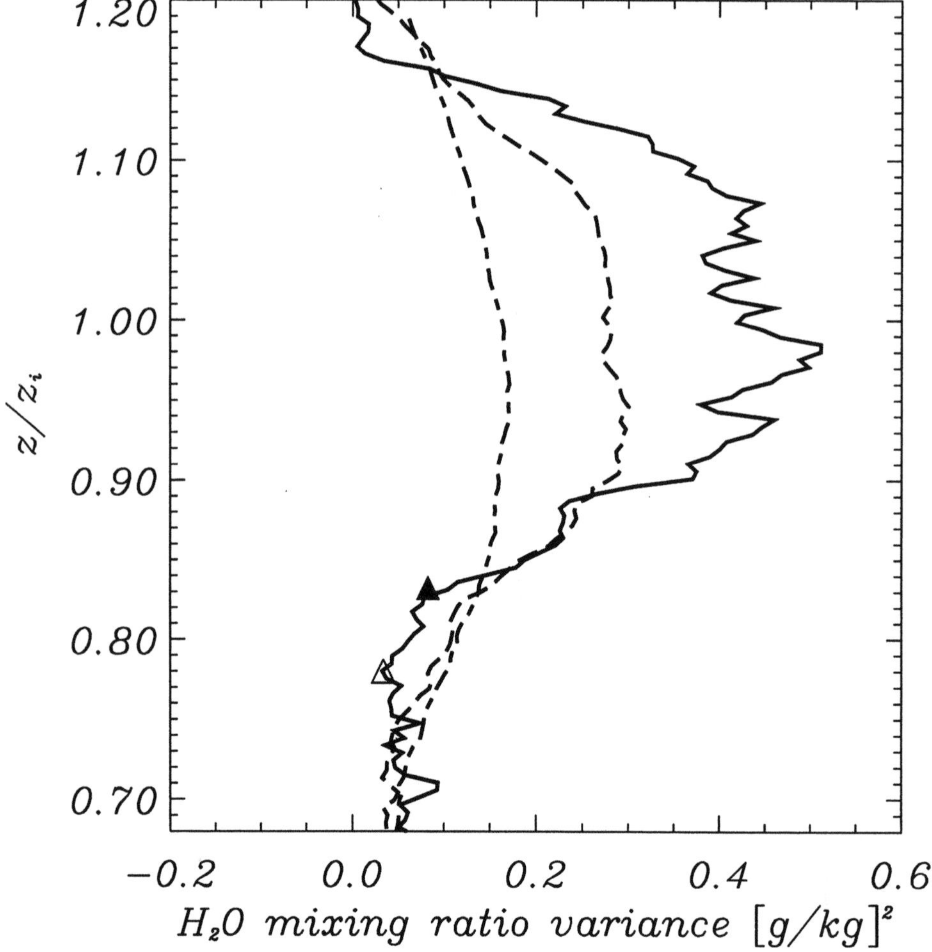

Figure 13. Variance profiles of DIAL water vapor mixing ratio for the flight leg of Fig.7 for three different vertical resolutions: ΔR=134m (solid), ΔR=224m (dashed) and ΔR=358m (dot-dashed). The black triangle stands for an in-situ level–leg measurement performed 15 minutes later, the white triangle for one performed 90 minutes later.

turbulence, only partly resolved by our measuring method whereas in the mixed layer, most of the variance lies at larger scales, fully resolved by our measurements. Between 0.75 and 0.85 z_i the variance computed using large ΔR values is higher than the variance computed for a ΔR of 134 m. Here the poor vertical resolution has smeared the variance signal in the vertical and spread out the relatively strong variance gradient located between 0.83 and 0.9 z_i. The in-situ leg values (triangles) agree with this, since they lie closest to the profile derived using a high vertical resolution, i.e. a small ΔR value. The coincidence of the black triangle at roughly 0.85 z_i with

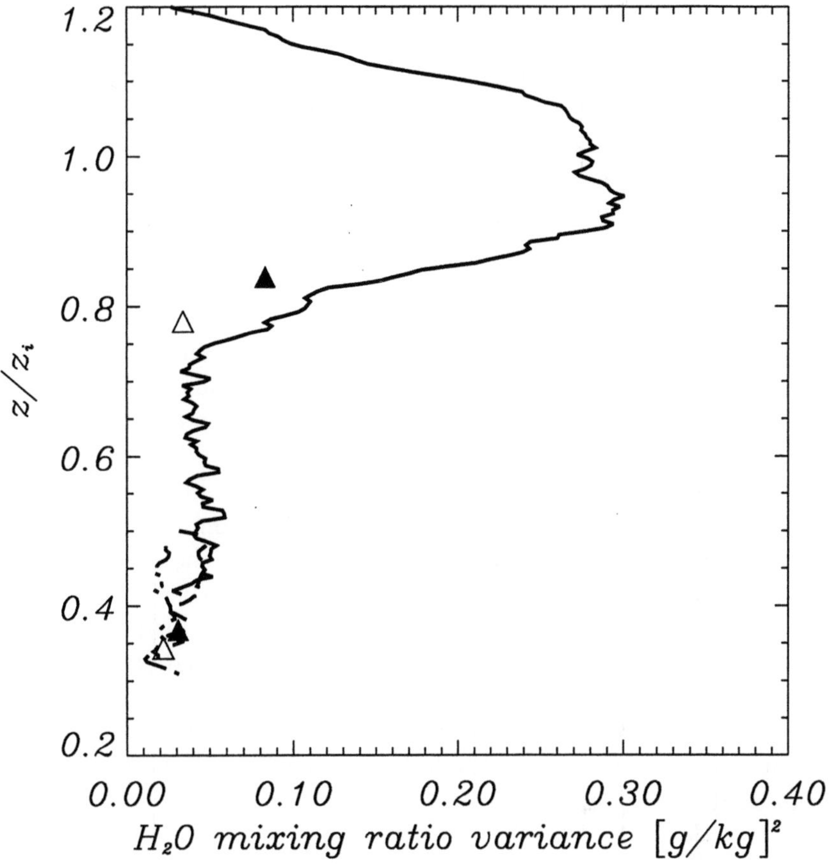

Figure 14. Variance profiles of DIAL water vapor mixing ratio for the flight leg of Fig.7 (solid, ΔR=224m), and two lower level DIAL legs, one performed 15 minutes later (dashed below 0.5 z_i), and the other one 75 minutes later (dot-dashed), both with ΔR=358m. As in Fig.13, the triangles represent in-situ level–leg values.

that DIAL profile is illustrated in Fig.12.

We have found that the EZ variances vary approximately inversely with ΔR for the values used here. Similarly, the variances were also found to vary approximately inversely with horizontal averaging length. Although extrapolation to zero averaging scales is risky, mostly because of the noisy data, the value of standard deviation we obtain by doing this is not very different from half the mixing ratio difference across the top of the CBL, which is the maximum possible standard deviation in the EZ. This gives us additional confidence in our technique for estimating variance.

Below 0.68 z_i, the ACF becomes too noisy when ΔR is 134 m. Figure 14 shows the variance profile down to 0.4 z_i using a ΔR of 224 m. Here, the DIAL variance is greater than the 0.8 z_i in-situ leg values (triangles) because of the smearing effect explained above. We see that the variance is nearly independent of altitude within the mixed layer. Variance profiles between 0.3 and 0.5 z_i, obtained from the two lower leg DIAL measurements, are consistent with both the upper leg DIAL measurement as well as the in-situ $0.35z_i$ variance. This further supports our method of estimating variance.

3.3.3. *Integral Scale Profiles*

Integral length scales for water vapor (λ_q) have been calculated using the method described in Lenschow and Stankov [10]. That is, λ_q is taken as the maximum value of the integral of the autocorrelation function, which is the autocovariance function normalized by the signal variance. Integral length scales for all DIAL water vapor measurements within the CBL in this case study, including time series in the EZ, lie between 1000 m and 2000 m. This is in good agreement with λ_q computed from the in-situ mixing ratio measurements which, on the upper–level flight legs (0.8 z_i) are found to be roughly 2000 m. An estimate for the random error in measured variance due to the finite length of flight legs has been derived [10]. This variance error is related to the integral scale of the second–order moment quantities, λ_{q^2}. From their measurements they found that λ_{q^2} is roughly 0.67 that of λ_q. Furthermore, they found the following relationship between λ_{q^2} and the normalized CBL height z/z_i:

$$\lambda_{q^2}/z_i = (z/z_i)^{1/2}. \tag{3}$$

Profiles of the DIAL–derived λ_{q^2}, plotted using the same format as Fig.4 of [10], are shown in Fig.15. While the bottom scale is λ_{q^2}/z_i, the top scale indicates the normalized leg length needed to obtain a 20 % water vapor variance accuracy. Note that the flight leg we are analyzing (Fig.1) is 90 km in length, that is, 56 × z_i. This z_i normalized length is represented by the dashed vertical line. The solid curve is λ_{q^2}/z_i measured for that leg. We see that it fluctuates about the 20 % variance accuracy (dashed) and about the empirical formulation of Lenschow and Stankov [10] (inclined dashed line). The dashed curve is λ_{q^2}/z_i obtained on the leg flown 1.5 hours later along the same track. Both profiles were obtained using $\Delta R = 224$ m. Reducing ΔR further is not possible, since the variance estimation technique fails below 0.7 z_i due to excessive scatter in the ACF. If ΔR is increased to twice the value above, no general shift in λ_{q^2}/z_i is observed, but the curves do become smoother. As a cross-check, a lower level leg DIAL measurement result is overplotted between 0.4 and 0.55 z_i. On the whole, the integral scales from the two flight legs agree well with each other, and with the

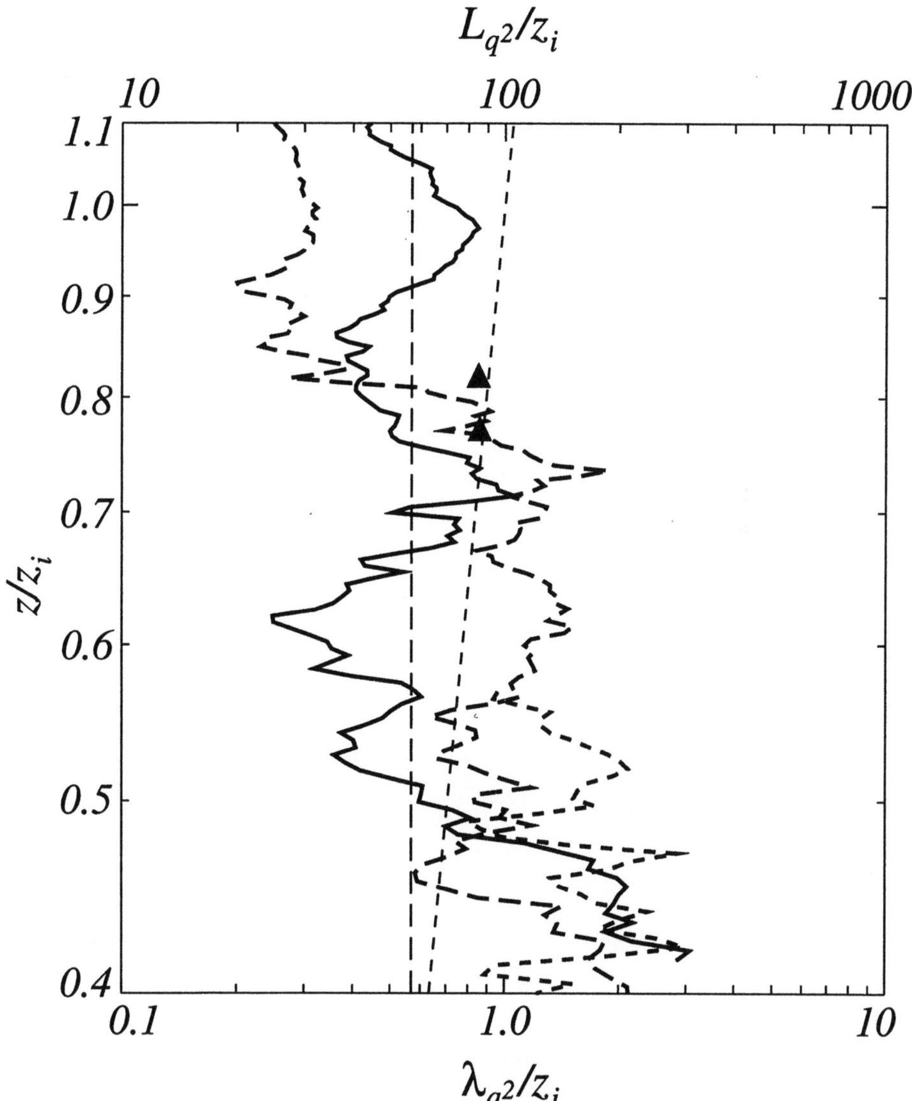

Figure 15. Integral scales profiles of DIAL water vapor mixing ratio for the flight leg of Fig.7 (solid), and the leg performed 1.5 hours later (dashed), both using a ΔR of 224m. The short-dashed curve below 0.6 z_i belongs to a lower–leg DIAL measurement. The triangles represent in-situ level–leg values. The bottom scale stands for the z_i normalized variance integral scale (explained in the text), the top scale indicates the z_i normalized leg length needed to obtain a 20 % water vapor variance accuracy. Referring to that scale, the dashed vertical line represents the z_i normalized length of the leg for the solid curve. The data of [10] are represented by the short-dashed inclined line which is a sketch of Equation 3.

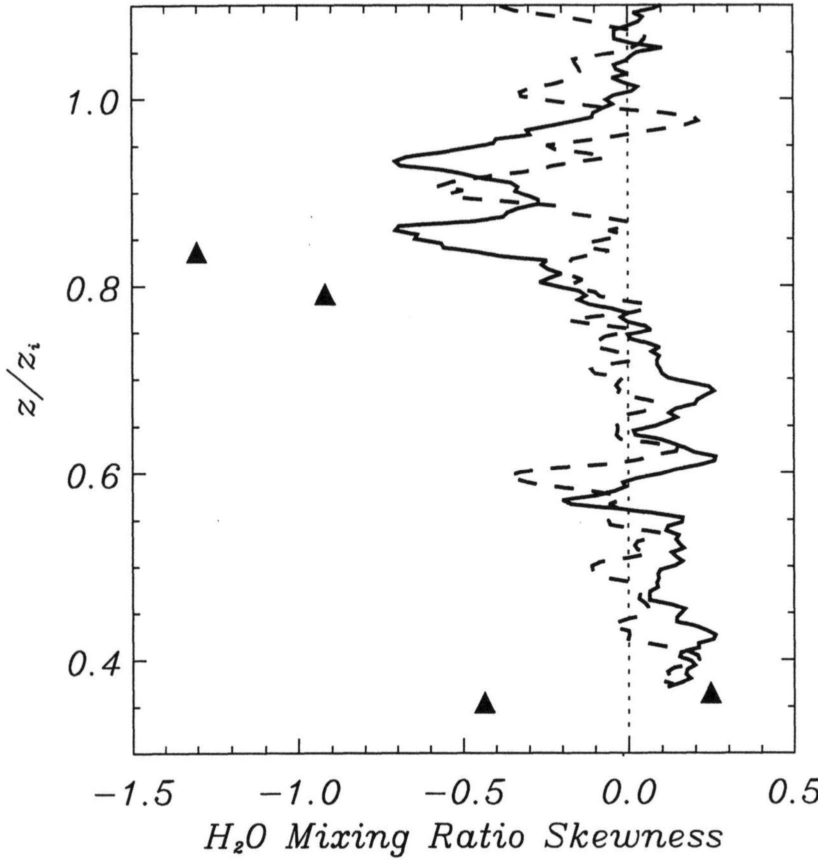

Figure 16. Skewness profiles of DIAL water vapor mixing ratio. Same data and line assignments as for Fig.15.

empirical formulation of [10] (equation 3). The good agreement with results from other experiments can be interpreted as another indirect confirmation of the DIAL variance retrieval method, since calculating λ_{q^2}/z_i requires a good estimation of the variance.

3.3.4. *Skewness Profiles*

Figure 16 shows a skewness profile calculated for this case using $\Delta R = 134$ m. The skewness seems to be significantly negative between 0.8 and 1.0 z_i, but considerably smaller in magnitude than the skewness calculated from in-situ data (triangles). The negative skewness is consistent with our conception of a CBL containing a random distribution of thermals pushing up over rather broad areas distorting the top of the boundary layer surrounded by sharp downward protrusions of subsiding air containing a mixture of entrained air from above and boundary–layer air from below. The relatively poor (as compared to variance) agreement between the in-situ and DIAL skewnesses is probably due mainly to the effect of averaging of the DIAL water vapor, and also to the increased uncertainty associated with calculations of higher–order moments such as skewness. We note that the skewness of the CBL top, the derivation of which is discussed below in the section on water vapor fluxes, is also found to be negative for both lidar over-passes. The effect of averaging is that skewness decreases in magnitude as ΔR increases. We therefore have limited confidence in the skewness values.

4. Water Vapor Fluxes

If the CBL is quasi–steady, horizontally homogeneous, and free from vertically varying horizontal advection, a relationship between the flux and the gradient of the mean vertical profile, proposed by Wyngaard and Brost [21], should apply. They suggested that the transport of scalars from the top of the CBL down, initiated by scalars entering the CBL from above (entrainment flux), and the transport of scalars from the bottom up, initiated by a surface flux (evaporation in the case of water vapor) could be treated independently and a separate flux–gradient relationship found for each. They also argued that the vertical gradient of a mean scalar mixing ratio in the CBL should depend only on the scalar surface and entrainment fluxes, a vertical velocity scale describing the convective mixing, the height of the potential temperature inversion (the top of the CBL), and two normalized dimensionless functions of height within the CBL, one for top–down and one for bottom–up transport. These gradient functions were determined empirically using LES ([16] and [15]). In addition, using a similar scaling argument they determined variance functions which describe a relationship between the surface and entrainment fluxes and the scalar variance profile. Given these mixed–layer gradient (MLG) or mixed–layer variance (MLV) relationships, if a mean scalar mixing ratio or variance profile, the height of the CBL, and the convective velocity scale can be measured, the scalar surface and entrainment fluxes can be uniquely determined.

The airborne DIAL provides all the data needed to estimate fluxes using the MLG and MLV methods, with the exception of the convective velocity scale, which we here estimate from in-situ measurements. The quality of the DIAL water vapor mixing ratios and variances degrades significantly in the lower CBL due to the large optical depth. In the upper CBL, where gradients and variances estimated from the DIAL data are much better, the relationships are dominated by entrainment. Therefore, we limit ourselves to deriving the entrainment flux from the DIAL data and use in-situ estimations of surface water vapor flux in the flux–gradient and flux–variance relationships.

We test these techniques using DIAL and in-situ water vapor and turbulence data from 4 August, 1994, a multi–level flight stack over the Northern Study Area. Two lidar passes on this day, the first of which is shown in figures 1 and 7, were used for estimating water vapor statistics in the CBL. Three–level in-situ flight stacks within the boundary layer accompanied each DIAL over-pass of the CBL. The in-situ flights were spaced at roughly 0.08, 0.35 and 0.80 z_i, allowing for a good linear extrapolation of the in-situ eddy–correlation water vapor and temperature fluxes to the surface and to z_i for estimating surface and entrainment fluxes.

4.1. MLG FLUXES

The mixed–layer flux–gradient relationship first presented by [21] can be solved for the entrainment flux:

$$\overline{wq}_{z_i} = -\frac{\partial \bar{Q}}{\partial z}\frac{z_i w_*}{g_t} - \overline{wq}_o \frac{g_b}{g_t}, \tag{4}$$

where $\bar{Q}$ is the horizontal mean water vapor mixing ratio, q the fluctuating component of the water vapor, w the vertical velocity, w_* the convective velocity scale, over-bars refer to horizontally–averaged values, and $g_b(z/z_i)$ and $g_t(z/z_i)$ are the normalized bottom–up and top–down gradient functions. In our case, we use lidar observations for z_i, and in-situ observations for w_* and $\overline{wq}_o$ (the surface flux of humidity). We use the LES results of [16] and [15] for the gradient functions.

Figures 17 and 18 show the results of using two lidar passes over the BOREAS northern study area on 4 August, 1994 to estimate water vapor entrainment by the MLG technique. Figure 17 shows the water vapor mixing ratio profile gradients derived from the DIAL, as well as the mixing ratio profile gradients predicted from the LES–derived gradient functions (g_b, g_t) given the in-situ surface and entrainment flux values. The DIAL observations show a change in mixing ratio across the EZ which is less steep than the LES prediction. Figure 18 plots the entrainment flux derived from the mean mixing ratio gradient profiles shown in Fig.17, applying equation 4 to each height. Since g_b, g_t and $\partial \bar{Q}/\partial z$ are functions of height, each estimate of the derivative of the mixing ratio profile with height yields an entrainment flux estimate. Thus, we obtain a "profile" of entrainment flux estimates. For the investigated altitude range between 0.8 and 0.9 z_i, the earlier DIAL leg (thick solid line) yields, on average, about 160 % larger entrainment flux estimates than the in-situ estimation (dashed vertical line).

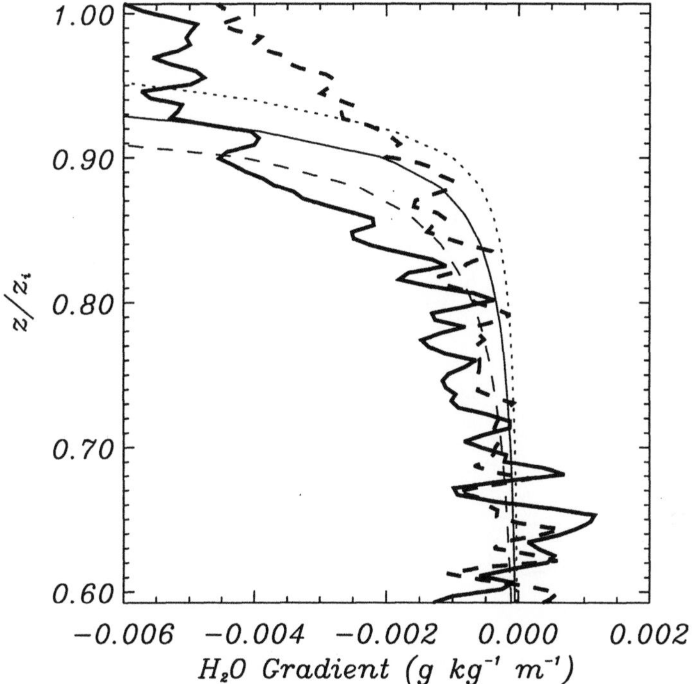

Figure 17. Illustration of the MLG method. DIAL water vapor vertical gradients (solid line for the leg of Fig.7, dashed line for the measurement 1.5 hours later) are compared to the LES–predicted gradient (thin solid), which was calculated using the in-situ surface and entrainment flux values. To check the sensitivity of the method, the LES predictions were re-run with an in-situ entrainment flux reduced by 50 % (thin dotted) and augmented by 100 % (thin dashed line).

The second flight leg (thick dashed line) is on average about 40 % larger than the in-situ estimation. The flux estimates obviously are not meaningful above about 0.9 z_i where both DIAL and LES begin to lose their ability to resolve large vertical gradients. Below about 0.8 z_i, the gradient becomes too noisy.

4.2. MLV FLUXES

The mixed–layer similarity theory of [21] has been extended to scalar variances by [16] and [15]. Solving their formulation for the entrainment flux yields,

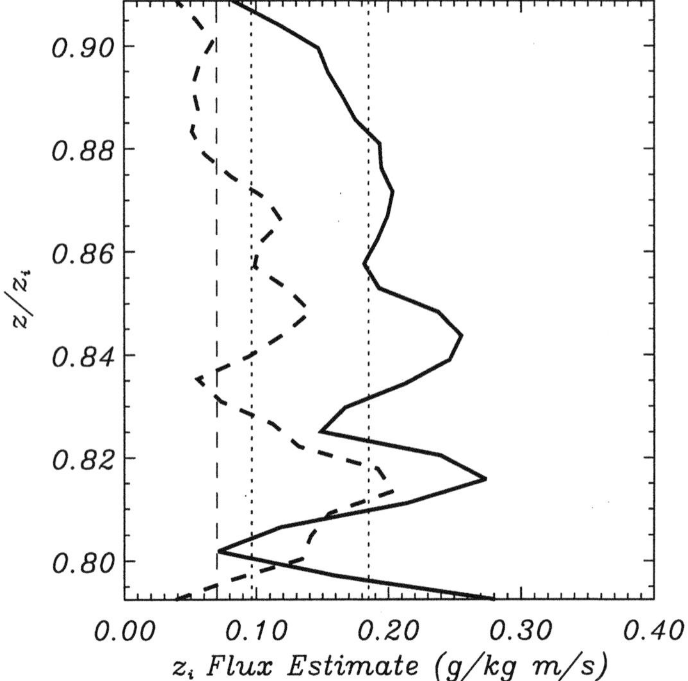

Figure 18. Estimates of entrainment flux using the MLG method and solving equation 4 for each height. Same DIAL line assignment as for Fig.17. The dotted thin vertical lines are the resulting mean entrainment flux estimates for both DIAL profiles. The dashed thin vertical line is the in-situ entrainment flux estimation.

$$\overline{wq}_{z_i} = \sqrt{(w_*^2\overline{q^2} - \overline{wq_o}^2 f_b)\frac{1}{f_t} + (\overline{wq_o}\frac{f_{tb}}{f_t})^2} - \overline{wq_o}\frac{f_{tb}}{f_t}, \qquad (5)$$

where f_t, f_b and f_{tb} are the dimensionless top–down, bottom–up and cross–correlated variance functions. By using the lagged autocovariance function to solve for water vapor variance $\overline{q^2}$ in the DIAL data, as described in the preceding section, we can use DIAL data to estimate the entrainment flux in nearly the same manner as the MLG method described above.

Figures 19 and 20, parallel to the MLG discussion above, present DIAL– and LES–derived variance profiles, as well as a comparison between the MLV–derived entrainment flux estimates and the in-situ eddy–correlation entrainment flux measurement. Here while the first DIAL leg (solid line) again yields a greater entrainment flux estimate than the second leg (dashed line), the MLV entrainment flux estimates are less noisy and agree more

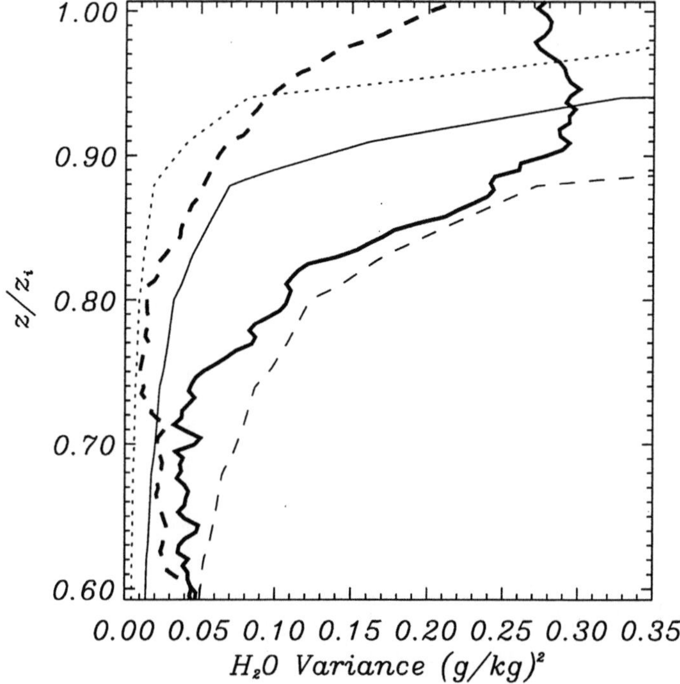

Figure 19. Illustration of the MLV method. DIAL water vapor variance profiles are compared to the LES–predicted variances (same line assignments as for Fig.17), which were calculated using the in-situ surface and entrainment flux values.

closely with the in-situ flux estimate than did the MLG–derived entrainment fluxes. For the altitude range between 0.6 and 0.9 z_i, the first leg yields, on average, about 60 % larger entrainment flux estimates than the in-situ estimation, and the second flight leg is on average 10 % smaller than in-situ. The latter agrees within roughly 30 % with in-situ. Again, the flux estimates are not meaningful above about $0.9z_i$, and become too uncertain below $0.6z_i$.

4.3. SENSITIVITY ANALYSES

The thin dashed lines of figures 17 and 19 illustrate the sensitivity of the LES relationship between the gradient or variance profile and the entrainment flux. The sensitivity to the entrainment flux is best when these lines are well separated. Below 0.75 z_i, there is little sensitivity to the entrainment flux. Above 0.9 z_i, the sensitivity is high, but the LES gradient and variance functions are less valid near the inversion due to the increasing

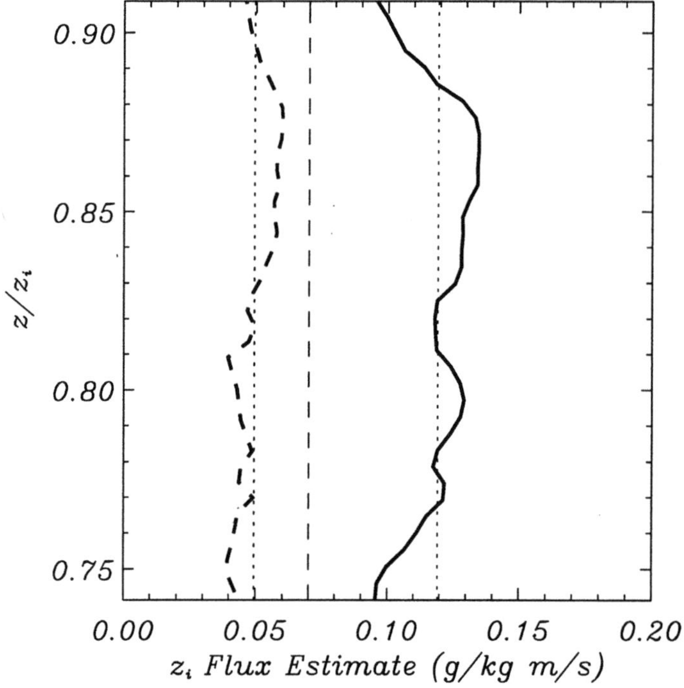

Figure 20. Estimates of entrainment flux using the MLV method and solving equation 5 for each height. Same line assignments as for Fig.18.

TABLE 1. Sensitivity of MLG and MLV entrainment flux estimates to various input parameters

Sensitivity to	Best estimate	Absolute change	Relative change	Entrainment flux relative change MLG	MLV
z_i	1612 m	± 100 m	± 6 %	+75 % - 50 %	± 23 %
w_*	2.21 m/s	± 0.2 m/s	± 10 %	± 9 %	± 10 %
$\overline{wq}_o$	0.06 g/kg m/s	± 0.06 g/kg m/s	± 100 %	± 1.7 %	∓ 5 %
ΔR	224 m	134 – 358 m		no effect	no effect
Δx	139 m	28 – 139 m			

predominance of small–scale turbulence. There the flux estimates are not meaningful. Thus, the optimal region for entrainment flux estimation lies between 0.75 and 0.9 z_i.

Table 1 describes the sensitivity of the entrainment flux derivations to the input parameters and the resolution of the DIAL data, for both MLG and MLV methods applied in the altitude range of 0.7 – 0.9 z_i. The CBL height (z_i) estimate has the greatest effect on the resulting entrainment flux estimate. The effect is higher for the MLG method. The ± 100 m chosen here as input parameter fluctuation correspond to the standard deviation of the CBL top undulations extracted from Fig.1 via a method referred to in the next section. Within the range of investigation, there is no apparent dependence of the fluxes on the DIAL data resolution, at least in the mid–CBL where the scales of turbulent eddies are large. The surface flux dependence is very low, since the MLG and MLV equations are rather insensitive to the surface flux in the investigated altitude range.

We also note that the MLV entrainment flux estimates seem to remain valid at higher altitudes within the CBL when using the coarser resolution LES variance functions [16]. We hypothesize that this is the case because the DIAL observations have similar resolution to this LES study. The higher–resolution LES runs resolve more of the water vapor structure in the EZ than do the DIAL observations. Hence solving for the entrainment flux using the DIAL entrainment zone variances together with the high–resolution LES EZ variance functions results in an underestimate of the entrainment flux. The results in the mid–CBL are not sensitive to the LES resolution as the eddies there are mostly resolved by both LES and the DIAL observations.

Our results show that estimates of the entrainment flux using LES–derived top–down, bottom–up variance formulations are more robust than those based on mean profiles. We believe this is due to the fact that the mixing ratio profile must be differentiated to derive MLG flux estimates. Differentiating the mean mixing ratio profile introduces noise in the MLG flux estimate, while no differentiation is needed for the MLV method. Furthermore, the lagged ACF approach used for the MLV method removes instrument noise.

4.4. MESOSCALE VARIABILITY

The DIAL–LES comparison would be most useful if the lidar measurements were high-pass filtered with a cut-off wavelength about equal to the horizontal domain of the LES to eliminate the mesoscale contributions to the water vapor field. We can get an idea of the extent of mesoscale contributions from plots of the z_i spectra obtained from the aerosol backscatter Fig.1. The estimates of z_i were obtained from the algorithm developed by Davis (personal communication) which uses a wavelet technique with a Haar basis function to estimate a maximum gradient in the backscatter.

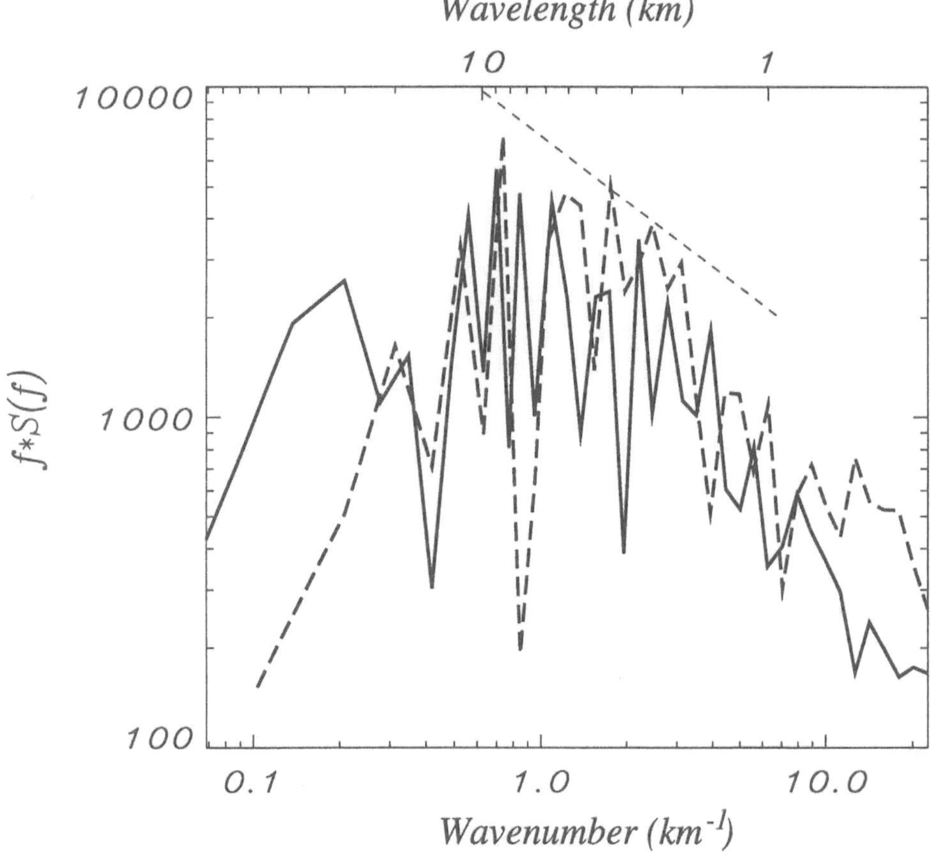

Figure 21. Spectra of the CBL top height of Fig.1 (solid), and of data obtained from the flight leg 1.5 hours later (dashed). The thin dashed line represents a -2/3 exponential slope for comparison.

The measured spectra (see Fig.21) shows a maximum at about 7 – 10 km, which is about the size of the LES domain. Here we have not removed the mesoscale contributions in order to emphasize the ability of the lidar to also measure the mesoscale. Fig.21 shows more mesoscale variations in the first DIAL leg than in the second one: Both spectra are, for wavelengths higher than 25 km, nearly an oder of magnitude apart. The overestimation of the flux, seen in the first DIAL leg data with both the MLG and the MLV method, is supposed to be due to these stronger mesoscale variations that are not accounted for in the LES.

5. Conclusions

We have shown several examples of boundary–layer structure delineated by the DIAL backscatter obtained from NCAR Electra over-flights of the BOREAS study areas. Smoke from forest fires enhanced boundary–layer backscatter, and was commonly observed in elevated stable layers above the boundary layer which reflect the variations in boundary–layer height. We also show examples of perturbations in boundary–layer structure induced by fair–weather cumulus.

We have used the two-dimensional DIAL water vapor fields from a particular flight to derive mean profiles, variances and skewnesses as functions of height within the CBL. A lagged autocorrelation method is used to remove instrument noise from our DIAL water vapor variance estimates. We see good agreement with in-situ observations for the mean mixing ratio and variance profiles.

We then use both the mean mixing ratio and variance profiles to estimate water vapor entrainment fluxes using the MLG and MLV methods first discussed in [1]. While the MLG method was successfully applied to in-situ profile observations in [2], this is the first application of the MLV method, and the first successful application of either method to remotely–sensed data. We find that the MLV method provides a more stable estimate of the entrainment flux. We also find that mesoscale variability, present in the observations but not the LES which was used to derive the mixed–layer gradient and variance functions [15], causes an overestimate of the entrainment flux in one case where a CBL–top spectrum shows that such variability is present. Whether or not mesoscale features such as these act to enhance entrainment is a question which remains to be answered.

References

1. Davis, K.J., (1992) Surface Fluxes of Trace Gases Derived from Convective-Layer Profiles, Ph.D. dissertation, University of Colorado. Available as NCAR CT-139.
2. Davis, K.J., Lenschow, D.H. and Zimmerman, P.R. (1994) Biogenic non-methane hydrocarbon emissions estimated from tethered balloon observations, *J. Geophys. Res.* **99**, 25587-25598.
3. Ehret, G., Kiemle, C., Renger, W. and Simmet, G. (1993) Airborne Remote Sensing of Tropospheric H_2O Using a Near Infrared DIAL, *J. Appl. Opt.* **32**, 4534-4551.
4. Ehret, G., Giez, A., Kiemle, C., Davis, K.J., Lenschow, D.H., Oncley, S.P. and Kelly, R.D. (1996) Airborne water vapor DIAL and in-situ observations of a sea–land interface, *Beitr. Phys. Atmosph.* **69**, 215-228.
5. Eichinger, W.E., Cooper, D.I., Holtcamp, D.B., Karl, R.R., Quick, C.R. and Tiee, J.J. (1993) Derivation of water vapor fluxes from lidar measurements, *Bound. Layer Meteor.* **63**, 39-64.
6. Giez, A., Ehret, G., Schwiesow, R.L., Davis, K.J., and Lenschow, D.H. (in press) Water vapor flux measurements from ground-based vertically-pointed water vapor differential absorption and Doppler lidars, *J. Atmos. Ocean. Tech.*
7. Kiemle, C., Kästner, M. and Ehret, G. (1995) The Convective Boundary Layer Structure from Lidar and Radiosonde Measurements during the EFEDA'91 Campaign, *J. Atmos. Ocean. Tech.* **12**, 771-782.
8. Kiemle, C., Ehret, G., Giez, A., Davis, K.J., Lenschow, D.H., and Oncley, S.P. (in press) Estimation of Boundary-Layer Humidity Fluxes and Statistics From Airborne DIAL, *J. Geophys. Res.*, BOREAS special issue, copyright by the AGU.
9. Lenschow, D.H. (1986) Aircraft Measurements in the Boundary Layer, in D.H. Lenschow (ed.), *Probing the Atmospheric Boundary Layer*, American Meteorological Society, Boston, MA, pp. 39–55.
10. Lenschow, D.H. and Stankov, B.B. (1986) Length scales in the convective boundary layer. *J. Atmos. Sci.* **43**, 1198-1209.
11. Meischner, P., Kiemle, C., Ehret, G., Kästner, M., Schreiber, H.G., Evtushenko, A., Kutuza, B., Petrenko, B. and Smirnov, M. (1994) Remotely Sensed Water Vapor Variations during CLEOPATRA '92, *Beitr. Phys. Atmosph.* **67**, 287-301.
12. Melfi, S.H., Spinhirne, J.D., Chou, S.H. and Palm, S.P. (1985) Lidar observation of vertically organized convection in the planetary boundary layer over the ocean, *J. Climate Appl. Meteor.* **24**, 806-821.
13. Michels, B.I. and Jochum, A.M. (1995) Heat and moisture flux profiles in a region with inhomogeneous surface evaporation, *J. Hydrol.* **166**, 383-407.
14. Moeng, C.-H. and Sullivan, P.P. (1994) A comparison of shear and buoyancy driven planetary-boundary-layer flows, *J. Atmos. Sci.* **49**, 2384-2401.
15. Moeng, C.-H. and Wyngaard, J.C. (1989) Evaluation of turbulent and dissipation closures in second-order modeling, *J. Atmos. Sci.* **46**, 2311-2330.
16. Moeng, C.-H. (1984) A large-eddy-simulation model for the study of planetary boundary-layer turbulence, *J. Atmos. Sci.* **41**, 2052-2062.
17. Oncley, S.P., Lenschow, D.H., Campos, T.L., Davis, K.J. and Mann, J. (in press) Regional-scale flux observations across the boreal forest during BOREAS, *J. Geophys. Res.*
18. Schmidt, H. and Schumann, U. (1989) Coherent structure of the convective boundary layer derived from large-eddy simulations, *J. Fluid. Mech.* **200**, 511-562.
19. Schotland, R.M. (1974) Errors in the lidar measurement of atmospheric gases by differential absorption, *J. Appl. Meteor.* **13**, 315-331.
20. Senff, C., Bösenberg, J. and Peters, G. (1994) Measurement of water vapor flux profiles in the convective boundary layer with lidar and radar–rass, *J. Atmos. Ocean. Tech.* **11**, 85–93.
21. Wyngaard, J.C. and Brost, R.A. (1984) Top-Down and Bottom-Up Diffusion of a Scalar in the Convective Boundary Layer, *J. Atmos. Sci.* **41**, 102-112.

EXPERIMENT, NUMERICAL MODELING, NUMERICAL SIMULATION, AND THEIR ROLES IN THE STUDY OF CONVECTION

J. C. WYNGAARD
Penn State University
University Park 16802, Pennsylvania, USA

1. Introduction

A casual observer of a turbulent flow, say a cigarette plume or the eddying motion downstream of a waterfall, can easily get the impression that turbulence behaves as it pleases locally in space and time. Any constraints on its behavior seem to come with a long leash.

The local behavior of turbulence has not traditionally been the focus of models and theories. Those have generally dealt with statistical properties of turbulence—what we call expected values, meaning ensemble averages. The casual observer does not see statistical properties, however; he sees local properties. Thus, we are poorly prepared by experience and by intuition for the challenge of modeling expected values. Perhaps it is like trying to imagine how things look in the infrared.

The need for distinguishing the local structure of turbulence from expected values is quite important but sometimes forgotten. A simple example comes to mind. Eddy-diffusivity models have been criticized on the basis of observations that are claimed to show "nondiffusive" behavior. Evidence of large eddies that transport material from one edge of a flow to another in single sweeps has been interpreted as indicating that the fluxes are not relatable to local gradients. The eddy-diffusivity model says nothing about local events, however, and it cannot legitimately be criticized on such grounds. In fact, the eddy-diffusivity model for expected values is usually found to be at least qualitatively valid.

Despite the conceptual difficulty of modeling expected values, that has been our most common approach to turbulence. Experimentalists have had the last word; models and theories survived only as long as they were not proven wrong by measurements.

With the supercomputer has come numerical simulation of turbulence and its outpouring of "data" on the local, three-dimensional structure of turbulent flows in the environment and in engineering. This has profoundly changed the nature of the turbulence enterprise. Today's new generation of turbulence researchers is apt to have been drawn to the field not by the challenges of measuring pressure or vorticity fluctuations or Reynolds stresses in turbulence, or of documenting a canonical turbulent flow in the laboratory, but by the magnetism of computer simulation. Turbulence fields produced by direct numerical simulation (DNS) and large-eddy simulation (LES) are now being used as "ground truth" in many research and applications areas. The once-vital role of the traditional experimentalist has been weakened and diminished. The turbulence pendulum swings.

E.J. Plate et al. (eds.), Buoyant Convection in Geophysical Flows, 239–251.

2. Experiment

Historically, observations have paced the growth in our understanding of turbulence. That growth has been slow, in large part because we cannot make reliable, complete measurements of turbulence fields, but also because of the inherent difficulty of the turbulence problem. The excitement surrounding the advent of acoustic sounding of the lower atmosphere (McAllister *et al.* [17]) soon sharpened our perceptions about the local structure of the atmospheric boundary layer, for example, but in retrospect probably also stimulated unrealistic expectations about the ability of acoustic sounding to produce detailed, three-dimensional data sets directly useful in the turbulence problem. In meteorology and oceanography today, acoustic sounders and other remote sensors are yielding multidimensional fields, but system performance limitations have made the data useful mostly in a qualitative sense.

Quantitative techniques for deducing two and three-dimensional fields of turbulence in laboratory flows are emerging. In general, however, the best we can routinely produce today is measurements of time series of variables at a point in the flow, as has been the case for much of the twentieth century.

Today there is a new and complicating dimension to the turbulence measurement problem. Turbulence simulation (LES and DNS) is producing tremendous quantities of three-dimensional, time-dependent, numerical "data" on turbulent flows. But seldom are the producers or the users of these data able to make more than primitive and cursory checks of their fidelity—nor are they called to. Nearly 20 years ago an obviously exasperated Hans Liepmann [13] decried the same situation in the turbulence modeling community. In brief, our turbulence-measurement ability falls even farther short of our "number-crunching" ability than it did in 1972, when Bradshaw [3] pointed to the "fact gap"—the gap between our abililty to produce calculations of the structure of turbulent flows and to test these calculations against data.

2.1. LABORATORY EXPERIMENT

Corrsin [5] speculated on the role that the new "large computing machines" could have in turbulence research. After estimating that about 10^{12} grid points would be required for a calculation of a turbulent flow at adequately large Reynolds number, he asked: "The foregoing estimate is enough to suggest the use of analog instead of digital computation; in particular, how about an analog consisting of a tank of water?"

In the 1970s Deardorff and Willis began a series of experiments with water in a laboratory convection tank 1 m on a side, using a layer of less dense fluid at typically 20 cm above the heated surface to cap the convection. (They worked deep in the sub-basement of the National Center for Atmospheric Research (NCAR), reportedly in part to isolate the studies from a management not supportive of laboratory convection experiments.) The turbulence structure they found [9, 27] bore a striking resemblance to that of the convective atmospheric boundary layer, despite the much lower Reynolds number ($Re = u\ell/\nu$ was about 4×10^3, four orders of magnitude less than in the atmosphere), the absence of mean shear, and the differences in geometry. They used the tank also to study the dispersion of turbulent plumes, both buoyant and nonbuoyant; their results caused a major revamping of short-range dispersion modeling in the lower atmosphere [26].

2.2. FIELD EXPERIMENT

Most of the field measurements of turbulence to date have been directed at statistical properties, with the time average serving as a surrogate for the ensemble average. The bulk of these measurements have been done with in-situ sensors measuring at a single point or at most a half-dozen points arrayed on a tower or on the tethering cable of a balloon.

Field measurements are complicated by the very long averaging times required for convergence to the ensemble average. We learned this in the 1968 Kansas experiment: one-hour average stresses and heat fluxes measured simultaneously at 5.7, 11.3, and 22.6 m height differed substantially, despite their being in the "constant flux" layer (Figure 1). This pervasive scatter can greatly limit the utility of in-situ measurements. The nature of the statistical convergence problem is such that lower-order moments converge faster than higher-order ones, and as a result our knowledge of mean profiles, say, tends to be better than that of second and third moments. The scatter in third moments, for example, can be so large as to render the data unusable for testing and refining turbulence models (Figures 2, 3).

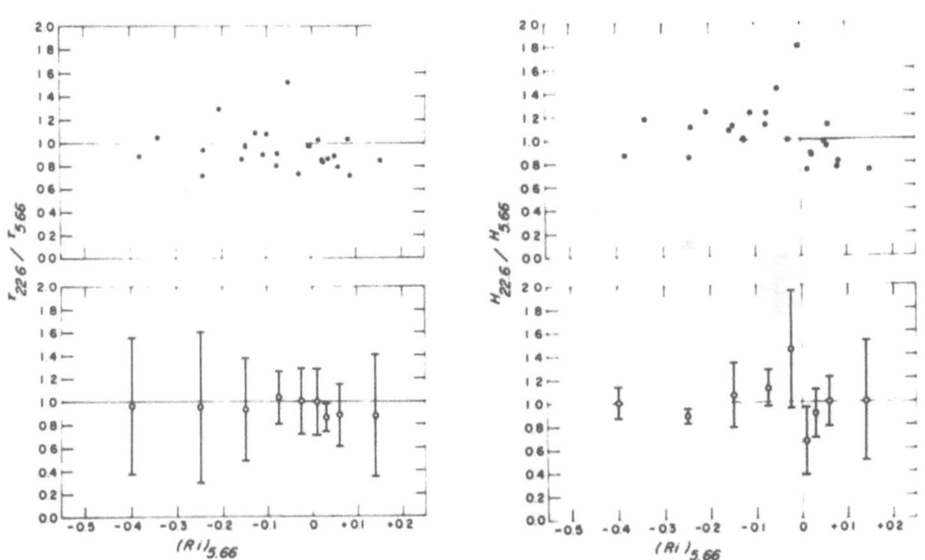

Figure 1. Ratios of the shearing stress τ (left panel) and heat flux H (right panel) at 22.6 m and 5.66 m measured in the 1968 Kansas experiment. Upper plots, 15-min data in 1-h blocks. Lower plots, data grouped into Richardson number intervals, with circle indicating the mean and bars indicating plus and minus one standard deviation. From Haugen *et al.* [10].

In the late 1970s NCAR scientists carried out a boundary-layer experiment with research aircraft over southeastern Colorado, their goal being to study the structure of the interfacial layer that caps the convective boundary layer. Turbulence signals in that region can be quite intermittent, making the averaging-time problem even more severe. The experiment yielded very little in the way of usable results.

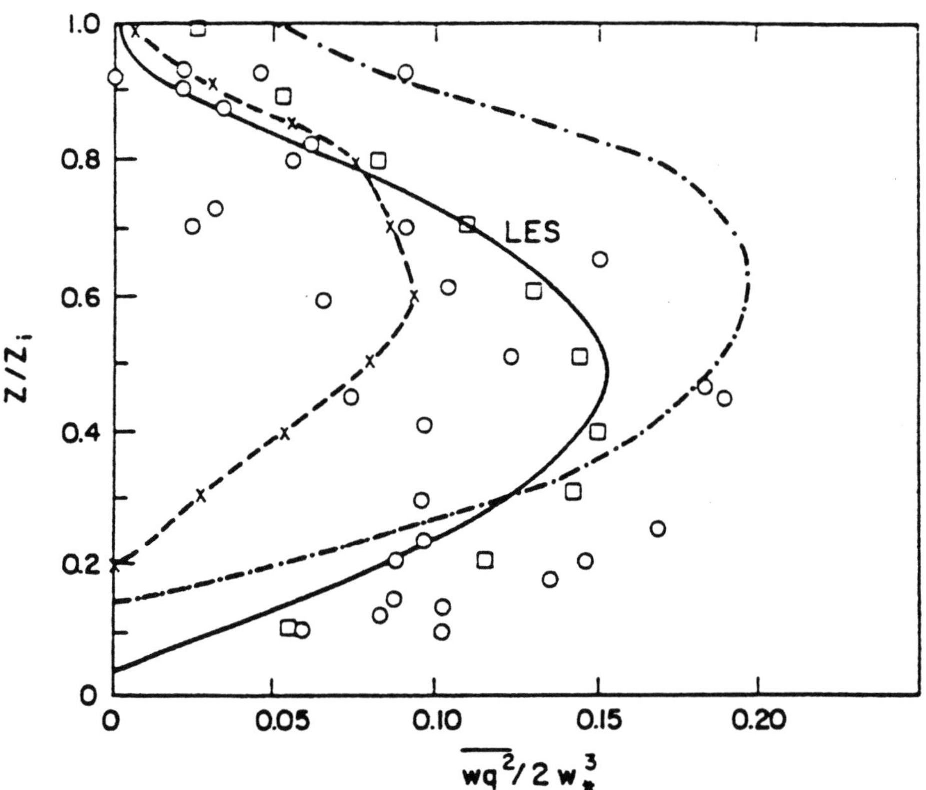

Figure 2. The vertical flux of energy from 96^3 LES (solid curve) compared with the parameterizations of Zeman and Lumley (dashed curve), Sun and Ogura (dash-cross curve), convection-tank data (squares) and AMTEX data (circles). From [18].

Although climate and mesoscale models contain representations or submodels of turbulence, these submodels are not generally rooted in the results of planetary boundary layer (PBL) experiments; instead, they are apt to be a mish-mash of closures for other flows. The exception is the surface-exchange submodel, which typically does use the hard-earned surface-layer results of micrometeorologists. The other turbulence submodels, such as those used to model turbulent flux divergences within the PBL, are generally not directly derived through experimental work in the atmosphere. Undoubtedly some of these submodels survive only because the data bases with which to test their performance have not existed. Numerical simulation data bases are beginning to be used for such testing, however [2].

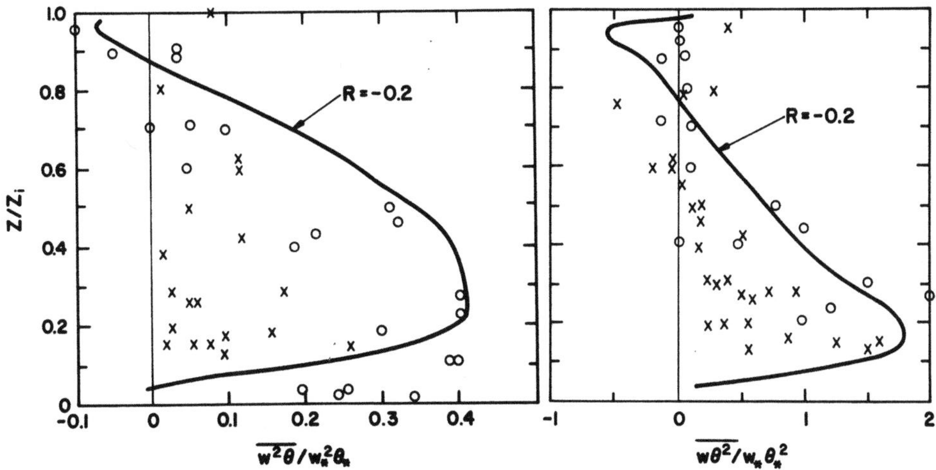

Figure 3. Comparisons of $\overline{w^2\theta}$ and $\overline{w\theta^2}$ computed through LES (solid curves) with data from the Minnesota (crosses) and AMTEX (circles) experiments. From [18].

3. Modeling and Simulation

I separate the many ways in which one can attempt to calculate turbulent flows into two categories: modeling and simulation. In *modeling* one represents turbulence through approximate equations that display behavioral similarities to turbulence. In *simulation* one uses equations that are derivable from the exact ones and, hence, remain faithful to the essential physics. This distinction between modeling and simulation is useful, I think, if not often made in the atmospheric and oceanic sciences.

Lumley [15] described a turbulence model as "a calibrated surrogate for turbulence." Regarding performance, he wrote: "We would thus expect that the models would work satisfactorily in situations not too far removed geometrically, or in parameter values, from the benchmark situations used to calibrate the model."

Within the turbulence community it now seems generally, if implicitly, agreed that turbulence models are not predictive tools. It was not always so. Quoting Lumley again: "Many of the initial successes of the models (in comparison to first order ones) have been in more complex flows, involving heat transfer, buoyancy and the like, because the relevant physical mechanisms are included. In addition, some of the successes have been in flows dominated by inertia or mean buoyancy, where the details of the turbulence model are irrelevant. Thus emboldened, the modelers have been overenthusiastic in promoting their models for other complex situations, often without considering at depth the difficult questions that arise. Consequently, there is some disillusionment with the models, a feeling that they embody too many *ad hoc* assumptions, and that they are unreliable as a result .... This reaction is probably justified, but it would be a shame if it resulted in a cessation of efforts to put a little more physics and mathematics into the models."

Because of their inherent simplicity turbulence models will probably always be important, even if and when the "turbulence problem" is solved. As we learn more about the structure of turbulent flows, however, we learn more about the weaknesses of our turbulence models and we can improve them. This "failure-improvement" cycle is an important part of turbulence modeling, even though it tends not to be part of its written record [28]. Numerical simulation is one path to improving models of turbulence.

In LES only the largest eddies are simulated; those smaller than the grid are modeled. Since it is believed to be more important to model the largest eddies reliably, I classify LES under simulation.

3.1. LARGE-EDDY SIMULATION

LES is much more popular today than it was 15 or so years ago. It is a good deal older than that; Deardorff did his first LES studies in the late 1960s. I recall meeting him for the first time in 1970, during a turbulence symposium in Boulder, Colorado organized by E. Plate and G. Hidy. Chandran Kaimal and I visited Deardorff in his office at NCAR. He was obtaining the first atmospheric results [7] from his code of $40 \times 40 \times 20$ grid points; his pioneering study [8] of turbulent channel flow had used a $24 \times 14 \times 20$ grid. (For comparison, today's largest LES codes are $512 \times 512 \times 512$.) For its subgrid-scale (SGS) closure his LES code used Smagorinsky's early notions as interpreted by Lilly [14] for inertial-range turbulence. I remember his excitedly showing us computational evidence of thermal plumes and eddies on the pages spewing from the huge, clattering line printer.

Before long, engineers appropriated the numerical simulation technique and, much to their credit, gave it the semantically precise name large-eddy simulation; that is much better than the vague term "three-dimensional numerical modeling" that Deardorff had used. The engineers also cleaned up the formal basis of the technique. Leonard [12] established clearly the role of spatial filtering in LES; that had been a bit fuzzy up to that point. Nothing else changed much, and the current practice of LES, as we now call it, differs mainly in details from that early, path-blazing work of Deardorff and Lilly.

LES is inherently an approximation: it deals explicitly with only the largest eddies and treats the remaining eddies in a way that is not yet entirely defensible from first principles. As one of the few ways to calculate time-dependent, three-dimensional turbulence fields, LES has to be taken seriously, however.

Except for the rare occasions when its predictions clearly disagree with experiment, as in the mean wind and mean scalar profiles near the surface [16], systematic experimental evaluation of LES results in micrometeorology has been conspicuously absent. Instead, LES seems increasingly to be taken as "ground truth." A recent intercomparison of LES results for the convective boundary layer [19] gave the reassuring result that all four codes behaved quite similarly despite their differences in formulation—but the authors were able to make almost no comparisons with experimental data. The parallel study of a neutral boundary layer [1] however, showed significant differences in the results from the four LES codes in the lowest third of the boundary layer. In this paper there were no comparisons with experimental data. Bradshaw's [3] "fact gap" seems wider than ever in micrometeorology.

Nonetheless, LES is now fairly routinely used for testing and developing models and for studying boundary-layer physics. Ayotte *et al.* [2], for example, used LES data bases in their study of the performance of contemporary boundary-layer models.

What are the problems with LES?

One is its SGS model, its approximate representation of the effects of the eddies removed by the spatial filter. Although LES is nearing the end of its third decade, this SGS modeling problem has not been solved satisfactorily. Many LES codes use essentially the SGS model used by Deardorff.

One of the strengths of LES is its relative insensitivity to the SGS model if the flow is well resolved. In some flows, though, the resolution is not uniformly adequate. For example, in the convective PBL the surface layer and the interfacial layer are apt to be poorly resolved. The horizontal scale of vertical velocity fluctuations near the surface goes as z, so that LES underresolves w at the first few grid points. In such regions the SGS model beomes more important and so it ought to perform well. We are just beginning to look at such details and, not surprisingly, we are finding that SGS models perform quite poorly within the surface layer [29].

If the SGS model of LES is inadequate in the surface layer there is a good chance that the resolvable-scale fields are inaccurate there. Given our limited ability to measure three-dimensional fields, even in the surface layer, it is very difficult to test this notion. One might suspect that LES has problems in the interfacial layer as well, where the SGS model must also carry a heavier load. That is even less accessible than the surface layer, however. This points out that the experimental testing of LES is in a bad state—we can do very little beyond testing expected values.

Some testing of LES can be done in laboratory facilities. Figure 4 shows the top-down and bottom-up diffusivites for scalars as determined through measurements in a replica of the Willis-Deardorff convection tank installed at the US EPA Fluid Modeling Facility. The results agree quite well with those determined through LES, even reproducing the mid-layer singularity in the bottom-up eddy diffusivity. To my knowledge comparable results have not yet been obtained in the atmosphere; to date the scatter has been too great.

3.2. DIRECT NUMERICAL SIMULATION

Direct numerical simulation is the straightforward numerical solution of the Navier-Stokes equation. It requires that one resolve the entire eddy size range, including the viscous eddies that keep turbulence near energetic equilibrium by dissipating kinetic energy (i.e., converting it into internal energy) at rate ϵ per unit mass through its kinematic viscosity ν. It follows that in DNS

$$\frac{\ell}{\eta} \sim N, \tag{1}$$

where ℓ is the integral scale, $\eta = \left(\nu^3/\epsilon\right)^{1/4}$ is the Kolmogorov microscale, and N is the

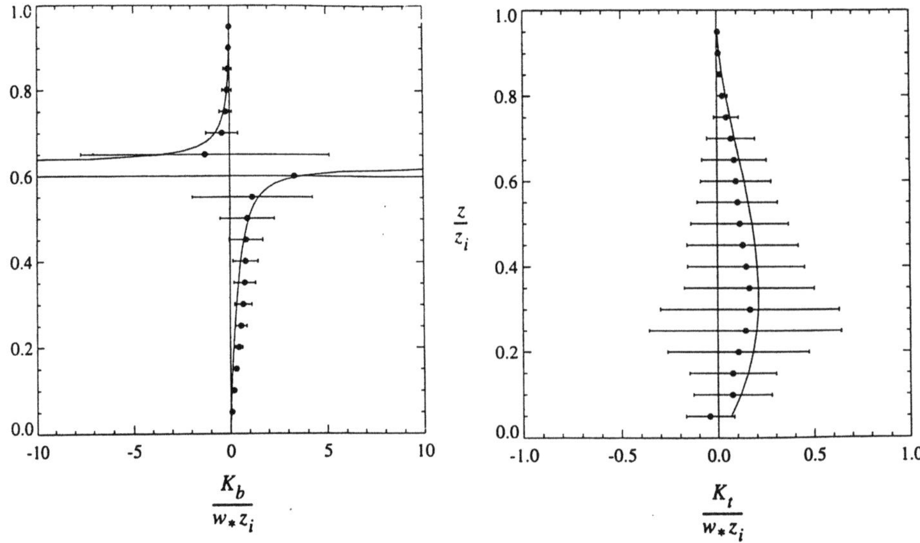

Figure 4. The profiles of top-down and bottom-up eddy diffusivities for scalars. Solid lines are the LES predictions; data points represent the average over 14 runs in the convection tank. Brackets denote the standard deviation of the data. From [21].

number of grid points per linear dimension. Since in any turbulent flow

$$\epsilon \sim \frac{u^3}{\ell}, \tag{2}$$

where u is the velocity scale of the energy-containing motion, we can rewrite this as

$$N \sim \left(\frac{u\ell}{\nu}\right)^{3/4} = Re^{3/4}, \tag{3}$$

so the number of grid points per linear dimension goes as the 3/4 power of the large-eddy Reynolds number Re. N can be a few hundred at most, so we see that DNS can only reach large-eddy Reynolds numbers of the order of 1000.

Can we do DNS of the convective PBL? Not literally, since $Re \sim 10^8$ there. We can do it only if we reduce its Re to 1000 or so—by changing $u, \ell,$ or ν.

The flow that Coleman *et al.* [4] calculated through DNS can thus be interpreted as a low speed, scaled down version of the PBL: it is about 1 m deep and has a mean wind speed on the order of 1 cm/sec. It takes strong faith to believe that such an abnormally shallow, low-speed boundary layer is representative of the PBL.

We could also visualize their flow as being of standard depth (1000 m, say) and speed (10 m/s, say) but of a fluid of 10^5 times the kinematic viscosity of air. This would be heavy molasses, for example.

For most of us the idea that a PBL of molasses would have the large-eddy structure of the real PBL is probably unreasonable. It is an example of what we implicitly assume when using DNS, however; we assume that the energy-containing structure of a high-Re flow is independent of Re so we can calculate it through DNS at much lower Re. This assumption is called Reynolds-number similarity; it means Reynolds-number independence. It has its roots in the essentially inviscid nature of the energy-containing range of turbulence. It is generally supported well by laboratory measurements of turbulent flows, but those measurements seldom span the five decades in Re that exist in this PBL application.

Where is Reynolds-number similarity not valid? When are DNS results not indicative of the structure of flows at much larger Re? The short answer is anywhere that viscosity is important—near walls, for example.

The longer answer is probably not beyond the energy-containing eddies—not in the dissipative range or in the inertial range (should it exist in the DNS). One implication is that any problem that depends on dissipative range structure could exhibit Re dependence.

A physical interpretation of this Re dependence starts with the observation that dissipative structure—e.g., time series of velocity derivatives—has long been known to be intermittent. The physical suggestion is that viscous dissipation is not uniformly distributed in space at any time, but is clustered in widely separated "hot spots" of activity. The shapes of these regions are of interest; Corrsin [6] speculated they were sheets of scale ℓ and thickness η. Tennekes [24] suggested instead they were tubes of diameter η and length λ, the Taylor microscale.

With some scaling relations in hand we can explore the Reynolds number trends implied by these fine-structure models. We start from the relations

$$\epsilon \sim \frac{u^3}{\ell} \sim \nu \frac{u^2}{\lambda^2}, \tag{4}$$

where λ is the Taylor microscale. These imply that

$$\frac{\ell}{\lambda} \sim Re^{1/2}. \tag{5}$$

We then compute for each model the volume fraction occupied by the dissipative regions:

$$\text{tube model: volume fraction} \sim \frac{\eta^2 \lambda}{\lambda^3} \sim \frac{\eta^2}{\lambda^2} \sim Re^{-1/2},$$

$$\text{sheet model: volume fraction} \sim \frac{\eta \ell^2}{\ell^3} \sim \frac{\eta}{\ell} \sim Re^{-3/4}, \tag{6}$$

where we have used (3). Each model shows that the dissipative regions occupy a decreasing volume fraction of the turbulent fluid as Re increases.

The models also predict that the velocity derivative signal grows in "spikiness" as *Re* increases. In the tube model, for example, the highest-amplitude velocity derivatives are of order

$$\frac{\partial u}{\partial x} \sim \frac{u}{\eta}. \tag{7}$$

It follows that the expected value of the fourth moment in the tube model is

$$\overline{\left(\frac{\partial u}{\partial x}\right)^4} \sim \frac{u^4}{\eta^4} \times \text{volume fraction} \sim \frac{u^4}{\eta^4} \times \frac{\eta^2}{\lambda^2} \sim \frac{u^4}{\eta^2\lambda^2}. \tag{8}$$

By definition the derivative variance is

$$\overline{\left(\frac{\partial u}{\partial x}\right)^2} \sim \frac{u^2}{\lambda^2}. \tag{9}$$

It follows from (8) and (9) that the "flatness factor" F of the velocity derivative for the tube model is

$$\text{tube model}: F \sim \frac{\overline{\left(\frac{\partial u}{\partial x}\right)^4}}{[\overline{\left(\frac{\partial u}{\partial x}\right)^2}]^2} \sim \left(\frac{\lambda}{\eta}\right)^2 \sim Re^{1/2}. \tag{10}$$

For the sheet model we find

$$F \sim Re^{1/4}. \tag{11}$$

Data (Figure 5) show clearly that F does increase with *Re*, although the trend is not as steep as these simple models imply.

The roots of these simple dissipation intermittency models models lie in part in the revised Kolmogorov hypotheses [11] for small-scale turbulence, which have also been discussed and interpreted by Obukhov [20]. These revised hypotheses, and the dissipation intermittency models they stimulated, indicate that the "landscape" of the dissipative range of turbulent flows depends continuously on *Re*, and therefore that the fine-structure landscapes are considerably different in low Reynolds number DNS and in the atmospheric boundary layer. In the atmospheric boundary layer the dissipative regions are predicted to be much rarer—i.e., much more intermittent. The flatness factor of derivatives of velocity or an advected scalar, a measure of the spikiness of the derivative signal, is observed to be an order of magnitude larger in the atmosphere than is calculated in DNS.

We conclude that in problems influenced by the fine structure of turbulence, DNS might not give results that are representative of the behavior in the much larger *Re* geophysical flows. Perhaps surprisingly, it seems that to date not much attention has been paid to the implications of this lack of *Re* similarity of turbulent fine structure. To my knowledge not many problems have been identified in which it is important. One exception could be cloud physics, where Tennekes and Woods [25] have shown that the coalescence of droplets is affected by the statistics of the velocity shear, a fine-structure property. It has recently been argued by Shaw *et al.* [22] that regions of spatially nonuniform cloud droplet concentration, which have been observed in nature, can be generated by the interactions of cloud droplets with turbulent fine structure. This raises the possibility that the relevant physics here could be quite different in direct numerical or laboratory simulations of the phenomena and in the vastly larger *Re* turbulence of precipitating clouds.

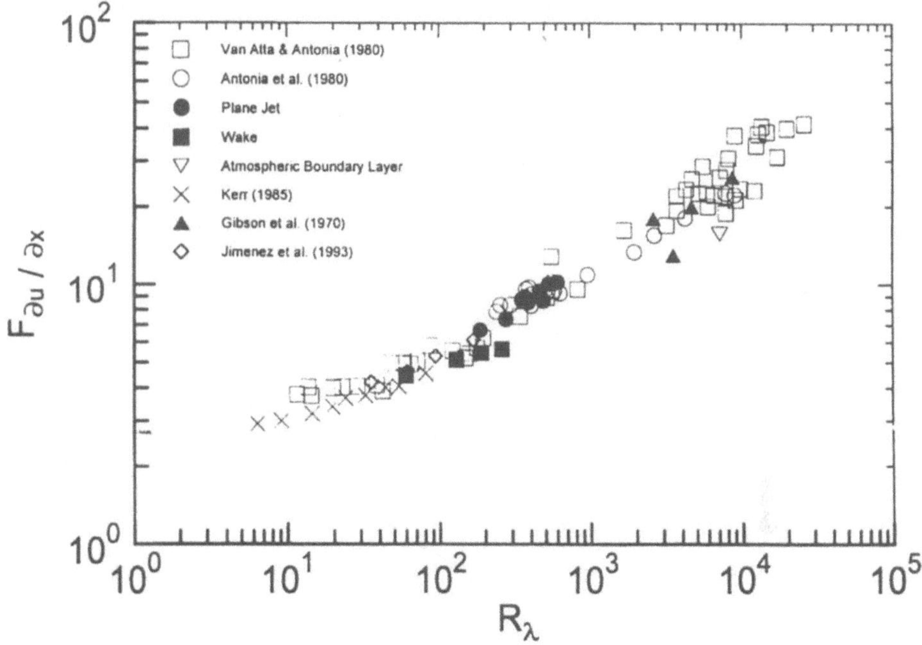

Figure 5. The Reynolds number (R_λ) dependence of the flatness factor of the velocity derivative as measured in various turbulent flows. $R_\lambda \sim Re^{1/2}$. From Sreenivasan and Antonia [23].

4. Summary

This is the era of turbulence simulation. LES, in particular, is increasing in popularity and is displacing traditional experiments and observations. On the one hand that is troubling because LES is not currently firmly tied to observational work. On the other hand, unlike traditional observations LES does provide a means of generating the three-dimensional fields that can be used as testbeds for developing turbulence models, and that represents a huge advance (Figures 2, 3, 6). A remaining and not widely acknowledged problem is the great difficulty of testing the fidelity and reliability of the LES results. Here laboratory flows can serve as ground truth, since they generally provide "cleaner", less scattered data than the atmosphere itself.

We need to focus on developing new experimental techniques that can be used in the atmosphere to produce data for testing LES. I think this is going to be difficult because I suspect part of the problem here is cultural. For an undoubtedly complex and subtle set of reasons, many students and young researchers today tend to find numerical simulation more attractive than observational work. It is clear that the student today does have an attractive option, numerical simulation, that those of previous generations did not have. Numerical simulation is undeniably a powerful tool, but it needs to be shaped, guided, and constrained by observations.

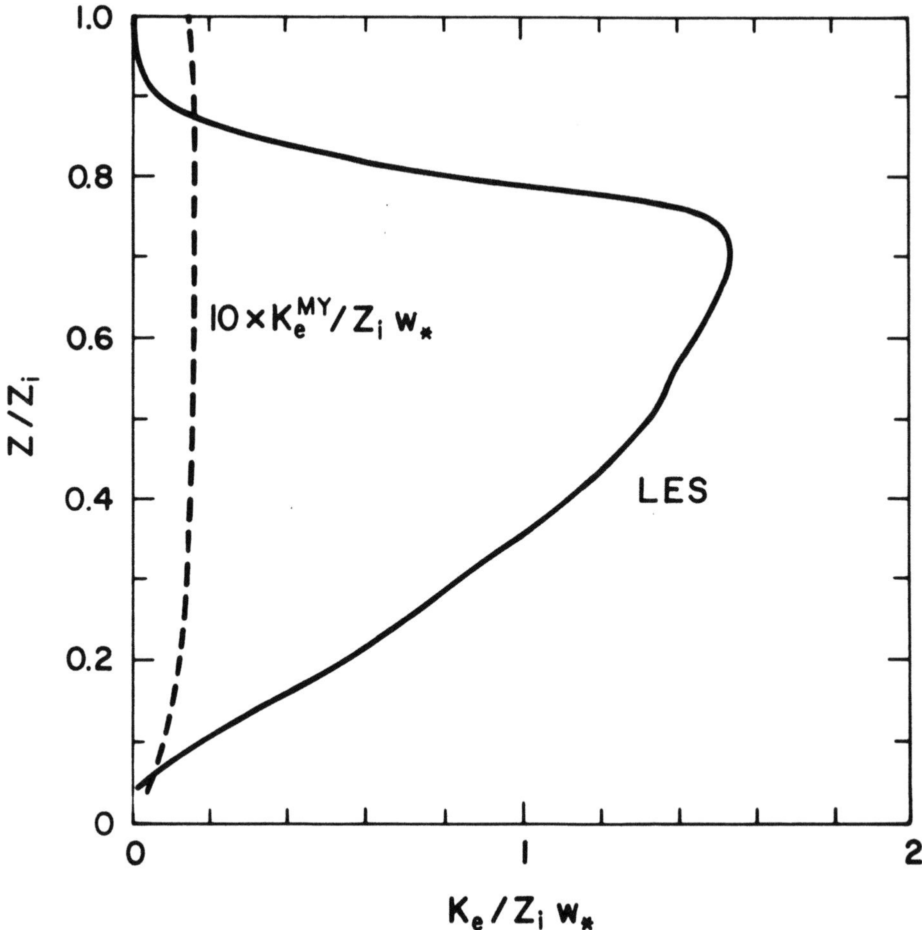

Figure 6. Profiles of the eddy diffusivity for energy flux from 96^3 LES (solid curve) and as used in a contemporary second-order model (dashed curve). The latter is magnified by a factor of ten. LES shows the model parameterization to be poor. From [18].

References

1. Andrén, A., Brown, A. R., Graf, J., Moeng, C.-H., Nieuwstadt, F. T. M., and Schumann, U. (1994) Large-eddy simulation of a neutrally stratified boundary layer: A comparison of four computer codes, *Quart. J. Roy. Meteor. Soc.* **120**, 1457–1484.
2. Ayotte, K., Sullivan, P., Andrén, A., Doney, S., Holtslag, A., Large, W., McWilliams, J., Moeng, C.-H., Otte, M., Tribbia, J., and Wyngaard, J. (1996) An evaluation of neutral and convective planetary boundary-layer parameterizations relative to large-eddy simulations, *Boundary-Layer Meteorol.* **79**, 131–175.
3. Bradshaw, P. (1972) The understanding and prediction of turbulent flow, *Aero. J.* **76**, 403–418.
4. Coleman, G. N., Ferziger, J. H., and Spalart, P. R. (1990) A numerical study of the turbulent Ekman layer, *J. Fluid Mech.* **213**, 313–348.
5. Corrsin, S. (1961) Turbulent flow, *Am. Sci.*, **49**, 300–325.

6. Corrsin, S. (1962) Turbulent dissipation fluctuations, *Phys. Fluids* **5**, 1301–1302.
7. Deardorff, J. W. (1970) Convective velocity and temperature scales for the unstable planetary boundary layer and for Rayleigh convection, *J. Atmos. Sci.* **27**, 1211–1213.
8. Deardorff, J. W. (1970) A numerical study of three-dimensional channel flow at large Reynolds numbers, *J. Fluid Mech.* **41**, 453–480.
9. Deardorff, J. W. and Willis, G. E. (1985) Further results from a laboratory model of the convective planetary boundary layer, *Bound.-Layer Meteor.* **32**, 205–236.
10. Haugen, D. A., Kaimal, J. C. , and Bradley, E. F. (1971) An experimental study of Reynolds stress and heat flux in the atmospheric surface layer, *Quart. J. Roy. Meteor. Soc.* **97**, 168–180.
11. Kolmogorov, A. N. (1962) A refinement of previous hypotheses concerning the local structure of turbulence in a viscous incompressible fluid at high Reynolds number, *J. Fluid Mech.* **13**, 82–85.
12. Leonard, A. (1973) Energy cascade in large eddy simulation of turbulent flows, *Adv. Geophys.* **18A**, 237–248.
13. Liepmann, H. W. (1970) The rise and fall of ideas in turbulence, *Am. Scientist*, **67**, 221–228.
14. Lilly, D. K. (1967) The representation of small-scale turbulence in numerical simulation experiments, *Proc. IBM Sci. Comput. Symp. Environmental Sci.*, IBM Form No. 320-1951, 195–210.
15. Lumley, J. L. (1983) Atmospheric modelling, *Mech. Eng. Trans.*, Inst. of Eng., Australia, ME8:153–159.
16. Mason, P. J. and Thomson, D. J. (1992) Stochastic backscatter in large-eddy simulations of boundary layers, *J. Fluid Mech.* **242**, 51–78.
17. McAllister, L. G., Pollard, J. R., Mahoney, A. R., and Shaw, P. J. R., (1969) Acoustic sounding: A new approach to the study of atmospheric structure, *Proc. IEEE* **57**, 579–587.
18. Moeng, C.-H. and Wyngaard, J. C. (1989) Evaluation of turbulent transport and dissipation closures in second-order modeling, *J. Atmos. Sci.* **46**, 2311–2330.
19. Nieuwstadt, F.T.M., Mason, P. J., Moeng, C.-H., and Schumann, U. (1993) Large-eddy simulation of the convective boundary layer: A comparison of four computer codes, in F. Durst, R. Friedrich, B.E. Launder, F.W. Schmidt, U. Schumann, and J.H. Whitelaw (eds.), *Turbulent Shear Flows 8*, Springer-Verlag, Berlin.
20. Obukhov, A. M. (1962) Some specific features of atmospheric turbulence, *J. Fluid Mech.* **13**, 77–81.
21. Piper, M., Wyngaard, J. C., Snyder, W. H., and Lawson, R. E. Jr. (1995) Top-down, bottom-up diffusion experiments in a water convection tank, *J. Atmos. Sci.* **52**, 3607–3619.
22. Shaw, R. A., Reade, W. C., Collins, L. R., and Verlinde, J. (1998) Preferential concentration of cloud droplets by turbulence: effects on the early evolution of cumulus cloud droplet spectra, to appear, *J. Atmos. Sci.*
23. Sreenivasan, K. R. and Antonia, R. A. (1997) The phenomenology of small-scale turbulence, *Annu. Rev. Fluid Mech.* **29**, 435–472.
24. Tennekes, H. (1968) A simple model for the small-scale structure of turbulence, *Phys. Fluids* **11**, 669.
25. Tennekes, H. and Woods, J. D. (1973) Coalescence in a weakly turbulent cloud, *Quart. J. Roy. Meteorol. Soc.* **99**, 758–763.
26. Weil, J. C. (1988) Dispersion in the convective boundary layer, in *Lectures on Air Pollution Modeling*, A. Venkatram and J. Wyngaard (eds.), AMS, Boston, 167–227.
27. Willis, G. E. and Deardorff, J. W. (1974) A laboratory model of the unstable planetary boundary layer, *J. Atmos. Sci.* **31**, 1297–1307.
28. Wyngaard, J. C. (1997) Review of *Simulation and Modeling of Turbulent Flows, Bull. Am. Meteorol. Soc.*, **78**, 1785–1787.
29. Wyngaard, J. C., Peltier, L. J., and Khanna, S. (1998) LES in the surface layer: Surface fluxes, scaling, and SGS modeling, to appear in *J. Atmos. Sci.* **55**.

TRANSILIENT TURBULENCE THEORY:
A NONLOCAL DESCRIPTION OF CONVECTION

R. B. STULL
Atmospheric Science Programme, Department of Geography
The University of British Columbia, 1984 West Mall
Vancouver, BC, V6T 1Z2 Canada

Abstract

Turbulent atmospheric boundary layers contain a range of eddy sizes, the largest of which can vertically advect air across large distances with little dilution. This nonlocal mixing can be described by a transilient matrix, where each element indicates what fraction of air at a destination height comes from any source height. Parameterization of this matrix allows transilient turbulence theory to be used to forecast boundary-layer evolution.

1. Introduction to Nonlocal Mixing

As discussed in the "Convective Transport Theory and the Radix Layer" chapter of this volume, the atmospheric convective mixed layer (ML) contains large, coherent turbulent structures called thermals. These thermals have undiluted cores that can nonlocally advect air across the whole depth of the ML. Such an advection-like process is poorly described by diffusion theories. This was the motivation for developing transilient turbulence theory (T3), which describes turbulence via nonlocal eddy-advections [8]. The word "transilient" is from a Latin root meaning "jump over", to suggest the nonlocal nature of this theory.

T3 is a first-order, nonlocal, ensemble-average turbulence closure theory [9]. It is usually applied within single vertical columns of air, where the column is vertically divided into discrete grid cells represented by grid points. The column is assumed to be sufficiently wide to cover a horizontal area of tens to hundreds of kilometers, within which there are many unresolved (subgrid) thermals and eddies of a wide range of sizes (Figure 1). T3 does not attempt to describe each individual eddy or thermal. Instead, it

E.J. Plate et al. (eds.), Buoyant Convection in Geophysical Flows, 253–264.

describes the vertically-resolved area-average net transport caused by an ensemble of horizontally unresolved thermals.

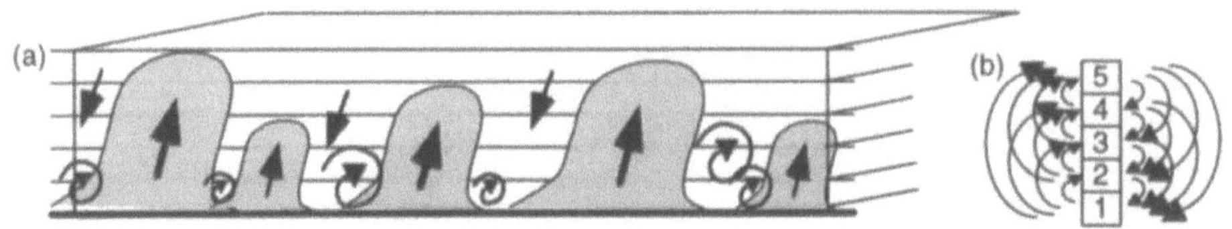

Figure 1. Illustration of (a) physical space showing one grid column with horizontally-unresolved thermals and eddies, and (b) a grid-point representation of the same processes, indicating all possible nonlocal and local vertical mixings that can be described by transilient turbulence theory.

The framework of T3 is a square matrix, called a transilient matrix, which describes the net transport from every possible source cell (with grid index j) to every possible destination cell (i) during a finite time step Δt, even if the sources and destinations are not adjacent in the vertical. The elements of this transilient matrix can be parameterized as a function of the mean flow and externally imposed forcings and constraints. This is a generic turbulence closure theory that can be applied over the whole column, regardless of whether there are subdomains of turbulence caused by boundary layers and/or clear-air turbulence (CAT). Forecasts made with T3 show that it can automatically capture the structure and growth of boundary layers, regardless of whether the air is statically unstable, neutral, or stable [12].

For comparison, local closure theories also have two components: a framework and a closure parameterization. One example of a local-closure framework is K-theory, which assumes that turbulent fluxes flow down local gradients of mean variables. To use such a closure, parameterizations of the value of K, the eddy diffusivity, must be made at every height in the column. K can be parameterized as a function of mean variables and external forcings. K-theory assumes that turbulent transport is dominated by small eddies, which is why it has difficulty in the convective ML where large eddies dominate. K-theory is just a subset of the more general T3, which includes the effect of all eddy sizes.

In section 2, the basic framework of T3 is summarized, and the need for parameterization is identified. Section 3 traces the evolution of T3 parameterizations during the past 15 years. New developments are introduced in section 4.

2. A Brief Overview of Transilient Turbulence Theory

2.1. CHARACTERISTICS OF TRANSILIENT MATRICES

A very thorough review of T3 is given by Stull [8]. The elements of a transilient matrix are $c_{ij}(t, \Delta t)$. The values of the elements indicate what fraction of air ending at destination i came from source j during time step Δt, such as illustrated in Figure 2.

The location of the elements in the matrix indicate the net size of the eddy; in other words, the amount of nonlocalness. The values of the elements change with time t for nonstationary turbulence, and at any instant in time they change with time step duration Δt.

To conserve air mass and state of the air (e.g., heat, moisture, etc.) each row and each column of the matrix must sum to one. This type of matrix is said to be doubly stochastic. Also, each element must be non-negative, which is required in our universe of increasing entropy or randomness; otherwise, negative matrix elements would describe unmixing. The net result of these restrictions is that each element must be in the range 0 to 1 inclusive.

Elements along the main diagonal of the matrix, where $i = j$, correspond to air having the same source and destination cell. In other words, elements along the main diagonal indicate how much of the air does NOT mix out of a grid cell during time step Δt. Values of 1 anywhere on the main diagonal indicate that there is no turbulent mixing to or from that grid cell. Elements adjacent to the main diagonal correspond to small-eddy turbulence, and give the amount of mixing between any grid cell and it's immediate vertical neighbor. A tridiagonal transilient matrix gives the same result as K-theory.

Elements further and further from the main diagonal indicate larger and larger net transport distances between more distant grid cells. This is where the nonlocal information is described. Because any one matrix corresponds to any one time step Δt, elements further from the main diagonal also indicate greater net vertical velocities of the turbulence. The elements in the upper triangle of the matrix indicate upward transport, while elements in the lower triangle indicate downward transport.

If all the off-diagonal elements of the transilient matrix are zero, then it equals the identity matrix, $\boldsymbol{I}$, which implies no turbulent mixing anywhere. The other extreme is where all of the elements are equal to $1/N$, where N is the number of grid points in the air column. This is a total-mixing matrix, $\boldsymbol{M}$, which completely homogenizes the column of air during one time step.

For very short time steps, one would anticipate that the actual transilient matrix should approach the identity matrix, because very little mixing is physically possible during very short time intervals.

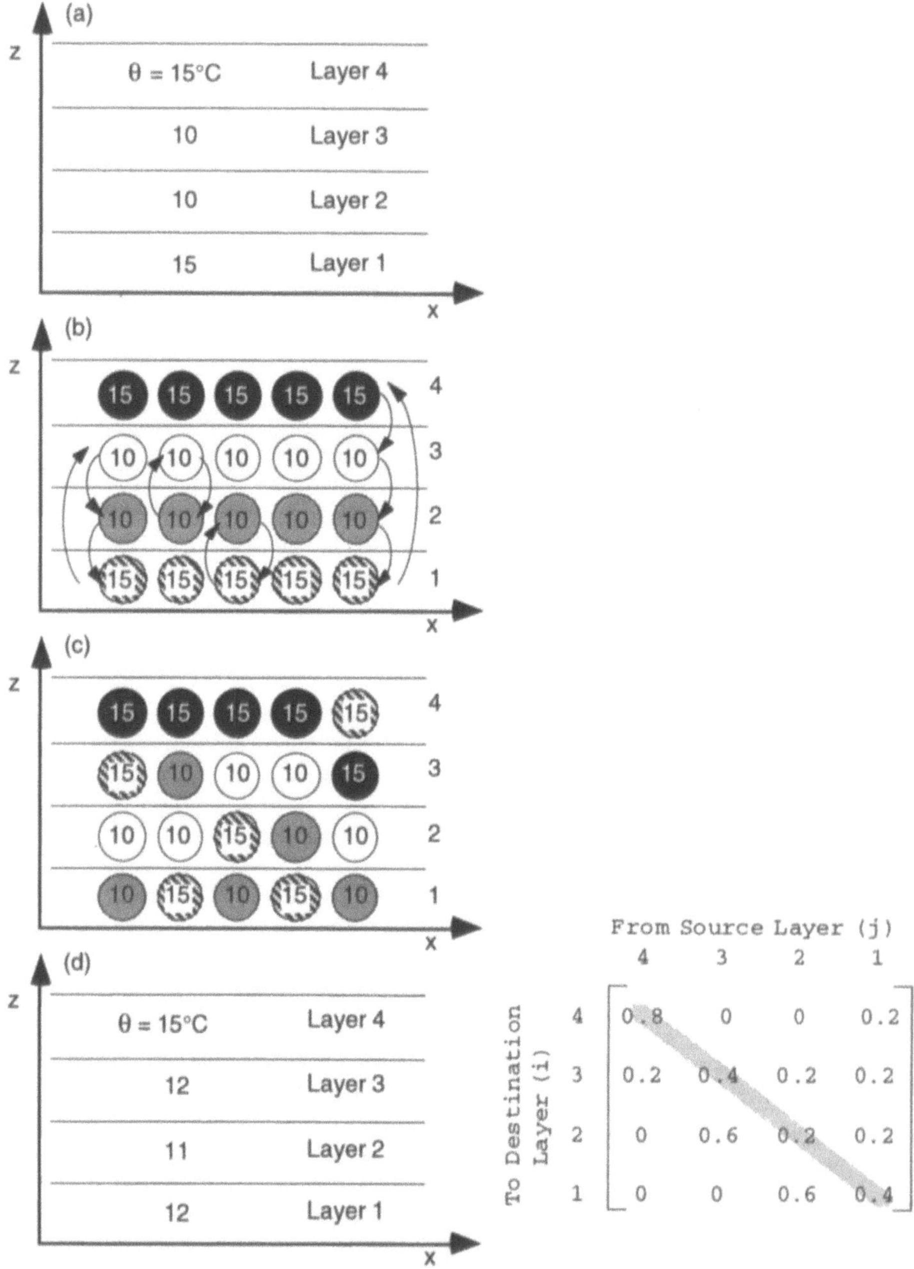

Figure 2. (a) Given four layers in a grid column, conceptually divide each layer into five identical air parcels (b). Suppose some of the parcels are exchanged by local and nonlocal mixing as sketched in (b) and (c). Averaging each layer gives the final mean potential temperatures (d) after one time step of mixing. The transilient matrix at right describes this mixing process [11], where each row and column sums to one.

On the other hand, for very long time steps, the transilient matrix should approach the well-mixed matrix within the subdomain of the matrix that is turbulent. Thus, the best time interval when using transilient turbulence theory is one that gives a transilient matrix midway between these two extremes. In most of the previous publications about transilient matrices, the matrix was flipped vertically relative to the normal mathematical notation, such that destination cells lower in the atmosphere (with lower values of grid index i), would correspond to the bottom of the matrix. This resulted in a main diagonal that went from lower left to upper right, and was somewhat confusing to new users of the theory. I now recommend that the matrix also be flipped left to right, such that lower grid cells (smaller i and j) appear at the lower right of the matrix, with the main diagonal in the usual mathematical direction from lower right to upper left [11]. For those of you reading the older transilient literature, just flip the matrix left to right to get a standard-looking matrix. Such a flip is NOT a transpose nor an inverse, so be careful.

2.2. APPLICATION OF TRANSILIENT MATRICES

The transilient matrix is used to make forecasts using simple matrix multiplication as the underlying framework:

$$\psi_i(t+\Delta t) = \sum_{j=1}^{N} c_{ij}(t,\Delta t)\cdot \psi_j(t) \tag{1}$$

where y represents any variable conserved during vertical advective movement, such as potential temperature, water vapor mixing ratio, pollutant concentration, and wind components. No matrix inversion is required, thereby making the method fast to compute.

A fundamental assumption of the previous equation is that turbulence consists of the undiluted nonlocal movement of air parcels, and that each air parcel carries with it the source characteristics of temperature, humidity, etc. In other words, the same transilient matrix is used for all variables. It is NOT like molecular diffusion, where different variables have different diffusivities.

Another fundamental assumption is that (1) describes only the turbulent advection part of the governing conservation equation for any variable. Other changes to the variable (e.g., radiative heating, chemical reactions, pressure correlations, advection by the mean wind, Coriolis force, latent heating) must be accounted for with separate forecast equations. Such a separation of physical computations is usually done in numerical weather forecast models anyway.

One outcome of the doubly-stochastic nature of the matrix is that (1) is absolutely numerical stable for any time step and any grid spacing [5]. This is a boon in weather forecast models. Nonetheless, it is usually wise to pick a time step that is small enough (roughly 15 minutes for convective mixed layers) that nonstationarity associated with the diurnal cycle or with other changes of the BL are captured.

Knowing the initial values of mean variables at the grid points at the start of each time step, and knowing the amount of air mixed between the various grid cells, it is easy to compute the net kinematic flux at any level k :

$$F_k(t) = \frac{\Delta z}{\Delta t} \sum_{i=1}^{k} \sum_{j=k+1}^{N} \left[c_{ji}\psi_i - c_{ij}\psi_j \right] \tag{2}$$

where Δz is the vertical grid spacing. This is a diagnostic equation that is not needed to make the forecast, but which can provide insight into the net effect of all the turbulent eddies. While i and j represent values in the center of grid cells, index k represents a location at the top of the grid cell. The ground is at $k = 0$, for which the turbulent flux F_0 must be zero by definition (unless clods of earth are jumping up and down in a turbulent dance). At the top of the model, the turbulent flux F_N must also be zero by definition.

When (1) is used to make forecasts, imposed bottom boundary conditions such as heat flux, moisture flux, and momentum flux from the earth's surface are applied via separate equations into only the bottom grid cell. The transilient matrix automatically takes care of distributing this boundary forcing to the other grid boxes during subsequent time steps. Any body-force terms such as radiative heating or Coriolis force must be applied via separate equations to every grid cell in the column.

3. History of Transilient Closures

3.1. GUIDING CONCEPTS

To use (1), the transilient matrix must be parameterized. To date, all of the parameterizations for this matrix have utilized first-order statistical closure [8]. Namely, the matrix is parameterized as a function of the mean flow (i.e., grid-averaged resolved values of potential temperature, wind components, etc.). As for other turbulence closures, there are no first-principles that directly give the parameterization. Instead, investigators can design closures to suit their tastes, subject to the constraints of any parameterization [7] such as physical realism, proper units, proper tensor characteristics, invariance with coordinate transform, etc.

To guide our formulation of nonlocal closure we evoke Le Chatelier's principle, which states that nature reacts in a way to partially undo any externally-imposed instabilities. For example, if the air becomes statically unstable, then we would expect turbulence to form to reduce the instability by mixing warm air upward and cool air downward. Another example, if the air becomes dynamically unstable, then turbulence will form to mix both winds and temperature such that the Richardson number rises above its critical value.

In other words, turbulence forms as a response to flow instabilities. This turbulence is temporary, and disappears once the instability has been eliminated. Usually the intensity of turbulence is proportional to the amount of instability. If external forcings act to continuously destabilize the atmosphere, such as by continually heating the bottom of the boundary layer during a sunny day, then turbulence also continuously exists during that period.

With these guiding concepts, a variety of closures are possible. In the next subsections, we trace the history of some of the closures that have been developed.

3.2. VERSION 1: TURBULENT ADJUSTMENT

The first transilient parameterization [4, 14] built on the concept of convective adjustment. For a convectively unstable atmosphere, using lapse rate as a measure of the instability, it is possible to solve the precise amount of mixing between vertically neighboring grid cells needed to eliminate the instability. Vertical mixing that is governed by this principle is called convective adjustment.

By extension, it was thought that measures of dynamic instability such as the Richardson number could be used to dynamically adjust neighboring grid cells toward an equilibrium. However, contrary to convective adjustment, the only way that dynamic equilibrium can be reached is by allowing nonlocal mixing. This was the original motivation for developing T3.

By defining a nonlocal Richardson number R_{ij} between source grid cell j and destination grid cell i, where i and j need not be vertical neighbors, a measure of the nonlocal instability was possible:

$$R_{ij} = \frac{g \cdot \Delta_{ij} z \cdot (\Delta_{ij} \theta_v)}{(T_{v\,i} + T_{v\,j}) \cdot [(\Delta_{ij} U)^2 + (\Delta_{ij} V)^2]} \tag{3}$$

where $\Delta_{ij} y = y_i - y_j$, θ_v is virtual potential temperature, T_v is virtual absolute temperature, z is height, g is gravitational acceleration, and (U, V) are the horizontal components of the mean wind. Details of how this Richardson can be used to find the transilient matrix are described by Stull [4].

This parameterization was applied to idealized boundary layers in the atmosphere and the ocean, and yielded good results. It was observed that boundary-layer depth need not be pre-specified in this approach; namely, the transilient scheme automatically determined the depth and growth of the turbulent boundary layer based on the initial conditions and the amount and history of instability.

Because the nonlocal Richardson number uses source and destination information equally to define temperature and wind differences, the resulting parameterization was unable to discriminate between upward and downward mixing. Namely, the parameterization yielded symmetric transilient matrices. Another drawback of this method was that the measure of instability and the constraints of mass and state conservation were rigidly tied together in a way that did not allow other destabilization factors to be accounted for. These drawbacks motivated version 2, described next.

3.3. VERSION 2: MIXING POTENTIAL

The Richardson number considers only buoyant and shear effects. However, turbulence is also affected by viscous dissipation, turbulent transport, storage, and pressure transport, such as described by the turbulence kinetic energy (TKE) equation [7]. The next logical step was to formulate a nonlocal approximation to the TKE equation, analogous to the nonlocal approximation to Richardson number, where differences between nonlocal grid cells are substituted in place of local gradients [6, 13].

However, it was impossible to get a transilient matrix that was perfectly doubly stochastic directly from the nonlocal TKE. So instead an intermediate step was devised to capture the potential for mixing in a matrix Y_{ij}, where the elements of this matrix need not satisfy any conservation of mass constraint.

$$Y_{ij} = (\Delta_{ij}U)^2 + (\Delta_{ij}V)^2 - \frac{g \cdot \Delta_{ij}z \cdot (\Delta_{ij}\theta_v)}{(T_{v\ i} + T_{v\ j}) \cdot R_c} - \varepsilon \cdot (\Delta_{ij}z)^2 \qquad (4)$$

where e is a dissipation parameter, and R_c is a critical Richardson number. This mixing potential matrix considers shear generation of turbulence (first two terms on the right), buoyant production or consumption, and viscous dissipation, but neglects the other turbulent and mean transport terms.

Although the elements of this matrix indicate the amounts of instability nonlocally, this instability might not be able to be totally released to cause turbulent mixing during any one time step, due to competing instabilities. A second step in the parameterization utilizes an L_∞ norm to transform the Y_{ij} matrix into a transilient matrix, where each row and column sums to one.

This transformation was not ideal, because it would give inappropriate weight to one of the rows of the matrix. Nonetheless, it proved a very successful and flexible method. It yielded fine results when used for more detailed verification against field experiments, for simulation of neutral and nocturnal boundary layers, and for cloud-topped boundary layers. A Richardson number version of the mixing potential method was also tested later [15], and compared against field data.

3.4. VERSION 3: BOUNDARY CONDITION SPLITTING

Recall from a previous subsection that surface boundary conditions are imposed only on the bottom grid point. This worked fine for time steps of about 15 minutes and vertical grid spacings of 200 km or so. However, as more simulations and verifications were performed, there were occasions where it was appropriate to take longer time steps with much thinner grid cells. One can imagine that if a heat flux is imposed into the bottom of a small grid cell for very long time, the temperature in that cell will rise to very high magnitudes. This would cause problems for the mixing potential algorithm, which would detect strong static instabilities associated with very warm air parcels that could potentially rise from the surface layer up into the stratosphere — an unrealistic situation for normal diurnal cycles of heating.

This problem was called the destabilization problem. To counteract this problem, a new version of the transilient parameterization was proposed where the application of boundary conditions was split into two parts [12]. First, based on observations in the real atmosphere of typical temperature gradients in the surface layer, an intermediate time step was taken that was appropriate for the vertical size of the bottom grid box, during which surface heating warmed that box. Next, the mixing potential and transilient matrices were calculated, but were temporarily not used. Then, the second portion of the boundary conditions was applied to the bottom grid cell, and finally the previously-calculated transilient matrix was used in (1) to cause the turbulent mixing.

This approach, while a bit more complicated, yield fine results in simulations of complex boundary layers having buoyant and shear forcings, and proved successful in simulating the dispersion of pollutants from both surface and elevated sources.

3.5. VERSION 4: MATRIX NORMALIZATION

As applications for the transilient method expanded, Inclan *et al.* [3] began applying the previous version to turbulent transport across forest canopies. To resolve the transpiration and radiative effects of the leaves and needles of the trees, it was necessary to utilize a very fine and complex vertical grid spacing, and very small time steps were taken because of constraints in the radiation algorithms.

It was discovered that the mixing-potential parameterization of the previous two versions yielded transilient matrices that did not approach the identity matrix, as was needed for the short time steps. To correct for this, an approximate method was developed to modify the mixing potential matrix depending on the discretization:

$$Yn_{ij} = \frac{\beta \cdot Y_{ij} \cdot (\Delta t / \Delta_{ij} z)^2}{1 + \beta \cdot Y_{ij} \cdot (\Delta t / \Delta_{ij} z)^2} \tag{5}$$

where b is an empirical constant roughly equal to 0.0005 , Y_{ij} is the intermediate mixing potential matrix (4), and Yn_{ij} is the desired normalized mixing potential matrix.

This procedure gives the proper structure to the normalized mixing-potential matrix Yn_{ij}, which when converted into a transilient matrix gives the desired results. Namely (***I*** --> ***c*** --> ***M***) as (0 --> Δt --> ∞). Unfortunately, as in all the previous versions it gives a symmetric transilient matrix.

3.6. VERSION 5: ASYMMETRIC

Meanwhile, Ebert *et al.* [2] and [Querzoli, 1997, personal communication] made large-eddy simulations and water-tank measurements of transilient matrices for special cases. They found that the transilient matrices are strongly asymmetric during free convection. That motivated a search for a measure of instability that does not have the same inherent symmetries as the Richardson number and TKE approaches.

A possible solution is to utilize convective available potential energy (CAPE), a concept that already receives widespread use in the study of thunderstorms. This approach integrates the buoyant energy of a rising or descending air parcel compared to the environmental sounding during its whole distance of travel, rather than considering just the buoyancy difference between source and destination locations [10].

Initial experiments with this approach suggest that it has the capability to capture the upward dominance of thermals heated from the surface, as well as the downward dominance of cold thermals sinking from a radiatively-cooled stratocumulus cloud top. By integrating parcel movement between all possible source and destination locations, an asymmetric buoyancy term can be used in the TKE parameterization in place of the buoyant term in (4). Even when the symmetric effects of wind differences and dissipation are included, the result is often an asymmetric mixing-potential matrix.

However, to utilize the asymmetric mixing potential, an alternative to the L_∞ method was needed to compute the final transilient matrix. Johnny Peterson [personal communication] proposed an iterative method that gradually relaxes the mixing potential matrix toward the doubly-stochastic transilient matrix, while faithfully preserving the general structure of the flow instabilities. The first guess of the transilient matrix is to set it equal to the mixing-potential matrix.

Then, each subsequent relaxation pass proceeds as follows. Divide each row of the original matrix by its own row sum to make the first intermediate matrix. Divide each column of the original matrix by its own column sum to get a second intermediate matrix. Finally average the two intermediate matrices element by element, to yield a new guess for the transilient matrix. Repeat the process using the new matrix as the starting point, and proceed until the row and column sums are within a small tolerance of summing to one.

While this method shows much promise for parameterizing asymmetric matrices, a robust application of this method has yet to be formulated. It appears that there are some special asymmetric mixing-potential matrices for which the Peterson method is unable to find any transilient matrix. This is a clue that there is some additional aspect of the physics of the real atmosphere that we are not properly accounting for in the parameterization. We are still experimenting. It should be noted that Peterson's method works excellently for symmetric matrices, and gives a much more reasonable transilient matrix for any given mixing-potential matrix. However, this improvement comes at the expense of added computations in the relaxation algorithm.

Finally, it should be noted that when the first measurements of asymmetric matrices appeared, it stimulated the definition of a wide range of new nonlocal turbulence statistics [12]. Many of these statistics are similar to traditional local turbulence statistics, but include nonlocal mixing effects that the traditional statistics cannot describe.

4. New Developments

We are currently continuing to test asymmetric parameterizations. Also, development of a higher-statistical-order nonlocal closure is beginning [Modzelewski, personal communication]. Bagliani [1] has recently devised new matrix invariants to use as measures of the asymmetry of transilient matrices. These are useful tools to test how faithfully the Peterson and other relaxation methods preserve asymmetries when converting from mixing-potential to transilient matrices. Water-tank and large-eddy simulation experiments are continuing by researchers around the world. Also, work is being done to develop more computationally efficient methods to allow transilient turbulence closure to be incorporated into operational numerical weather prediction models [Modzelewski].

5. Conclusions

In this paper, transilient turbulence theory was presented as a way of parameterizing nonlocal vertical mixing within a column of air. Development and refinement of this parameterization during the past one and a half decades was summarized, and future challenges were presented. There is tremendous opportunity for other investigators to add new ideas to the transilient framework.

In the companion paper by Stull in this volume, two other nonlocal parameterizations were presented: convective-transport theory for flux at the surface of the earth, and radix-layer theory for the vertical profiles of mean variables in the bottom fifth of the ML. Taken together with T3, these works define a unified philosophy for the nonlocal nature of turbulence throughout the depth of the atmosphere, and serve as a solid basis for modeling the atmospheric boundary layer.

References

1. Bagliani, M. (1998) Indices of asymmetry in the context of transilient turbulence theory. *J. Atmospheric Sciences* **55**, (in press).
2. Ebert, E.E., Schumann, U., and Stull, R.B. (1989) Nonlocal turbulent mixing in the convective boundary layer evaluated from large-eddy simulation. *J. Atmospheric Sciences* **46**, 2178-2207.
3. Inclan, M.G., Forkel, R., Dlugi, R., and Stull, R.B. (1996) Application of transilient turbulent theory to study interactions between the atmospheric boundary layer and forest canopies. *Boundary-Layer Meteorology* **79**, 315-344.
4. Stull, R.B. (1984) Transilient turbulence theory. Part I: The concept of eddy-mixing across finite distances. *J. Atmospheric Sciences* **41**, 3351-3367.
5. Stull, R.B. (1986) Transilient turbulence theory. Part III: Bulk dispersion rate and numerical stability. *J. Atmospheric Sciences* **43**, 50-57.
6. Stull, R.B. (1987) Transilient algorithms to model mixing across finite distances. *Environmental Software* **2**, 4-12.
7. Stull, R.B. (1988) *An Introduction to Boundary Layer Meteorology*, Kluwer Academic Publishers, Dordrecht.
8. Stull, R.B. (1993) Review of nonlocal mixing in turbulent atmospheres: transilient turbulence theory, *Boundary-Layer Meteorology* **62**, 21-96.
9. Stull, R.B. (1995) A review of parameterization schemes for turbulent boundary-layer processes, *Seminar Proceedings: Parameterization of sub-grid scale physical processes, 5-9 September 1994.* European Centre for Medium Range Weather Forecasts (ECMWF), 163-174.
10. Stull, R.B. (1995) Parameterization of asymmetric transilient matrices for convective PBLs, *Preprints of the 11th Symposium on Boundary Layers and Turbulence, March 27-31, 1995, Charlotte, NC,* American Meteorological Society, Boston, 57-58.
11. Stull, R.B. (1995) *Meteorology Today for Scientists and Engineers*, Wadsworth (West) ITP, Belmont, CA.
12. Stull, R.B. and Bartnicki, J. (1998) Transport using transilient matrices, in Fox, P. and Kerr, R. (Eds) *Geophysical and Astrophysical Convection.* Gordon & Breach Science Publ., New York.
13. Stull, R.B. and Driedonks, A.G.M. (1987) Applications of the transilient turbulence parameterization to atmospheric boundary layer simulations, *Boundary-Layer Meteorology* **40**, 209-239.
14. Stull, R.B. and Hasagawa, T. (1984) Transilient turbulence theory. Part II: Turbulent adjustment. *J. Atmospheric Sciences* **41**, 3368-3379.
15. Zhang, Q. and Stull, R. (1992) Alternative nonlocal descriptions of boundary-layer evolution, *J. Atmospheric Sciences* **49**, 2267-2281.

BULK MODELS OF THE ATMOSPHERIC CONVECTIVE BOUNDARY LAYER

E. FEDOROVICH
Institute of Hydrology and Water Resources Planning
Karlsruhe University
Kaiserstrasse 12, 76128 Karlsruhe, Germany

Abstract

The paper presents an overview of modeling the atmospheric convective boundary layer (CBL) using bulk parameterizations for the vertical structure of the layer. Such parameterizations are constructed based on empirical knowledge about vertical distributions of meteorological variables in the CBL.

Two main types of CBL bulk models are presented and discussed. The first model considered is the so-called zero-order jump model, which implies a vertical homogeneity of meteorological variables in the bulk of the CBL, and zero-order discontinuities of variables at the interfaces of the layer. Integral budgets of momentum and heat in the zero-order jump model of the dry atmospheric CBL are considered. A general version of the equation describing the CBL growth rate (the entrainment rate equation) is obtained through integration of the turbulence kinetic energy balance equation, evoking basic assumptions of the zero-order representation of the CBL vertical structure. The developed theory is generalized for the case of the CBL over an irregular terrain.

The second model considered is the general-structure CBL bulk model, which incorporates a self-similar representation of the buoyancy profile within the capping inversion layer. This representation has been examined against the data from atmospheric measurements, laboratory experiments with buoyancy agitated turbulence, and large eddy simulations. The growth rate equations for mixed and inversion layers are derived using the turbulence kinetic energy balance equation and the Deardorff scaling hypothesis refined to account for the inversion-layer structure. The model is found to be able to reproduce transition regimes of the CBL development affected by nonstationarity of the entrainment zone.

1. Introduction

Studies of the atmospheric convective boundary layer (CBL) within the framework of bulk-model methodology have passed through several historical stages. Most extensively, the CBL was studied with the so-called zero-order jump models originating from the pioneering works of Ball [1] and Lilly [28].

E.J. Plate et al. (eds.), Buoyant Convection in Geophysical Flows, 265–290.

According to the zero-order jump model, the potential temperature within the CBL is presumed to be height constant. Its changes with height in the surface layer, and in the entrainment layer at the CBL top are reduced to the zero-order discontinuities of the temperature profile. The analogous zero-order parameterization is employed for the wind velocity profile (Garrat *et al.* [22]). Wind velocity in the CBL is taken to be uniform with height. Velocity shears across the surface layer and in the entrainment layer are represented by steps in the wind profile. The vertical structure of the velocity and temperature fields in the stably stratified free-atmosphere layer above the CBL is assumed to be known.

The breakthrough in the zero-order jump modeling of the CBL happened in the seventies when Plate [38], Betts [4], Carson [7], Tennekes [47], Stull [42-44], Carson and Smith [8], Zilitinkevich [52], and Zeman and Tennekes [51] proposed a set of zero-order jump approaches for modeling a horizontally homogeneous, shear-free CBLs. Later, Zilitinkevich [54] suggested a generalized zero-order jump CBL model, comprising the aforementioned models as asymptotic cases. Several applied mesometeorological models, *e.g.* those of Kraus and Leslie [26], Brutsaert [6], Batchvarova and Gryning [2], and Zilitinkevich *et al.* [55] were developed based on the zero-order parameterization of the CBL vertical structure. Most of those models dealt with a horizontally homogeneous CBL and used simplifying assumptions concerning the effects of wind shear.

In an attempt to reproduce the CBL vertical structure in a more detailed way, higher-order bulk models of the CBL were proposed. In the first-order jump model, Betts [5] introduced the interfacial layer of finite thickness between the mixed layer and the free atmosphere. The potential temperature profile was taken to be linear in this layer, undergoing first-order discontinuities at its upper and lower boundaries. The general-structure CBL bulk models of Deardorff [13], and Fedorovich and Mironov [21] provided for realistic representation of the temperature / buoyancy profile in the entrainment zone, accounting for nonstationarity of the entrainment. All mentioned models of higher orders were proposed for shear-free cases of the atmospheric CBL.

Within the last two decades, modeling activities in the CBL area have switched almost entirely to large eddy numerical simulations. Although the first large eddy simulation (LES) of the atmospheric CBL has been already performed by J. Deardorff during the early seventies (Deardorff [9, 11-12]), progress in the field was slow due to the lack of computer power and insufficient capacity of data storage devices. In the eighties, the situation had improved, and significant results in LES studies of the atmospheric CBL were achieved. Most notable are the works of Moeng [31-33], Nieuwstadt and Brost [37], Moeng and Wyngaard [35], Mason [30], Schmidt and Schumann [41], and Moeng and Sullivan [34]. Different types of the CBL and various features of its structure have been investigated with the aid of LES during the last fifteen years.

However, although being very valuable and efficient tools for fundamental studies of the atmospheric convection, and possessing the same ability in reproducing fine features of the flow structure as high-resolution laboratory experiments, the LES models can hardly be used at present for applied purposes due to the enormous computer resources they still demand.

This is one of the reasons why the bulk approach remains to be an attractive model framework for applied studies of the atmospheric CBL.

2. Zero Order Jump Model

A zero-order jump parameterization for the vertical structure of the atmospheric CBL is presented in Fig. 1. It is based on the observed features of vertical distributions of meteorological variables in the CBL which are extensively discussed in a number of contributions to this ASI volume, see *e.g.* papers by Lenschow [27], Plate [39], and Fedorovich and Kaiser [20]. The main feature is the strong turbulent mixing embracing the whole layer and causing the approximate uniformity of the vertical distribution of physical substances within the CBL. Thus, within the zero-order approach, the CBL is represented by the convectively mixed layer with two interfaces, the upper one (at $z=h$) and the lower one (at $z=0$), across which the potential temperature and wind velocity change in a jumplike way.

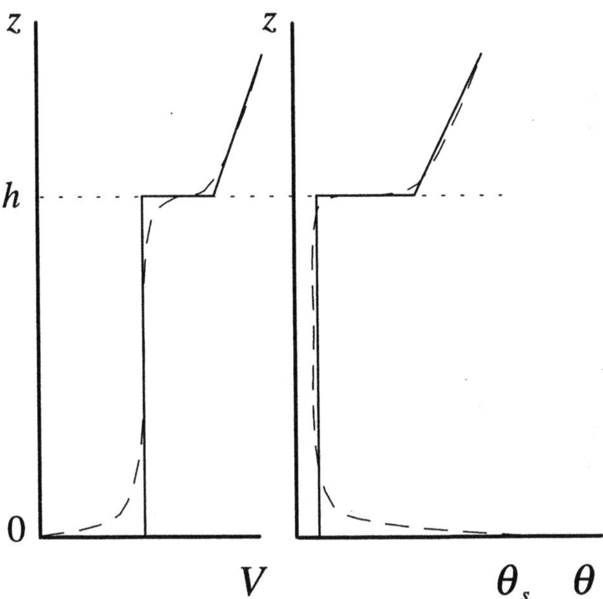

Figure 1. Actual (dashed lines) and zero-order jump (solid lines) profiles of wind velocity V, and potential temperature θ in the convective boundary layer; θ_s is the near-surface value of θ.

In historical retrospective, the ideology of zero-order jump modeling was mainly developed for the case of horizontally homogeneous, shear-free atmospheric CBL. For the CBL with wind shear, quite a few parameterizations have been suggested. Some of them are reviewed in Stull [46]. The effects of baroclinicity (thermal wind), advection, and horizontal diffusion, which can be important for forming the meteorological regime of the atmospheric CBL, were left beyond the scope of traditional zero-order jump model studies. A theoretical framework for applied zero-order jump CBL modeling, which takes into account the above effects, has been proposed by Fedorovich [19]. In the following subsections, particular features of this model framework will be considered.

2.1. HEAT AND MOMENTUM BUDGETS IN ZERO ORDER CBL MODEL

The following initial equations written in the Reynolds form describe the flow in the nonstationary, horizontally inhomogeneous atmospheric CBL.

The momentum balance equations:

$$\frac{\partial u}{\partial t}+\frac{\partial uu}{\partial x}+\frac{\partial vu}{\partial y}+\frac{\partial wu}{\partial z}=f(v-v_{g0}-\Gamma_v z)+\frac{\partial \tau_x}{\partial z}, \tag{1}$$

$$\frac{\partial v}{\partial t}+\frac{\partial uv}{\partial x}+\frac{\partial vv}{\partial y}+\frac{\partial wv}{\partial z}=-f(u-u_{g0}-\Gamma_u z)+\frac{\partial \tau_y}{\partial z}, \tag{2}$$

where u, v and w are the components of mean wind velocity along axes x, y and z, respectively; $\tau_x=-<w'u'>$ and $\tau_y=-<w'v'>$ are components of the vertical turbulent flux of momentum normalized by density, the sign <...> denotes the operation of Reynolds averaging; f is the Coriolis parameter; u_{g0} and v_{g0} are the near-surface values of the geostrophic wind components, Γ_u and Γ_v are the vertical gradients of these components. The last four quantities at this point are presumed to be prescribed functions of time and horizontal coordinates. Later we shall show a way of determining them from pressure and temperature spatial distributions. Equations (1) and (2) imply that contribution of the horizontal turbulent transport to the momentum balance is negligible compared to vertical turbulent transport, and conditions of hydrostatic equilibrium are satisfied in the CBL (Qi *et al.* [40]). These assumptions hold when horizontal scales in the flow dominate over the vertical ones which is typical for most of the CBL cases observed in nature and simulated in the laboratory.

The mass conservation equation is employed in the form:

$$\frac{\partial u}{\partial x}+\frac{\partial v}{\partial y}+\frac{\partial w}{\partial z}=0. \tag{3}$$

In the heat transfer equation

$$\frac{\partial \theta}{\partial t}+\frac{\partial u\theta}{\partial x}+\frac{\partial v\theta}{\partial y}+\frac{\partial w\theta}{\partial z}=-\frac{\partial Q}{\partial z}, \tag{4}$$

where θ is the mean potential temperature, and $Q=<w'\theta'>$ is the turbulent kinematic heat flux, the following assumption, which is analogous to the one used in the momentum balance equations, is made: the advection and the vertical turbulent transport ultimately dominate the horizontal turbulent exchange. In Eq. (4), we also neglected the contributions of molecular and radiation heat transfer to the heat balance of the CBL. The first one is very small in the atmosphere compared to the turbulent heat transfer, whereas the radiation effects are omitted for simplicity. Principally, they can be incorporated in the CBL zero-order jump model, see *e.g.* Zilitinkevich *et al.* [55].

The above equations correspond to the general case of the CBL over a flat underlying surface. To obtain the equations of the momentum and heat budget of the layer, we integrate Eqs. (1), (2), and (4) over the boundary-layer depth, *i.e.* over z from 0 to h, taking into account the zero-order jump representations of the temperature and velocity profiles, see Fig. 1. As we integrate, we will include the zero-order discontinuity surface in the integration domain, thus regarding the upper side of this surface as the CBL edge.

Calculating the integrals from each term of the first equation of motion, we come to the equation of the integral budget of momentum along the x axis, (see derivation in Fedorovich [19]):

$$h\left(\frac{\partial \bar{u}}{\partial t}+\bar{u}\frac{\partial \bar{u}}{\partial x}+\bar{v}\frac{\partial \bar{u}}{\partial y}\right)=\Delta u\frac{Dh}{Dt}+h\left[f\left(\bar{v}-v_{g0}\right)-\frac{h}{2}f\Gamma_v\right]-\tau_{xs}\,. \tag{5}$$

where $\bar{u}$ is the average value of the x component of wind velocity within the mixed layer; $\Delta u = u_h - \bar{u}$ is the increment of this component across the mixed-layer upper interface; u_h is the value of u in the stable layer, at the upper side of the zero-order interface represented by the surface of zero-order discontinuity; $\frac{Dh}{Dt}\equiv\frac{\partial h}{\partial t}+\frac{\partial \bar{u}h}{\partial x}+\frac{\partial \bar{v}h}{\partial y}$ is the substantial (total) variation of h. The operation of averaging over the mixed-layer depth is defined here as $\overline{(\,)}=\frac{1}{h}\int_0^h(\,)dz$.

The variation Dh/Dt in fact describes the combined effect of three mechanisms determining the evolution of the CBL depth. The first of them, represented by $\partial h/\partial t$, is associated with local changes of h (nonstationarity). Horizontal advection constitutes the second mechanism. It contributes to the total variation by $\bar{u}(\partial h/\partial x)+\bar{v}(\partial h/\partial y)$. The third mechanism is the subsidence related to the horizontal divergence of the flow in the CBL: $h[(\partial\bar{u}/\partial x)+(\partial\bar{v}/\partial y)]=-w_{hs}$, where w_{hs} is the subsidence velocity at the CBL top. The subsidence velocity is a part of

$$w_h=-\int_0^h\frac{\partial u}{\partial x}dz-\int_0^h\frac{\partial v}{\partial y}dz=-\frac{\partial \bar{u}h}{\partial x}+u_h\frac{\partial h}{\partial x}-\frac{\partial \bar{v}h}{\partial y}+v_h\frac{\partial h}{\partial y},$$

which is the total vertical velocity at $z=h$.

Integration of the second equation of motion (2), and of the heat balance equation (4) gives (Fedorovich [19])

$$h\left(\frac{\partial \bar{v}}{\partial t}+\bar{u}\frac{\partial \bar{v}}{\partial x}+\bar{v}\frac{\partial \bar{v}}{\partial y}\right)=\Delta v\frac{Dh}{Dt}-h\left[f\left(\bar{u}-u_{g0}\right)-\frac{h}{2}f\Gamma_u\right]-\tau_{ys} \tag{6}$$

and

$$h\left(\frac{\partial\bar{\theta}}{\partial t}+\bar{u}\frac{\partial\bar{\theta}}{\partial x}+\bar{v}\frac{\partial\bar{\theta}}{\partial y}\right)=\Delta\theta\frac{Dh}{Dt}+Q_s, \tag{7}$$

respectively. The latter expression, in which Q_s is the near-surface value of the kinematic heat flux, presents the heat budget of the CBL.

For now we shall consider Q_s, as well as the components τ_{xs} and τ_{ys} of the near-surface shear stress to be known functions of time and the horizontal coordinates. In section 2.5, some approaches towards specifying these characteristics in the zero-order jump models of the CBL will be presented.

2.2. PROFILES OF TURBULENT FLUXES

The expressions of the momentum and heat-flux profiles are derived by integrating Eqs. (1), (2), and (4) over the vertical co-ordinate from 0 to z. From the Eq. (1), we have for the x component of the momentum flux:

$$\tau_x=\tau_{xs}+\int_0^z\left(\frac{\partial u}{\partial t}+\frac{\partial uu}{\partial x}+\frac{\partial vu}{\partial y}+\frac{\partial wu}{\partial z}\right)dz-\int_0^z f(v-v_{g0}-\Gamma_v z)dz.$$

Employing the zero-order jump representation of the velocity profile and evaluating w by integration of the continuity equation (3) from 0 to z, we obtain τ_x as the quadratic function of z:

$$\tau_x=\tau_{xs}+z\left(\frac{\partial\bar{u}}{\partial t}+\bar{u}\frac{\partial\bar{u}}{\partial x}+\bar{v}\frac{\partial\bar{u}}{\partial y}\right)+z\left[f(\bar{v}-v_{g0})+\frac{z}{2}f\Gamma_v\right],$$

which can be written in the form (Fedorovich [19]):

$$\tau_x=\tau_{xs}(1-\zeta)+\Delta u\frac{Dh}{Dt}\zeta+\frac{h^2}{2}f\Gamma_v\zeta(\zeta-1), \tag{8}$$

where $\zeta=z/h$ is the dimensionless height.

Similarly, the expressions of the y component of the momentum flux

$$\tau_y=\tau_{ys}(1-\zeta)+\Delta v\frac{Dh}{Dt}\zeta-\frac{h^2}{2}f\Gamma_u\zeta(\zeta-1), \tag{9}$$

and of the heat flux

$$Q=Q_s(1-\zeta)-\Delta\theta\frac{Dh}{Dt}\zeta \tag{10}$$

profiles as functions of dimensionless height can be obtained. It is seen from Eq. (10) that kinematic heat flux in the zero-order jump CBL model changes linearly with height.

Equations (8), (9), and (10) indicate that both components of the momentum flux, as well as the heat flux, undergo the zero-order discontinuities at the mixed-layer interface. Just below this surface they reach the values $\Delta u\frac{Dh}{Dt}$, $\Delta v\frac{Dh}{Dt}$, and $-\Delta\theta\frac{Dh}{Dt}$, respectively. From the upper side of the discontinuity surface all turbulent fluxes vanish.

2.3. ENTRAINMENT RATE EQUATION

The CBL depth h is one of the most important variables characterizing the process of convection. Due to the penetration of thermals into the stably stratified flow above the mixed layer, the heat and momentum from the stable region are entrained or mixed down into the bulk of the turbulized convective layer. The entrainment is thus the principal mechanism responsible for the CBL growth dynamics.

To describe variations of h in time and space we depart from the turbulent kinetic energy (TKE) balance equation. In the case under consideration, this equation has the form

$$\frac{\partial e}{\partial t}+\frac{\partial ue}{\partial x}+\frac{\partial ve}{\partial y}+\frac{\partial we}{\partial z}=\tau_x\frac{\partial u}{\partial z}+\tau_y\frac{\partial v}{\partial z}+\beta Q-\frac{\partial \Phi}{\partial z}-\varepsilon, \tag{11}$$

where e is the TKE per unit mass, ε is the energy dissipation rate, and Φ is the vertical transport of energy due to turbulent exchange and pressure fluctuations.

To obtain the entrainment rate equation, we integrate Eq. (11) over z from 0 to h. The integration of the left-hand part, representing the temporal variations of e and its transformations due to advection, yields

$$\int_0^h\left(\frac{\partial e}{\partial t}+\frac{\partial ue}{\partial x}+\frac{\partial ve}{\partial y}+\frac{\partial we}{\partial z}\right)dz=\frac{\partial}{\partial t}\overline{e}h+\frac{\partial}{\partial x}\overline{eu}h+\frac{\partial}{\partial y}\overline{ev}h.$$

While deriving the above expression, we set $e_h=0$, since the zero-order jump model assumes that turbulence vanishes at $z=h$.

The integration of the shear production terms in the right-hand part of (11) cannot be carried out directly because we have to integrate the products of the shear stress components (which are discontinuous at $z=h$) and vertical derivatives of the velocity components (they are infinite at $z=0$ and $z=h$ within the framework of the zero-order parameterization). The following approach for calculating the integrals can be applied (we shall demonstrate it using the first term as an example).

In the vicinity of h we isolate a thin layer with the depth δh. Then we approximate the velocity gradient across this layer with $\Delta u/\delta h$, and the increment of the x component of momentum flux with the linear function $\Delta u\frac{Dh}{Dt}\frac{h-z}{\delta h}$, multiply them, and carry out the integration over the layer δh. This yields $\frac{1}{2}\Delta u^2 \frac{Dh}{Dt}$, the value of the integral being independent on δh. Therefore it holds true when δh tends to zero. While calculating the integral over the rest of the mixed layer we should take into account that in the bulk of the layer velocity is height constant, and therefore there is no shear production of turbulent kinetic energy in this region. The contribution of shear in the thin near-surface layer, where velocity sharply increases from zero to the value characteristic of the mixed layer, and where variation of the momentum flux with height is negligibly small, can be evaluated analogous to the way used in the vicinity of *h*. It is easy to show that such integration results in $\bar{u}\tau_{xs}$.

Thus, for the integral TKE production by shear we obtain:

$$\int_0^h \left(\tau_x \frac{\partial u}{\partial z} + \tau_y \frac{\partial v}{\partial z} \right) dz = \bar{u}\tau_{xs} + \bar{v}\tau_{ys} + \frac{1}{2}\left(\Delta u^2 + \Delta v^2\right)\frac{Dh}{Dt}.$$

The integral TKE production by buoyancy forces is expressed as

$$\int_0^h \beta Q dz = \beta \frac{h}{2}\left(Q_s - \Delta\theta \frac{Dh}{Dt} \right).$$

We assume that there is no transport of energy through the underlying surface. Therefore the integral of the transport term yields $-\Phi_h$, which is the negative flux of energy at the CBL top. Within the employed model framework, the energy drain at *z=h* should be associated solely with the internal gravity waves (Zilitinkevich [54]), (see also section 3.2), since the zero-order jump model postulates vanishing of turbulent transport above *h*. It was noted by Stull [45] that for typical atmospheric CBL cases the energy drain by gravity waves is relatively small. On the other hand, Fedorovich and Mironov [21] have found a pronounced effect of the energy transport by waves in the CBL case with rather moderate stable stratification in the turbulence-free flow region aloft.

Summarizing the above expressions for the different components of the integral TKE balance, we come to the entrainment equation

$$\left[\bar{e} - \frac{1}{2}\left(\Delta u^2 + \Delta v^2 - \beta h \Delta\theta\right)\right]\frac{Dh}{Dt} + \left(\frac{\partial \bar{e}}{\partial t} + \bar{u}\frac{\partial \bar{e}}{\partial x} + \bar{v}\frac{\partial \bar{e}}{\partial y} - \frac{\beta}{2}Q_s + \bar{\varepsilon} \right) h = \bar{u}\tau_{xs} + \bar{v}\tau_{ys} - \Phi_h . \quad (12)$$

To solve this equation, one should specify a way of evaluating the mixed-layer means $\bar{e}$ and $\bar{\varepsilon}$, and also a method of calculating the transport of energy Φ_h at the mixed-layer top. Within the zero-order approach these variables are commonly determined using parameterizations based on similarity arguments.

In the most studied case of the nonsteady, horizontally quasi-homogeneous, shear-free CBL, the entrainment equation (12) reduces to

$$\left(\bar{e}+\frac{\beta h\Delta\theta}{2}\right)\frac{dh}{dt}+\left(\frac{d\bar{e}}{dt}+\bar{\varepsilon}-\frac{\beta}{2}Q_s\right)h=-\Phi_h\,. \tag{12a}$$

For the shear-free CBL, traditional parameterizations for $\bar{e}$ and $\bar{\varepsilon}$ result from the Deardorff [10] self-similarity hypothesis, which states that profiles of e and ε normalized using h as the height scale and $w_*=(\beta Q_s h)^{1/3}$ as the velocity scale are universal functions of the dimensionless height $\zeta=z/h$. This allows the representations $\bar{e}=C_e(\beta Q_s h)^{2/3}$ for the mean TKE, and $\bar{\varepsilon}=C_\varepsilon \beta Q_s$ for the mean dissipation rate, where C_e and C_ε are universal constants.

From the analysis of water tank and atmospheric experimental data, Zilitinkevich [54] found these constants to be 0.5 and 0.4, respectively. Fedorovich and Mironov [21], who additionally employed data from several LES studies of the shear-free CBL, has shown that these estimates are apparently exaggerated by 25%. This could be due to the disturbing presence of shear during field measurements and parasite enhancement of horizontal velocity fluctuations by bottom temperature variations in the water tank experiments.

For the vertical transport of energy from the CBL top, the two most known parameterizations were suggested, the first of Kantha [25], and the second of Zilitinkevich [54]. Both are based on the relationship of Thorpe [49], who expressed the energy drain from the CBL through the parameters of the waves propagating in the nonturbulent fluid above the layer. Parameterization of Kantha, accompanied by the geometric formula of Stull [44] relating the so-called entrainment coefficient $A=\dfrac{\Delta\theta(dh/dt)}{Q_s}$ to the ratio $(\Delta h/2)/(h-\Delta h/2)$, where Δh is the entrainment-zone depth, results in

$$\Phi_h=C_{Nh}N^3h^3\left(\frac{A}{1+A}\right)^2,$$

whereas the Zilitinkevich parameterization gives

$$\Phi_h=C_{Nh}'N^3h^3\left(\frac{A}{1+A}\right)^3.$$

In the above expressions, C_{Nh} and C_{Nh}' are dimensionless constants. Zilitinkevich [54] estimated C_{Nh}' to be 0.02. From the experiments with a general-structure CBL model, Fedorovich and Mironov [21] found that C_{Nh} is about one order of magnitude smaller than C_{Nh}'.

Further simplification of the entrainment equation commonly involves omitting the $\bar{e}$ containing terms in the left-hand side of Eq. (12a), and neglecting the wave-related energy flux in its right-hand side. This provides a closure relationship

$$A = \frac{\Delta\theta \dfrac{dh}{dt}}{Q_s} = 1 - 2C_\varepsilon = const\,, \tag{12b}$$

where the value of constant is set equal 0.2. The last expression works decently for the CBL cases when the simplifications made are justified.

For the CBL with wind shear, a general parameterization of Φ_h has not yet been proposed. In this case, the employment of the geometric formula [44] for determination of Δh is becoming questionable because it does not take into account the influence of shear on the entrainment-zone dynamics. One of the feasible solutions is to use the diagnostic relationship between the normalized entrainment zone depth, $\Delta h / h$, and a dimensionless parameter of entrainment $\mathrm{Ri}_E = \beta\Delta\theta(dh/dt)^{-2}h$, suggested by Gryning and Batchvarova [23] and tested later in their applied model for the height of the daytime mixed layer and the entrainment zone (Batchvarova and Gryning [3]): $\Delta h / h = 3.3\mathrm{Ri}_E^{-1/3} + 0.2$.

Applied CBL models based on the bulk approach usually employ a stationary version of the TKE balance equation (Stull [43], Zilitinkevich *et al.* [55], and Gryning and Batchvarova [23]). This enables one to avoid the problem of parameterizing $\overline{e}$, but leaves the problem of $\overline{\varepsilon}$ determination open. In most of suggested applied models, $\overline{\varepsilon}$ is merely set equal to combination of the TKE integral production components, each taken with an empirical proportionality coefficient (Stull [43], Tennekes and Driedonks [48], Driedonks [17], Driedonks and Tennekes [18], and Batchvarova and Gryning [2]). The shear forcing at the CBL bottom is parameterized in these models through u_*^3, and the effect of the elevated shear is expressed in terms of $(\Delta\overline{V})^3$, where $u_* = \sqrt[4]{\tau_{xs}^2 + \tau_{ys}^2}$ is the friction velocity, and $\Delta\overline{V}$ is the wind shear across the CBL top. As one can infer from Eq. (12), the direct integration of TKE balance equation, evoking the assumptions of zero-order jump approach, yields quite different expressions for the shear production terms: $\overline{u}\tau_{xs} + \overline{v}\tau_{ys}$ (at the bottom), and $\frac{1}{2}\left(\Delta u^2 + \Delta v^2\right)\frac{Dh}{Dt}$ (at the upper interface). Generally, even for average conditions, it is not as evident whether $\overline{u}\tau_{xs} + \overline{v}\tau_{ys}$ can be taken proportional to u_*^3, and $\frac{1}{2}\left(\Delta u^2 + \Delta v^2\right)\frac{Dh}{Dt}$ to $(\Delta\overline{V})^3$. It is also easy to notice that the first of the last two terms always remains positive during the convective-layer growth, whereas the second has the sign of $\Delta\overline{V}$ and thus can be a sink of TKE in the growing CBL.

2.4. ATMOSPHERIC CBL OVER IRREGULAR TERRAIN

Under common atmospheric conditions, the convective boundary layer develops over an underlying surface with variable topography, roughness, and thermal properties. If these variations are not very sharp in space, it is possible to generalize the above theory for the case when lower interface of the CBL is represented by some known function

$H=H(x, y)$ describing the topography, and the spatial distribution of aerodynamic roughness is given by $z_{0u}(x,y)$, see Fig. 2.

The initial equations for the case under consideration slightly differ from the model equations corresponding to the case of the CBL over the flat surface. In the momentum balance equations (1) and (2) the first terms in the right-hand parts must be modified in the following way:

$$f\left[v - v_{g0} - \Gamma_v(z - H)\right],$$

$$-f\left[u - u_{g0} - \Gamma_u(z - H)\right],$$

since the near-surface values of the geostrophic wind are now prescribed at the level $H(x, y)$. Functions Γ_v and Γ_u become the internal parameters of the model. We shall indicate in section 2.5 the method of evaluating them from the temperature field. The mass consistency and the heat balance are expressed, as previously, by relationships (3) and (4), respectively.

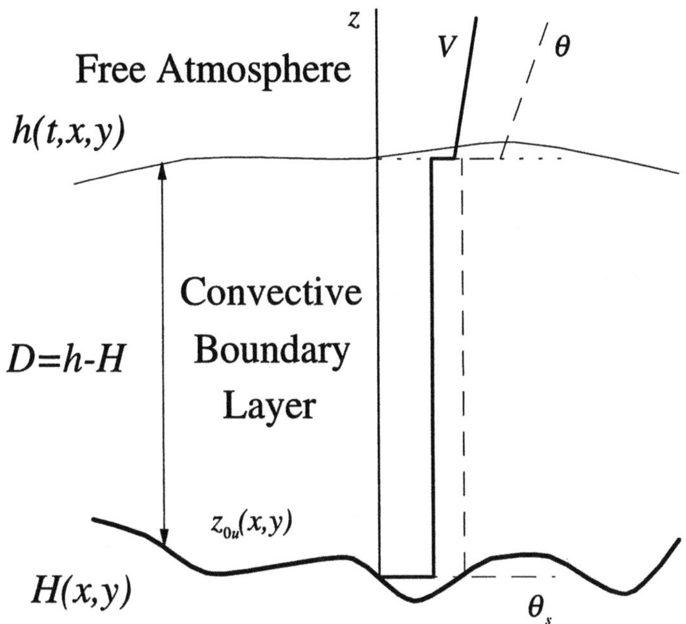

Figure 2. Schematic of the atmospheric convective boundary layer over irregular terrain.

We transform the co-ordinate system by introducing new time and space variables

$$t_n = t, \quad x_n = x, \quad y_n = y, \quad z_n = \frac{z - H(x,y)}{h(t,x,y) - H(x,y)},$$

where new variables are denoted by subscript *n*.

In the transformed co-ordinates, the first equation of motion can be integrated over z_n, see Fedorovich [19], which yields the momentum-budget equation

$$D\left(\frac{\partial\bar{u}}{\partial t}+\bar{u}\frac{\partial\bar{u}}{\partial x}+\bar{v}\frac{\partial\bar{u}}{\partial y}\right)=\Delta u\left(\frac{\partial h}{\partial t}+\frac{\partial\bar{u}D}{\partial x}+\frac{\partial\bar{v}D}{\partial y}\right)+D\left[f\left(\bar{v}-v_{g0}\right)-\frac{D}{2}f\Gamma_v\right]-\tau_{xs}. \quad (13)$$

In the above expression, $D(t, x, y)=h-H$ is the relative CBL depth, and the averaging is defined through integration over the dimensionless vertical coordinate. Equation (13) is quite similar to Eq. (5) presenting the integral balance of the x component of momentum in the CBL over the flat surface.

The equation of the momentum balance along the y axis

$$D\left(\frac{\partial\bar{v}}{\partial t}+\bar{u}\frac{\partial\bar{v}}{\partial x}+\bar{v}\frac{\partial\bar{v}}{\partial y}\right)=\Delta v\left(\frac{\partial h}{\partial t}+\frac{\partial\bar{u}D}{\partial x}+\frac{\partial\bar{v}D}{\partial y}\right)-D\left[f\left(\bar{u}-u_{g0}\right)-\frac{D}{2}f\Gamma_u\right]-\tau_{ys}, \quad (14)$$

and the integral heat balance equation

$$D\left(\frac{\partial\bar{\theta}}{\partial t}+\bar{u}\frac{\partial\bar{\theta}}{\partial x}+\bar{v}\frac{\partial\bar{\theta}}{\partial y}\right)=\Delta\theta\left(\frac{\partial h}{\partial t}+\frac{\partial\bar{u}D}{\partial x}+\frac{\partial\bar{v}D}{\partial y}\right)+Q_s \quad (15)$$

can be obtained analogously.

Integration of the momentum balance, and heat transfer equations over the normalized vertical coordinate from 0 to z_n yields expressions of the momentum flux components and turbulent heat flux as functions of z_n:

$$\tau_x=\tau_{xs}\left(1-z_n\right)+\Delta u\left(\frac{\partial h}{\partial t}+\frac{\partial\bar{u}D}{\partial x}+\frac{\partial\bar{v}D}{\partial y}\right)z_n+\frac{D^2}{2}f\Gamma_v z_n\left(z_n-1\right), \quad (16)$$

$$\tau_x=\tau_{xs}\left(1-z_n\right)+\Delta u\left(\frac{\partial h}{\partial t}+\frac{\partial\bar{u}D}{\partial x}+\frac{\partial\bar{v}D}{\partial y}\right)z_n+\frac{D^2}{2}f\Gamma_v z_n\left(z_n-1\right), \quad (17)$$

$$Q=Q_s\left(1-z_n\right)-\Delta\theta\left(\frac{\partial h}{\partial t}+\frac{\partial\bar{u}D}{\partial x}+\frac{\partial\bar{v}D}{\partial y}\right)z_n. \quad (18)$$

Substituting the above expressions in the TKE balance equation (11) written in the new coordinate system, and carrying out the termwise integration of this equation in a way similar to that shown in section 2.3, we obtain a version of the entrainment rate equation for the atmospheric CBL over irregular terrain:

$$\left[\overline{e}+\frac{1}{2}\left(\Delta u^2+\Delta v^2-\beta h\Delta\theta\right)\right]\left(\frac{\partial h}{\partial t}+\frac{\partial \overline{u}D}{\partial x}+\frac{\partial \overline{v}D}{\partial y}\right)$$

$$+\left(\frac{\partial \overline{e}}{\partial t}+\overline{u}\frac{\partial \overline{e}}{\partial x}+\overline{v}\frac{\partial \overline{e}}{\partial y}-\frac{\beta}{2}Q_s+\overline{\varepsilon}\right)D=\overline{u}\tau_{xs}+\overline{v}\tau_{ys}-\Phi_h\,, \tag{19}$$

This equation differs in merely a few details form the entrainment rate equation for the CBL over a flat surface, see Eq. (12).

2.5. EXTERNAL PARAMETERS

In atmospheric boundary-layer modeling, the value of the near-surface heat flux Q_s is usually derived from the equation of the surface heat balance, provided the methods of evaluating the other balance components are known. The simple algorithm for this purpose was proposed in Zilitinkevich *et al.* [55]. This algorithm incorporates the expression relating $\overline{\theta}$ to the near-surface value of temperature θ_s, and a formula for calculating the roughness parameter with respect to the temperature.

Departing from the value of Q_s, and from the roughness parameter with respect to the wind, $z_{0u}(x,y)$, the components of the near-surface momentum flux τ_{xs} and τ_{ys} can be evaluated. In the boundary-layer model of Zilitinkevich *et al.* [55], the following relationships between the momentum flux and velocity components in the CBL have been employed:

$$\overline{V}=u_*\frac{a_u+\ln\left(|L|/z_{0u}\right)}{k},\quad \overline{u}=\overline{V}\cos\alpha_h,\quad \overline{v}=\overline{V}\sin\alpha_h\,,$$

$$\sin\left(\alpha_s-\alpha_h\right)=\frac{a_\alpha}{k}\left(\frac{h}{|L|}\right)^{-1/3}\frac{u_*}{\overline{V}}\operatorname{sign}f,\quad \tau_{xs}=u_*^2\cos\alpha_s,\quad \tau_{ys}=u_*^2\sin\alpha_s\,,$$

where α_h is the angle between the x axis and the velocity vector $\overline{\mathbf{V}}$ with components $\overline{u}$ and $\overline{v}$, α_s is the angle between the x axis and the surface shear stress, $L=-u_*^2/(\beta Q_s)$ is the Monin-Obukhov length scale, k=0.4 is the von Karman constant, and a_u and a_α are dimensionless parameters for which Zilitinkevich [53] obtained estimates 1 and 3, respectively.

The above expressions constitute a closed set of equations for determination of the components of the near-surface momentum flux, provided Q_s, $\overline{u}$, $\overline{v}$, and z_{0u} are known.

The near-surface values of the geostrophic wind components in the atmospheric boundary layer, u_{g0} and v_{g0}, can be expressed through the horizontal gradients of the near-surface pressure field p_s by the well known formulas:

$$u_{g0} = -\frac{1}{f\rho_s}\frac{\partial p_s}{\partial y}, \quad v_{g0} = \frac{1}{f\rho_s}\frac{\partial p_s}{\partial x},$$

where ρ_s is the near-surface density. Both p_s and ρ_s can be evaluated from the weather forecast data, or from the output of some larger-scale atmospheric model.

The vertical variations of the geostrophic wind are usually related to the effects of the so-called thermal wind, which is one of the manifestations of the baroclinicity in the atmosphere. In the CBL, the thermal wind appears as a result of the thermal horizontal nonuniformity of the layer. Under these conditions, the vertical gradients of the geostrophic wind components can be represented as first approximations by

$$\Gamma_u = -\frac{\beta}{f}\frac{\partial \bar{\theta}}{\partial y}, \quad \Gamma_v = \frac{\beta}{f}\frac{\partial \bar{\theta}}{\partial x},$$

following the traditional thermal wind relationship given, *e.g.* in Holton [24], and using the CBL potential temperature averages.

The theory presented in the previous sections of the paper deals with different cases of dry CBL. To account for the effects of air humidity in the unsaturated atmosphere, one should complement the model with the equation of moisture transfer. It is shown in Zilitinkevich *et al.* [55] that the integral humidity budget, and profile of the turbulent flux of moisture in the zero-order CBL model can be expressed in a way analogous to the heat budget and to the heat flux. In the equation terms representing the buoyancy effects, the potential temperature θ should be replaced by the virtual potential temperature, $\theta_v = \theta(1+0.61q)$, where q is the specific air humidity, and the buoyancy flux has to be written as $\beta Q + 0.61gE$, where E is the kinematic turbulent flux of moisture.

3. General Structure Model of the Shear Free CBL

In many instances, the buoyant production of TKE in the atmospheric CBL ultimately dominates the TKE production due to wind shears, so the CBL can be taken as shear-free.

Figure 3 shows the vertical profiles of buoyancy b and vertical turbulent buoyancy flux B in the shear-free CBL heated from below. The buoyancy is defined here as $b=g(\rho_0-\rho)/\rho_0 \approx g(\theta-\theta_0)/\theta_0$ where ρ_0 is a reference density value. As a result of heating, a well-mixed layer forms in the lower part of the fluid. In the near vicinity of the heated surface, the buoyancy drops sharply from a surface value b_s to the mixed-layer value b_m. The homogeneous mixed layer is separated from the nonturbulent layer aloft (which is linearly stratified with the vertical buoyancy gradient $db/dz = N^2$) by the interfacial (entrainment) layer, where the buoyancy increases with height.

The vertical turbulent buoyancy flux B decreases linearly with height in the main portion of the CBL. Its zero-crossing height roughly defines the mixed-layer depth h_0.

(Note that the mixed-layer depth from the general point of view is different to the CBL depth; only in the zero-order CBL model do these depths coincide.) Being negative over the interfacial layer, B reaches a minimum within this layer and vanishes towards its upper boundary $z = h_0 + \Delta h$. Thus, four regions can be distinguished in the shear-free CBL: the surface layer (whose depth is typically small compared to h_0), the mixed layer, the interfacial (entrainment) layer, and the nonturbulent layer.

As can be inferred from the comparison of Fig. 3 and Fig. 1, the buoyancy jump in the zero-order CBL model does not correspond to the actual buoyancy difference across the entrainment layer. It is also possible to show (Fedorovich and Mironov [21]) that the zero-order model value of buoyancy (temperature) flux at the CBL upper edge is not equal to the most negative flux of entrainment presented in Fig. 3.

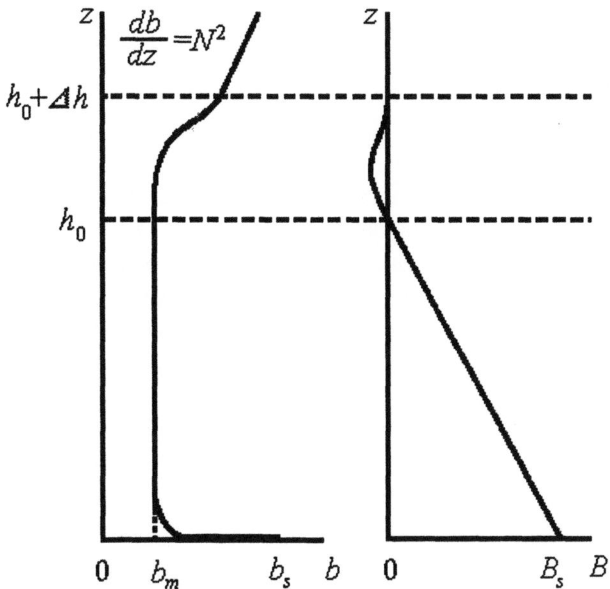

Figure 3. Vertical profiles of buoyancy b and turbulent buoyancy flux B in the shear-free CBL.

Based upon the experimental evidence that the buoyancy increment in the upper portion of the CBL occurs over a layer of significant thickness, Betts [5] proposed the so-called first-order jump CBL model. The model assumes that the mixed layer extends up to the height of the most negative buoyancy flux of entrainment, and that b and B increase linearly with height throughout the interfacial layer, undergoing the first-order discontinuities at its upper and lower boundaries. However, one can note from Fig. 3 that the profiles of b and B in the interfacial layer are hardly the linear ones in reality, and that the minimum buoyancy flux of entrainment occurs within the interfacial layer, not at its bottom. Consequently, the interfacial-layer thickness in the first-order jump model is significantly smaller than the actual one.

The next step towards more realistic parameterization of the entrainment zone was made by Deardorff [13] who allowed all negative buoyancy flux of entrainment to take place within the entrainment layer. The entrainment-layer buoyancy profile in the Deardorff model is represented as

$$b = b_m + F\Delta b \, , \tag{20}$$

where F is a dimensionless function of height. The normalized integral of this function over the entrainment layer (the integral shape factor),

$$C_b = \Delta h^{-1} \int_{h_0}^{h_0+\Delta h} F dz \, , \tag{21}$$

is assumed to be dependent on relative stratification $G = N^2\left(\Delta h / \Delta b\right)$, *i.e.* on the square of the ratio of N and the mean buoyancy frequency in the IL, $\left(\Delta b / \Delta h\right)^{1/2}$. The empirical approximation of $C_b(G)$ was suggested in Deardorff [13], however, the form of the function F was not determined.

The Deardorff general-structure CBL model performed well in the cases of steady-state entrainment with large-scale subsidence ($dh_0/dt = d\Delta h/dt = d\Delta b/dt = 0$) and pseudoencroachment ($d\Delta h/dt = d\Delta b/dt = 0$ and $G = 1$), when no closure assumption on dh_0/dt or $d\Delta h/dt$ was required. For unsteady entrainment, a quite arbitrary closure assumption was made concerning $d\Delta h/dt$ which looks more like a fitting relation than a theoretically substantiated equation. Furthermore, this relation provides an ambiguous expression for $d\Delta h/dt$ when Δb and Δh become small.

Following the model line of Deardorff [13], Fedorovich and Mironov [21] derived rate equations for h_0 and Δh departing from the TKE balance equation and using the Deardorff [10] hypothesis of similarity for convective boundary layers revised to account for the entrainment-layer structure. They also attempted to find a reasonable approximation for the buoyancy profile within the IL.

The general-structure CBL model of Fedorovich and Mironov [21] and some results from their study are presented in the following subsections of the paper.

3.1. BUOYANCY BUDGET

We simplify the discussion by considering a horizontally homogeneous boundary layer without a large-scale subsidence. Let us assume that with the development of convection the vertical buoyancy profile in the CBL keeps the following form:

$$b = \begin{cases} b_m & \text{at} \quad 0 \le z \le h_0, \\ b_m + \Delta b F(\zeta, G) & \text{at} \quad h_0 \le z \le h_0 + \Delta h, \\ b_m + \Delta b + N^2(z - h_0 - \Delta h) & \text{at} \quad h_0 + \Delta h \le z. \end{cases} \tag{22}$$

The quiescence of the turbulence-free layer above the CBL is adopted in the model. This implies that in the free atmosphere some initial linear buoyancy profile $b_0(z)$ is conserved, so that $b(z,t)=b_0(z)$ at $z \geq h_0 + \Delta h$.

In the representation (22), F is a function of the dimensionless coordinate $\zeta=(z-h_0)/\Delta h$ and stratification parameter G. It satisfies the following boundary conditions:

$$F\Big|_{\zeta=0} = \frac{\partial F}{\partial \zeta}\Big|_{\zeta=0} = 0, \quad F\Big|_{\zeta=1} = 1, \quad \frac{\partial F}{\partial \zeta}\Big|_{\zeta=1} = G \ . \tag{23}$$

The parameterization (22) omits the difference between b_m and the buoyancy in the thin near-surface layer, (see Fig. 3). Its small contribution to the total buoyancy budget in the shear-free CBL can be neglected.

The evolution of the buoyancy profile (3) should satisfy the buoyancy transfer equation

$$\frac{\partial b}{\partial t} = -\frac{\partial B}{\partial z} \ . \tag{24}$$

Integrating Eq. (24) over z from 0 to h_0 with due regard to the representation (22) and taking into account the definition of the mixed-layer depth as the buoyancy flux crossover height, *i.e.* B=0 at $z=h_0$, we obtain the equation of the mixed-layer buoyancy budget, (see Fedorovich and Mironov [21] for details of derivation),

$$\frac{d}{dt}\left[N^2(h_0+\Delta h)-\Delta b\right]=\frac{B_s}{h_0}. \tag{25}$$

Integration of Eq. (24) over z from 0 to $h_0+\Delta h$ gives the equation of total buoyancy budget in the CBL:

$$\frac{d}{dt}\left\{\frac{1}{2}N^2(h_0+\Delta h)^2-\Delta b\left[h_0+(1-C_b)\Delta h\right]\right\}=B_s\,, \tag{26}$$

where $C_b(G)=\int_0^1 F(\zeta,G)d\zeta$ is the integral shape factor, see Eq. (21).

Equations (25) and (26) are the two ordinary differential equations for three unknowns: h_0, Δh and Δb. An additional expression relating these variables is needed to close the problem.

3.2. ENTRAINMENT RATE EQUATION

The closure equation can be derived departing from the TKE balance in the shear-free CBL:

$$\partial e / \partial t = B - \partial \Phi / \partial z - \varepsilon \, . \tag{27}$$

We adopt the hypothesis of similarity of the convective regime considered. This hypothesis states that the basic turbulence parameters, being normalized by the length scale $h_0 + \Delta h$ and the velocity scale $w_* = \left[B_s(h_0 + \Delta h)\right]^{1/3}$, cease to depend on time in their explicit form and depend on it only through these scales, *i.e.* they become universal functions of the dimensionless height $z/(h_0 + \Delta h)$. Hence, the vertical profiles of turbulent energy and its dissipation rate can be presented in the form:

$$e = w_*^2 F_e\left(\frac{z}{h_0 + \Delta h}\right), \qquad \varepsilon = \frac{w_*^3}{h_0 + \Delta h} F_\varepsilon\left(\frac{z}{h_0 + \Delta h}\right), \tag{28}$$

where F_e and F_ε are dimensionless functions satisfying the boundary conditions $F_e(1) = F_\varepsilon(1) = 0$.

The employed closure hypothesis is very similar to the one proposed by Deardorff [10] which has been widely used in the CBL zero-order jump models. However, instead of an arbitrary height within the limits of the interfacial layer (usually corresponding to the elevation of buoyancy flux minimum), we use $h_0 + \Delta h$, *i.e.* the whole CBL depth, as an appropriate length scale. Fedorovich and Mironov [21 have shown that employment of $h_0 + \Delta h$ as the length scale allows one to decrease the range of empirical estimates of the universal functions for ε and e in the upper portion of the CBL.

The termwise integration of Eq. (27) over z from 0 to $h_0 + \Delta h$ leads to the following equation relating h_0, Δh and Δb (Fedorovich and Mironov [21]):

$$\begin{aligned} \frac{10}{3} C_e \left(1 + \frac{\Delta h}{h_0}\right)^{2/3} (E_h + E_\Delta) = (1 - 2C_\varepsilon) - 2C_\varepsilon \frac{\Delta h}{h_0} - \left(1 - 2C_{bb} + 2G\frac{dC_{bb}}{dG}\right)\left(\frac{\Delta h}{h_0}\right)^2 + \\ +2\left(C_b - GC_{bb} + G^2 \frac{dC_{bb}}{dG}\right)\frac{\Delta h}{h_0} \mathrm{Ri}_b E_h + \\ +2\left[C_b - 2C_{bb} - GC_{bb} - G(1-G)\frac{dC_{bb}}{dG}\right]\frac{\Delta h}{h_0} \mathrm{Ri}_b E_\Delta - \\ -\frac{4}{3} C_e \left(1 + \frac{\Delta h}{h_0}\right)^{5/3} \mathrm{De} - 2\frac{\Phi(h_0 + \Delta h)}{B_s h_0}, \end{aligned} \tag{29}$$

where $E_h = (B_s h_0)^{-1/3} dh_0 / dt$ is the dimensionless rate of h_0 change; $E_\Delta = (B_s h_0)^{-1/3} d\Delta h / dt$ is the dimensionless rate of Δh change; $\mathrm{Ri}_b = B_s^{-2/3} h_0^{1/3} \Delta b$ is the Richardson number based on h_0 and buoyancy increment Δb across the IL;

$\mathrm{De} = B_s^{-4/3} h_0^{2/3} dB_s / dt$ is the nonstationarity parameter introduced in Deardorff *et al.* [16]; $C_e = \int_0^1 F_e(x)dx$ and $C_\varepsilon = \int_0^1 F_\varepsilon(x)dx$ are dimensionless constants; $C_{bb} = \int_0^1 d\zeta \int_0^\zeta F(\zeta', G)\zeta'$ is a dimensionless function of G; $\Phi(h_0 + \Delta h)$ is the energy flux at the boundary layer top.

The latter can be evaluated evoking the relationships proposed either by Kantha [25] or by Zilitinkevich [54], see section 2.3. In terms of the general-structure CBL model, the first of these relationships yields

$$2\frac{\Phi(h_0 + \Delta h)}{B_s h_0} = C_N \mathrm{Ri}_N^{3/2} \left(\frac{\Delta h}{h_0}\right)^2, \tag{30}$$

and the second results in

$$2\frac{\Phi(h_0 + \Delta h)}{B_s h_0} = C_N' \mathrm{Ri}_N^{3/2} \left(\frac{\Delta h}{h_0}\right)^3, \tag{31}$$

where $\mathrm{Ri}_N = B_s^{-2/3} h_0^{4/3} N^2$ is the Richardson number based on h_0 and N, and C_N and C_N' are dimensionless constants.

Equations (25), (26), (29), (30) or (31) constitute a closed set. The dimensionless constants C_ε, C_e, C_N or C_N', and the shape factors $C_b(G)$ and $C_{bb}(G)$ have been determined by Fedorovich and Mironov [21] from data of atmospheric, oceanic, laboratory, and LES studies of the shear-free CBL.

An approximation of the function $F(\zeta, G)$, fitting experimental and model data reasonably well, was obtained in [21] from geometrical arguments. To match the continuity conditions (23), the required function was represented in the form of a fourth-order polynomial. Coefficients of this polynomial were set to be functions of C_b in order to satisfy the additional integral condition (21). This provided the following approximation for the dimensionless buoyancy profile in the entrainment zone:

$$F(\zeta, G) = \left(\frac{3}{2}G - 12 + 30C_b\right)\zeta^2 + (28 - 4G - 60C_b)\zeta^3 + \left(\frac{5}{2}G - 15 + 30C_b\right)\zeta^4, \tag{32}$$

where the dependence of the integral shape factor C_b on the relative stratification G was given by

$$C_b = 0.55\exp(-0.27G). \tag{33}$$

The last expression, which had been originally proposed by Deardorff [13], was verified by Fedorovich and Mironov [21] against data collected from a variety of CBL experimental and model studies.

3.3. MODELING LABORATORY CBL

The pure case of the shear-free CBL was simulated in the series of water-tank experiments by Deardorff *et al.* ([15], [16], Willis and Deardorff [50], Deardorff and Willis [14]). The most comprehensive data set was presented in Deardorff *et al.* [16] hereafter referred to as DWS. It was used by Fedorovich and Mironov [21] to test their general-structure CBL model.

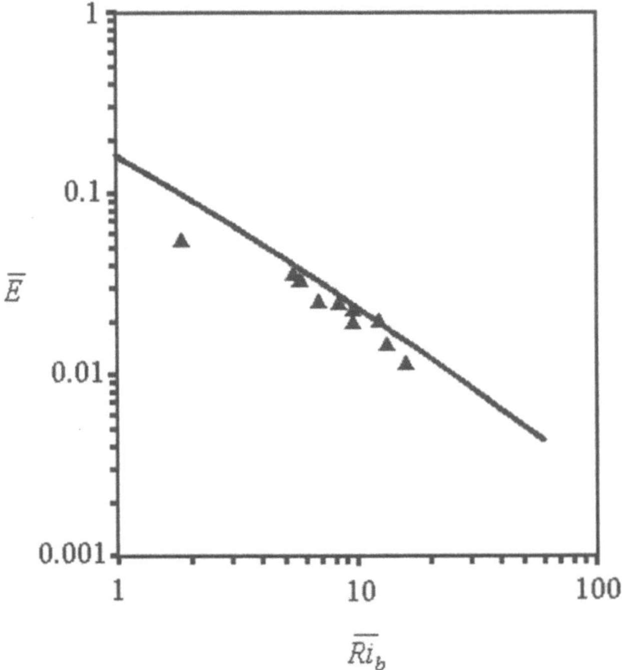

Figure 4. Dimensionless entrainment rate versus the Richardson number for a two-layer fluid system. The model curve from [21] is shown by the solid line; the points are the DWS results.

Following the DWS notation originating from the zero-order jump approach, we will express parameters of entrainment in terms of $\overline{h}$, which is the elevation of buoyancy-flux minimum within the entrainment zone:

$$\overline{E} = (B_s\overline{h})^{-1/3} d\overline{h} / dt, \quad \overline{\mathrm{Ri}_b} = B_s^{-2/3}\overline{h}^{1/3}\Delta b, \quad \overline{\mathrm{Ri}_N} = \frac{1}{2} B_s^{-2/3}\overline{h}^{4/3} N^2 . \tag{34}$$

The general-structure model allows the determination of $\overline{h}$ directly from the shape of the buoyancy flux profile.

Predictions of entrainment rate by the general-structure CBL model agree well with the data from the DWS two-layer fluid experiments, see Fig. 4. The two-layer fluid case corresponds to neutral stratification in the free atmosphere above the CBL ($\overline{\mathrm{Ri}_N}$ =0). Here we use the traditional terminology, speaking about a "two-layer fluid", although in reality it is a three-layer one: mixed layer - interfacial layer - nonturbulent neutral layer. Within the zero-order approach, this is really a two-layer fluid because the interfacial layer is reduced to the zero-thickness surface. The calculated dependence of $\overline{E}$ on $\overline{\mathrm{Ri}_b}$ practically coincides with the basic relation of the zero-order model for convection in the two-layer fluid: $\overline{E}\cdot\overline{\mathrm{Ri}_b}$ =*const.*

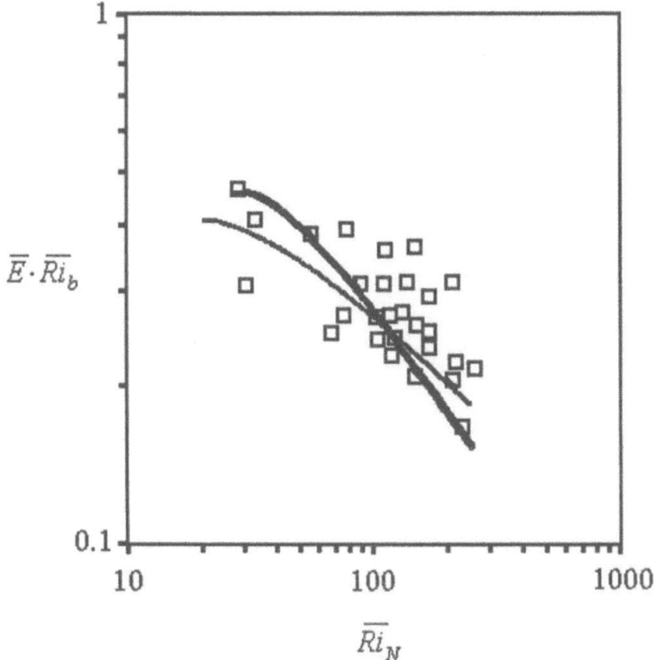

Figure 5. Entrainment in the linearly stratified fluid. The heavy solid curve is calculated using Eq. (30), the solid line shows the results of calculation with Eq. (31); the points are the DWS data.

If the nonturbulent layer above the CBL is stably stratified, the entrainment law relates $\overline{E}$ to two Richardson numbers, $\overline{\mathrm{Ri}_b}$ and $\overline{Ri_N}$. In Fig. 5, the product $\overline{E}\cdot\overline{\mathrm{Ri}_b}$ is plotted against $\overline{\mathrm{Ri}_N}$. Model curves presented in the plot were obtained using two parameterizations, (30) and (31), for wave-related energy flux at the boundary-layer top. The dimensionless constants C_N=0.001 and C_N'=0.012 were estimated in [21] as the best fit to empirical data from the DWS experiments with the linearly stratified fluid. The overall difference between the two parameterizations in Fig. 5 is rather small considering the data scatter. Still, the DWS data gives more proof to Eq. (31), than to Eq. (30).

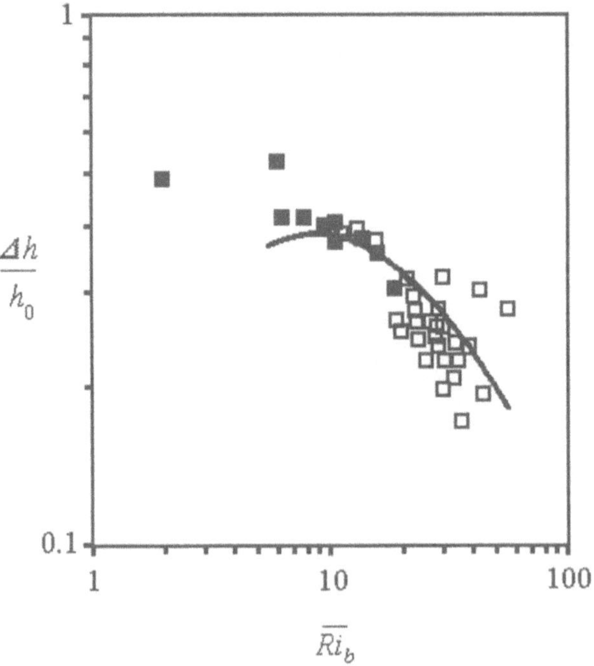

Figure 6. Normalized entrainment layer depth versus the Richardson number in the linearly stratified fluid. The curve is calculated by the general-structure CBL model with the energy flux parameterized after (31). The open squares represent the DWS data referring to a linearly stratified fluid. The DWS data from experiments intended to treat a two-layer fluid system are shown by filled squares.

The DWS experiments with a two-layer fluid were actually performed in the presence of weak stable density stratification in the quiescent layer. This can be inferred from parameters of the experimental setup tabulated in [16]. Calculations with the general-structure model [21] showed that the effect of this stratification could be, nevertheless, strong enough to modify the regime of entrainment compared to the pure case of convection in the two-layer fluid. The $\Delta h / h_0$ values from the DWS experiments are plotted against $\overline{\mathrm{Ri}_b}$ in Fig. 6 with due regard to non-zero values of

N^2. They conform fairly well with the theoretical curve for a linearly stratified fluid at small $\overline{\mathrm{Ri}}_b$.

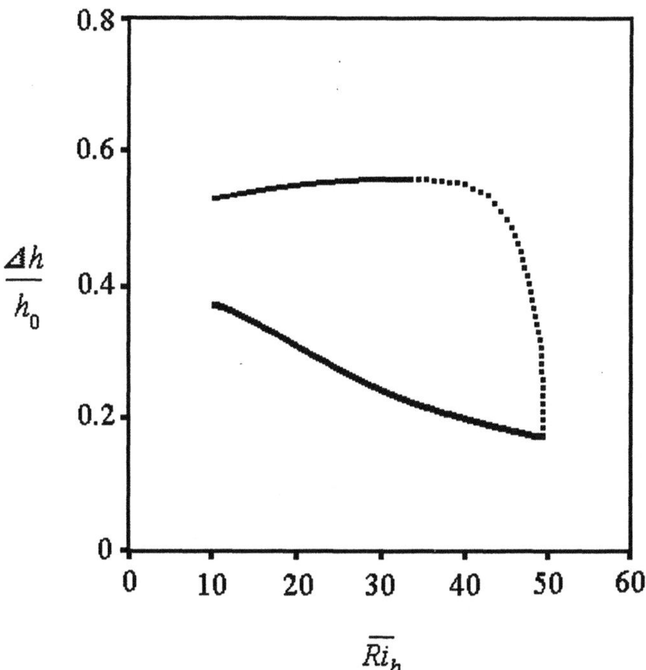

Figure 7. Convective boundary layer development in a composite stratified flow. The heavy solid curve corresponds to the entrainment in a linearly stratified layer. The dashed curve represents the transition stage of the boundary layer development when the entrainment adjusts to a regime characteristic of a two-layer system. The solid line displays the quasi-equilibrium entrainment in a two-layer fluid.

With the aid of the general-structure CBL model, Fedorovich and Mironov [21] found that nonstationarity of the entrainment zone can essentially modify relationships between the entrainment parameters in the shear-free CBL. Figure 7 shows the numerical results for the CBL growing through a stratified flow composed of linearly stratified lower layer and neutrally stratified upper layer. The entrainment occurs first in a linearly stratified fluid, where $\Delta h / h_0$ decreases monotonously with $\overline{\mathrm{Ri}}_b$ (heavy solid curve). After the entrainment-layer top has reached the base of the neutral layer, the entrainment zone passes through a transition stage (dashed curve) towards a quasi-equilibrium state characteristic of a two-layer fluid. At the transition stage of entrainment, $\Delta h / h_0$ reveals inverse dependence on $\overline{\mathrm{Ri}}_b$ as compared with the quasi-equilibrium entrainment regime in a nearly neutral fluid (solid curve).

The model curve in Fig. 7 forms nearly a closed loop. A similar behavior of the interdependencies of the entrainment parameters was observed in the day-time atmospheric boundary layer by Nelson *et al.* [36], who called this phenomenon the hysteresis of the entrainment zone.

4. Summary

We have presented two typical CBL bulk models: the zero-order jump model and the general-structure CBL model.

The zero-order CBL model proves to be a useful tool for applied studies of the atmospheric CBL. Wind engineering and pollutant-dispersion modeling can be mentioned as prospective areas for the application of the zero-order jump approach.

Despite the simplicity of zero-order jump model approach, it provides the opportunity of accounting for a variety of physical mechanisms determining the temporal and spatial structure of the atmospheric CBL. The model allows a generalization for the CBL over irregular terrain. Further progress in zero-order CBL modeling essentially depends on success in developing scalings and parameterizations for different turbulence regimes in the atmospheric CBL.

The general-structure CBL model is based on the realistic representation of the buoyancy profile throughout the entrainment zone of the shear-free CBL. The explicit self-similar representation of this profile was proposed. The integral shape factor of the profile was shown to be a universal function of the dimensionless stratification parameter relating buoyancy gradients above and across the entrainment zone.

The general-structure model approach allows the simulation of entrainment-zone dynamics and description of transition entrainment regimes in multilayer flows. The model successfully reproduces basic shear-free CBL cases simulated in the laboratory experiments of Deardorff *et al.*, and provides theoretical explanations for particular cases of the atmospheric shear-free convection affected by nonstationarity of the entrainment zone.

References

1. Ball, F. K. (1960) Control of inversion height by surface heating, *Quart. J. Roy. Meteorol. Soc.*, **86**, 483-494.
2. Batchvarova, E. and Gryning, S.-E. (1991) Applied model for the growth of the daytime mixed layer, *Bound.-Layer Meteor.*, **56**, 261-274.
3. Batchvarova, E. and Gryning, S.-E. (1994) An applied model for the height of the daytime mixed layer and the entrainment zone, *Bound.-Layer Meteor.*, **71**, 311-323.
4. Betts, A.K. (1973) Non-precipitating cumulus convection and its parameterization. *Quart. J. Roy. Meteorol. Soc.*, **99**, 178-196.
5. Betts, A. K. (1974) Reply to comment on the paper "Non-precipitating cumulus convection and its parameterization", *Quart. J. Roy. Meteorol. Soc.*, **100**, 469-471.
6. Brutsaert, W. (1987) Nearly steady convection and the boundary layer budgets of water vapor and sensible heat, *Bound.-Layer Meteor.*, **39**, 283-300.
7. Carson, D. J. (1973) The development of dry inversion-capped convectively unstable boundary layer, *Quart. J. Roy. Meteorol. Soc.*, **99**, 450-467.
8. Carson, D. J., and Smith, F. B. (1974) Thermodynamic model for the development of a convectively unstable boundary layer, in H. E. Landsberg and J. Van Mieghem (eds.), *Advances in Geophysics*, **18A**, Academic Press, pp. 111-124.

9. Deardorff, J. W. (1970) Preliminary results from numerical integration of the unstable boundary layer, *J. Atmos. Sci.*, **27**, 1209-1211.
10. Deardorff, J. W. (1970) Convective velocity and temperature scales for the unstable planetary boundary layer and for Raleigh convection, *J. Atmos. Sci.*, **27**, 1211-1213.
11. Deardorff, J. W. (1972) Numerical investigation of neutral and unstable planetary boundary layers, *J. Atmos. Sci.*, **29**, 91-115.
12. Deardorff, J. W. (1974) Tree dimensional numerical study of turbulence in an entraining mixed layer, *Bound.-Layer Meteor.*, **7**, 199-226.
13. Deardorff, J. W. (1979) Prediction of convective mixed-layer entrainment for realistic capping inversion structure, *J. Atmos. Sci.*, **36**, 424-436.
14. Deardorff, J. W., and Willis, G. E. (1985) Further results from a laboratory model of the convective planetary boundary layer, *Bound.-Layer Meteor.*, **32**, 205-236.
15. Deardorff, J. W., Willis, G. E., and Lilly, D. K. (1969) Laboratory investigation of non-steady penetrative convection, *J. Fluid Mech.*, **35**, 7-31.
16. Deardorff, J. W., Willis, G. E., and Stockton, B. H. (1980) Laboratory studies of the entrainment zone of a convectively mixed layer, *J. Fluid Mech.*, **100**, 41-64.
17. Driedonks, A. G. M. (1982) Models and observations of the growth of the atmospheric boundary layer, *Bound.-Layer Meteor.*, **23**, 283-306.
18. Driedonks, A. G. M. and Tennekes, H. (1984) Entrainment effects in the well-mixed atmospheric boundary layer, *Bound.-Layer Meteor.*, **30**, 75-103.
19. Fedorovich, E. (1995) Modeling the atmospheric convective boundary layer within a zero-order jump approach: an extended theoretical framework, *J. Appl. Meteor.*, **34**, 1916-1928.
20. Fedorovich, E. and Kaiser, R. (1998) Wind tunnel model study of turbulence regime in the atmospheric convective boundary layer. See this volume.
21. Fedorovich, E. E. and Mironov, D. V. (1995) A model for a shear-free convective boundary layer with parameterized capping inversion structure, *J. Atmos. Sci.*, **52**, 83-95.
22. Garrat, J. R., Wyngaard, J. C., and Francey, R. J. (1982) Winds in the atmospheric boundary layer – prediction and observation, *J. Atmos. Sci.*, **39**, 1307-1316.
23. Gryning, S.-E. and Batchvarova, E. (1994) Parameterization of the depth of the entrainment zone above the daytime mixed layer, *Quart. J. Roy. Meteorol. Soc.*, **120**, 47-58.
24. Holton, J. R. (1972) *An Introduction to Dynamic Meteorology*, Academic Press.
25. Kantha, L. H. (1977) Note on the role of internal waves in thermocline erosion, in E. B. Kraus (ed.), *Modelling and predictions of the upper layer of the ocean*, Pergamon Press, pp. 173-177.
26. Kraus, E. B. and Leslie, L. D. (1982) The interactive evolution of the oceanic and atmospheric boundary layers in the source regions of the trades, *J. Atmos. Sci.*, **39**, 2760-2772.
27. Lenschow, D. (1998) Observations of clear and cloud-capped convective boundary layers, and techniques for probing them. See this volume.
28. Lenschow, D. H., Wyngaard, J. C., and Pennel, W. T. (1980) Mean-field and second-momentum budgets in a baroclinic, convective boundary layer, *J. Atmos. Sci.*, **37**, 1313-1326.
29. Lilly, D. K. (1968) Models of cloud-topped mixed layers under a strong inversion, *Quart. J. Roy. Meteorol. Soc.*, **94**, 292-309.
30. Mason, P. J. (1989) Large-eddy simulation of the convective atmospheric boundary layer, *J. Atmos. Sci.*, **46**, 1492-1516.
31. Moeng, C.-H. (1984) A large-eddy simulation for the study of planetary boundary layer turbulence, *J. Atmos. Sci.*, **41**, 2052-2062.
32. Moeng, C.-H. (1986) Large-eddy simulation of a stratus-topped boundary layer. Part I: Structure and budgets, *J. Atmos. Sci.*, **43**, 2886-2900.
33. Moeng, C.-H. (1987) Large-eddy simulation of a stratus-topped boundary layer. Part II: Implications for mixed-layer modeling, *J. Atmos. Sci.*, **44**, 1606-1614.
34. Moeng, C.-H. and Sullivan, P .P. (1994) A comparison of shear- and buoyancy-driven planetary boundary layer flows, *J. Atmos. Sci.*, **51**, 999-1022.
35. Moeng, C.-H. and Wyngaard, J. C. (1988) Spectral analysis of large-eddy simulations of the convective boundary layer, *J. Atmos. Sci.*, **45**, 3574-3587.
36. Nelson, E., Stull, R., and Eloranta, E. (1989) A prognostic relationship for entrainment zone thickness, *J. Appl. Meteorol.*, **28**, 885-903.
37. Nieuwstadt, F. T. M. and Brost, R. A. (1986) Decay of convective turbulence, *J. Atmos. Sci.*, **43**, 532-546.

38. Plate, E. J. (1971) *Aerodynamic Characteristics of Atmospheric Boundary Layers*, US Atomic Energy Commission, Oak Ridge, Tennessee.
39. Plate, E. J. (1998) Convective boundary layer: a historical introduction. See this volume.
40. Qi, Y., Zhou, J., and Fu, B. (1994) Airflow over a mountain and the convective boundary layer, *Bound.-Layer Meteor.*, **68**, 301-318.
41. Schmidt, H. and Schumann, U. (1989) Coherent structure of the convective boundary layer derived from large-eddy simulations, *J. Fluid. Mech.*, **200**, 511-562.
42. Stull, R. B. (1973) Inversion rise model based on penetrative convection, *J. Atmos. Sci.*, **30**, 1092-1099.
43. Stull, R. B. (1976) The energetics of entrainment across a density interface, *J. Atmos. Sci.*, **33**, 1260-1267.
44. Stull, R. B. (1976) Mixed-layer depth model based on turbulent energetics, *J. Atmos. Sci.*, **33**, 1268-1278.
45. Stull, R. B. (1976) Internal gravity waves generated by penetrative convection, *J. Atmos. Sci.*, **33**, 1279-1286.
46. Stull, R. B. (1988) *An Introduction to Boundary Layer Meteorology*, Kluwer, Dordrecht.
47. Tennekes, H. (1973) A model for the dynamics of the inversion above a convective boundary layer, *J. Atmos. Sci.*, **42**, 558-567.
48. Tennekes, H. and Driedonks, A. J. M. (1981) Basic entrainment equations for the atmospheric boundary layer, *Bound.-Layer Meteor.*, **20**, 515-531.
49. Thorpe, S. A. (1973) Turbulence in stably stratified fluids: a review of laboratory experiments, *Bound.-Layer Meteor.*, **5**, 95-119.
50. Willis, G.E. and Deardorff, J. W. (1974) A laboratory model of the unstable planetary boundary layer, *J. Atmos. Sci.*, **31**, 1297-1307.
51. Zeman, O. and Tennekes, H. (1977) Parameterization of the turbulent energy budget at the top of the daytime atmospheric boundary layer, *J. Atmos. Sci.*, **34**, 111-123.
52. Zilitinkevich, S. S. (1975) Comments on "A model for the dynamics of the inversion above a convective boundary layer", *J. Atmos. Sci.*, **32**, 991-992.
53. Zilitinkevich, S. S. (1975) Resistance laws and prediction equations for the depth of the planetary boundary layer, *J. Atmos. Sci.*, **32**, 741-752.
54. Zilitinkevich, S. S. (1991) *Turbulent Penetrative Convection*, Avebury Technical, Aldershot.
55. Zilitinkevich, S. S., Fedorovich, E. E., and Shabalova, M. V. (1992) Numerical model of a non-steady atmospheric planetary boundary layer, based on similarity theory, *Bound.-Layer Meteor.*, **59**, 387-411.

PARAMETERIZATIONS OF THE CONVECTIVE BOUNDARY LAYER IN ATMOSPHERIC MODELS

C.-H. MOENG
National Center for Atmospheric Research
P.O.Box 3000, Boulder, CO 80307 USA

1. Introduction

The horizontal grid mesh in atmospheric forecast models is almost always larger than several kilometers, and hence all turbulent eddies within the Planetary Boundary Layer (PBL) are subgrid-scale motions. Therefore the net effect of PBL turbulence has to be parameterized as an ensemble mean through an ensemble-mean turbulence model. There exist many PBL parameterization schemes. Based on a survey Wyngaard and Moeng performed in 1990 (Taylor and Wyngaard [24]), we found these PBL schemes can be categorized into six generic types:

1. Stablility Bounds model
2. Bulk model
3. Transilient turbulence model
4. Richardson-number dependent K model
5. K-profile model
6. Higher-order closure approach

The first scheme was developed by Price *et al.* [20] and is perhaps the simplest kind of all. It is a stability adjustment scheme; the main mixing rule is that whenever the stratification between two adjacent layers becomes unstable, mixing is allowed to occur to adjust the sounding back to neutral. So far this scheme has been applied only to the oceanic mixed layer, and I will not discuss it further. I will also not talk about bulk and transilient turbulence models in this lecture because they are covered in the Fedorovich [8] and Stull [22] papers, respectively. Only the last three modeling types will be covered in this lecture.

E.J. Plate et al. (eds.), Buoyant Convection in Geophysical Flows, 291–311.

2. Richardson-number-dependent K Model

In this type of modeling approach, the turbulence fluxes of momentum, heat, and other scalars are assumed to relate to the vertical gradients of the mean fields through eddy viscosity and eddy diffusivity coefficients K_M and K_H as

$$\overline{uw} = -K_M \frac{\partial U}{\partial z}, \tag{1}$$

$$\overline{vw} = -K_M \frac{\partial V}{\partial z}, \tag{2}$$

$$\overline{w\theta} = -K_H \frac{\partial \Theta}{\partial z}, \tag{3}$$

where these Ks are made to be functions of the gradient Richardson number,

$$R_i \equiv \frac{g}{\theta_0} \frac{\partial \Theta / \partial z}{(\partial V / \partial z)^2}. \tag{4}$$

The most-used Richardson-number-dependent K model was developed by Louis [12] and is used in the ECMWF medium-range forcast model. This scheme extends the surface-layer similarity formulae to the whole PBL, as described in the following. Based on the surface similarity theory, the mean wind speed and temperature at any height z within the surface layer for the convective PBL can be expressed as

$$|\vec{V}(z)| = \frac{u_*}{\kappa}[\ln\frac{z}{z_0} - \Psi_1(\frac{z}{L})], \tag{5}$$

$$\Theta(z) - \Theta_0 = \frac{0.74\theta_*}{\kappa}[\ln\frac{z}{z_0} - \Psi_2(\frac{z}{L})], \tag{6}$$

where Ψ_1 and Ψ_2 are stability functions, $|\vec{V}(z)| = \sqrt{U(z)^2 + V(z)^2}$, and

$$L \equiv \frac{\theta_0}{g} \frac{u_*^2}{\kappa\theta_*} \tag{7}$$

is the Monin-Obukhov length scale, g/θ_0 is the thermal coefficient, $u_* \equiv (\overline{uw}_0^2 + \overline{vw}_0^2)^{1/4}$ is the friction velocity, $\theta_* \equiv -\overline{w\theta}_0/u_*$, and $\overline{w\theta}_0$ is the surface heat flux. z/L is referred to as the stability parameter; it is negative for convective, positive for stable, and zero for neutral PBL. The Monin-Obukhov length can be expressed as a function of the Richardson number by substituting (5) and (6) into (7):

$$L = 0.74 \frac{\theta_0}{g} \frac{|\vec{V}|^2}{\Theta - \Theta_0} \frac{[\ln(z/z_0) - \Psi_2(z/L)]}{[\ln(z/z_0) - \Psi_1(z/L)]^2}, \tag{8}$$

thus z/L could be expressed explicitly as

$$\frac{z}{L} \propto \mathrm{F}_{\mathrm{m,h}}(\frac{z}{z_0}, R_{iB}), \tag{9}$$

where the surface Richardson number is expressed as

$$R_{iB} \equiv \frac{g}{\theta_0} \frac{z(\Theta - \Theta_0)}{|\vec{V}|^2}. \tag{10}$$

Thus, the surface fluxes can be expressed as

$$\overline{uw}_0 \equiv -u_*^2 \mathrm{cos}\alpha_0 = \{C_{DN} F_m(\frac{z}{z_0}, R_{iB})\}\ |\vec{V}|U(z), \tag{11}$$

$$\overline{vw}_0 \equiv -u_*^2 \mathrm{sin}\alpha_0 = \{C_{DN} F_m(\frac{z}{z_0}, R_{iB})\}\ |\vec{V}|V(z), \tag{12}$$

$$\overline{w\theta}_0 \equiv -u_*\theta_* = \{\frac{C_{DN}}{0.74} F_h(\frac{z}{z_0}, R_{iB})\}\ |\vec{V}|(\Theta(z) - \Theta_0), \tag{13}$$

where $C_{DN}(\equiv \kappa^2/(\ln\frac{z}{z_0})^2)$ is the drag coefficient in neutral stratification conditions, the quantities inside the brackets are surface drag coefficients C_D and C_H, and α_0 is the angle between the surface wind and the x (east-west) axis.

Louis computed the functions F_m and F_h as functions of R_{iB} for different z/z_0, shown in Fig. 1, using Businger's empirical formulae of Ψ_1 and Ψ_2, and proposed the following curve fits to these functions:

$$F = 1 - \frac{bR_{iB}}{1 + c|R_{iB}|^{1/2}} \quad \text{for convective cases,} \tag{14}$$

and

$$F = \frac{1}{(1 + b'R_{iB})^2} \quad \text{for stable cases,} \tag{15}$$

for both F_m and F_h, where $b = 2b' = 9.4$ and $c = C_* C_{DN} b(z/z_0)^{1/2}$ with $C_* = 7.4$ for momentum and $C_* = 5.3$ for scalar fluxes. These formulae are derived from the surface-layer data.

For the outer region of the PBL—above the surface layer— Louis assumed that the same Richardson-number dependence applies to K_M and K_H as

$$K_M = \ell^2 |\frac{\Delta\vec{V}}{\Delta z}| F(R_i), \tag{16}$$

$$K_H = \frac{1}{0.74} \ell^2 |\frac{\Delta\vec{V}}{\Delta z}| F(R_i). \tag{17}$$

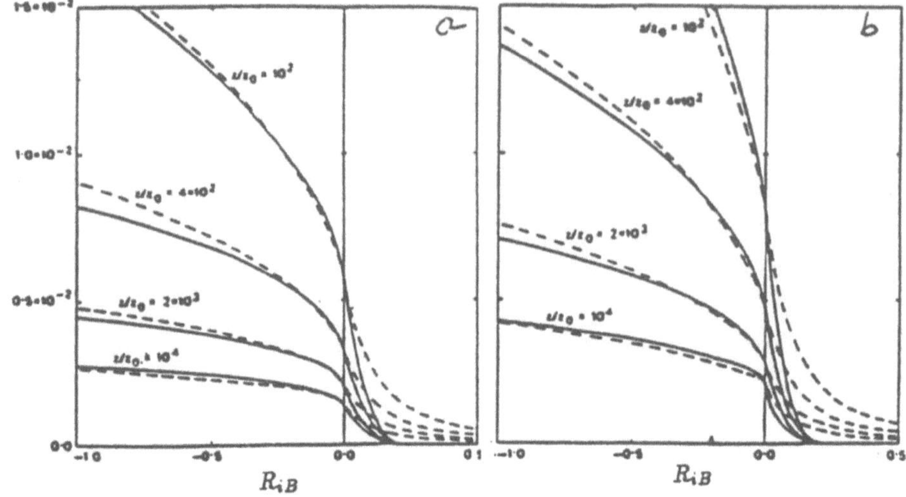

Figure 1. Drag coefficients for a) momentum and b) heat in terms of the bulk Richardson number and the roughness length. Solid curves are obtained by iterations of (9) and dashed curves are from the analytical formulae (14) and (15). (From Louis [12])

Here $\Delta\vec{V}/\Delta z$ is a discrete form of $\partial\vec{V}/\partial z$, where Δ denotes the difference between any two adjacent grid levels. Above the surface layer, the gradient Richardson number in discrete form is

$$R_i = \frac{g}{\Theta_0}\frac{\Delta\Theta/\Delta z}{(\Delta\vec{V}/\Delta z)^2}. \tag{18}$$

The constant c in (14) for the outer layer is

$$c = C^* \frac{\ell^2 b\{[(z+\Delta z)/z]^{1/3} - 1\}^{3/2}}{z^{1/2}\Delta z^{3/2}}. \tag{19}$$

Louis computed the length scale ℓ in (16) and (17) using a formula similar to Blackadar's length scale (Blackadar [4]),

$$\ell = \frac{\kappa z}{1 + \frac{\kappa z}{\ell_0}}, \tag{20}$$

but for the asymptotic value ℓ_0 he assumed an adjustable constant. As it turns out the model results, especially near the PBL top, are very sensitive to the specification of ℓ_0. Louis suggested $\ell_0 = 100$ m in his 1979 paper but later changed it to 150 m for momentum flux and 450 m for heat flux (Louis *et al.* [13]).

3. K-profile Model

The K-profile approach differs from the Richardson-number-dependent K approach in that the vertical shape of K is prescribed in the former but is determined by the local wind and temperature gradients in the latter. We know that the shape of K—when well behaved—typically takes the form of $z(1 - z/z_i)^p$, which is zero at the surface and at the PBL top z_i, and reaches a maximum somewhere near the mid-PBL. The most widely used K-profile model today was developed by Troen and Mahrt [23] and its modified version is now used in the NCAR climate model (Holtslag and Boville [9]). Troen and Mahrt set the exponent $p = 2$ and expressed the eddy viscosity as

$$K_M = \kappa \; w_s \; z \; (1 - z/z_i)^2, \tag{21}$$

where κ is the von Karman constant.

The velocity scale w_s in (21) depends on the PBL regime. Troen and Mahrt set

$$w_s = u_*/\phi_m \qquad \text{for stable PBL,} \tag{22}$$

and

$$w_s = (u_*^3 + c_1 w_*^3)^{1/3} \qquad \text{for convective PBL,} \tag{23}$$

where $\phi_m = 1 + 4.7z/L$ is from Businger *et al.* [6], $w_* \equiv (g\overline{w\theta}_0 z_i/T_0)^{1/3}$ is the convective velocity scale, and $c_1 \sim 0.28$ based on Businger's empirical function. Thus, K in this model has a fixed shape, but its magnitude is determined by the surface friction velocity and the surface heat flux. The Troen-Mahrt K formula for stable PBLs is similar to that proposed by Brost and Wyngaard [5] except that the exponent p in Brost and Wyngaard was 1.5.

For scalar fluxes, Troen and Mahrt added a countergradient term to the flux-K relationship as

$$\overline{wc} = -K_H(\frac{\partial C}{\partial z} - \gamma), \tag{24}$$

where $\gamma = C_1\overline{wc}_0/(z_i w_s)$ and $C_1 = 6.5$. Note that this countergradient term does not reduce to zero in neutral or stable conditions when the surface heating vanishes. Holtslag and Moeng [10] modified this term to $\gamma \propto w_*\overline{wc}_0/(\overline{w^2}z_i)$ based on a LES study, so the countergradient effect reduces to zero when the surface heating vanishes (i.e., when $w_* = 0$), as we anticipate.

Troen and Mahrt relate K_H to K_M using the following formula:

$$P_r \equiv \frac{K_M}{K_H} = \frac{\phi_h}{\phi_m}\frac{z_r}{L} + \kappa\frac{z_r}{z_i}C_1, \tag{25}$$

which is obtained by matching the surface fluxes, of those computed by the K formula (21) and by the Monin-Obuhkov surface similarity theory, at the top of the surface layer z_r where z_r is taken to be $0.1z_i$.

To use the K formula (21), we need to know the PBL height z_i. Troen and Mahrt proposed the following formula for the PBL height:

$$z_i = R_{ic} \frac{T_0}{g} \frac{|\vec{V}(z_i)|^2}{\Theta_v(z_i) - \Theta_s}, \tag{26}$$

where $\vec{V}(z_i)$ and $\Theta_v(z_i)$ are the mean wind and virtual potential temperature at the PBL top. The critical Richardson number R_{ic} is set to 0.5 in that study. Θ_s is the air virtual temperature near the surface and is defined as

$$\Theta_s = \Theta_v(z_1) \quad \text{for stable PBL} \tag{27}$$

and

$$\Theta_s = \Theta_v(z_1) + C_1 \frac{\overline{w\theta_0}}{w_s} \quad \text{for convective PBL.} \tag{28}$$

Here C_1 is the same constant as that used in defining γ, w_s is given in (23), and $\Theta_v(z_1)$ is the virtual potential temperature at the first grid level, which should be within the surface layer.

The formula (26) can be rewritten as

$$\Theta_v(z_i) = \Theta_s + R_{ic} \frac{T_0}{g} \frac{|\vec{V}(z_i)|^2}{z_i}. \tag{29}$$

This, along with Fig. 2, describes how the PBL top is obtained: For convective cases, the PBL top is at the height where a lifting surface air parcel becomes neutrally buoyant, which is the first term in (29). The second term adds a small modification to the PBL height due to the effect of wind at that height. For the stable PBL, z_i is defined at the height where virtual potential temperature exceeds $\Theta_v(z_1)$ by the amount of $R_{ic}\frac{T_0}{g}\frac{|\vec{V}(z_i)|^2}{z_i}$; in other words, the stable PBL top is where the bulk Richardson number $R_i \equiv \frac{g}{T_0}\frac{z_i(\Theta(z_i)-\Theta_s)}{|\vec{V}(z_i)|^2}$ reaches its critical value R_{ic}.

4. Higher-order Closure Approach—Single-point Closure

4.1. CONCEPT AND GOVERNING EQUATIONS

The above K models assume a closure relationship of second moments (i.e., fluxes) to the first moments (i.e., the vertical gradient of the means), i.e., Eqs. (1)-(3), and hence are called the first-order closure models. The higher-order closure approach prognostically solves the governing equations

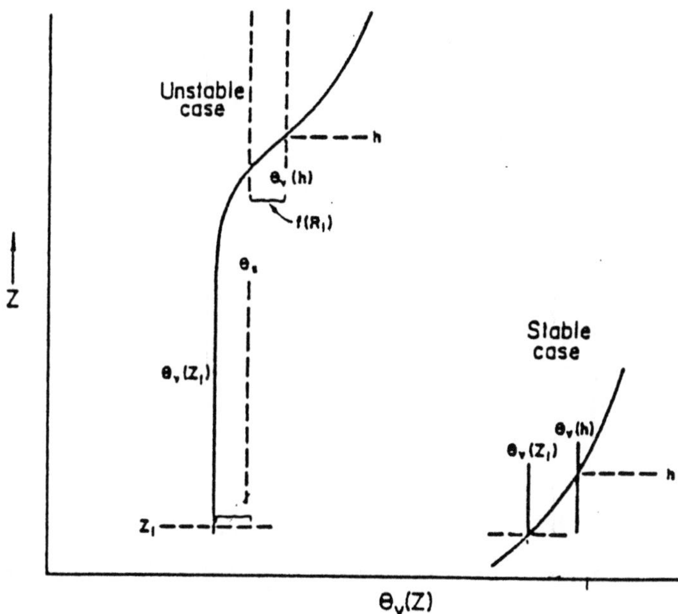

Figure 2. Geometric sketch of the boundary-layer depth relationship to the profile of potential temperature above the surface layer (solid profile). (From Troen and Mahrt [23])

of the higher moment statistics, and assumes closure relationships among the higher moment statistics, e.g., relating third moments to second and/or first moments. The basis behind this approach is a hypothesis that "if a crude assumption for second moments predicts first moments adequately, perhaps a crude assumption for third moments will predict second moments adequately" (Lumley and Khajeh-Nouri [15]).

The governing equations for the higher moment statistics can be derived as follows (ref: C. duP. Donaldson [7]): The Navier-Stokes equation for a Boussinesq fluid reads

$$\frac{\partial u_i}{\partial t} + u_j \frac{\partial u_i}{\partial x_j} = \frac{g_i}{T_0}\theta - \frac{1}{\rho}\frac{\partial p}{\partial x_i} + \nu \frac{\partial^2 u_i}{\partial x_j^2}, \tag{30}$$

where $g_i = g\delta_{i3}$ is the gravitational acceleration rate, p is the pressure field, and ν is the molecular kinetic viscosity. By decomposing (30) into the Reynolds mean (represented by an overline) and fluctuations (denoted by the prime), we have

$$(\frac{\partial \overline{u_i}}{\partial t} + \frac{\partial u_i'}{\partial t}) + (\overline{u_j}\frac{\partial \overline{u_i}}{\partial x_j} + \overline{u_j}\frac{\partial u_i'}{\partial x_j} + u_j'\frac{\partial \overline{u_i}}{\partial x_j} + u_j'\frac{\partial u_i'}{\partial x_j})$$

$$= \frac{g_i}{T_0}(\overline{\theta} + \theta') - \frac{1}{\rho_0}\frac{\partial \overline{p}}{\partial x_i} - \frac{1}{\rho_0}\frac{\partial p'}{\partial x_i} + \nu\frac{\partial^2 \overline{u_i}}{\partial x_j^2} + \nu\frac{\partial^2 u_i'}{\partial x_j^2}, \tag{31}$$

Applying the Reynolds averaging to Eq. (31) yields

$$\frac{\partial \overline{u_i}}{\partial t} + \overline{u_j}\frac{\partial \overline{u_i}}{\partial x_j} = +\frac{g_i}{T_0}\overline{\theta} - \frac{1}{\rho_0}\frac{\partial \overline{p}}{\partial x_i} - \frac{\partial \overline{u_i'u_j'}}{\partial x_j} + \nu\frac{\partial^2 \overline{u_i}}{\partial x_j^2}, \tag{32}$$

which is Reynolds mean equation for incompressible fluids where $\partial u_i/\partial x_i = 0$. The $\overline{u_i'u_j'}$ term in (32) is called the "Reynolds stress". For high R_e turbulence, such as that in the PBL, the molecular term in (32) is much smaller than the Reynolds stress term and can be neglected.

Next the equations for velocity fluctuations are obtained by subtracting the mean, Eq. (32), from the total, Eq. (31),

$$\frac{\partial u_i'}{\partial t} + \overline{u_j}\frac{\partial u_i'}{\partial x_j} + u_j'\frac{\partial \overline{u_i}}{\partial x_j} + u_j'\frac{\partial u_i'}{\partial x_j} - \overline{u_j'\frac{\partial u_i'}{\partial x_j}} = +\frac{g_i}{T_0}\theta' - \frac{1}{\rho_0}\frac{\partial p'}{\partial x_i} + \nu\frac{\partial^2 u_i'}{\partial x_j^2}. \tag{33}$$

Multiplying (33) by u_k' and taking ensemble averages (which have the properties of Reynolds averaging) leads to second moment equations for $\overline{u_i'u_j'}$: (For simplicity we will omit the primes of the moment statistics under the overline signs from here on.)

$$\begin{aligned} \frac{\partial \overline{u_i u_k}}{\partial t} &= -\frac{\partial \overline{w u_i u_k}}{\partial z} - (\overline{u_i w}\frac{\partial U_k}{\partial z} + \overline{u_k w}\frac{\partial U_i}{\partial z}) + \frac{g}{T_0}\overline{w\theta_v} \\ &\quad - \frac{1}{\rho_0}(\overline{u_i\frac{\partial p}{\partial x_k}} + \overline{u_k\frac{\partial p}{\partial x_i}}) + \nu\frac{\partial^2 \overline{u_i u_k}}{\partial x_j^2} - 2\nu\overline{\frac{\partial u_i}{\partial x_j}\frac{\partial u_k}{\partial x_j}}. \end{aligned} \tag{34}$$

In deriving the above equation we further assumed horizontal homogeneity, so $\partial\overline{()}/\partial x = \partial\overline{()}/\partial y = 0$. This is referred to as the "PBL assumption". Here we use U_i and $\overline{u_i}$ interchangeably to denote the mean wind. The mean vertical velocity W is assumed to be zero. The last two terms are molecular diffusion and dissipation. In these second-moment equations one finds third-moment terms, $-\partial\overline{wu_iu_k}/\partial z$, $-\overline{u_i\partial p/\partial x_k}$, and $-\overline{u_k\partial p/\partial x_i}$. (Pressure fluctuations p are second-moment statistics since it has units of u^2.) To solve (34), one must obtain these third moments as well as the molecular diffusion and dissipation terms.

There are two ways to obtain $\overline{wu_iu_k}$. One is to make closure assumptions on $\overline{wu_iu_k}$ by diagnostically relating it to some second or first moments; this method closes the system of equations at the second-order moment level and is thus called second-order closure modeling. The other is to solve the prognostic equations of $\overline{wu_iu_k}$, which can be easily derived from Eq. (33), and make closures on the fourth moments. The latter approach is third-order closure modeling. [Higher-order closure modeling is also called single-point closure modeling because it solves the equations of single-point correlation quantities, $\overline{u_i(x_l,t)u_k(x_l,t)}$, not correlations between two different location points like $u_i(x_l,t)u_k(x_m,t)$.]

4.2. CLOSURE ASSUMPTIONS

The basic principles of making closure assumptions were described in duP. Donaldson [7]. Each closure formula has to be invariant under both coordinate transformation and Galilean transformation, have the right dimensional properties on both sides of the formula, and satisfy all the conservation relationships known to govern the variables in question. For example, a third moment term, $\overline{u_iu_ju_k}$, can be expressed as

$$\overline{u_iu_ju_k} \propto \frac{\partial}{\partial x_i}\overline{u_ju_k} + \frac{\partial}{\partial x_j}\overline{u_iu_k} + \frac{\partial}{\partial x_k}\overline{u_iu_j}, \tag{35}$$

which is invariant under coordinate transformation. To have the same dimensions, we then must have

$$\overline{u_iu_ju_k} \propto -\ell q[\frac{\partial}{\partial x_i}\overline{u_ju_k} + \frac{\partial}{\partial x_j}\overline{u_iu_k} + \frac{\partial}{\partial x_k}\overline{u_iu_j}], \tag{36}$$

where ℓ is a characteristic length scale and q is a characteristic velocity scale. The minus sign is added to insure that turbulent energy will be diffused down the gradient. This is the well-known downgradient diffusion assumption. However, this assumption is now known to be inappropriate for buoyancy-driven turbulence (Zeman and Lumley [25]; Moeng and Wyngaard [19]).

Molecular dissipation terms are typically written in a decay form, for example,

$$-\nu\overline{\frac{\partial u_i}{\partial x_j}\frac{\partial u_k}{\partial x_j}} = -\nu\frac{\overline{u_i\,u_k}}{\lambda^2}, \tag{37}$$

where λ is another characteristic length scale.

The pressure term can be written as

$$-\overline{u_i\frac{\partial p}{\partial x_j}} = -\frac{\partial\overline{u_ip}}{\partial x_j} + \overline{p\frac{\partial u_i}{\partial x_j}}, \tag{38}$$

where the first term on the right-hand side is pressure transport, which is zero for the $\overline{u^2}$ and $\overline{v^2}$ equations since it is an x or y derivative. Thus, in the TKE equation the only pressure transport term left is $-\partial\overline{wp}/\partial z$.

The second term on the right-hand side of (38) drops out of the TKE equation, since $\partial u_i/\partial x_i = 0$ for incompressible fluid. In other words, the effect of this pressure term is only to *redistribute* energy among the $\overline{u^2}$, $\overline{v^2}$ and $\overline{w^2}$ components; it doesn't generate any turbulent kinetic energy. For example, in buoyancy-driven turbulence, buoyancy directly generates vertical-velocity fluctuations and the pressure effect redistributes energy from $\overline{w^2}$ to $\overline{u^2}$ and $\overline{v^2}$. In shear-driven turbulence, the along-shear velocity component gains energy directly from the mean shear, and pressure fluctuations redistribute it to the other two components. This redistribution of turbulent kinetic energy is the so-called return-to-isotropy effect. With this concept Rotta [21] developed a closure for this term as:

$$-\overline{u_i\frac{\partial p}{\partial x_j}} - \overline{u_j\frac{\partial p}{\partial x_i}} = -\frac{1}{\tau}(\overline{u_i u_j} - \frac{2}{3}\delta_{ij}\overline{E}), \tag{39}$$

where τ is the return-to-isotropic time scale.

Later researchers (e.g., Lumley [14]; Launder *et al.* [11]) showed that (39) represents only the effect of pressure fluctuations due to nonlinear, turbulence-turbulence interaction, i.e., the "slow" part of pressure effect. When mean shear (or buoyancy) exists, there should be an additional pressure fluctuation effect ("rapid" part) added to the closure formula (39), such as a term proportional to $\partial U_i/\partial x_j + \partial U_j/\partial x_i$ due to the mean-shear-induced pressure fluctuation effect. Readers who are interested in this subject are referred to Lumley [14] or Andren and Moeng [2].

So far the highest closure scheme used to predict PBL statistics is third-order closure modeling (e.g., Andre *et al.* [1]; Moeng and Randall [18]); there, the well-known quasi-normal approximation, which can properly relate fourth moments to second moments as $\overline{abcd} = \overline{ab}\,\overline{cd} + \overline{ac}\,\overline{bd} + \overline{ad}\,\overline{bc}$, is used as the closure for fourth moments. This normal (i.e., Gaussian) assumption however does not apply to the lower moments. Most PBL modelers, however, stay on the second-order closure approach because third-order closure modeling requires solving many more prognostic equations.

4.3. CHARACTERISTIC LENGTH SCALE

The most difficult and uncertain part of higher-order closure approach is determining the characteristic scales of length, velocity, and time that appear in the above closure formulae. These scales are supposed to represent the net effect of *all* turbulent motions—large and small eddies. For most models, all length scales show up in (36) and (37) are assumed to be proportional to a master length scale L_0, i.e., $\ell \sim \lambda \sim L_0$. This master length scale is then determined by either diagnostic formulae or prognostic methods. In the former approach, the Blackadar formula (20) is often adopted for the convective PBL, which yields a master length scale that increases linearly with height near the surface and reaches an asymptotic value ℓ_0 in and above the mid-PBL where

$$\ell_0 = \alpha \frac{\int_0^\infty qz dz}{\int_0^\infty q dz}. \tag{40}$$

Moeng and Wyngaard [19] showed that $\alpha \sim 0.2$ is an appropriate constant based on their LES. For the stable PBL, Brost and Wyngaard [5] parameterized the master length scale as

$$L_0 = \frac{z}{1 + z/\ell_B}, \tag{41}$$

where ℓ_B is given as

$$\ell_B^2 = \frac{\overline{w^2}}{\frac{g}{T_0}\frac{\partial \Theta}{\partial z}}, \tag{42}$$

which came from the assumed balance between inertial force ($\sim \overline{w^2}/\ell_B$) and buoyancy force [$\sim (g/T_0)(\partial\Theta/\partial z)\ell_B$]. Thus, $L_0 \sim z$ near the surface and $L_0 \sim \ell_B$ in the upper part of the PBL.

For the prognostic approach, some studies solve for the governing equation of the molecular dissipation rate and relate the length scale to the dissipation rate through (37) (e.g., Zeman and Lumley [25]), while others solve for the governing equation of $q^2 L_0$ where q^2 is twice of the turbulent kinetic energy (Mellor and Yamada [17]).

All models present the characteristic velocity scale as the root-mean-square of twice the total TKE, $q = \sqrt{2E}$, and relate the time scale to the length and velocity scales as $\tau \propto L_0/q$.

4.4. THE MELLOR-YAMADA MODELS

A full second-order closure model thus consists of a set of prognostic equations for all second moment statistics, and a set of closure formulae for all third moment statistics and molecular dissipation rates. It is clearly too computationally expensive for use in any mesoscale or climate forecast model. For that reason, Mellor and Yamada [16] developed a hierarchy of turbulence closure models for PBLs by systematically simplifying the full second-order closure model based on the order of a non-dimensional measure of anisotropy, a^2, where $\overline{u_i u_j} = (\delta_{ij}/3 + a_{ij})\overline{u_i u_i}$, as shown in the following table.

The so-called Level 4 model is a full second-order closure model, which solves for the full set of prognostic equations of all second moments. The Level 3 neglects all the time-rate-of-change and the turbulent transport

TABLE 1. Hierarchy of Mellor-Yamada Models

Level 4	All second moment equations are prognostic
Level 3	Only TKE and $\overline{\theta^2}$ equations remain prognostic
Level 2	All second moment equations are algebraic
Level 1	Further simplification from Level 2 model

terms except those in the equations of TKE and temperature variance ($\overline{\theta^2}$). The Level 2 further neglects the time-rate-of-change and the turbulent transport terms in the TKE and $\overline{\theta^2}$ equations. Hence in the Level 2 model, the set of second-moment equations reduces to a set of diagnostic (algebraic) equations, and can be shown to have the following K-form for the fluxes:

$$\overline{uw} = -K_M \frac{\partial U}{\partial z}, \tag{43}$$

$$\overline{vw} = -K_M \frac{\partial V}{\partial z}, \tag{44}$$

and

$$\overline{w\theta} = -K_H \frac{\partial \Theta}{\partial z}. \tag{45}$$

We will show next that these K_M and K_H are Richardson-number dependent. The Level 1 has even simpler K_M and K_H, but this model is seldom if ever used.

The most popular single-point closure scheme used in GCMs is the Level 2.5, with complexity between the Level 2 and the Level 3, in which a further assumption is made to the Level 3 so that the only prognostic equation retained is that of TKE. Therefore it is also known as one-equation closure modeling.

Analytically solving the set of the second-moment algebraic equations, Mellor-Yamada obtained

$$\overline{uw} = -\{\ell^2 |\frac{\partial \vec{V}}{\partial z}| S_M\} \frac{\partial U}{\partial z}, \tag{46}$$

$$\overline{vw} = -\{\ell^2 |\frac{\partial \vec{V}}{\partial z}| S_M\} \frac{\partial V}{\partial z}, \tag{47}$$

$$\overline{w\theta} = -\{\ell^2 |\frac{\partial \vec{V}}{\partial z}| S_H\} \frac{\partial \Theta}{\partial z}, \tag{48}$$

for their Level 2 model, where S_M and S_H are functions of Richardson number as shown in Fig. 3. The quantities inside the brackets in the above

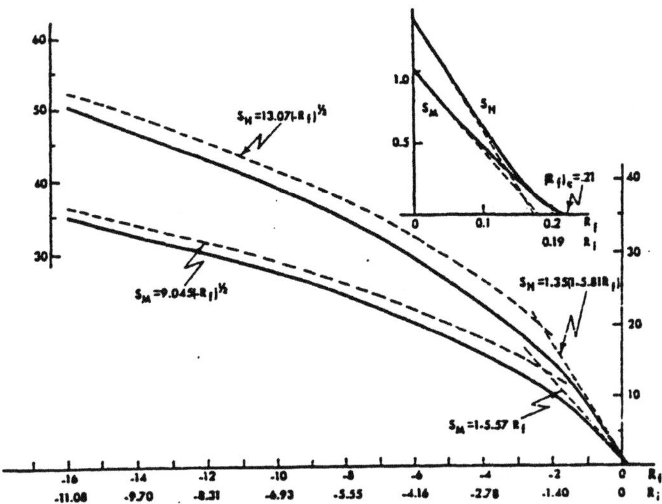

Figure 3. The stability functions S_M and S_H as functions of flux Richardson number or gradient Richardson number. (From Mellor and Yamada [16])

eqs. are K_M and K_H; they have the same form as those given in (16) and (17) except now S_M and S_H are used instead of F.

It is interesting to compare Louis' K formula (see section 2) with that of Mellor-Yamada Level 2. Using the curve fits, $S_M = 9.045(-R_f)^{1/2}$ for $R_f < -2$, $S_M = 1 - 5.57R_f$ for $R_f > -2$, $F = 1 - bR_i/[1 + c|R_i|^{1/2}]$ for $R_i < 0$ and $F = 1/[(1 + b'R_i)^2]$ for $R_i > 0$ as proposed by the authors, and assuming that the flux Richardson number is the same as the gradient Richardson number, we compare these two K models in Fig. 4. For the c value [see (19)] appeared in the F formula, we use $\ell = 100$ m, $z = 500$ m, and two different grid sizes, $\Delta z = 50$ m and $\Delta z = 200$ m, respectively. Figure 4 indicates that in the middle of PBL [i.e., use $z = 500$ m in (19)] both K_M values are quite similar even though these two K_M formulae were derived from very different sources: one based on surface-layer similarity formulae and the other on second-moment budgets.

5. System Performance of the Existing PBL Models

To see how each PBL model performs on key PBL parameters such as turbulent fluxes in different PBL regimes, Ayotte *et al.* [3] carried out an evaluation project to examine the six generic PBL models listed in section 1 against an large eddy simulation (LES) database. The LES database consists of ten PBL regimes covering a range of shear and buoyancy flows, as summarized in Tables 2 and 3. At this first stage of the evaluation project,

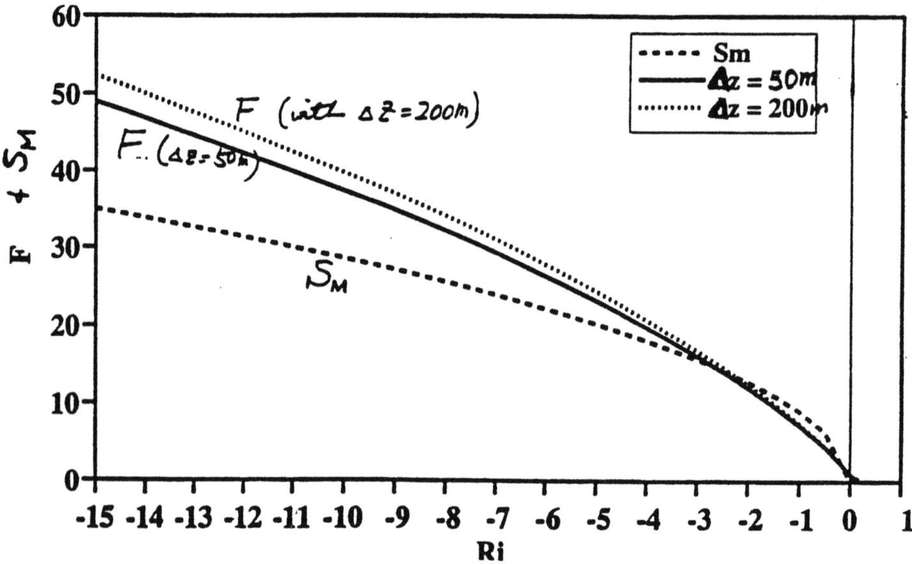

Figure 4. Comparison of the Mellor-Yamada stability function S_M with the Louis' F_M.

only cloud-free, neutral and convective PBL regimes are considered.

To perform this study, a PBL Model Evaluation Software package was developed. This software, which is interactive on a UNIX Workstation, consists of the above six generic model codes, the mean and fluxes of the LES fields (at some earlier time t_0 to use as the initial condition for the PBL model integration and at some later time t_f to use as a database for comparison as sketched in Fig. 5), and a graphic package that compares all PBL and LES solutions. All PBL models share the same main routine, input format, and output graphical display.

Detailed descriptions of the model comparison are given in Ayotte *et al.* [3]. Here we summarize certain key findings. We found that different PBL models produce quite different entrainment rates. This is shown in Fig. 6. A_1 is defined as follows: performing time and height integrations to the governing equation of a conservative scalar ϕ

$$\frac{\partial \phi}{\partial t} = -\frac{\partial \overline{w\phi}}{\partial z}, \tag{49}$$

from t_0 to t_f and from the PBL top at t_0 to the top of the model domain H yields

$$\frac{-\int_{z_i(t_0)}^{H}(\phi(z,t_f)-\phi(z,t_0))dz}{t_f-t_0} = -\frac{1}{t_f-t_0}\int_{t_0}^{t_f}\overline{w\phi}(z_i(t_0),t)dt = A_1, \tag{50}$$

TABLE 2. Case/Parameter matrix

Case	L_x, L_y, L_z (km)	Q_*(K m/s)	U_g (m/s)	V_g (m/s)
EKM	2x2x3.6	0.00	10.0	0.0
00WC	5x5x2	0.00	15.0	0.0
05WC	3x3x1	0.05	15.0	0.0
00SC	3x3x1	0.00	15.0	0.0
03SC	3x3x1	0.03	15.0	0.0
05SC	3x3x1	0.05	15.0	0.0
24SC	5x5x2	0.24	15.0	0.0
24F	5x5x2	0.24	0.0	0.0
15B	5x5x2	0.15	10.0	0-20
24B	5x5x2	0.24	10.0	0-20

TABLE 3. Case/Parameter matrix

Case	$-z_i/L$	L (m)	z_i (m)	w_* (m/s)	u_* (m/s)	τ (s)	z_i Method
EKM	0.0	∞	1500	0.00	0.390	3600	II
00WC	0.0	∞	1065	0.00	0.667	1596	III
05WC	2.26	-480.7	1087	1.20	0.680	905	III
00SC	0.0	∞	448	0.00	0.500	896	III
03SC	1.30	-379.4	493	0.79	0.530	624	I
05SC	1.30	-369.7	480	0.93	0.623	516	I
24SC	18.4	-55.9	1025	2.00	0.566	512	I
24F	∞	0.00	1033	2.00	0.00	515	I
15B	7.73	-125.3	968	1.68	0.627	576	I
24B	12.1	-83.47	1011	1.99	0.640	508	I

where t_0 and t_f are intial and final times of the PBL model integration, respectively. Thus A_1 represents the averaged flux of ϕ across $z_i(t_0)$ level over the whole integration time period. It is representative of the time-averaged entrainment flux of ϕ.

Figure 6 plots the predicted A_1 values from different PBL models for a top-down scalar C, which means the surface flux of c is zero. It shows that most of the PBL models underpredict the entrainment flux of a top-down scalar, especially in the free convection limit, except for the mixed-layer model. The mixed-layer model predicts the entrainment flux quite well compared to the LES solution, particularily in the free convective limit. This is because the entrainment-rate formula in the mixed-layer model is

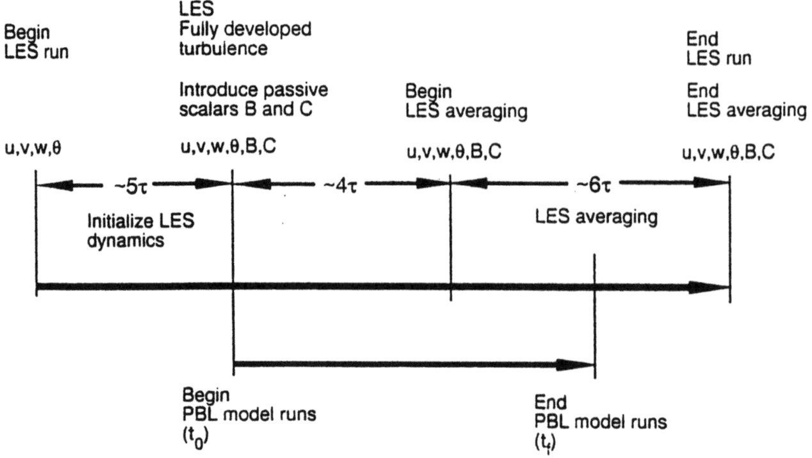

Figure 5. The integration sequence of the LES and the PBL models. (From Ayotte *et al.* [3])

based on an empirical formula which has a limit of the well-known −0.2 value of the ratio between the entrainment buoyancy flux and the surface buoyancy flux in the free-convection case (ref. Fedorovich's paper [8]).

All PBL models yield reasonable surface momentum flux compared to the LES, as shown in Fig. 7. The differences between the PBL models and the LES solutions are less than 10% $\sim$ 15%.

The sensitivity of the PBL model results to the vertical grid resolution is shown in Fig. 8. The x-axis indicates the number of vertical grid levels within the PBL, and the y-axis is the aggregate error averaged over all cases studied here and over the whole PBL depth, for the potential temperature, two passive scalars, and the wind speed. The results show that a minimum of 10 grid levels is needed for most of the PBL models to obtain reasonable mean statistics.

6. Summary and Conclusions

There exist many different kinds of PBL parameterization schemes for use in large-scale meteorological models. These models are simple representations of ensemble-mean statistics for the complicated PBL turbulent motion.

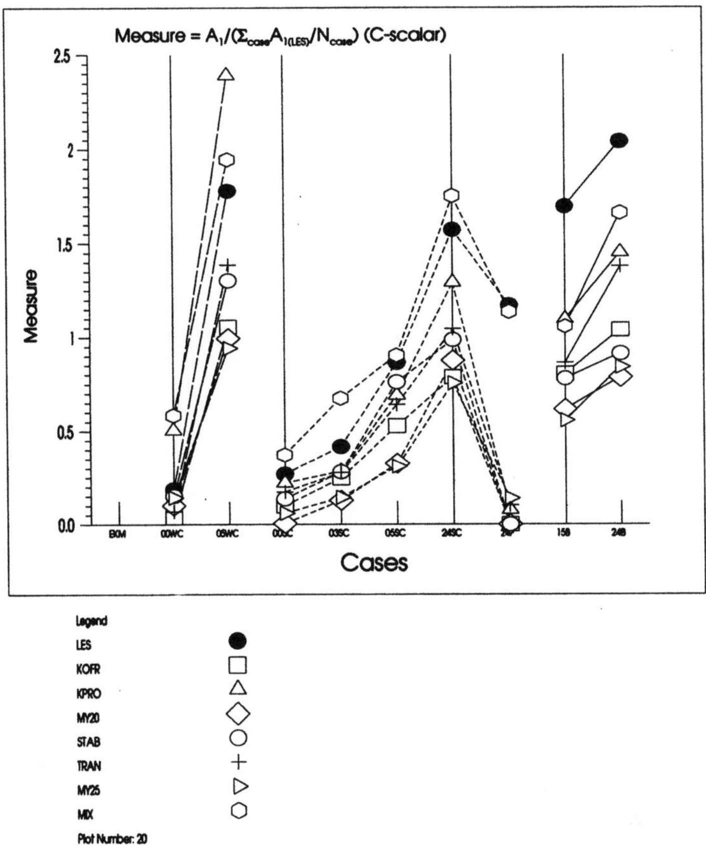

Figure 6. The A_1 measure of the top-down C-scalar, normalized by the LES A_1 value averaged over all PBL cases. (From Ayotte *et al.* [3])

Therefore, we should not expect them to perform well on all key PBL parameters over all PBL regimes. Certain models (such as bulk models) were developed to satisfy the statistical behavior of the highly convective

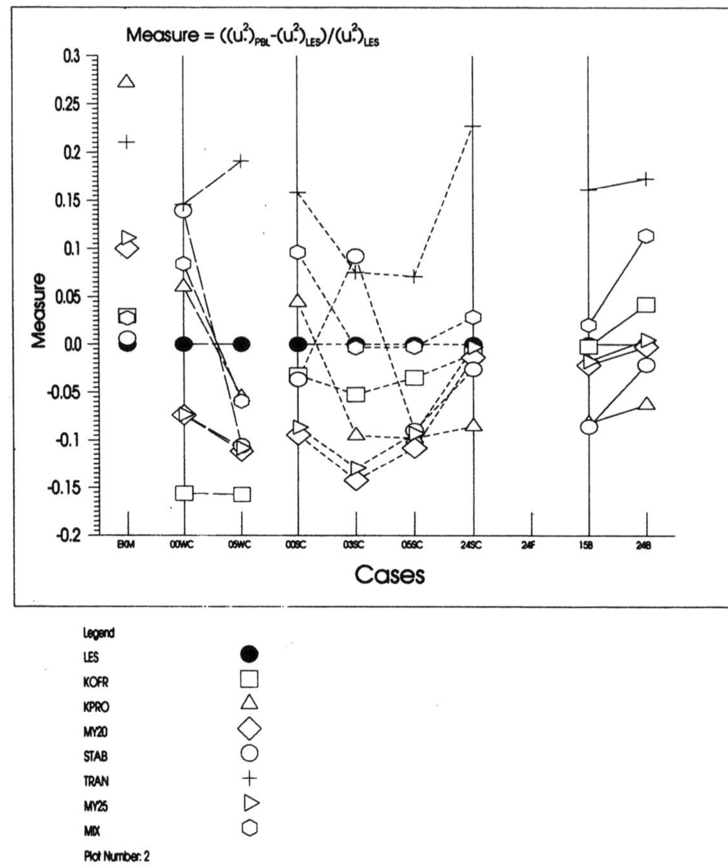

Figure 7. Difference between PBL models and LES for surface momentum flux normalized by the LES value. (From Ayotte *et al.* [3])

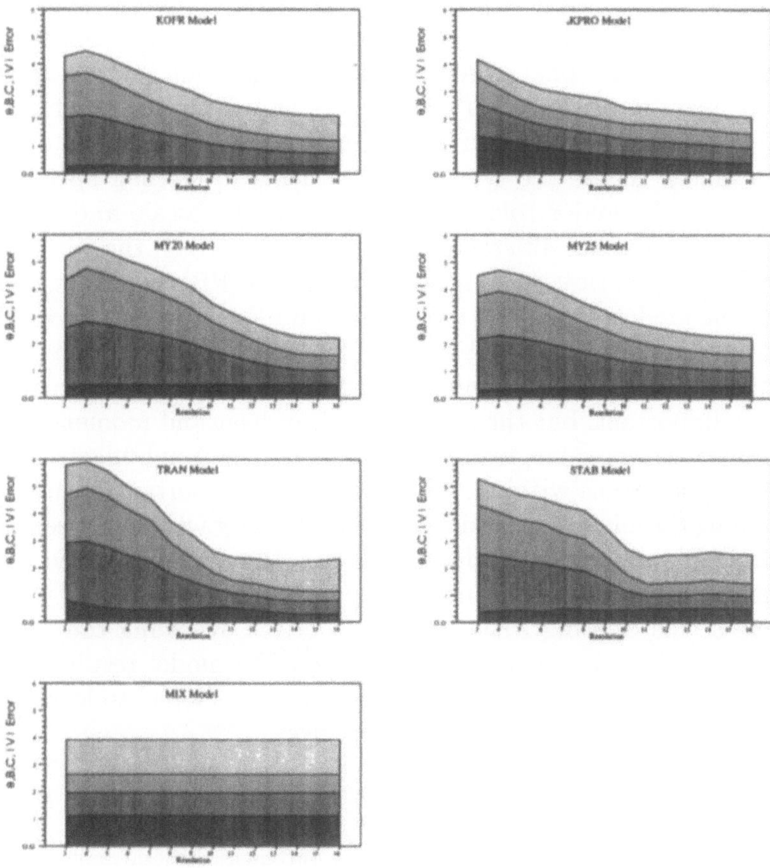

Figure 8. Aggregate error over all cases for the layer-averaged θ, $B-$scalar, $C-$ scalar and wind speed (bottom to top) plotted against resolution. Numbers along abscissa are number of grid levels in the PBL. (From Ayotte *et al.* [3])

PBL, while others (such as local eddy diffusivity models) were designed mainly for the shear-driven type of turbulent flows. Higher-order closure models are computational expensive, but they provide information on the higher-moment statistics, such as variances and covariances, which can be useful for studying, e.g., subgrid-scale condensation or chemical dispersion.

One should choose the type of PBL scheme to fit their specific needs. For example, the major role the PBL plays in mesoscale and climate applications can be quite different. In mesoscale modeling, the vertical distributions of temperature and moisture within the PBL could be important because the models may need to resolve cloud formations. The wind direction within the PBL can also be crucial for mesoscale system development.

In climate modeling, however, the above details of PBL structure may not be so important, but the surface fluxes of heat and momentum can be crucial in determining long-term energy and momentum budgets. Persistent stratocumulus regime within the PBL is another important phenomenon in determining the climate radiation budget. The interaction between the PBL and the land surface, between the PBL and the ocean, and the transport of chemical species all play crucial role in the climate system.

None of the existing PBL models is perfect. Knowing the deficiencies of the PBL model in use can help interpret the model results. The PBL Model Evaluation Software described here was developed to learn the PBL model performance at different PBL regimes.

References

1. Andre, J.C., De Moore, G., Lacarrere, P., Therry, G., Du Vachat, R. (1978) Modelling the 24 h Evolution of the Mean and Turbulent Structures of the Planetary Boundary Layer, *J. Atmos. Sci.*, **35**, 1861–1883.
2. Andren, A. and Moeng, C.-H. (1993) Single-Point Closures in a Neutrally Stratified Boundary Layer. *J. Atmos. Sci.*, **50**, 3366–3379.
3. Ayotte, K. and co-authors (1996) An Evaluation of Neutral and Convective Planetary Boundary-Layer Parameterizations Relative to Large Eddy Simulations. *Boundary Layer Meteorol.*, **79**, 131–175.
4. Blackadar, A.K. (1962) The Vertical Distribution of Wind and Turbulent Exchange in a Neutral Atmosphere. *J. Geophys. Res*, **67**, 3095–3102.
5. Brost, R.A. and Wyngaard, J.C. (1978) A Model Study of the Stably Stratified Planetary Boundary Layer. *J. Atmos. Sci.*, **35**, 1427–1440
6. Businger, J.A., Wyngaard, J.C., Izumi, Y. and Bradley, E. F. (1971) Flux-Profile Relationships in the Atmospheric Surface Layer. *J. Atmos. Sci.*, **28**, 181–189.
7. Donaldson, C. duP. (1973) Construction of a Dynamic Model of the Production of Atmospheric Turbulence and the Dispersal of Atmospheric Pollutions, *Workshop on Micrometeorology*, Ed. D.A. Haugen, AMS 313–392.
8. Fedorosich, E.E. (1998) Bulk models of the atmospheric convective boundary layer. See this volume.
9. Holtslag, A.A.M. and Boville, B.A. (1993) Local Versus Nonlocal Boundary-Layer Diffusion in a Global Climate Model. *J. Climate*,i **6**, 1825–1842.
10. Holtslag, A.A.M. and Moeng, C.-H. (1991) Eddy Diffusivity and Countergradient Transport in the Convective Atmospheric Boundary Layer. *J. Atmos. Sci.*, **48**, 1690–1698

11. Launder, B.E., Reece, G. T. and Rodi, W. (1975) The Development of a Reynolds Stress Turbulent Closure. *J. Fluid Mech*, **68**, 537–566.
12. Louis, J.-F. (1979) A Parametric Model of Vertical Eddy Fluxes in the Atmosphere. *Boundary Layer Meteorol*, *17*, 187–202.
13. Louis, J.-F., Tiedtke, M. and Geleyn, J.-F. (1981) *A Short History of the PBL Parameterization at ECMWF*, Workshop on Planetary Boundary Layer Parameterization, 25–27 November 1981, ECMWF.
14. Lumley, J.L. (1978) Computational Modelling of Turbulent Flows. *Adv. Appl. Mech.*, **18**, 123–176.
15. Lumley, J.L. and Khajeh-Nouri, B. (1974) Computational modeling of turbulent transport. *Andances in Geophysics*, **18A**, 169–192
16. Mellor, G.L. and Yamada, T. (1974) A Hierarchy of Turbulence Closure Models for Planetary Boundary Layers. *J. Atmos. Sci.*, **31**, 1791–1806
17. Mellor, G.L. and Yamada, T. (1982) Development of a Turbulence Closure Model for Geophysical Fluid Problems. *Rev. Geophys. Space Phys.*, **20(4)**, 851–875
18. Moeng, C.-H. and Randall, D.A. (1984) Problems in Simulating the Stratocumulus-Topped Boundary Layer with a Third-Order Closure Model. *J. Atmos. Sci.*, **41**, 1588–1600.
19. Moeng, C.-H. and Wyngaard, J.C. (1989) Evaluation of Turbulent Transport and Dissipation Closures in Second-Order Modeling. *J. Atmos. Sci.*, **46**, 2311–2330
20. Price, J.F., Weller, R.A. and Pinkel, R. (1986) Diurnal Cycling: Observations and Models of the Upper Ocean Response to Diurnal Heating, Cooling and Wind Mixing. *J. Geophys. Res.*, **91(C7)**, 8411–8427.
21. Rotta, J.C. (1951) Statistische theorie nichthomogener turbulenz. *Z. Phys.*, **129**, 547–572.
22. Stull, R.B. (1998) Transilient turbulence theory: a nonlocal description of convection. See this volume.
23. Troen, I and Mahrt, L. (1986) A Simple Model of the Atmospheric Boundary Layer: Sensitivity to Surface Evaporation. *Boundary Layer Meteorol.*, **37**, 129–148.
24. Taylor, P. and Wyngaard, J.C. (1990) *PBL Model Evaluation Workshop*, European Center for Medium-Range Forecasts, 14–15 August 1989, Reading, World Climate Research Programme, Series 42, WMO/TD 378.
25. Zeman, O. and Lumley, J.L. (1976) Modelling Buoyancy Driven Mixed Layers. *J. Atmos. Sci.*, **33**, 1974–1988

WIND TUNNEL SIMULATION OF CONVECTIVE BOUNDARY LAYER PHENOMENA

Simulation Criteria and Operating Ranges of Laboratory Facilities

R. N. MERONEY
Department of Civil Engineering, Colorado State University
Fort Collins, Colorado, 80523, USA

Abstract

The operating ranges of meteorological wind tunnels for convective boundary-layer (CBL) simulation are defined in this paper based on a review of the theoretical and practical limitations of the flow phenomena and the facilities available. Wind-tunnel operating ranges are limited by the dimensions of the simulated circulations and of the tunnel itself, the tunnel flow speed and turbulence processes, and the characteristics of the measurement instrumentation.

1. Introduction

Convective boundary layer (CBL) circulations are surface temperature-forced atmospheric circulations observed almost daily over most of the earth's surface. An understanding of the convective boundary layer has only begun to emerge from laboratory, field, and numerical research in the last thirty years. Recent studies have focused on the characteristics of the large and central mixed layer, but still unresolved are the details of how the various CBL sublayers are linked together. Much of the remaining difficulties relate to the fact that definitive measurements in regions above the meteorological tower heights but below the mixed layer and above the mixed layer within the capping entrainment layer are difficult to acquire. Much of today's understanding of the CBL has come from convection tank and laboratory studies. Conventional type and size water-tank and wind-tunnel facilities provided data which were used to interpret the mixed-layer atmospheric physics and to validate analytic and numerical models for plume behavior (Thorpe [29]). Routine engineering investigations of CBL impact on power-plant plumes or other activities of man will be constrained by the inherent limitations of the size and type of simulations used. This paper examines some of the limitations of fluid modeling.

2. Laboratory Simulation of the CBL

Much of the understanding of the CBL has come from convection tank and large eddy simulation (LES) studies. The original work by Willis and Deardorff [31] is now

E.J. Plate et al. (eds.), Buoyant Convection in Geophysical Flows, 313–325.

complemented by wind-tunnel measurements taken at Colorado State University (Arya and Plate [1]; Poreh and Cermak [24]), University of Karlsruhe (Poreh *et al.* [26]; Rau *et al.* [27]; Fedorovich *et al.* [9]; Fedorovich and Kaiser [8]; Kaiser and Fedorovich [11]), and the work at the Atmospheric Research Division, CSIRO Aspendale, Australia, who are using a salt-water analog to the thermal CBL process to simulate mixing layer phenomena (Hibberd and Sawford [10]).

Thermally stratified wind tunnels currently exist at Colorado State University, USA (1.8 m × 1.8 m cross-section, 24 m long) and at the National Institute for Environmental Studies, Japan (2 m × 3 m cross-section, 24 m long). New wind-tunnel facilities specifically designed to study the CBL are under construction or evaluation at the Central Electricity Generating Facility, Leatherhead, UK (1.5 m × 3.5 m cross-section, 20 m long), the Ecole Centrale de Lyon, France (2 m × 3.7 m cross-section, 15 m long), University of Karlsruhe, Germany (1.5 m × 1.5 m cross-section, 10 m long), and the very ambitious facility at Monash University, Australia (5 m × 10 m cross-section, 40 m long).

2.1. WATER TANK EXPERIMENTS

Laboratory experiments in water tanks have contributed substantially to a better understanding of convectively mixed layers. Experiments with such facilities continue to contribute important information about the mixed layer and its rates of growth. Visualization experiments performed with laser sheets and chemically fluorescing plumes yields detailed information about the mixing characteristics of buoyant plumes in the CBL. These studies clearly demonstrate that the process is non-Gaussian.

Small size limits the spatial resolution characteristics of most water-tank simulations of the CBL. Typical mixed-layer depths examined are from 0.1 m to 0.5 m; hence, the equivalent atmospheric surface-layer regions are below 1 cm, and the stability length may be at most of the order of millimeters. It is unlikely then that CBL water tank measurements will contribute substantially to a better understanding of the atmospheric surface layer (ASL) where the three sublayers reside.

Deeper entrainment layers may be produced in larger water-tank facilities. For example, the Stratified Mixing Facility at Colorado State University consists of a 2 m deep, 10 m long tank provided with thick side wall insulation, floor heating, recirculation pumps, a salt water filling system, visualization windows, and temperature and concentration measurement instrumentation.

2.2. WIND TUNNEL EXPERIMENTS

Deardorff [7] showed for the CBL that when $-z_i/L_{mo}>10$, (note: Arya [2] suggests >5), where z_i is the mixed-layer depth and L_{mo} is the Monin-Obukhov scaling length, then the CBL turbulence scales with z_i and the convective velocity w_*. Characteristic distributions of temperature T, velocity U, and heat flux $<wt>$, are sketched in Fig. 1. Time is expected to scale with z_i/w_*, vertical distance with z_i, and horizontal distance with $(z_i U)/w_*$. For the admixture concentration C the scaling is $Cz_i^2 U/Q_s$, where Q_s is the source strength. Assuming undistorted vertical and horizontal scales, this means one must maintain model and prototype w_*/U equivalent.

Once the mixing layer depth and convective velocity scales are stipulated for the laboratory experiment, it is also necessary that the rate of mixing layer growth

$(dz_i / dt) / w_*$ be similar. Unfortunately, for convenient model velocities, e.g. $U \approx 1$ to 2 m/s, very large surface heat fluxes may be required, e.g. 200 to 15000 W/m^2, and intense stable temperature gradients at the top of the mixing layer, e.g. $dT/dz \approx 50$ to 2000 °C/m.

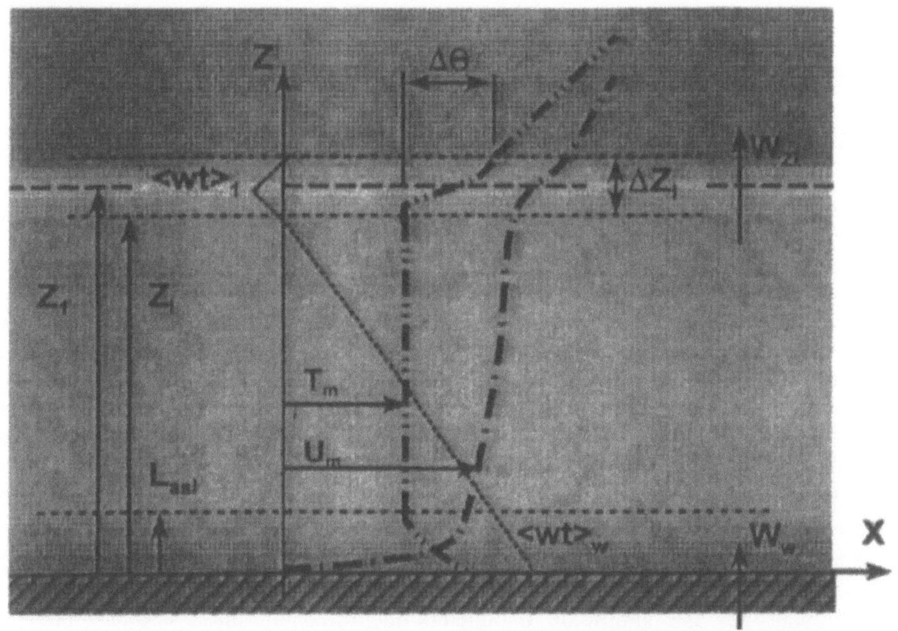

Figure 1. Schematic of the structure of a typical convective boundary layer (CBL).

An alternative approach to the creation of an elevated inversion which might be adequate is to use the wind-tunnel roof as the effective inversion height. The validity of this approach must be examined carefully, however, since a downward flux of heat due to interfacial entrainment plays an important role in many mixed-layer characteristics.

3. Wind Tunnel Operating Range for CBL Simulations

Meteorological wind tunnels are, in effect, analog computers with "near-infinitesimal" resolution and "near-infinite" memory (Snyder [28]). They employ real fluids, not mathematical models of fluids, and produce inherently viscous, turbulent, nonhydrostatic, non-Boussinesq, and compressible flows with nonslip boundary conditions. However, flows in scaled physical models are also only partially similar and cannot at present include all processes present in the atmosphere such as Coriolis acceleration, exchange of energy by radiation, conduction into the soil, and phase changes of water.

Simulation of atmospheric motions by wind-tunnel flows has occurred for almost 100 years since Professor Le Cour constructed a wind-mill test facility in Askov, Denmark, in 1895, and Gustave Eifel designed his exhibition tower in 1889 (Eifel used his tower as a drop site, a meteorological tower, and constructed two wind tunnels at the site). Background reviews about laboratory simulation were prepared by Cermak [4], Davenport and Isyumov [6] and Melbourne [12, 13]. Meroney [15, 18] considered the simulation of complex terrain and valley drainage situations. Snyder [28] suggested similarity criteria for the study of air-pollution meteorology in near neutral situations. Meroney [17] extended the discussion to the simulation of dense-gas plumes in the surface layer. Meroney *et al.* [19] and Avissar *et al.* [3] proposed simulation criteria and operating ranges for the simulation of sea and land breezes.

3.1. GENERAL SIMILARITY REQUIREMENTS

During physical model simulations of atmospheric flows, scale-model replicas of observed ground-level buildings and terrain are constructed and inserted into a laboratory flow facility. The flow characteristics and stratifications of the air in the wind tunnel are adjusted to be similar as possible to the atmospheric conditions. Complete equivalence of the laboratory model and atmospheric prototype flow fields requires geometric, kinematic, dynamic, and thermal similarity. In addition, boundary conditions upstream, downstream, at the lower surface, and near the top of the physical model must be similar to those at the corresponding boundaries of the modeled atmospheric domain. These multiple similarity requirements, the characteristics of the wind tunnel and its instrumentation, and the nature of the atmospheric phenomenon to be modeled all help to determine the operating range (OR) for a wind-tunnel simulation.

Similarity characterization of stratified atmospheric flows are summarized by Avissar *et al.* [3] and Meroney [14]. Specific characterization of CBL phenomena are discussed by Meroney and Melbourne [20], Poreh [23], Poreh *et al.* [26], and Rau *et al.* [27]. Equality of the similitude parameters identified by these authors must be supplemented by the requirements that the surface boundary conditions and the approach-flow characteristics also be similar for model and prototype. Boundary-condition similarity requires similar values of

- @ surface roughness,
- @ topographic relief,
- @ surface temperature distribution,
- @ upstream distribution of mean and turbulent velocities,
- @ upstream distribution of mean and turbulent temperatures, including inversion height, and
- @ longitudinal pressure gradient.

If all of the above conditions are met simultaneously, then all scales of motion ranging from the atmospheric microscale to mesoscale could be simulated exactly by the laboratory model. Unfortunately, not all conditions can be satisfied simultaneously by a scaled model since some are incompatible or conflicting; hence, only partial or approximate similarity can be achieved. This suggests that a laboratory model for a particular meteorological situation must be designed to simulate most accurately those scales of motion which are of greatest significance for the application. In the case of a CBL with undistorted horizontal and vertical scales primary parameters of interest will be w_*/U and inversion height z_i.

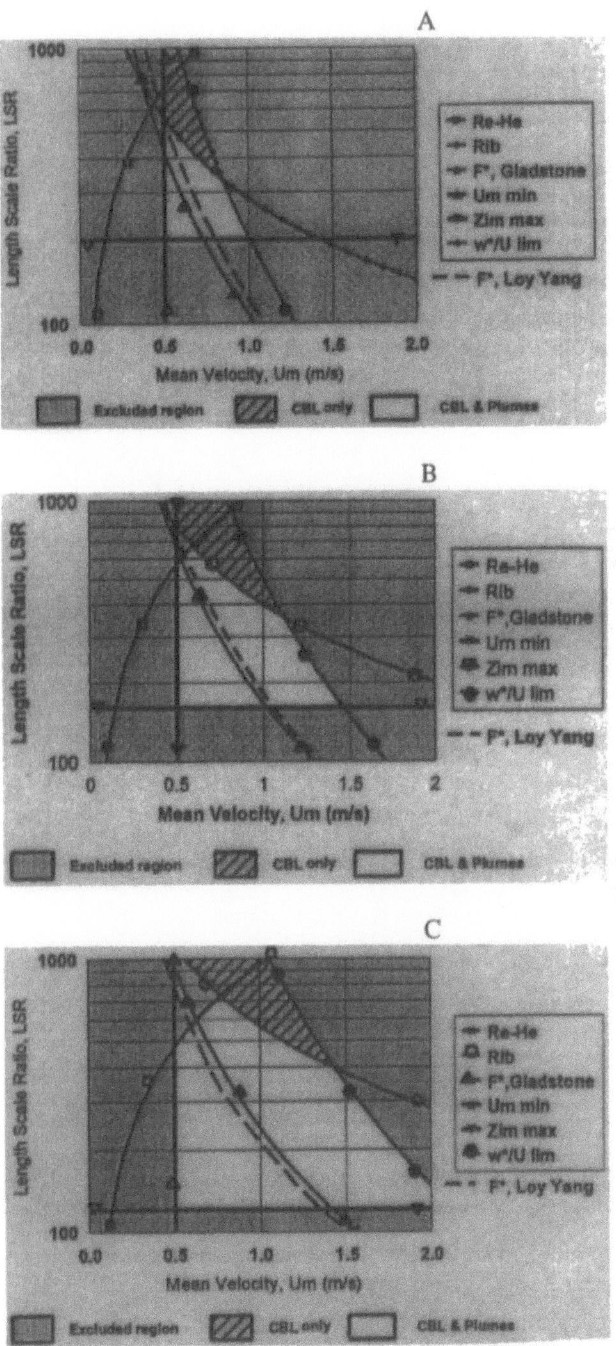

Figure 2. Operating ranges for CBL simulations of Pasquill-Gifford categories A, B, and C conditions in Monash EWT.

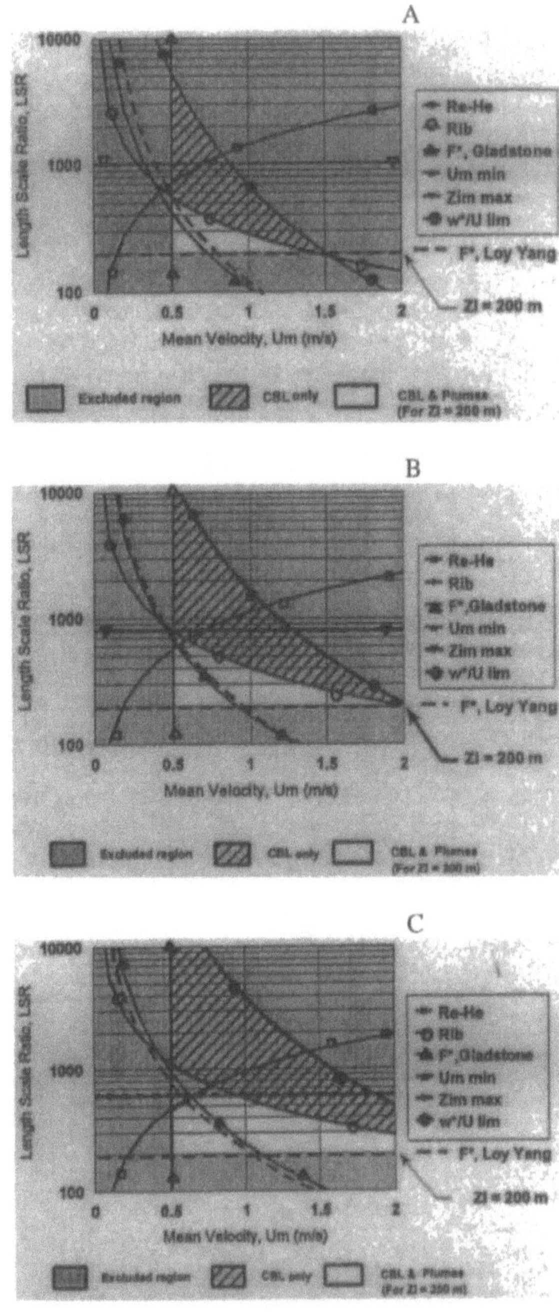

Figure 3. Operating ranges for CBL simulations of Pasquill-Gifford categories A, B, and C conditions in Colorado State University MWT.

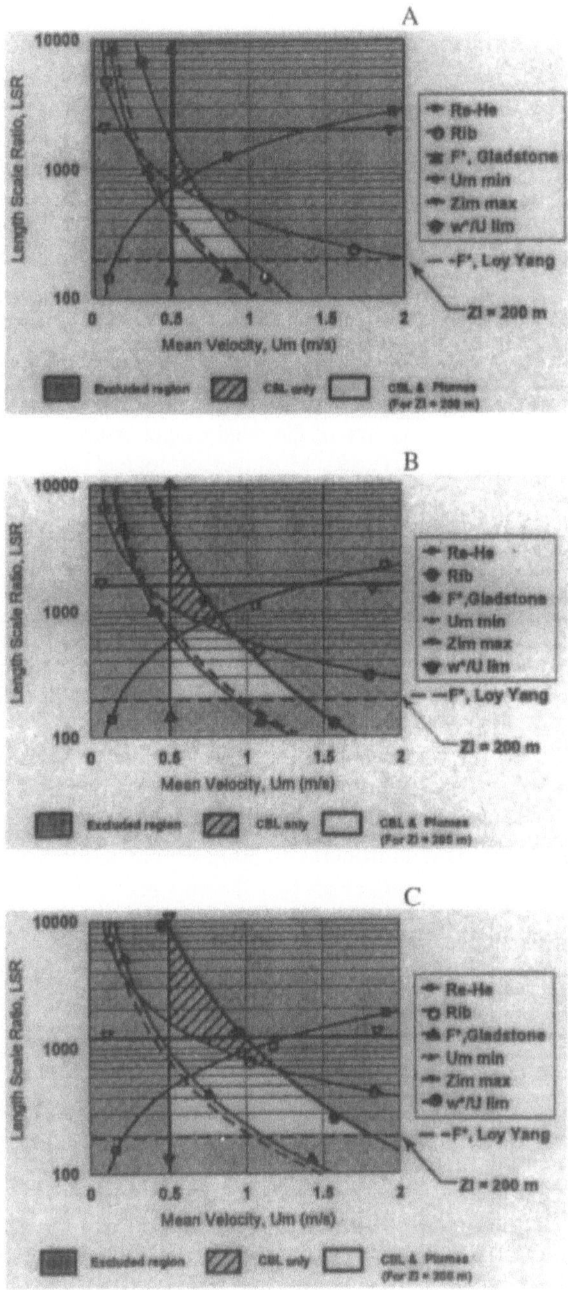

Figure 4. Operating ranges for CBL simulations of Pasquill-Gifford categories A, B, and C conditions in University of Karlsruhe SBLWT.

3.2. INSTRUMENTATION CHARACTERISTICS

Avissar *et al.* [3] reviewed wind-tunnel instrumentation characteristics, and they suggested that measurement accuracy was enhanced for model flow speeds of 0.5 m/s and greater. Averaging times and sampling rates may be expected to produce fractional errors of 1, 5 and 10% for averaging times of 400, 16, and 4 seconds, respectively. Conventional sized hot-film instrumentation should produce only a 4% spatial resolution error.

3.3. WIND TUNNEL CHARACTERISTICS

Simulation of the atmospheric CBL is not only a function of the governing flow physics but also depends on the availability of a suitable simulation facility and its instrumentation. In particular the size of the wind-tunnel test section will determine the smallest model length scale ratio (LSR). Most meteorological tunnels range in size from 0.5 m × 0.5 m to those with working cross-sections of 3 m × 4 m. An exception is the new environmental wind tunnel at Monash with a cross-section of 5 m × 10 m. Density stratification can be induced by use of heat exchangers, injection of heated air, gases of different molecular weight, or latent heat absorption or release during phase change (e.g. Ogawa *et al.* [22] and Meroney [16]). By using vortex generators, fences, roughness elements, grids, screens or jets, a wide range of turbulence integral scales can be introduced into the tunnel boundary layer. Choice of model surface roughness or stratification permits control of surface turbulence intensity, dimensionless wall shear, and velocity profile shape.

Three tunnels have been chosen to represent the characteristics of such facilities. The meteorological wind tunnel (MWT) at Colorado State University is a large, closed-circuit facility with a 1.8 m high by 1.8 m wide by 24 m long test section. Wind speeds are continuously variable from 0.1 to 30 $m \cdot s^{-1}$ and ambient air temperatures can be varied from 5 to 205°C (*e.g.* Cermak [4]). Ten meters of the upstream floor can be cooled between 1°C and ambient temperature while 12 m of the downstream test-section floor can be heated from 1 to 200°C.

The environmental wind tunnel (EWT) at Monash University is a large, open-circuit facility with a 5 to 7 m high by 10 to 12 m wide, and 40 m long test section. Wind speeds are continuously variable from 0.1 to 18 $m \cdot s^{-1}$ (or to 45 $m \cdot s^{-1}$ with cross-section reduced). Ambient air temperatures are drawn into the entrance section, but wall heaters may be inserted to regulate wall temperatures. The wall heaters have a surface heat capacity of 5 $kW \cdot m^{-2}$.

The stratified boundary layer wind tunnel (SBLWT) at the University of Karlsruhe is a unique configuration closed-circuit facility with a 1.5 m by 1.5 m, and 10 m long test section. Wind speeds are variable from 0.5 to 8 $m \cdot s^{-1}$. Air layers are heated and circulated through ten separate contiguous ducts, and additional surface heating is provided over the test section floor. The wall heaters have a surface heat capacity of 1.5 $kW \cdot m^{-2}$.

Coriolis force considerations also limit the maximum acceptable LSR. Snyder [28] suggests a 5 km maximum cut-off point for horizontal length scales for modeling atmospheric diffusion. Mery *et al.* [21] suggests a 15 km limit, Ukeguchi *et al.* [30] suggest a 40 to 50 km limit, and Cermak *et al.* [5] proposed a 150 km limit. Given the strong mixing present in CBL situations and time scales of about 3 hours then surface generated stresses should dominate most situations for at least 2 to 5 km.

The following criteria have been identified which limit the MWT, EWT, and SBLWT operational ranges:

TABLE 1. Operational range limitations of typical meteorological tunnels

	Colorado State MWT	Monash EWT	Karlsruhe SBLWT
@ Maximum model inversion height	< 1 m	< 5 m	< 0.5 m
@ Maximum inversion strength	< -50°Cm^{-1}	< -50°Cm^{-1}	< -100° Cm^{-1}
@ Minimum model inversion height	0	0	0
@ Maximum model blockage	< 5 %	< 5 %	< 5 %
@ Minimum plume Reynolds number	> 300	> 300	> 300
@ Minimum model wind speed	> 0.5 ms^{-1}	> 0.5 ms^{-1}	> 0.5 ms^{-1}
@ Maximum model heat flux	< 5 kWm^{-2}	< 5 kWm^{-2}	<1.5 kWm^{-2}

3.4. WIND TUNNEL PERFORMANCE ENVELOPES FOR CBL SIMULATIONS

The most severe restrictions on the wind-tunnel operating range (or performance envelope) result from geometric similarity constraints. For example Deardorff [7] proposed that laboratory conditions should be sought where $-z_i/L_{mo}>10$ $(z_i/L>25)$ to simulate the mixed layer. Yet the height of most facilities (1-2 m) would limit the dynamic layer to 2 mm, the dynamic convective layer to 2 cm, and the free convection layer to less than 20 cm. Presuming wind-tunnel facilities with model z_i = 0.5, 1.0 and 2.5 m, respectively, then the various characteristic model depths might be:

TABLE 2. Operating ranges of typical meteorological wind tunnels

	Karlsruhe SBLWT	Colorado State MWT	Monash EWT
z_i	0.5 m	1.0 m	2.5 m
L_*	0.002 m	0.004 m	0.01 m
L	0.02 m	0.02 m	0.10 m
L_{**}	0.04-0.10 m	0.08-0.20 m	0.20-0.50 m
L_{ASL}	0.04-0.12 m	0.08-0.24 m	0.20-0.60 m.

In the table above, L=$-kL_{mo}$ is the stratification length scale (k is the von Kármán constant), L_* is the dynamic sublayer characteristic lenth scale, L_{**} is the dynamic-convective characteristic length scale, and L_{ASL} is the atmospheric surface-layer characteristic length scale. See Meroney and Melbourne [20] for definitions of scaling lengths.

An additional criterion required to avoid secondary circulations imposed by facility size is that the aspect ratio of width to inversion height should be greater than 4. Willis and Deardorff [31] found that at values of 2, results were no longer consistent with an infinitely large homogenous layer.

Based on the previous discussions, the following four similarity criteria appear pertinent to the physical simulation of CBL surface layer circulations:

1. $[Ri_B]_m = [Ri_B]_p$
2. $[w_*/U]_m = [w_*/U]_p$ in the mixed layer;
3. $[h_s/z_i]_m = [h_s/z_i]_p$;
4. Similar upwind velocity, temperature, and turbulence profiles.

In the criterion 1, $Ri_B = \beta \frac{\Delta\theta}{w_*^2} \Delta z_i$ is the bulk Richardson number across the inversion, where β is the buoyancy parameter, $\Delta\theta$ is the temperature increment across the inversion, and Δz_i is the inversion-layer depth. In the criterion 3, h_s is the stack height.

TABLE 3. Prototype conditions for convective boundary layers (summarized from various sources)

Pasquill-Gifford Stratification Category		A	B	C
G (m/s)	Geostrophic wind speed	10	10	10
f_c (s^{-1})	Coriolis constant	0.0001	0.0001	0.0001
Z_{ref} (m)	Reference height	10	10	10
Z_t (m)	Troposphere height	10000	10000	10000
T_{air} (°K)	Air temperature	300	300	300
L_{mo} (m)	Monin-Obukhov Length	-5	-10	-20
Ri_f	Flux Richardson Number	-5	-2	-1
Ri	Gradient Richardson Number	-2	-1	-0.5
μ	Ekman parameter	-4000	-120	-60
z_o (m)	Roughness length	0.01	0.01	0.01
z_o/L_{mo}		-0.002	-0.001	-0.0005
z_i (m)	Inversion height	1000	800	600
z_i/L_{mo}		-200	-80	-30
u_* (m/s)	Friction velocity	0.30	0.30	0.30
w_* (m/s)	Convective velocity	2.33	2.01	1.65
$\Phi_m(Z_{ref}/L_{mo})$	Dimensionless shear	0.42	0.50	0.59
u_*/U		0.073	0.069	0.065
w_*/U		0.58	0.40	0.28
U (m/s)	Wind speed at ref. height	4	5	6
$d\Theta/dz$ (°C/km)		-10.0	-7.8	-5.6
L_p (m)	Integral length scale	313	313	313
σ_u/w_*	x-speed variance	0.4	0.4	0.4
σ_w/w_*	z-speed variance	0.4	0.4	0.4
σ_θ/T_*	Temperature variance	0.2	0.2	0.2
σ_u/u_*	x-speed variance	4	3.3	3
σ_w/u_*	z-speed variance	2.39	1.98	1.70
σ_w/σ_u		0.60	0.60	0.57
σ_T/Θ_*	Temperature variance	0.75	3.54	4.08
α (z_o=0.01 m)	Power law coefficient	0.05	0.08	0.1
λ_w/h	Peak wave length at ref. height	0.2	0.2	0.2
$\varepsilon h/w_*^3$	Normalized dissipation	1	1	1

In a CBL situation, mechanical and thermal turbulence influence one another in the surface layers. In turn, this alters the velocity and temperature profiles throughout the CBL. The bulk Richardson number equality (criterion 1) requires that above the CBL inversion where LSR denotes the prototype-to-model vertical length-scale ratio and subscripts m and p refer to model and prototype conditions, respectively,

$$LSR = [\frac{(\partial T/\partial z)_m}{(\partial \theta/\partial z)_p}]^{1/2} [\frac{\theta_p}{T_m}]^{1/2} [\frac{U_p}{U_m}]. \tag{1}$$

The requirement that the convective velocity to advection velocity scale ratio (criterion 2) is also satisfied implies that

$$LSR = \frac{\beta(z_i)_p\, Q_m}{(w_*)_p^3} [\frac{U_p}{U_m}]^3. \tag{2}$$

Similarity of the depth of the CBL (criterion 3) requires that

$$LSR = \frac{z_{ip}}{z_{im}}. \tag{3}$$

Combining the similarity constraints given by the equations for LSR above with the characteristic sizes and flow capacities of the wind tunnels described in section 3.3 and typical atmospheric CBL conditions (see Table 3) provides the relationships and data needed to construct the wind-tunnel operating range and to identify reasonable simulation scenarios. Figures 2 through 4 display performance envelopes for the wind-tunnel sizes that are required for CBL surface layer stratification, respectively. The shaded areas are excluded per criteria limitations for CBL simulations. Model plumes produced under CBL conditions in the cross-hatched region may not be turbulent.

Although the smaller MWT can marginally reproduce CBL conditions (Poreh and Cermak [24]: $z_i/L_{mo} \approx 5.6$-10.7), simultaneous simulation of the surface layer circulations will be unlikely if not impossible. Even for the EWT joint simulation of the CBL and the surface layer regions can only be performed over a limited LSR and U_m range.

It is concluded that an extensive range of wind-tunnel modeling of the three surface layers in a highly unstable atmospheric boundary layer could be achieved with a wind-tunnel working section 5 m high by 10 m wide by 40 m long with 1 MW of floor heating. Model length scale ratios would center about 1/200, with the smallest being about 1/400 and the largest 1/100. The limitations at the small scale are Reynolds number and the practicability of controlling very low wind-tunnel speeds. The limitations at the large scale are mainly the enormous heat requirements and size of the facility needed to model adequate downstream distances from the source.

4. Conclusions

This paper has examined the characteristics, capabilities, and limitations of meteorological wind tunnels to simulate the atmospheric convective boundary layer. A significant conclusion is that such facilities provide an opportunity to explore CBL sublayer and entrainment layer behavior.

Extensive experience with stratified water tanks and wind tunnels definitely suggests that the important turbulence characteristics of the convective boundary layer can be simulated in the laboratory. Nonetheless, laboratory simulation is often not automatic or convenient. Examination of the characteristics of mixing layer, entrainment, and surface layer behavior determines that:

1. Useful CBL simulations may be obtained in sufficiently large stratified wind-tunnel facilities without augmentation techniques.
2. The use of sufficiently large stratified wind-tunnel facilities will provide a means to study the atmospheric sublayers associated with the CBL.

References

1. Arya, S. P. S. and Plate, E. J. (1969) Modeling of the stably stratified atmospheric boundary layer, *J. Atmos. Sci.*, **26**, 656-665.
2. Arya, S. P. S. (1982) Atmospheric boundary layers over homogeneous terrain, in E. J. Plate (ed.), *Engineering Meteorology*, Elsevier, New York, pp. 233-268.
3. Avissar, R., Moran, M. D., Wu, G., Meroney, R. N., and Pielke, R. A. (1990) Operating ranges of mesoscale numerical models and meteorological wind tunnels for the simulation of sea and land breezes, *Boundary-Layer Meteorol.*, **50**, 227-275.
4. Cermak, J. E. (1975) Applications of fluid mechanics to wind engineering - A Freeman Scholar lecture, *J. Fluids Eng.*, **97**, 9-38.
5. Cermak, J. E., Sandborn, V. A., Plate, E. J., Binder, G. H., Chuang, H., Meroney, R. N. and Ito, S. (1966) Simulation of atmospheric motion by wind-tunnel flows, Report No. CER66-17, May, Department of Civil Engineering, Colorado State University, 101 pp.
6. Davenport, A. and Isyumov, N. (1967) The application of the boundary layer wind tunnel to the prediction of wind loading, *Proc. Int. Seminar on Wind Effects on Buildings and Structures*, Ottawa, Canada.
7. Deardorff, J. W. (1972) Numerical investigation of neutral and unstable planetary boundary layers, *J. Atmos. Sci.*, **29**, 91-115.
8. Fedorovich, E. and Kaiser, R. (1998) Wind tunnel model study of turbulence regime in the atmospheric convective boundary layer, see this volume.
9. Fedorovich, E., Kaiser, R., Rau, M., and Plate, E. (1996) Wind tunnel study of turbulent flow structure in the convective boundary layer capped by a temperature Inversion, *J. Atmos. Sci.*, **53**, 1273-1289.
10. Hibberd, M. F. and Sawford, B. L. (1994) A saline model of the planetary convective boundary layer for diffusion studies: scaling criteria, design and validation, *Bound. Layer Meteor.*, **67**, 229-250.
11. Kaiser, R. and Fedorovich, E. (1998) Turbulence spectra and dissipation rates in a wind tunnel model of the atmospheric convective boundary layer, *J. Atmos. Sci.*, **55**, in press.
12. Melbourne, W. H. (1977) Development of natural wind models at Monash University, *6th Aust. Hydraulics and Fluid Mechanics Conference*, Adelaide, pp. 190-194.
13. Melbourne, W. H. (1982) Atmospheric dispersion and wind-wave test facility, Monash University, 8 pp.
14. Meroney, R. N. (1976) An algebraic stress model for stratified turbulent shear flows, *Computer and Fluids,* **4**, 93-107.
15. Meroney, R. N. (1981) Physical simulation of dispersion in complex terrain and valley drainage flow situations, in C. De Wispelaere (ed.), *Air Pollution Modeling and Its Application*, Plenum Publishing Co.
16. Meroney, R. N. (1986) Wind-tunnel modeling of convective boundary-layer phenomena, Colorado State University Memorandum CEM86-87-RNM49, 37 pp.
17. Meroney, R. N. (1987) Guidelines for fluid modeling of dense gas cloud dispersion, *Journal of Hazardous Materials*, **17**, 23-46.
18. Meroney, R. N. (1990) Fluid dynamics of flow over hills / mountains - insights obtained through physical modeling, in W. Blumen (ed.), *Atmospheric Processes Over Complex Terrain*, American Meteorological Society Monograph No. 45, pp. 145-171.
19. Meroney, R. N., Cermak, J. E., and Yang, B. T. (1975) Modeling of atmospheric transport and fumigation at shoreline sites, *Boundary-Layer Meteorol.*, **9**, 69-90.

20. Meroney, R. N. and Melbourne, W. H. (1991) Operating ranges of meteorological wind tunnels for the simulation of convective boundary layer phenomena, *Boundary-Layer Meteorol.*, **61**, 145-174.
21. Mery, P., Schon, J. P., and Solal, J. (1974) Comparison of thermally neutral and unstable shear flows in the wind tunnel and the atmosphere, *Adv. Geophys.*, **18B**, 273-287.
22. Ogawa, Y., Diosey, P. G., Uehara, K., and Ueda, H. (1985) Wind tunnel observation of flow and diffusion under stable stratification, *Atmos. Environ.*, **19**, 65-74.
23. Poreh, M. (1995), Small scale modelling of combined wind and buoyancy effects, *9th Int. Conf. on Wind Engineering. State of the Art Volume*, New Delhi, India, 9-13 January, 1995, pp. 393-417.
24. Poreh, M. and Cermak, J. E. (1984) Wind tunnel simulation of diffusion in a convective boundary layer, *29th OHOLO Biological Conference*, Ya'Acov, Israel 25-28 March, *Boundary-Layer Meteorol.*, **30**, 431-455.
25. Poreh, M. and Cermak, J. E. (1985) Study of neutrally buoyant plumes in a convective boundary layer with mean velocity and shear, *Proceedings of the 7th Symposium on Turbulence and Diffusion*, November 12-15, 1985, Boulder, Colorado, 119-121.
26. Poreh, M., Rau, M., and Plate, E. J. (1991) Design considerations for wind tunnel simulations of diffusion within the convective boundary layer, *Atmos. Environ.*, **25A**, No. 7, 1251-1256.
27. Rau, M., Bachlin, W., and Plate, E. J. (1991) Detailed design features of a new wind tunnel for studying the effects of thermal stratification, *Atmos. Environ.*, **25A,** No. 7, 1257-1262.
28. Snyder, W. (1982) Guideline for fluid modeling of atmospheric diffusion, U. S. Environmental Protection Agency Report EPA-600/8-81-009, 185 pp.
29. Thorpe, S. A. (1973) Turbulence in stably stratified fluids: a review of laboratory experiments, *Boundary-Layer Meteorol.*, **5,** 95-119.
30. Ukeguchi, N., Sakata, H., Okamoto, H. and Ide, Y. (1967) Study on stack gas diffusion, *Mitsubishi Tech. Bull.*, **52,** 1-13.
31. Willis, G. E. and Deardorff, J. W. (1974) A laboratory model of the unstable planetary boundary layer, *J. Atmos. Sci.*, **31**, 1297-1307.

WIND TUNNEL MODEL STUDY OF TURBULENCE REGIME IN THE ATMOSPHERIC CONVECTIVE BOUNDARY LAYER

E. FEDOROVICH AND R. KAISER
Institute of Hydrology and Water Resources Planning
Karlsruhe University
Kaiserstrasse 12, 76128 Karlsruhe, Germany

Abstract

Results from a wind tunnel model of the sheared atmospheric convective boundary layer (CBL) are presented and discussed. The model is realized in the thermally stratified wind tunnel of the Institute of Hydrology and Water Resources Planning (IHW), University of Karlsruhe. Design of the wind tunnel, and the technique employed for velocity and temperature measurements are described. The performed experiments reproduce convective conditions in the atmosphere when both capping inversion and hydrostatic stability of the free-atmosphere air are not very strong. Under such conditions, the turbulence regime in the CBL is mainly determined by the bottom-up convective and shear forcings, whereas the heat flux of entrainment and temperature fluctuation magnitude within the inversion layer are comparatively small. Temporal averaging has been employed for deriving turbulence statistics from the velocity and temperature time series measured. These statistics are compared with their counterparts from atmospheric measurements, and from numerical and laboratory model studies.

Surface shear and elevated shear were found to be essential factors modifying the CBL turbulence regime. The elevated shear turned out to be an inhibitor of entrainment at the CBL top. Turbulence enhancement has been observed in the transition zone between the unmixed convective layer in the neutral environment and the well-mixed CBL capped by a temperature inversion.

The turbulence spectra obtained display short but pronounced inertial intervals in large-wavenumber ranges. The characteristic value of the turbulent Reynolds number Re_l in the simulated CBL is $3 \cdot 10^3$. The vertical to horizontal velocity spectral ratio is slightly higher than the local-isotropy value of 4/3. Velocity spectra measured in the lower portion of the CBL show the buoyancy to be the dominant factor of turbulence production at smaller wavenumbers, while the shear contribution increases with the wavenumber growth. The combined effect of buoyant and shear forcings apparently leads to elongation and flatness of the production ranges in the velocity spectra. The inertial-subrange theory relationships have been used for evaluation of the turbulence dissipation rates from the velocity spectra. The turbulence kinetic energy dissipation rates in the CBL with shear were found to be comparatively large over the whole turbulized region.

E.J. Plate et al. (eds.), Buoyant Convection in Geophysical Flows, 327–370.

1. Introduction

The principal feature of the atmospheric convective boundary layer (CBL) is the strong turbulent mixing occupying the main portion of the layer in the vertical. This mixing is chiefly maintained by buoyant plumes and thermals (warm updrafts), rising from the warm underlying surface, and associated descending motions (cool downdrafts). Wind shear is a secondary mechanism of turbulence generation in the atmospheric CBL.

The CBL mixed core, where horizontally (or temporally) averaged meteorological variables like potential temperature, specific humidity, and wind velocity are nearly height-constant, is commonly called the mixed layer. This layer is separated from the underlying surface by a relatively shallow surface layer, within which meteorological variables change sharply from their near-surface values to the mixed-layer ones. The interfacial (inversion) layer, also called the entrainment layer (zone), is located above the mixed layer. In the entrainment layer, the energy of thermals is expended for their penetration into the stably stratified atmospheric air aloft and for mixing (entraining) it downwards.

Entrainment is accompanied by the downward transport of heat from the warmer outer flow region into the mixed layer. Thus, the entrainment zone corresponds to the range of heights with negative kinematic heat flux. Entrainment of more buoyant air into the mixed layer leads to the increase of the CBL depth.

Typically, the interfacial (entrainment) layer is characterized by strong vertical gradients of averaged meteorological quantities. Wind shear across the interfacial layer (the so-called elevated shear) contributes to the turbulence production in the upper part of the CBL and, presumably, can modify the regime of entrainment. The entrainment-layer depth is quite variable, but usually smaller than the mixed-layer depth. Towards the top of the interfacial layer, turbulence decays and turbulent transport becomes negligible.

Observations of the atmospheric CBL are presented in this volume by D. Lenschow [26].

When the buoyant (convective) production of turbulence in the CBL dominates the mechanical (shear) production, the CBL can be considered as shear-free. This CBL case has been simulated in numerous laboratory water tank experiments pioneered by J. Deardorff and his colleagues (Deardorff *et al.* [10,11], Willis and Deardorff [56], Deardorff and Willis [9]; see also reports from the water tank studies of Kumar and Adrian [25] Adrian *et al.* [1], and Cenedese and Querzoli [6]). Meanwhile the sheared CBL, which is a much more common case in the atmosphere, was left beyond the scope of water tank experiments.

Systematic wind tunnel studies of the atmospheric convective turbulence have been carried out by only a few research teams over the world (Rey *et al.* [45], Ogawa *et al.* [36], and Poreh and Cermak [41]). The analysis of the operating ranges of the existing wind tunnel facilities for simulating the atmospheric CBL, (Meroney and Melbourne [32], see also the paper by Meroney [31] in this volume), gives an idea of the multitude of technical problems awaiting investigators attempting to reproduce details of the CBL turbulence in wind tunnels. Inspection of available data from wind tunnel experiments makes evident that in most of them the flow structure was not appropriately resolved for making quantitative conclusions concerning the CBL turbulence regime. Among the other problems of wind tunnel CBL modeling are size limitations (usually, a very long tunnel is needed to generate a sufficiently deep CBL), and difficulties with matching the model to prototype scale requirements for a number of important integral parameters of the CBL, namely, for the inversion strength, the boundary-layer depth, the temperature gradient in the turbulence-free layer, and the surface heat flux.

Analysis of the above problems, and the experience accumulated from water tank and wind tunnel model studies of atmospheric convection at the Institute of Hydrology and Water Resources Planning (IHW), Karlsruhe University, (see paper by E. Plate [40] in this volume), enabled the development of a concept of the thermally stratified wind tunnel with preshaped vertical temperature and velocity profiles at the inlet of the test section. The preshaped profiles roughly correspond to vertical distributions of temperature and velocity in the developed atmospheric CBL. Adjustment of the preshaped boundary layer to the buoyant and shear forcings takes place while the flow proceeds downstream over the heated, aerodynamically rough floor of the test section. This method minimizes the fetch, over which a quasi-equilibrium and quasi-homogeneous turbulence regime in the simulated CBL is achieved. By varying the shapes of the temperature and velocity profiles at the test section inlet, different CBL flow regimes with combined buoyant and shear forcings can be investigated. This is a step forward compared with the pure shear-free CBL cases reproduced in the water tank models.

Study of the mean-flow characteristics and turbulence statistics in the IHW wind tunnel was reported by Fedorovich *et al.* [15]. This study was followed by investigation of spectral properties of the CBL turbulence by Kaiser and Fedorovich [23].

The simulation approach employed in the IHW tunnel allows generation of a quasi-stationary CBL, which develops longitudinally and grows by the advection-type entrainment of warmer air from upper turbulence-free layers into the convectively mixed layer. A CBL of similar type develops in the atmosphere when a neutrally stratified air mass is advected over a heated terrain. Preceding model studies of the CBL, both numerical and laboratory, dealt mostly with cases of nonstationary, horizontally quasi-homogeneous CBL. Under general conditions, in the presence of mean wind and over a nonuniform underlying surface, the atmospheric CBL grows by both advection and nonstationary mechanisms.

Due to the quasi-stationarity of the simulated CBL, the turbulence statistics in the wind tunnel model are calculated through temporal averaging. In studies of the nonstationary CBL the statistics are commonly obtained through area averaging.

The turbulence regime in the wind tunnel model of the sheared atmospheric CBL is analyzed and discussed in the present paper. Before showing the simulation results, we will give a brief description of the technique and methods employed for velocity and temperature measurements in the tunnel. The technique for the flow visualization will be considered as well, and some visualized flow patterns will be shown. Particular attention will be devoted to the comparison of the turbulence statistics and spectra obtained in the tunnel with their counterparts derived from atmospheric measurements, and from numerical and water tank models of the atmospheric CBL.

2. Experimental Setup and Measurement Technique

The thermally stratified wind tunnel of the IHW is specially designed for simulating turbulent flow and dispersion in the sheared atmospheric CBL capped by a temperature inversion. The design features of the tunnel are considered in detail by Rau *et al.* [43], Poreh *et al.* [42], and Rau and Plate [44].

The tunnel is of the closed-circuit type, (see Fig. 1), which allows minimization of the energy supply necessary for maintaining a stable thermal regime in the flow. The test section of the tunnel is 10 m long, 1.5 m wide, and 1.5 m high. The return section is subdivided into 10 individually insulated layers. Each 15-cm deep layer is operated by its own fan and heating element. The system of fans and heating elements is driven by a feedback control system. In this way, the vertical velocity and temperature distributions at the inlet of the test section can be preset and kept constant. The test section floor, which is constructed of aluminum plates, is also heated in a controlled way to ensure a constant heat flux through the bottom. The floor roughness may be regulated by employing variable-size roughness elements mounted on the floor.

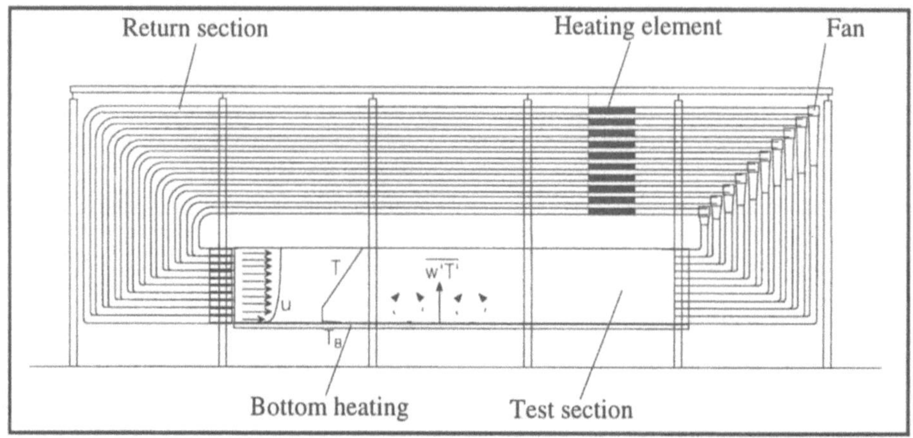

Figure 1. Schematic cut of the thermally stratified wind tunnel along the central longitudinal plane.

The design of the tunnel gives the opportunity of reproducing convective regimes with different extents of shear contribution to the turbulence production and generating a developed CBL over comparatively short fetches.

At the same time, there are certain limitations imposed on the flow parameters in the tunnel and, consequently, on the simulated turbulence regimes. In particular, the maximum inversion elevation that can be presently achieved in the tunnel is about 0.5 m, (remember that the height of the test section is 1.5 m). Maximum values of the temperature gradient in the outer flow above the inversion, and of the temperature increment across the inversion layer are 100 K/m and 100 K, respectively. The flow velocity at the inlet can be varied within the interval from 0.5 m/s to 2 m/s. The upper limit for the heat flux through the bottom is 1.5 K·m/s.

2.1. SCALING CONSIDERATIONS

To relate the simulated CBL to its atmospheric prototype, a set of nondimensional similarity parameters must be elaborated. Following Deardorff *et al.* [11], Zilitinkevich [60], and Fedorovich and Mironov [16] we will use for this purpose two

Richardson numbers, $\mathrm{Ri}_{\Delta T}$ and Ri_N. The first of them is based on the temperature increment ΔT across the inversion layer: $\mathrm{Ri}_{\Delta T} = \beta w_*^{-2} z_i \Delta T$. The second represents the effect of stable stratification (characterized by the value of buoyancy frequency N) above the interfacial layer: $\mathrm{Ri}_N = N^2 z_i^2 w_*^{-2}$. The other two scaling parameters entering $\mathrm{Ri}_{\Delta T}$ and Ri_N are the CBL depth z_i and the Deardorff [8] convective velocity scale $w_* = (\beta Q_s z_i)^{1/3}$, in which $\beta = g / T_0$ is the buoyancy parameter (g is the acceleration due to the gravity, T_0 is the reference temperature value), and Q_s is the near-surface value of the turbulent kinematic heat flux. The above Richardson numbers represent only the buoyant forcing in the CBL. To account for the effects of wind shear at the surface, these parameters must be complemented by the ratio u_* / w_*, which characterizes the proportion between dynamic (u_* is the friction velocity) and buoyant contributions to the turbulence production in the CBL (Meroney and Melbourne [32]).

Another important dimensionless parameter of the CBL turbulence regime is the Reynolds number, $\mathrm{Re} = U_c L_c / \nu$, where U_c and L_c are characteristic velocity and length scales of the flow, respectively, and ν is the kinematic viscosity of the air.

In the atmospheric CBL, typical $\mathrm{Ri}_{\Delta T}$ and Ri_N values are within the range of 0 to 100, and Re is of the order 10^8. In the IHW wind tunnel, a maximum $\mathrm{Ri}_{\Delta T}$ of about 10, and a maximum Ri_N of about 20 can presently be achieved. This corresponds to atmospheric convection cases with comparatively weak damping effects of stable stratification in the upper portion of the CBL and above the convection zone. In the wind tunnel CBL, Re varies within the range of 10^3 to 10^5, depending on the length and velocity scales chosen. With such Re values, the reproduced turbulent regime can be roughly considered Re independent.

Turbulence production in the CBL is buoyancy-dominated when the ratio u_* / w_* is sufficiently small. Based on classification proposed by Holtslag and Nieuwstadt [19], u_* / w_* =0.34 may be obtained as a conventional boundary value separating the regimes of sheared and shear-free convection. Typical values of u_* / w_* in the wind tunnel are within the range of 0.2 to 0.5, that is near and above the boundary value for shear effects.

Temporal and spatial scales of turbulent motions in the simulated CBL may be related using the Taylor translation hypothesis (Willis and Deardorff [57]). It was shown in Fedorovich *et al.* [15] that this hypothesis is satisfied fairly well in the wind tunnel CBL where velocity turbulent fluctuations are typically small compared to the mean-flow velocity.

2.2. MEASUREMENT AND VISUALIZATION TECHNIQUE

2.2.1. *Laser Doppler Velocimetry*

The laser Doppler velocimetry (LDV) provides a method for measuring flow velocity with high accuracy and spatial resolution. The LDV measuring signal is independent on temperature fluctuations which is of particular importance in high-temperature wind tunnel flow. All constants of the LDV system are usually known and thus time-consuming calibrations, characteristic of the hot-wire anemometry, can be avoided in the LDV measurement procedures. Comprehensive explanations of the LDV principles are given in literature, see *e.g.* Drain [13], Wiedemann [55], Durst *et al.* [14], and Ruck [46].

The two-beam configuration is the most common LDV setup. Its schematic is shown in Fig. 2. All velocity measurements in the wind tunnel CBL were performed with this setup.

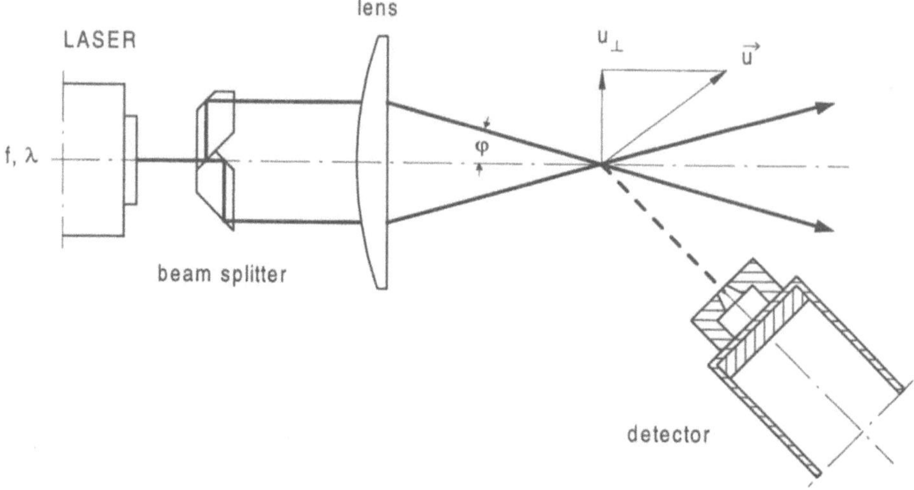

Figure 2. A dual-beam laser Doppler velocimeter, after Ruck [46].

The flow velocity is determined by the LDV from the speed of small particles transported by the flow. Such particles are produced using a 5% aqueous solution of Na_2SO_4. This solution is injected into the flow at the test section inlet by means of jet atomizers. In the test section, the water component of the spray droplets evaporates and the flow is seeded by small salt crystals. The Na_2SO_4 crystals have a refraction index of 1.5, and thus scatter light well. The Na_2SO_4 compound is neither toxic nor corrosive. Besides that, it is not hygroscopic, and comparatively cheap.

The frequency response of the velocity amplitudes of the particle and of the flow can be roughly estimated as (Domnik [12])

$$\frac{\hat{u}_P}{\hat{u}_F} = \frac{1}{\sqrt{1+(\omega\tau)^2}} . \tag{1}$$

Here, the indices P and F refer to the particle and to the flow, respectively, $\hat{u}_P / \hat{u}_F$ is the frequency response of the velocity amplitudes, ω is the circular frequency, $\tau = (2/9)(r^2/\nu)(\rho_P/\rho_F)$ is the time constant of the particle (relaxation time), r is the radius of the particle, and ρ_P, ρ_F are the densities of the particle material and of the transporting fluid, respectively.

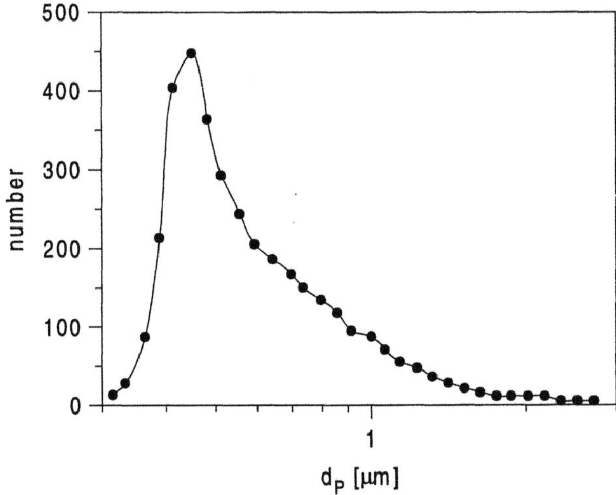

Figure 3. Size distribution of the seeding particles.

Figure 3 shows the size (diameter) distribution of the Na_2SO_4 seeding particles. Most of the particles have a diameter between 0.3 µm and 1.5 µm which gives a relaxation time within the range of 1 µs to 20 µs. The particle sinking velocity in a calm fluid, $u_E = (1-\rho_F/\rho_P)g\tau$, is therefore between 10 µm/s and 200 µm/s. For the cut-off frequency $f_c = 1/(2\pi\tau)$ we obtain an estimate between 8 KHz and 160 KHz.

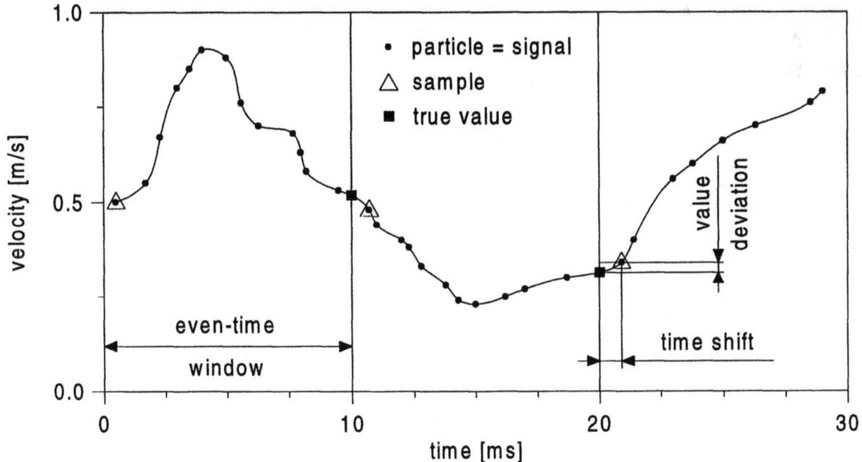

Figure 4. Quasi-equidistant sampling with LDV.

To perform quasi-equidistant velocity field sampling, the LDV processor was operated in the even-time mode, see Fig. 4. In each even-time window corresponding to the sampling time interval the average amount of arriving particles (and therefore the amount of signals) was typically about 10. The signal of the first arriving particle in the window is taken as a sample and assigned to the time moment of the window beginning. The shift between the sample time and the real time of the particle arrival (signal) causes a deviation of the registered velocity value from the real value corresponding to the beginning of the window. This can be a source of high-frequency noise in velocity spectra calculated from the quasi-equidistantly sampled time series. The velocity component statistics are not influenced by the deviations mentioned. When calculating cross-correlations between the velocity components, special measurements were carried out in the even-time coincident mode.

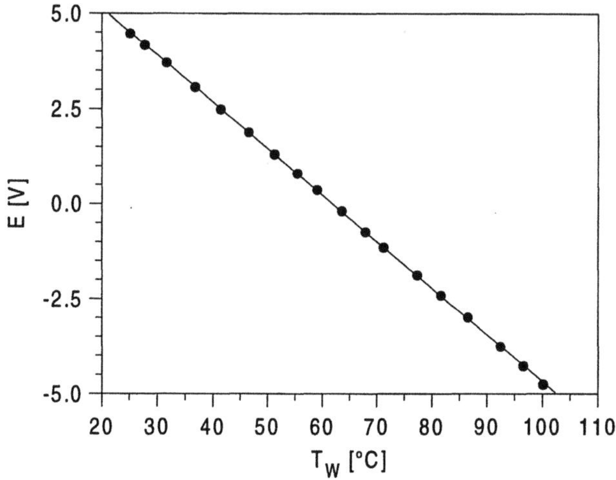

Figure 5. Calibration function of the temperature probe with platinum-plated tungsten wire.

2.2.2. *Temperature Measurements*

For the temperature measurements in the wind tunnel the resistance-wire technique is applied. The temperature probe is made up of a thin wire suspended between the tips of two prongs. When the probe is placed in the flow, the temperature of the ambient air can be determined from the temperature-dependent electrical resistance of the wire. The resistance-wire technique, which is a special branch of hot-wire anemometry, is described in detail in literature, see *e.g.* Sandborn [47], Strickert [51], Perry [39], Lomas [29], and Bruun [4].

In the IHW wind tunnel, the flow temperature is measured with a 2.5 μm-thick and 1.25 mm-long platinum-plated tungsten wire probe. Figure 5 presents a characteristic linear calibration function of the probe.

The electric current through the wire causes a spurious heating, which should be taken into account in the temperature measurements. The employed current of 1 mA leads to an excess temperature value of about 0.2 K.

In addition, the heat inertia of the wire must be accounted for in the measurement procedure. If the time constants of the wire and of the prongs are known, it is possible to compute the transfer function of the resistance-wire probe and correct the temperature time series for the inertia effects of the probe.

For the experimental determination of the time constant of the wire, its middle point was heated by a focused laser beam. The laser beam was switched on and off by means of a rotating disc, one half of which was transparent. The circular velocity of the disk was 4000 cycles per minute. The signal of the wire placed in a constant-velocity air flow was digitized with a temporal resolution of 40 KHz. In Fig. 5, the measured cooling rates of the wire corresponding to different flow velocities are shown. The time constant of the wire was evaluated as the period during which the temperature of the wire drops by the factor of *e* from its initial value.

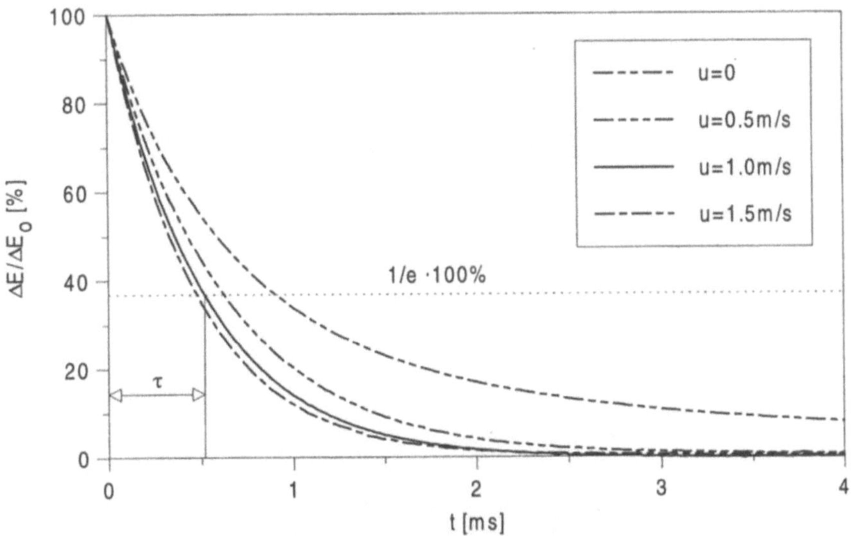

Figure 6. Cooling rate of the resistance wire.

TABLE 1. Time constants and cut-off frequencies of the platinum-plated tungsten wire (d=2.5 μm, l=1.25 mm) experimentally determined at different flow velocities.

Flow velocity, m/s	Time constant, ms	Cut-off frequency, Hz
0	0.90	177
0.5	0.62	257
1.0	0.52	306
1.5	0.47	339

TABLE 2. Time constants and cut-off frequencies of the prongs experimentally determined at different flow velocities.

Flow velocity, m/s	Time constant, s	Cut-off frequency, Hz
0.5	5.09	0.03
1.0	1.10	0.14
1.5	0.85	0.19

The time constant of the prongs was estimated in a similar way. In this case, the whole probe was heated by means of a hot-air jet. After switching off the heating, the probe was exposed to a cold air flow and the cooling rate was measured. The time constant of the wire is several orders of magnitude smaller than that of the prongs. Therefore, the influence of the wire could be neglected while determining the cooling rate of the prongs.

The experimentally determined time constants of the wire and the prongs are listed in Tables 1 and 2 for different flow velocities.

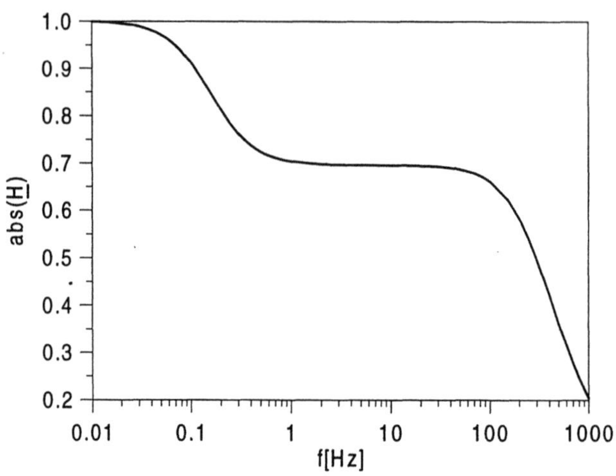

Figure 7. Transfer function of the resistance-wire probe computed with the experimentally determined time constants of the wire and of the prongs.

The transfer function of the temperature probe was calculated based on the experimentally determined time constants of the wire and the prongs. An empirical approach verified by Paranthoen *et al.* [38] was employed for the calculation. The calculated transfer function for flow velocity of 1 m/s is shown in Fig. 6. In the low-frequency range, this function exhibits the fast decay associated with the inertia of the prongs. Within the range of 1 Hz to 100 Hz, the transfer function passes through a plateaulike interval. With the frequency increasing beyond this interval, the shape of the transfer function is becomes closer to that of an infinitely long wire.

2.2.3. *Parallel Velocity and Temperature Measurements*

The width of the LDV even-time window is 0.01 s which provides a sampling frequency of 100 Hz. Each velocity time series includes values obtained in 16384 even-time windows. Thus, its length is 163.84 seconds.

Discretization of the analog signal of the resistance wire is triggered by the LDV signal in order to synchronize the velocity and temperature time series. In this way, the temperature time series have the same length as the velocity time series. The temperature probe is placed at the same level as the LDV measuring volume, and shifted 1 mm downstream. This shift desynchronizes velocity and temperature measurements at frequencies of 1000 Hz and more, (the typical mean-flow velocity in the tunnel is 1 m/s). As will be seen from the spectra shown in section 3.3, at such frequencies the energy of both velocity and temperature turbulent fluctuations is small compared with the energy in the resolved frequency range. Therefore, the velocity and temperature measurements in the tunnel can be regarded as coincident in time and space.

2.2.4. *Flow Visualization*

The laser light sheet technique is employed to visualize flow and turbulence structure in the wind tunnel. An argon-ion laser is used as the source of light. The laser beam is embedded into a multimode glass fiber of 25-mm diameter. From the exit of the fiber the beam passes through a spherical biconvex lens of 10-mm focal length. The lens provides a divergent light bunch beyond the focal point of the lens. An additional convex cylindrical lens of 80-mm focal length is used to bundle up the bunch in the horizontal plane while in the vertical it is kept divergent. As a result, a light sheet is generated in front of the lens system. The distance between the sheet and the lenses may be varied in the range of 1 m to 7 m depending on the disposition of the lenses. The typical thickness of the light sheet is about 1 cm, and its height is about 1 m.

A white 1,2-propyleneglycol fog is used as the visualizing material. It is admixed into the lowest layer at the test section inlet. Buoyant thermals rising from the heated floor of the tunnel are thus colored white. They are clearly seen in the light sheet against the background of more transparent air entrained down from the upper layers of the flow.

The visualized flow patterns are captured by a television camera and recorded on a video tape. Afterwards, typical flow patterns are selected from the tape and digitized for further image processing.

3. Model Results

3.1. CBL DEVELOPMENT IN THE WIND TUNNEL

3.1.1. *Evolution of Temperature and Turbulence Fields*

The inlet velocity, and temperature profiles corresponding to the basic experimental setup of the present study are schematically shown in Fig. 8. The velocity is held constant in the vertical with a value of 1 m/s. The temperature is constant (20°C) over the two lower layers, 30-cm deep in total. Between the second and the third layer, a temperature jump of 30°C is imposed. Above the jump, the temperature is grows with height at a rate of 67°C/m. The heat flux of 1250 W/m^2 through the floor of the tunnel is kept constant in the horizontal. The presented experimental configuration differs slightly from the one considered in Fedorovich *et al.* [15], where the CBL case with a weaker stratification (33°C/m) above the inversion was investigated.

Measurements of the flow parameters in the tunnel are carried out at five fixed locations corresponding to the positions of the windows, through which the laser light penetrates in the tunnel. If the origin of the coordinate system is placed at the floor midpoint near the test section inlet, and the x axis is oriented parallel to the flow, then the measurement locations (windows) are found at x=0.68 m (I window), x=2.33 m (II window), x=3.98 m (III window), x=5.63 m (IV window), and x=7.28 m (V window). Accordingly, the y axis is oriented perpendicular to the flow, and the z axis is directed upwards. All measurement data presented in this section refer to the central longitudinal plane of the tunnel.

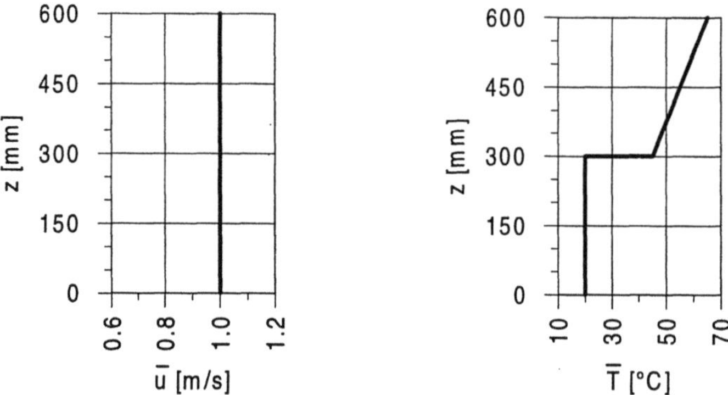

Figure 8. Idealized velocity and temperature profiles at the test section inlet.

The evolution of the mean temperature field in the tunnel is shown in Fig. 9. Corresponding changes of the turbulence field can be seen in Fig. 10. At the first measurement location, x=0.68 m, the shape of temperature profile is quite similar to that of the inlet profile, see Fig. 8. Beyond the unstably stratified near-surface region occupying roughly one quarter of the depth of the initially neutrally stratified layer, the changes in the temperature field are not significant. In the unstable region, fluctuations of both plotted velocity components are exaggerated due to the turbulence production by buoyancy forces. However, the resulting turbulent exchange is not strong enough to mix up the air near the surface. The vertical velocity excess in the turbulized region vanishes towards the surface due to the damping effect of the surface on the vertical turbulent motions.

While the flow proceeds further over the heated floor of the tunnel, the depth of the zone warmed by the convective heat transfer from the surface increases. At x=2.33 m, only the uppermost portion of the initial neutral layer preserves its original temperature. The turbulence intensity in the heated region is markedly higher than at x=0.68 m, whereas the overall temperature lapse rate is smaller due to the stronger mixing. The surface velocity shear also contributes to the intensification of turbulent exchange and to the transport of heat upwards. The beginning of temperature inversion erosion can be identified in the temperature profile at x=2.33 m. It is nevertheless clear that at this stage of flow evolution the turbulence is not developed enough to provide a strong turbulent mixing typical for the atmospheric CBL.

Over the range of distances between x=2.33 m and x=3.98 m, the temperature profile drastically changes its shape. The kneelike minimum in the profile under the capping inversion at x=2.33 m is replaced in the profile at x=3.98 m by much higher temperature values, which grow with height. This comparatively rapid heating of the flow in the under-inversion region is caused by two main mechanisms: (i) heat transport upwards from the surface by strong convective motions, (note turbulence intensities at x=2.33 m and at x=3.98 m in Fig. 10), and (ii) downward transport of heat from the warm region above the inversion due to entrainment. The latter is maintained by convective thermals, which are sufficiently developed at x=3.98 m and thus capable of overshooting in the stably stratified environment above the inversion. The increase of the turbulence intensity within the range of fetches from x=2.33 m to x=3.98 m is associated with the release of potential energy accumulated in the unstable sublayer below the inversion with a superadiabatic temperature gradient. This potential energy is transformed into the kinetic energy of turbulent fluctuations when the under-inversion layer overturns, (see the temperature profile transformation from x=2.33 m to x=3.98 m in Fig. 9).

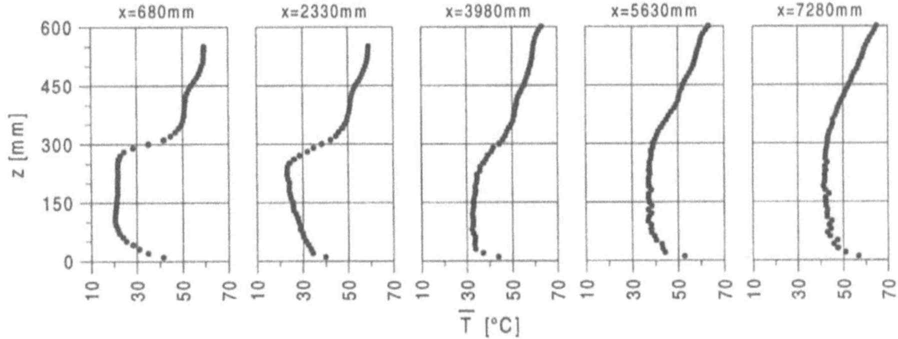

Figure 9. Evolution of the mean-temperature field along the tunnel.

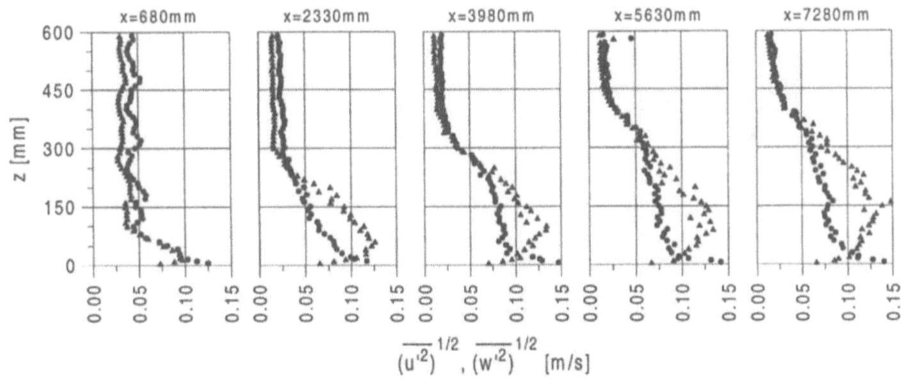

Figure 10. Development of turbulence along the tunnel. Dots: *u* fluctuations, triangles: *w* fluctuations.

As a result of the above processes, the convectively mixed layer is formed in the lower portion of the wind tunnel flow.

Downstream from x=3.98 m, at x=5.63 m and at x=7.28 m, the mean-flow temperature profile retains the shape characteristic of the developed atmospheric CBL: a sharp decrease of temperature in the near-surface layer; an approximate thermal uniformity within the mixed layer; a temperature increase with height across the inversion; and a quiescence of temperature field in the stably stratified flow aloft. The CBL depth, which can be roughly estimated from the elevation of the capping inversion midpoint in Fig. 9, and also from the vertical extension of the turbulized zone in

Fig. 10, increases progressively with distance. This growth is accompanied by cooling of the stably stratified layer due to the entrainment of warmer air from above the inversion into the bulk of the mixed layer.

3.1.2. *Visualization of Turbulence Structure*

In Fig. 11, a visualized flow pattern over the central longitudinal plane of the tunnel is presented. The fog in the plot is colored black. Large towerlike convective structures (thermals) transporting the fog upwards are clearly seen over the whole range of distances shown. The vertical dimensions of the thermals increase downwind. The CBL growth rate along the tunnel can be estimated from longitudinal changes of the heights of thermals. This rate is about 5 cm per 1 m length. The thermals penetrate deeply into the stably stratified outer flow region. Their diameters decrease with height. This is also a feature of thermals observed in the atmospheric CBL (Hooper and Eloranta [20]).

The maximum height of thermals in the simulated CBL is about 0.5 m. According to Nelson *et al.* [34], this height can be taken as the upper edge of the entrainment zone, and hence as the CBL top. A typical horizontal scale of rising thermals is approximately 0.2 m. The downdrafts may be identified in the plot as lighter colored areas. They are typically twice as wide as the thermals. Proportions between the horizontal scales of updrafts and downdrafts in the wind tunnel CBL correspond well to atmospheric observations by Lenschow and Stephens [27]. The upper parts of the wind tunnel thermals do not display a significant sideward transport of air from the thermals at the capping inversion level.

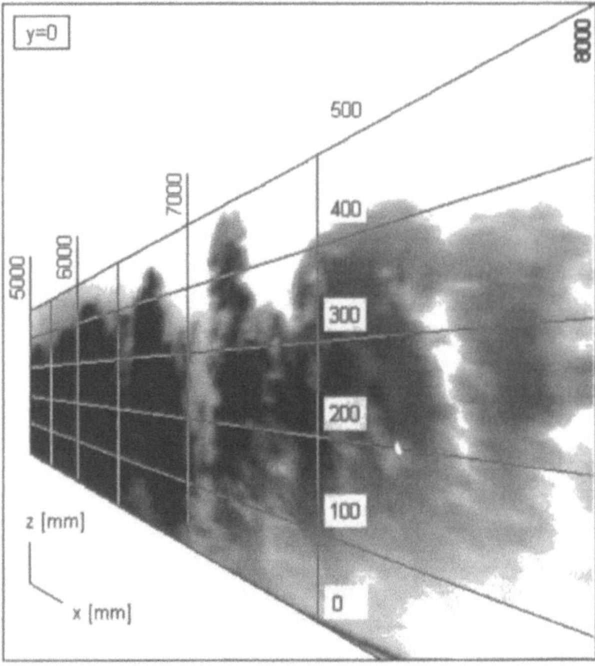

Figure 11. Light sheet image of the CBL turbulence structure along the central longitudinal plane of the tunnel.

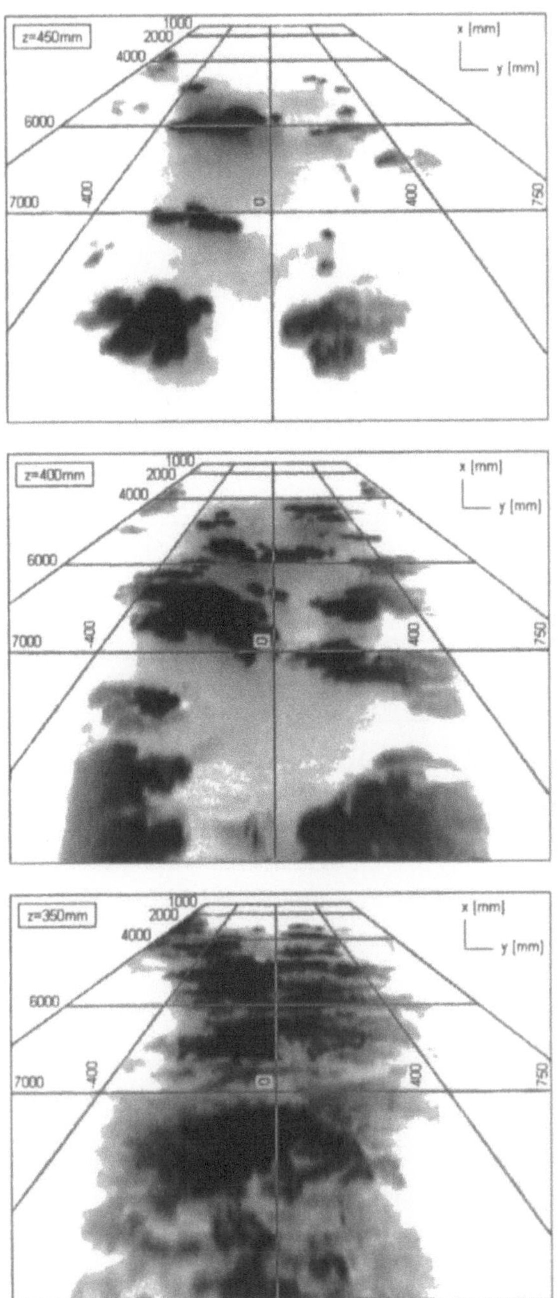

Figure 12. Changes of turbulence structure throughout the upper portion of the CBL. The lowest light sheet image refers to the lowest registration plane.

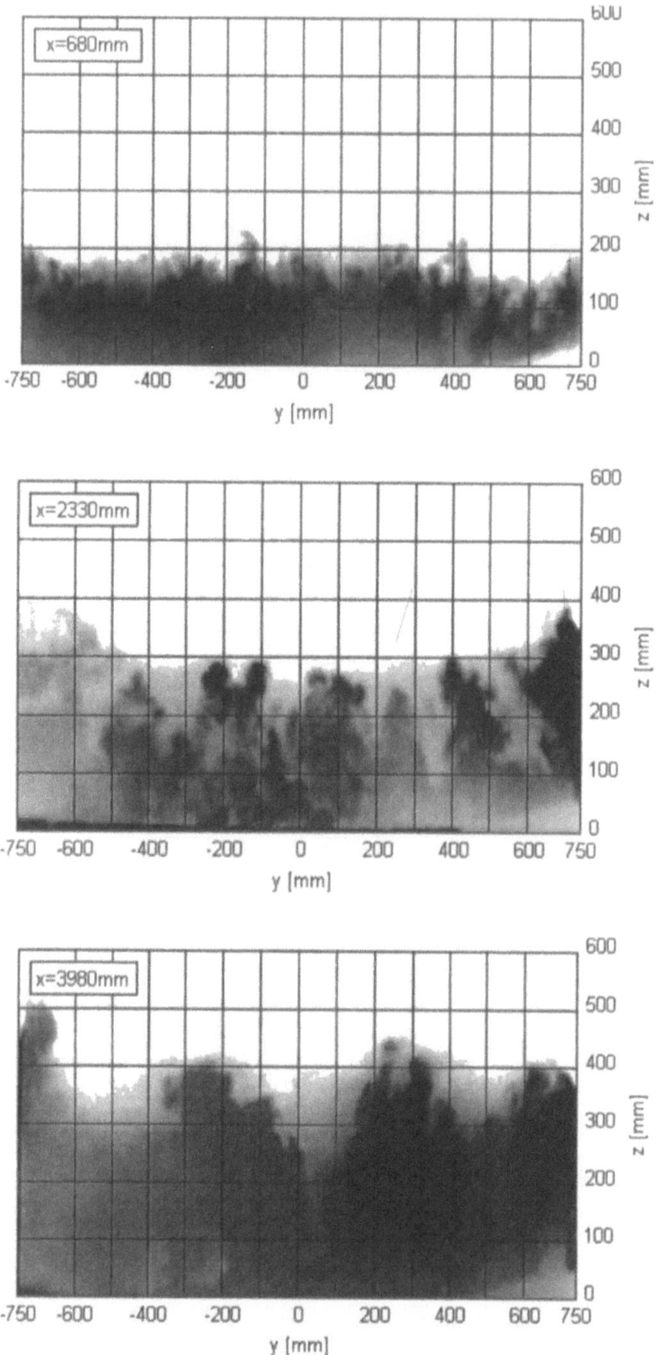

Figure 13. Light sheet images of the CBL turbulence structure perpendicular to the wind tunnel flow.

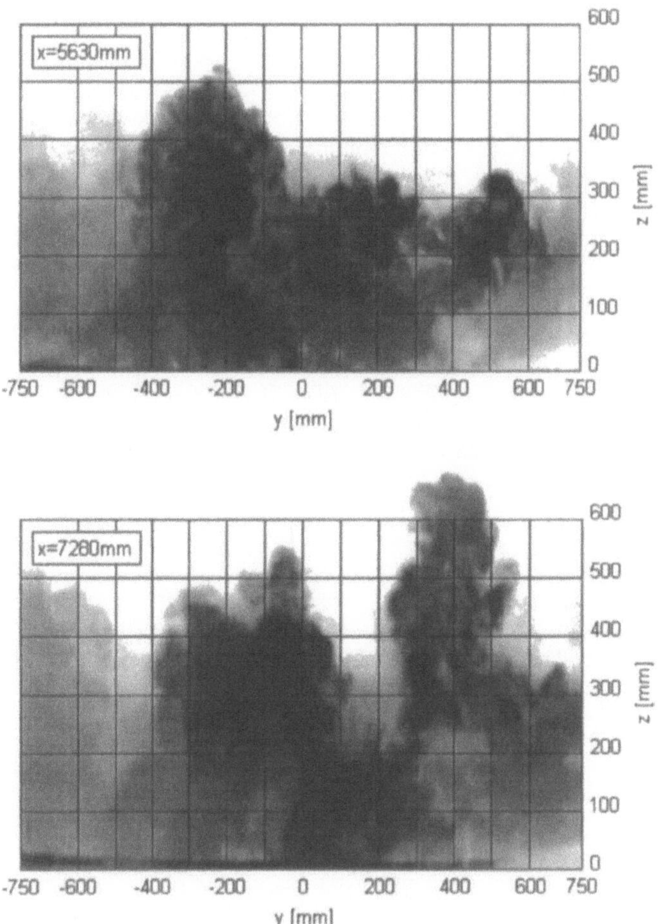

Figure 13 (continued). Light sheet images of the CBL turbulence structure perpendicular to the wind tunnel flow.

The visualized patterns presented in Fig. 12 display the changes of turbulence structure in the vicinity of the capping inversion. As can be inferred from the temperature plots in Fig. 9, the inversion elevation within the fetch range of 4 m to 7 m changes from 0.35 m to 0.45 m. The patterns in Fig. 12 visualize three flow slices at different heights within the inversion layer. At z=0.35 m, which is at the lower edge of the inversion layer, large-scale turbulence structures are dominant in the flow. The turbulence pattern at the next level, z=0.40 m, which is in the core of the inversion layer, is much more patchy. Typical sizes of rising thermals here are smaller than under the inversion. At the highest elevation shown, z=0.45 m, the tops of only a few thermals are observed which appear as black spots on the background of white ambient air. The shape of these spots indicates that the horizontal cross-sections of the thermals are rugged due to the lateral entrainment of surrounding air inside the thermals.

Light sheet images of the flow in the cross-sections corresponding to the measurement windows are presented in Fig. 13. At the first location, x=0.68 m, small buoyant plumes rising from the heated floor into the neutrally stratified environment are clearly seen. They are too small to reach the inversion base. At x=2.33 m, the size of the plumes increases, but only a few of the plumes extend up to the inversion, which blocks them. Peculiar fine-structure convective elements, which look like fingers, can be observed at the tops of the plumes.

Further development of the turbulence regime in the tunnel is associated with a merging of plumes into the buoyant thermals, which are bigger and stronger turbulence structures than the plumes. Larger-scale thermals are well seen in the plots corresponding to x=3.98 m, x=5.63 m, and x=7.28 m. The thermals penetrate far into the stably stratified flow above the inversion. Their horizontal dimensions typically decrease with height which is consistent with our previous observations of thermals in Figs. 11 and 12. It is remarkable that the upper portions of some thermals are composed of fingerlike buoyant elements similar to the tops of convective plumes at the earlier stages of the CBL development. These fingers penetrate into the stably stratified flow ahead of a thermal itself.

3.1.3. *Stages of CBL Evolution*

Departing from temperature and turbulence evolution patterns, examples of which are shown in Figs. 9 and 10, and based on the analysis of visualized flow patterns, (see Figs. 11 - 13), a subdivision of the CBL evolution into stages can be suggested. This subdivision is schematized in Fig. 14.

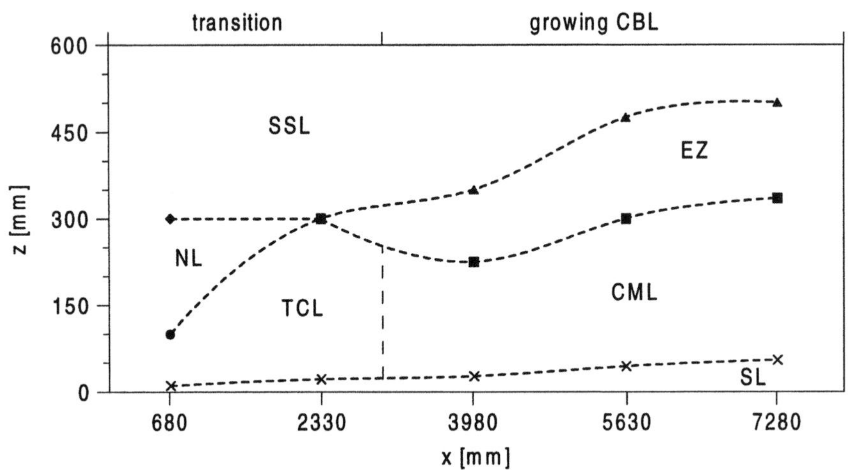

Figure 14. Development of the CBL in the wind tunnel. Abbreviations: SSL – stably stratified layer, EZ – entrainment zone, NL – neutrally stratified layer, TCL – transitional convective layer, CML – convectively mixed layer, and SL – surface layer.

The transitional convective region is located within the first three meters of the test section length. Throughout this region, the adjustment of the preshaped flow to the bottom-up buoyant and shear forcings takes place. The transitional convective boundary layer grows on the background of the neutrally stratified flow. The depth of this layer is determined by the average height of convective plumes growing into the neutral layer. The overall temperature gradient in the transitional convective layer is superadiabatic. After buoyant plumes rising from the heated surface reach the inversion base, the convectively mixed layer develops within a comparatively short range of fetches. Turbulence pattern in this layer is primarily constituted of the convective thermals (warm updrafts) and the associated cold downdrafts, which both are stronger and larger convective structures than plumes in the transitional convective layer.

Above the convectively mixed layer, the interfacial layer (entrainment zone) is formed. The intermittent turbulence structure in this layer is composed of penetrating thermals, whose horizontal dimensions decrease with height, and of wide downdrafts transporting warm air down from the outer flow region. Due to the entrainment mechanism, the CBL progressively grows along the tunnel. The typical depth of the entrainment layer in the wind tunnel flow is from one quarter to one third of the total CBL depth.

Across the comparatively shallow surface layer adjacent to the heated floor, the mean-flow temperature and velocity characteristics sharply change with height from their near-wall values (zero for velocity) to the mixed-layer values.

TABLE 3. Integral parameters of the simulated CBL.

Fetch, m	3.98	5.63	7.28
z_i , m	0.30	0.35	0.40
w_* , m/s	0.163	0.185	0.198
T_* , K	3.86	4.32	4.45

TABLE 4. Dimensionless characteristics of the simulated CBL.

Fetch, m	3.98	5.63	7.28
$\mathrm{Ri}_{\Delta T}$	3.9	3.0	2.5
Ri_N	5.2	5.4	6.0
u_* / w_*	0.43	0.38	0.35

In Tables 3 and 4, integral parameters of the convective regime and values of dimensionless combinations corresponding to three measurement locations within the developed CBL are given. The Deardorff [8] convective temperature scale T_* in Table 3 is defined as $T_* = Q_s / w_*$. The CBL depth scale z_i is determined from the elevation of the heat flux minimum within the entrainment zone.

3.2. RESULTS OF TURBULENCE MEASUREMENTS

3.2.1. *Turbulence Statistics*

Dimensionless profiles of turbulence statistics in the wind tunnel CBL are shown in Figs. 15 - 17. Hereafter, we shall exclusively consider data obtained at $x \geq 3.98$ m which thus refer to the stage of the developed CBL. The plotted statistics are normalized with the traditional Deardorff [8] convective scales: z_i for the height, w_* for the velocity, and T_* for the temperature. Despite the well-known limitations, (discussed, for instance, in Wyngaard [58]), the Deardorff mixed-layer scaling presents a convenient common background for the comparison of measurement and model results from different CBL studies. Some of these results are shown in Figs. 15 - 17.

The heat flux profiles are presented in Fig. 15a. In the interfacial layer, the negative heat flux of entrainment from the wind tunnel model is much smaller compared to the negative Q values from the other studies. The weaker inversion in the wind tunnel CBL leads to the smaller magnitude of temperature fluctuations in the interfacial layer. In the water tank and large eddy simulation (LES) studies, data from which we consider for comparison, the characteristic values of $\mathrm{Ri}_{\Delta T}$ and Ri_N were respectively 15 and 150 (Deardorff and Willis [9]), and 25 and 100 (Schmidt and Schumann [48]).

The effects of capping inversion and stable stratification in the outer flow are most vividly displayed in the normalized profiles of the temperature fluctuations (Fig. 15b). In the lower portion of the CBL, where the turbulence regime is mainly governed by bottom-up forcings, all data sources provide similar values of the temperature variance. The variances from the wind tunnel experiments are smallest near the CBL top, where water tank and LES studies with stronger stabilities inside and above the inversion layer predict much higher values of the normalized temperature variance.

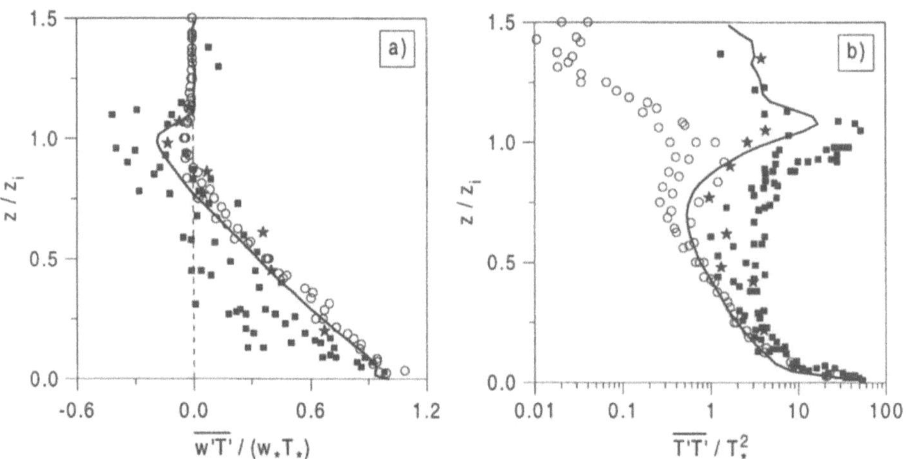

Figure 15. The vertical turbulent heat flux (a) and the temperature variance (b) in the CBL. The wind tunnel data (circles) are compared with the results of atmospheric measurements (Caughey and Palmer [5], and Lenschow *et al.* [28]), squares; water tank simulations (Deardorff and Willis [9]), stars; and LES (Schmidt and Schumann [48]), lines.

The divergence of $\overline{u'u'}$ data from the different sources in Fig. 16a is rather large. There are indications (Schmidt and Schumann [48]) that the water tank simulations of Deardorff and Willis [9] and the atmospheric measurements of Caughey and Palmer [5], which both give rather high $\overline{u'u'}$ values, were affected by horizontal variations of the surface heat flux in the first case, and by wind shear in the second.

Surface velocity shear in the wind tunnel CBL leads to the enhancement of horizontal velocity variance in the lower portion of the CBL. The turbulence enhancement in the downwind region of the transition zone, at x=3.98 m (see section 3.1.1), is represented in the plot by the group of wind tunnel data points shifted to the right from the main profile.

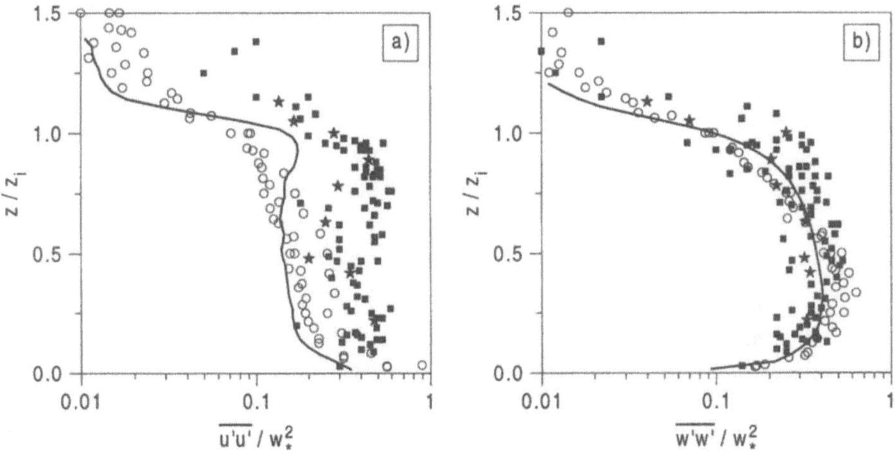

Figure 16. Variances of the longitudinal (a) and vertical (b) velocity components in the CBL. The notation is the same as in Fig. 15.

Two main mechanisms are presumably responsible for the enhancement of the longitudinal velocity fluctuations near the CBL top. The first is the sideward transport of air from the thermals in the inversion layer. This is an explanation for the marked maxima in the $\overline{u'u'}$ profiles from the water tank and LES studies with the relatively strong capping inversions. The wind tunnel data obtained in the CBL with a weak capping inversion merely give a slope change (bend) in the $\overline{u'u'}$ profile at $z/z_i \approx 1$. This points to a strong correlation between the magnitude of the $\overline{u'u'}$ variance in the upper portion of CBL and the inversion strength. The second mechanism of the $\overline{u'u'}$ enhancement at $z/z_i \approx 1$ is associated with the velocity shear across the inversion layer. Such an elevated shear may be one of the reasons for the large $\overline{u'u'}$ variance values in the upper part of the atmospheric CBL.

Figure 16a suggests one more conclusion: with a stronger damping effect of the buoyancy force above the inversion (larger Ri_N, see the water tank and LES data in the plot), the horizontal velocity variance in the upper portion of the CBL decays faster with height than in the wind tunnel case of small Ri_N.

Within the main portion of the CBL, the wind tunnel data on $\overline{w'w'}$ (Fig. 16b) are in fair agreement with the water tank, atmospheric measurements, and the LES results. Still, the $\overline{w'w'}$ variances obtained in the wind tunnel at z/z_i <0.5 are slightly larger than the variances from the other studies. Like in the case of the $\overline{u'u'}$ variance, the enhancement of $\overline{w'w'}$ in the wind tunnel flow can be caused by the surface shear and, at x=3.98 m, by the transition effects. Additionally, the vertical velocity fluctuations in the core of the wind tunnel CBL may be exaggerated due to weak quasi-stationary cellular circulations, which are discussed in Fedorovich *et al.* [15]. The behavior of $\overline{w'w'}$ at $z/z_i > 1$ is analogous to that of $\overline{u'u'}$: with larger Ri_N (*e.g.* in the LES experiments) the decay of the velocity variance with height is steeper.

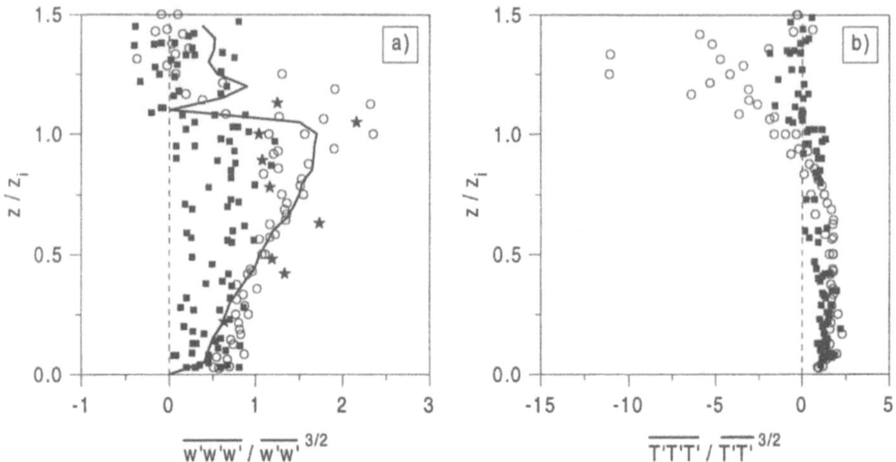

Figure 17. Skewness of (a) vertical velocity and (b) temperature in the CBL. The wind tunnel data (circles) are compared with the results of atmospheric measurements (Sorbjan [50]), squares; water tank simulations (Deardorff and Willis [9]), stars; and LES (Schmidt and Schumann [48]), line.

Vertical distributions of the vertical velocity and temperature fluctuation skewness are shown in Fig. 17. The wind tunnel data on the temperature skewness are compared only with the atmospheric measurements.

Positive values of the vertical velocity skewness in the upper portion of the CBL reveal a wider range of values occupied by updrafts (positive w') than by downdrafts (negative w'). The skewness observed in the atmosphere varies only slightly in the vertical, contrary to the wind tunnel, water tank and LES data, which clearly indicate the growth of skewness with height in the main portion of the layer. Narrow buoyant fingerlike elements rising from the tops of the thermals in the simulated CBL may be the main cause of the large skewness values from the wind tunnel. We have observed these elements in the visualized flow patterns, see section 3.1.2.

Distinctions between the atmospheric and wind tunnel data on the temperature skewness (Fig. 17b) are large inside and above the inversion layer, where the regime of temperature fluctuations is essentially determined by the thermals penetrating into the stably stratified environment. Comparatively narrow and cool thermals in the wind tunnel CBL provide more asymmetric distribution of temperature fluctuations and thus larger negative values of the temperature skewness at $z/z_i \geq 1$.

3.2.2. *Turbulence Kinetic Energy Budget*

We consider the turbulence kinetic energy (TKE) balance in a quasi-stationary CBL, which slowly evolves in the x direction and which is quasi-homogeneous in the y direction (Stull [52]):

$$\underbrace{\bar{u}\frac{\partial e_k}{\partial x}+\bar{w}\frac{\partial e_k}{\partial z}}_{\text{advection}} = \underbrace{-\overline{u'w'}\frac{\partial \bar{u}}{\partial z}}_{\substack{\text{shear}\\\text{production}}} \underbrace{+\frac{g}{T_0}\overline{w'T'}}_{\substack{\text{buoyant}\\\text{production}}} \underbrace{-\frac{1}{\rho_0}\frac{\partial \overline{p'w'}}{\partial z}}_{\substack{\text{pressure}\\\text{correlation}}} \underbrace{-\frac{\partial}{\partial z}\overline{e_k w'}}_{\substack{\text{turbulent}\\\text{diffusion}}} - \underbrace{\varepsilon}_{\text{dissipation}} = 0, \qquad (2)$$

where $e_k = 0.5(u'^2 + v'^2 + w'^2)$ stands for the turbulence kinetic energy; p is the pressure; g is the acceleration due to gravity; the ensemble (Reynolds) averages are denoted by overbars; the primes signify turbulent fluctuations; T_0 and ρ_0 are reference values of temperature T and density ρ, respectively. The TKE integral budget in the wind tunnel CBL was evaluated and analyzed in Fedorovich *et al.* [15].

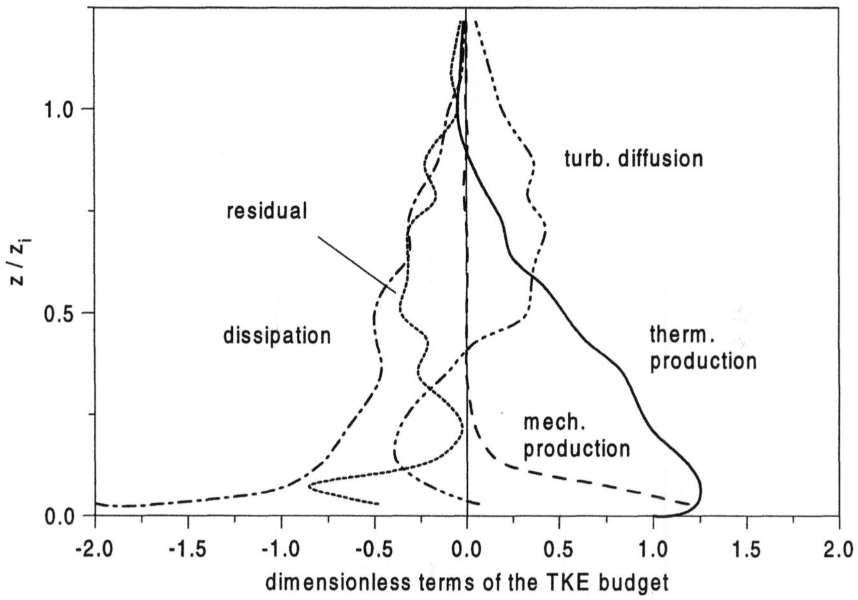

Figure 18. Normalized components of the TKE budget in the wind tunnel CBL.

Only two components of velocity can be simultaneously registered by the LDV system employed for velocity measurements in the tunnel. According to a regular measurement program, these are the longitudinal (u) and vertical (w) velocity components. Special measurements performed for the estimation of the transversal velocity component v have shown that v'^2 values are typically very weakly correlated with w'. The estimate $-\partial\overline{e_k w'}/\partial z = -0.5(\partial\overline{u'^2 w'}/\partial z + \partial\overline{w'^3}/\partial z)$ has been employed for the evaluation of TKE turbulent diffusion from the velocity time series. The pressure correlation term has not been calculated due to the absence of data on pressure fluctuations in the simulated CBL. The advection term has also been omitted due to the inability of determining it accurately from the wind tunnel data. We may expect, however, that it is a comparatively small component of the TKE balance in the wind tunnel (Fedorovich *et al.* [15]). The TKE dissipation rate ε was determined from the velocity spectra (the procedure is described in section 3.3.3). The shear and buoyancy production terms were directly evaluated from the measured data.

The TKE balance in the wind tunnel flow at x=5.63 m is presented in Fig. 18. Before being plotted, the balance components were normalized by w_*^3 / z_i. The residual term represents the net contribution to the TKE balance by advection, pressure transport, and other smaller terms omitted in the budget considerations.

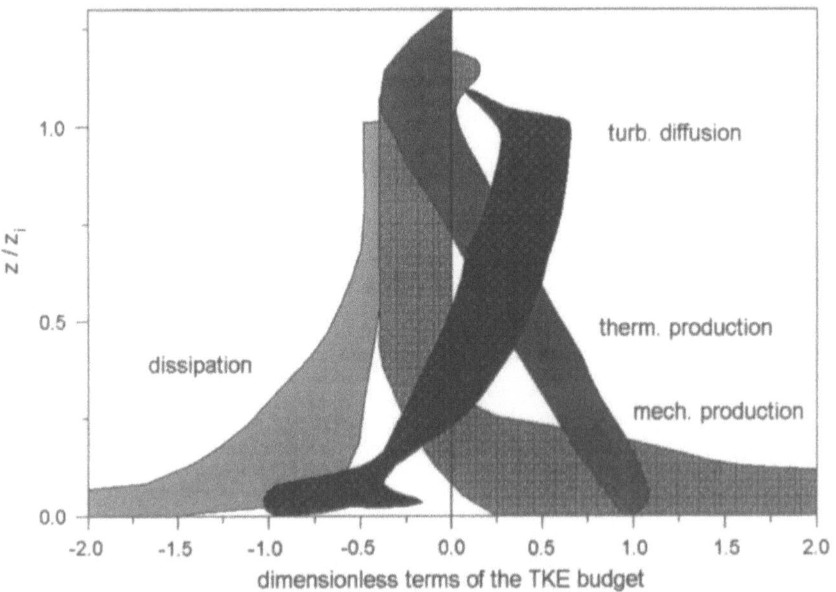

Figure 19. The TKE budget in the CBL. The shaded areas indicate ranges of values obtained by merging data from different experimental and model sources, after Stull [52].

The mechanical (shear) production of the TKE is maximum near the surface, where it exceeds even the thermal (buoyant) production. The total TKE production in the lower portion of the CBL is compensated largely by high TKE dissipation rate. Also, some energy is extracted from this flow region due to the TKE turbulent transport (diffusion) upwards. The transport term changes its sign slightly below the CBL middle. In the upper portion of the layer, the TKE grows at the expense of the energy diffusion from below. Across the lower third of the CBL, the TKE shear production steeply drops to very small values, and remains close to zero up to the CBL top. The thermal production gradually decays with height throughout the main portion of the CBL. The proportion between the mechanical and buoyant TKE production terms shows that buoyant production is dominant in the bulk of the wind tunnel CBL. Close to the CBL top, the thermal production of the TKE is negative within a certain range of heights due to the heat exchange across the capping inversion.

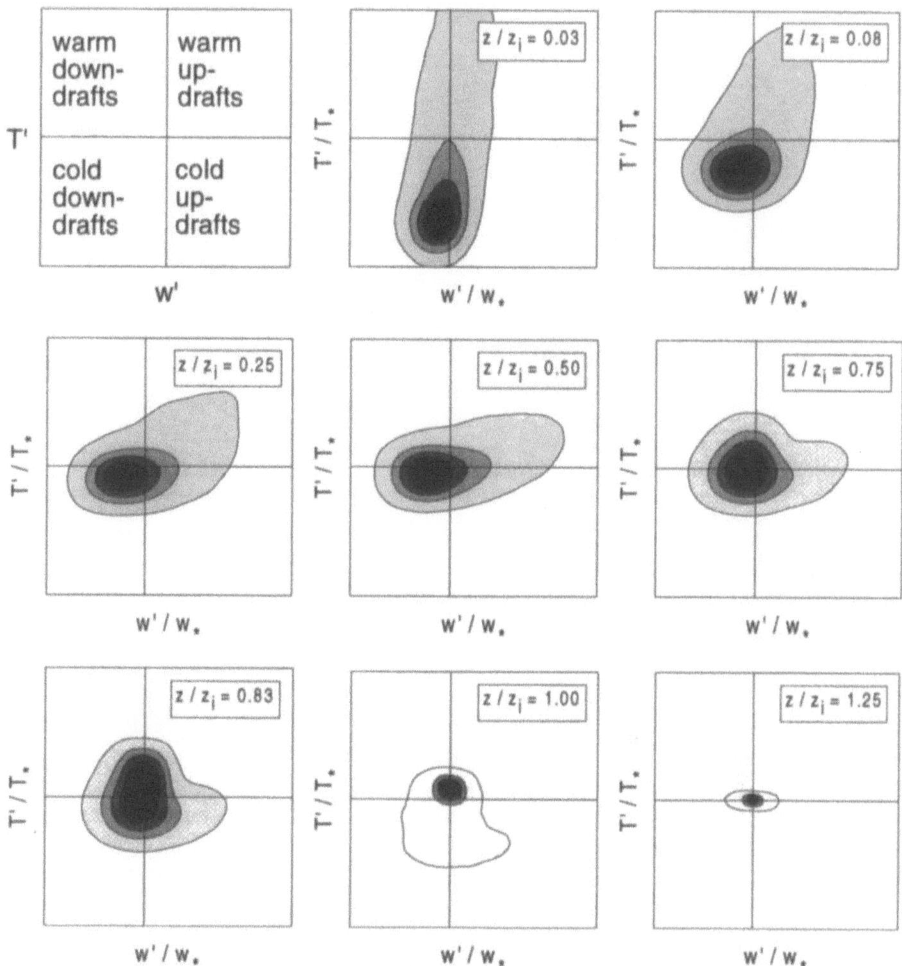

Figure 20. Joint probability density distribution of the vertical velocity and temperature fluctuations at different elevations in the wind tunnel CBL at x=3.98 m.

Figure 19 presents the normalized components of the TKE balance compiled from a variety of experimental and model studies of the atmospheric CBL. The comparison with Fig. 18 certifies that the TKE balance components in the simulated CBL are well within the range of values provided by other data sources.

3.2.3. *Probability Density Distributions*

Joint probability density distributions of the vertical velocity and temperature fluctuations shown in Fig. 20 provide information about the statistical structure of the vertical heat transport by convective motions. The four quadrants of the distribution base (the plot in the upper left corner of Fig. 20) can be related to the warm and cold up- and downdrafts, which constitute the internal turbulence pattern of the CBL, (see the visualized flow images in section 3.1.2).

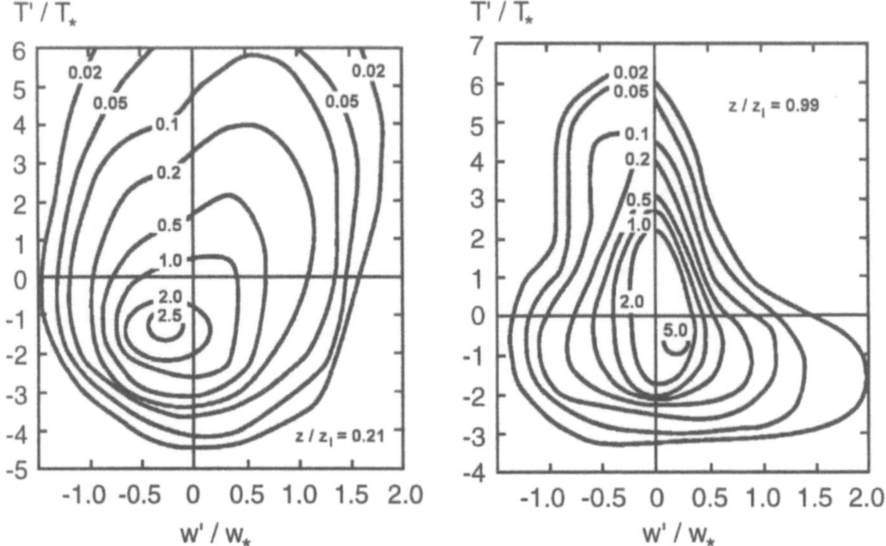

Figure 21. Joint probability density distributions of the vertical velocity and temperature fluctuations at two dimensionless elevations in the water tank CBL model of Deardorff and Willis [9].

At the lowest elevation shown (z / z_i =0.03), which is located inside the convective surface layer, the joint distribution provides evidence of strong vertical turbulent heat flux and large, positively skewed temperature fluctuations. Whilst the height increases within the range of 0.08 to 0.5, the temperature fluctuation magnitude decreases and the magnitude of the vertical velocity fluctuations increases, (see also Figs. 15b and 16b). This is the reason for the clockwise turn of the distribution pattern. From the probability distributions referring to the lower half of the CBL one can deduce that the vertical heat transport in this flow region is maintained mainly by hot updrafts and cold downdrafts. In the upper half of the convectively mixed layer, at z / z_i =0.75-0.83, the updrafts become cooler while the probability of warm downdrafts transporting hot air from the region above the inversion into the mixed layer increases. Inside the inversion layer, at z / z_i =1.00, a characteristic bulge in the distribution pattern may be noted which is caused by relatively cold updrafts penetrating in the stably stratified layer. The pattern also gives an impression that these updrafts do not mix with the ambient warm air, but merely sink back into the convectively mixed layer as cool downdrafts. This observation is consistent with the previously discussed smallness of the heat flux of entrainment in the simulated CBL. At the upper edge of the inversion layer (z / z_i =1.25), the distribution shows almost no correlation between w and T fluctuations. This is a sign of rather weak turbulent heat transport in this portion of the flow.

The joint probability patterns shown in Fig. 20 can be compared with their counterparts from the water tank study of Deardorff and Willis [9], see Fig. 21. In the lower portion of the CBL, at z/z_i =0.21, the probability pattern from the water tank reveals joint skewness of the velocity and temperature fluctuations associated with the upward heat transport by the thermals. The shape of the pattern is quite similar to that of the probability distribution from the wind tunnel model at z/z_i =0.25. Like the probability pattern from the wind tunnel CBL at z/z_i =0.83, the water tank distribution at z/z_i =0.99 exhibits asymmetry associated with the entrainment flux. Such asymmetry has been also observed by Mahrt and Paumier [30] in the probability pattern derived from measurements in the atmospheric CBL.

3.2.4. *Conditional Sampling*

Further information on the convective turbulence structure in the wind tunnel can be obtained through conditional sampling of the measured velocity and temperature time series. The updraft/downdraft conditional sampling implies sorting data from the time series into categories according to a selection criterion based on the sign and magnitude of the vertical velocity fluctuations. These categories are defined by the so-called indicator functions (Stull [52]). The indicator functions impose the thresholds which must be passed by an indicator variable (w' in our case) in order to be attributed to a particular category.

TABLE 5. Parameters of indicator function I (IF I).

	Threshold for w'
Updraft	$0 \le w'$
Downdraft	$w' < 0$

TABLE 6. Parameters of indicator function II (IF II).

	Threshold for w'	Threshold for L_x
Updraft	$0.04\text{m/s} < w'$	$0.04\text{m} < L_x$
Environment	$\lvert w' \rvert \le 0.04\text{m/s}$	$L_x \le 0.04\text{m}$
Downdraft	$w' < -0.04\text{m/s}$	$0.04\text{m} < L_x$

A time interval within which w' keeps a value beyond the threshold may be used as an additional selection criterion. This time interval can be expressed in terms of the horizontal length L_x of a rising / descending portion of the air, provided the mean longitudinal velocity of the flow is known. Using both the magnitude and the duration (length) criteria, the data points in the time series can be sorted into updraft and downdraft categories and then separately analyzed. Data that do not belong to any of the imposed categories may be associated with the environment embedding the ascending and descending motions.

Table 5 presents the parameters of the first indicator function (IF I) employed for conditional sampling. This indicator function merely distinguishes between upward and downward motions. The second indicator function (IF II), which parameters are given in Table 6, attributes some of the data points to the category of environment.

As seen from the examples of conditional sampling applied to the vertical turbulent heat flux and longitudinal velocity, (Fig. 22a and Fig. 22b, respectively), the sampling results are quite sensitive to the indicator function employed. With the imposed thresholds for the magnitude and size of upward / downward motions, the features of turbulent transport generally become more pronounced.

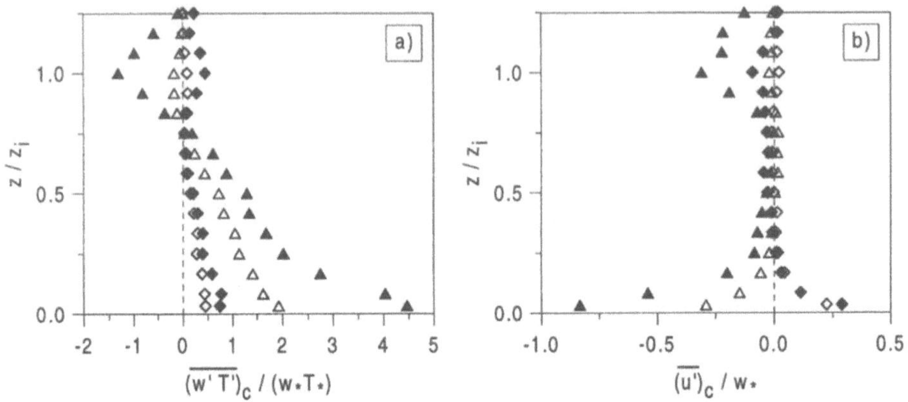

Figure 22. Conditionally sampled values of (a) vertical turbulent heat flux and (b) longitudinal velocity in the wind tunnel CBL at x=3.98 m. Open symbols: IF I; filled symbols: IF II. Triangles: updrafts; rhombi: downdrafts.

It may be concluded from Fig. 22a that in the lower portion of the CBL the heat is mainly transported by the warm updrafts. In the entrainment zone, the negative heat flux is primarily associated with the cold updrafts. This flux is partially compensated by the positive heat flux resulting from correlated negative temperature and vertical velocity fluctuations in the cold downdrafts falling back down into the convectively mixed layer. From conditionally sampled u fluctuations (Fig. 22b), one can deduce that the horizontal velocity fluctuations with a negative sign near the CBL bottom are mainly attributed to updrafts, while downdrafts transport positive momentum fluctuations, which are smaller in magnitude. In combination, this provides a negative flux of the horizontal momentum in the lower portion of the CBL, (see Fig. 5b in Fedorovich *et al.* [15]). The sampling pattern with IF II gives evidence of the negative momentum transport across the inversion by large-scale motions. Nevertheless, the net vertical transport of momentum in the inversion layer is rather small.

Figure 23a testifies that in the lower portion of the simulated CBL the temperature excess of the warm updrafts is larger than that of the cool downdrafts. This feature of the CBL turbulence structure we have already pointed out while discussing the joint probability patterns in Fig. 20. Within the inversion layer, the net values of the temperature fluctuations are quite small due to the relatively weak inversion in the wind tunnel CBL, (see discussion to Fig. 15b). On the other hand, the temperature sampling with IF II provides noticeably larger magnitudes of negative temperature fluctuations in up- and downdrafts, which are both cooler than the surrounding air. This allows to infer that the heat flux of entrainment in the wind tunnel CBL is maintained by comparatively small-scale and slow vertical motions, meanwhile the air from the larger-scale and faster updrafts is only slightly heated above the inversion before it sinks back into the mixed layer, (see also Fig. 22a).

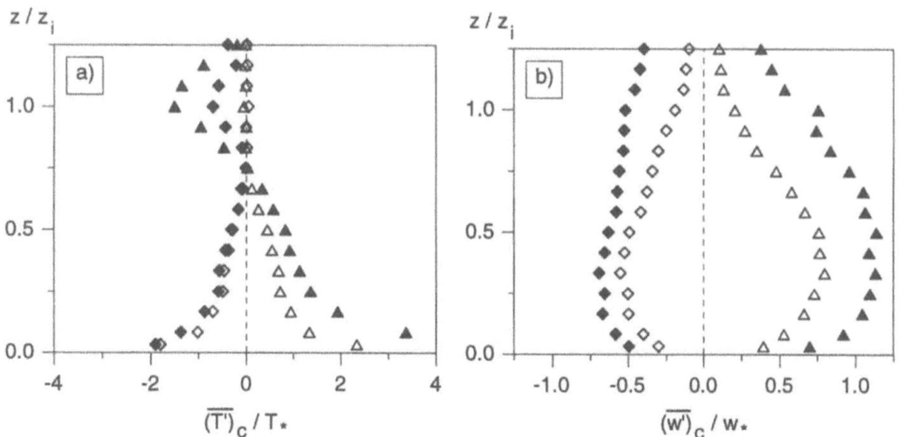

Figure 23. Conditionally sampled values of (a) temperature and (b) vertical velocity in the wind tunnel CBL at x=3.98 m. For notation see Fig. 22.

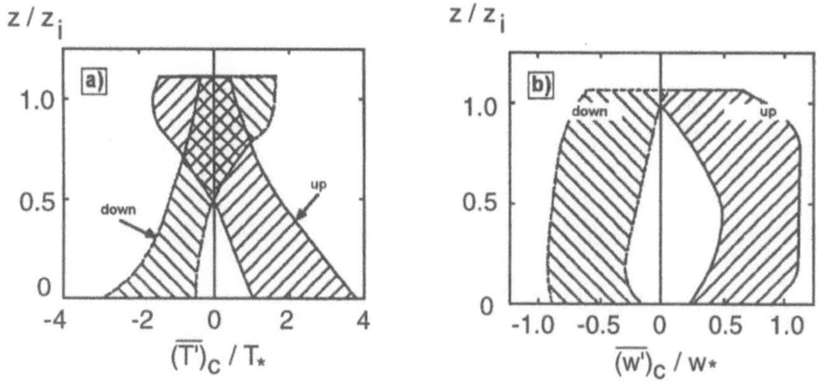

Figure 24. Conditionally sampled values of (a) temperature and (b) vertical velocity in the atmospheric CBL, after Greenhut and Khalsa [18], and Young [59].

Results of temperature sampling in the upper portion of the atmospheric CBL (Fig. 24a) show both positive and large temperature fluctuations in the atmospheric downdrafts. Apparently, this corresponds to CBL cases with strong capping inversions. The wide range of values indicated in Fig. 24a are related, in part, to different indicator functions used by different investigators (Stull [52]).

Conditionally sampled values of the vertical velocity from the wind tunnel (Fig. 23b) are well within the range of *w* sampled in the atmosphere (Fig. 24b). It is seen in both graphs that updrafts in the bulk of the CBL are characterized by higher absolute vertical velocity values than downdrafts. Near the CBL top, the magnitudes of *w* fluctuations in updrafts and downdrafts are equalized.

3.3. SPECTRAL ANALYSIS

3.3.1. *Spectra Derivation from Measurement Data*

We calculate spectral densities by the finite Fourier transforms of the original time series based on the Wiener-Khinchin relationship. This relationship equates the spectral densities to the one-sided spectral density functions defined in terms of Fourier transforms of correlation functions (Bendat and Piersol [3]). Since the introduction of algorithms for fast Fourier transforms by Cooley and Tukey [7], this approach towards the discrete spectra calculation has become dominant.

To average the spectra, an original time series consisting of 16384 data values was separated into 31 overlapping records, each containing 1024 measurement values. The calculated spectra were subdivided into 400 equal intervals with respect to the logarithm of frequency, and the average spectral density value within each interval was calculated. This was done merely to improve the layout of the spectra when plotted in logarithmic scales. For the details of spectra calculation see Kaiser and Fedorovich [23].

Since the Taylor hypothesis was found to be valid for the simulated CBL (Fedorovich *et al.* [15]), it was used to transfer the spectra from the frequency domain to the wavenumber one. Provided the mean-flow horizontal velocity u in the measurement location of spectrum is known, the following relationship between the wavenumber k, the wavelength λ, and the frequency f may be employed:

$$k = \frac{2\pi}{\lambda} = \frac{2\pi f}{u}. \tag{3}$$

The wavenumber spectrum $P_\varphi(k)$ and the frequency spectrum $P_\varphi(f)$ of a quantity φ are related by

$$P_\varphi(k) = \frac{u}{2\pi} P_\varphi(f). \tag{4}$$

Experimental spectra are usually presented with logarithmic scales of frequency or wavenumber (Jensen and Busch [21], Panofsky and Dutton [37]). If the ordinates are $f P_\varphi(f)$ or $k P_\varphi(k)$, then the area in a log-frequency or log-wavenumber interval represents the contribution of the fluctuations in this interval to the variance of φ. Another advantage of these ordinates is their independence on the selected units of frequency or wavenumber, because $f P_\varphi(f) = k P_\varphi(k)$. Most of the wind tunnel spectra are shown below in such a logarithmic frequency / wavenumber form.

3.3.2. *Velocity and Temperature Spectra*

We begin consideration of the turbulence spectra with the evaluation of the vertical to horizontal velocity spectral ratio P_w / P_u in the wind tunnel CBL. This characteristic is important for making assumptions concerning the local isotropy of turbulence and the existence of inertial subranges in the measured spectra. Both features should be taken into account when the TKE dissipation rate and the destruction rate of temperature fluctuations are evaluated from the measured spectra.

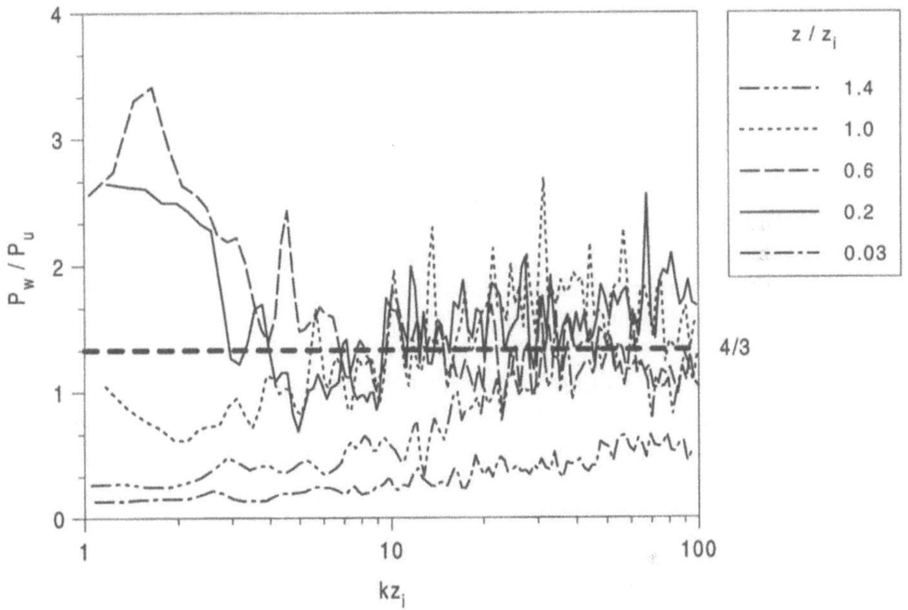

Figure 25. Spectral ratio P_w / P_u at different dimensionless elevations in the wind tunnel flow (x=3.98 m).

Kolmogorov [24] introduced the concept of inertial subrange as that of the energy cascade interval in the turbulence spectrum where no energy enters the system from the outside and no energy dissipates. Within this interval, the energy spectrum is isotropic, it does not depend on viscosity, and its shape is determined only by the energy flux over the spectrum which is equal to the energy dissipation rate ε. Thus, in the inertial subrange

$$P_i(k) \sim \varepsilon^{2/3} k^{-5/3},$$

where $P_i(k)$ is the spectral density of the velocity component u_i. Obukhov [35] has shown that the assumptions of the Kolmogorov [24] theory lead also to a -5/3 law for the inertial-subrange spectrum of temperature fluctuations. Local isotropy of turbulence and the presence of the inertial subrange in spectra are conditioned by a sufficiently high turbulent Reynolds number of the flow, $Re_t = ul/\nu$, where u is the representative turbulence velocity scale, l is the integral turbulence length scale, and ν is the kinematic air viscosity. According to the analysis of Tennekes and Lumley [54], Re_l has to be of the order 10^3 and more to justify the employment of the Kolmogorov theory relationships for spectra in the inertial subrange.

In the wind tunnel, $\boldsymbol{u}$ is of the order 0.1 m/s, and $\boldsymbol{l}$ is approximately 0.5 m (Fedorovich *et al.* [15]) which gives $Re_l \approx 3 \cdot 10^3$ (adopting $\nu = 1.5 \cdot 10^{-5}$ m^2/s). The characteristic wind tunnel value of the Reynolds number $Re_\lambda = \boldsymbol{u}\lambda/\nu$ based on the Taylor microscale λ is about 50. Here, the estimate $\lambda = 10^{-2}$m is obtained from $\lambda^2 = 15\nu\boldsymbol{u}^2 / \varepsilon$, where ε is taken to be equal $10^{-2}\text{m}^2/\text{s}^3$, (see section 3.3.3).

Under conditions of local isotropy, the Kolmogorov theory predicts the P_w / P_u ratio in the inertial subrange to be equal 4/3 (Mestayer [33]). The wind tunnel values of P_w / P_u are plotted against the dimensionless wavenumber kz_i in Fig. 25. With $kz_i > 10$, the P_w / P_u ratio in the main portion of the CBL is close to the local-isotropy value. It is possible to notice, however, that the average of P_w / P_u slightly exceeds the 4/3 criterion. Such a small, but persistent anisotropy of turbulent motions at comparatively large wavenumbers may be due to the insufficiency of the pressure fluctuations to isotropize turbulence in the presence of dominant buoyant forcing, which favors vertical motions (Schmidt and Schumann [48]). Nevertheless, it is seen from Fig. 25 that turbulence in the simulated CBL may be considered as approximately isotropic in the wavenumber range $10 < kz_i < 80$, and in the range of heights $0.2 \le z / z_i \le 1$.

Above the CBL, at $z / z_i = 1.4$, the local isotropy is observed only over a very short range of wavenumbers. Close to the surface, at $z / z_i = 0.03$, the turbulence is essentially anisotropic over the whole range of scales. In this portion of the flow, the horizontal motions dominate the vertical ones due to the damping effect of the surface. This is also a reason why the turbulence becomes more isotropic with decreasing scales of motion.

With dimensionless wavenumbers smaller than 10, the behavior of P_w / P_u differs in the CBL mixed core and in the inversion layer. In the mixed-layer, at $z / z_i = 0.2$ and $z / z_i = 0.6$, the P_w / P_u ratio reveals a strong turbulence anisotropy associated with the dominance of vertical motions. Within the inversion layer, at $z / z_i = 1$, the dependence of P_w / P_u on k supports our earlier findings concerning the vertical motion suppression at the CBL top, and the enhancement of the horizontal velocity fluctuations due to the sideward transport of air from the blocked thermals.

Comparison of the wind tunnel velocity spectra with the spectral data from the water tank study of Deardorff and Willis [9], and from the LES study by Schmidt and Schumann [48] is presented in Figs. 26a and 26b. In both studies considered here for comparison, the shear-free atmospheric CBL was modeled. The wind tunnel turbulence statistics have been compared with data from these studies in section 3.2.1. It has been shown there that u and w velocity component variances in the lower and middle portions of the wind tunnel CBL are relatively large. The spectral distribution of divergence between the wind tunnel, water tank, and LES velocity variances can be traced in the presented velocity component spectra.

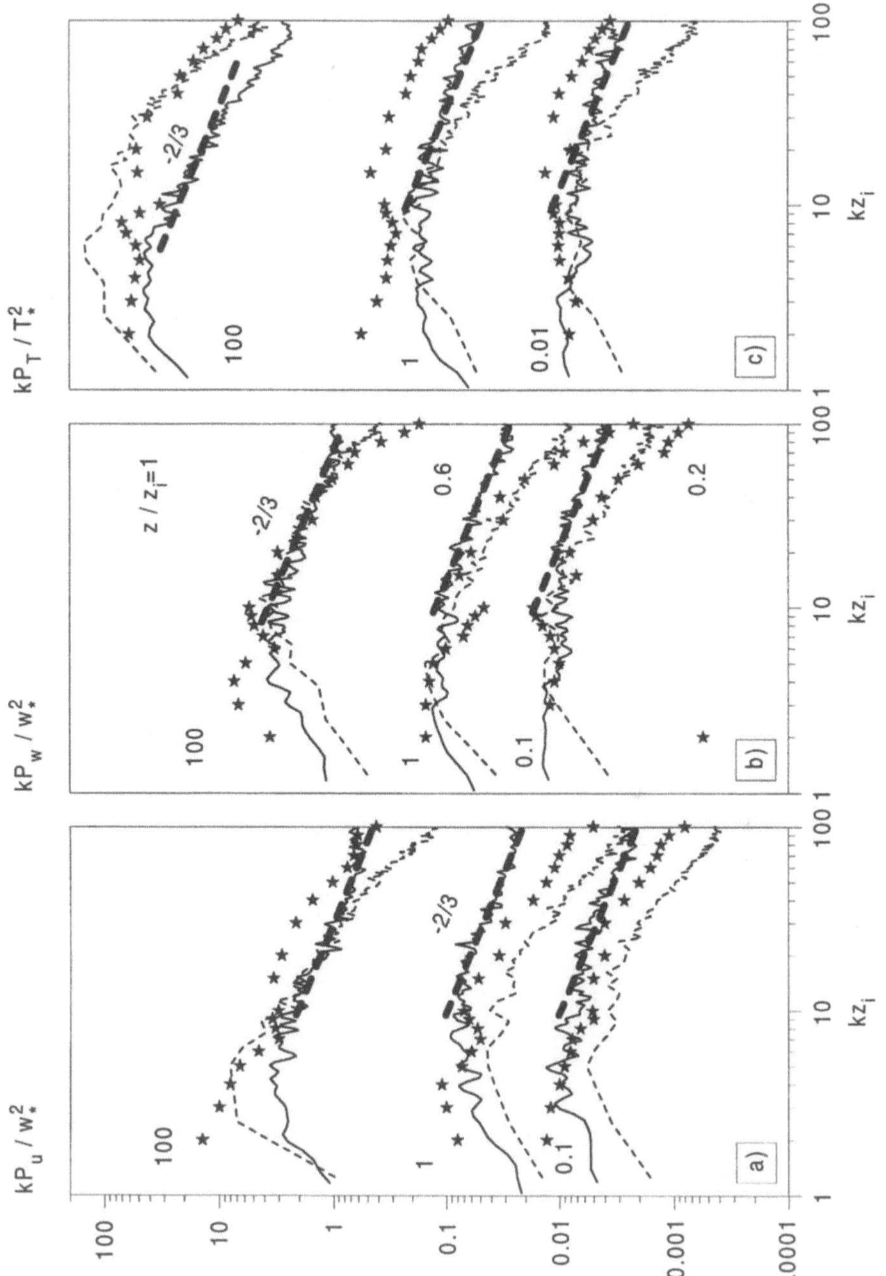

Figure 26. Spectra of (a) longitudinal velocity, (b) vertical velocity, and (c) temperature in the wind tunnel CBL, solid lines. Corresponding spectra from the water tank model of Deardorff and Willis [9], and from the LES study of Schmidt and Schumann [48] are given by stars and dashed lines, respectively.

At smaller wavenumbers (kz_i <8), the wind tunnel u spectra (Fig. 26a) are positioned between the water tank and LES spectra. Comparatively large water tank spectra values in this wavenumber range presumably exhibit the effect of horizontal inhomogeneity of the surface heat flux first pointed out by Schmidt and Schumann [48], (see also discussion in section 3.2.1).

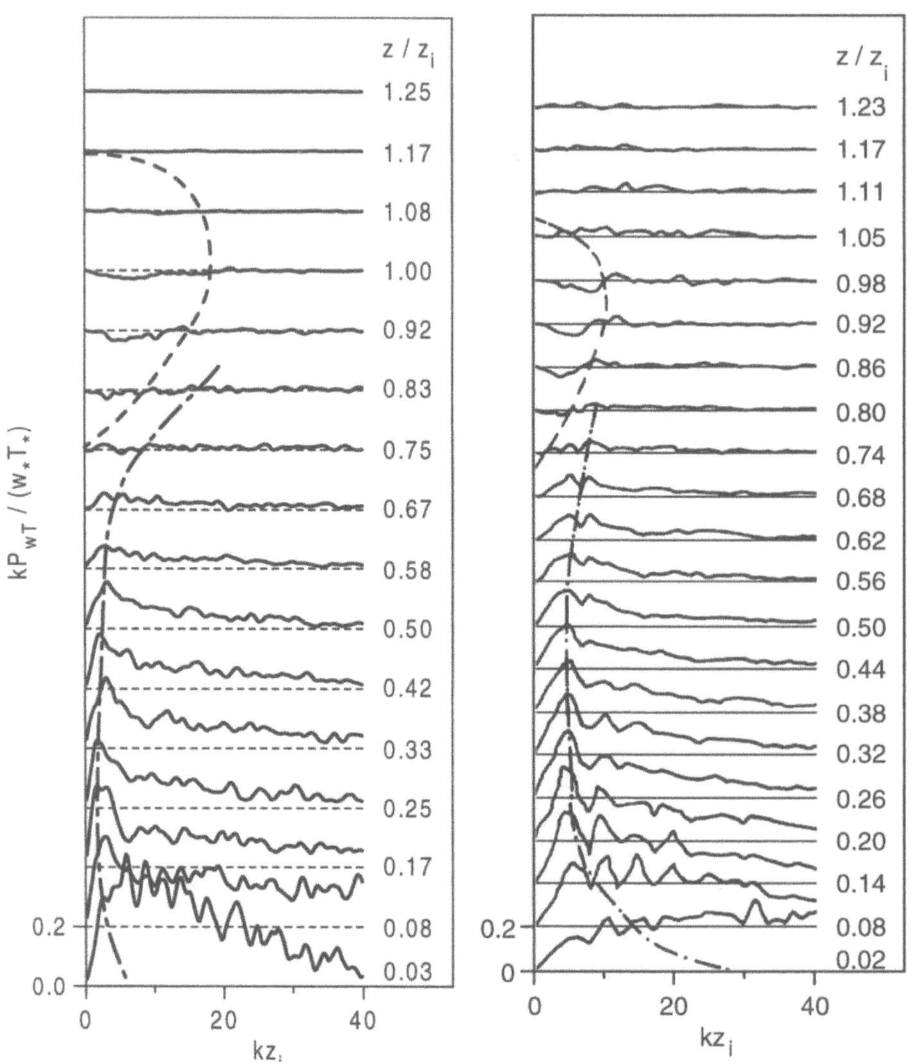

Figure 27. Spectra of the vertical turbulent heat flux from the wind tunnel model, left plot, and from the LES study of Schmidt and Schumann [48], right plot.

Larger w spectral values in the wind tunnel model at small wavenumbers may be explained by the shear contribution, and by the quasi-steady cellular circulations, which are both sources of the velocity variance enhancement in the wind tunnel CBL (Fedorovich *et al.* [15]). Near the CBL top, at $z/z_i=1$, where the turbulence regime is practically unaffected by the surface shear, the wind tunnel spectra are close to the spectral curves from the other model studies.

At $z/z_i=0.2$ and $z/z_i=0.6$, the LES spectra decrease with an increasing wavenumber, while the wind tunnel spectral curves remain relatively flat. As a result, the wind tunnel spectra of both velocity components are larger than their water tank and LES counterparts with $kz_i>10$. This suggests that in the lower portion of the CBL, the buoyancy acts as the dominant factor of turbulence production at smaller wavenumbers, while the contribution of shear becomes important with increasing wavenumbers. The joint effect of buoyant and shear forcings leads to the elongation of the production range and to the relative flatness of the normalized spectra in the high-wavenumber portion of this range. Such behavior of the velocity spectra in the sheared CBL is in accordance with the -1 power law predicted by Tschen [53] and Gisina [17].

At all altitudes represented in Fig. 26a, the normalized velocity spectra from the wind tunnel model display linear intervals at $kz_i>10$, which can be approximated fairly well by the -5/3 (for the normalized spectra by -2/3) power laws of the Kolmogorov [24] inertial-subrange theory. These linear intervals, although clearly seen in the spectra, are rather short compared with the typical extensions of inertial subranges observed in the velocity spectra from atmospheric measurements (Kaiser and Fedorovich [23]).

The temperature spectra from the wind tunnel model are compared in Fig. 26c with *T* spectra from the water tank and LES studies. At the two lower elevations in the wind tunnel CBL, the temperature spectra shape is similar to those of *u* spectra shown in Fig. 26a. This bears evidence of a vast range of temperature fluctuation scales that almost equally contribute to the spectral energy. In the CBL core, the wind tunnel spectra are located close to the LES spectra, but are not as convex as the latter ones.

In the inversion layer, at $z/z_i=1$, the decay of the wind tunnel temperature spectrum in the high-wavenumber range is faster than in the spectra measured at lower levels. Here, the effect of inversion strength on the temperature fluctuation magnitude is clearly seen. The smaller inversion strength in the wind tunnel experiments is the reason for the lower position of the corresponding spectrum curve in Fig. 26c as compared with temperature spectra from the water tank and LES studies.

At larger wavenumbers, linear -5/3 intervals are observed in the wind tunnel temperature log-spectra. However, they are even shorter than the inertial subranges in the velocity spectra, (compare Fig. 26c with Figs. 26a and 26b).

Figure 27 presents the wind tunnel heat flux spectra (cospectra of the vertical velocity and temperature fluctuations) compared with the heat flux spectra from the LES study of Schmidt and Schumann [48]. At $z/z_i=0.03$, the influence of the surface shear in the wind tunnel CBL can be a reason for the comparatively weak correlation between *w* and *T* fluctuations at larger wavenumbers. We have already noted that under the influence of shear, the production ranges of *w* spectra are elongated towards higher frequencies. Velocity fluctuations produced by shear are not coupled with temperature fluctuations, rather they are merely stochastically mixed with them. This results in the comparatively fast decay of the heat flux spectrum with increasing wavenumbers.

In the middle portion of the CBL, the wind tunnel cospectra consistently exhibit comparatively broad spectral maxima related to the thermals (w' and T' are both positive), and to cool downdrafts (w' and T' are both negative). These maxima are located within the wavenumber range from kz_i=3 to kz_i=8. The LES cospectra in the CBL core have their maxima approximately within the same range. This corresponds to the observation of Kaimal *et al.* [22], who found that motions with scales of the order of z_i are the main contributors to the turbulent transport of heat in the atmospheric CBL.

The magnitude of the cospectrum decreases towards the upper portion of the CBL, and the spectral maximum shifts to larger wavenumbers, (remember the decrease with height of the horizontal sizes of thermals in the visualized patterns of section 3.1.2). In the inversion layer, at z/z_i >0.8, the heat flux spectra become negative with smaller wavenumbers. These negative spectral values mainly represent large-scale cool updrafts, which are the characteristic structures in the upper portion of the wind tunnel CBL. There are also warm downdrafts of the entrained air which are partially responsible for the negative cospectra values within the inversion layer. Higher-frequency components of the heat flux in the entrainment zone retain positive values and gradually decay towards the CBL top, (see the cospectrum at z/z_i =1.2).

3.3.3. *Dissipation Rates of TKE and Temperature Fluctuations*

For the evaluation of the TKE dissipation rate ε from the wind tunnel velocity spectra we employ the following relationships of the Kolmogorov inertial-subrange theory (Tennekes and Lumley [54]):

$$P_u(k) = \alpha\varepsilon^{2/3}k^{-5/3}, \tag{5}$$

$$P_w(k) = \frac{4}{3}\alpha\varepsilon^{2/3}k^{-5/3}, \tag{6}$$

where the value of the dimensionless parameter α is taken equal 0.52 (Andreas [2], and Schmidt and Schumann [48]). From a practical point of view, the procedure of the determination of ε is based on the least-square approximation of the inertial-subrange spectrum using Eqs. (5) and (6). The applicability of these equations to spectra obtained near the CBL top, where spectral inertial subranges are very short, seems to be rather questionable.

Vertical profiles of ε from the wind tunnel model are shown in Fig. 28a. We have seen in section 3.3.2 that the assumption of local isotropy is only approximately satisfied in the wind tunnel CBL. Thus we cannot expect horizontal and vertical velocity spectra to provide exactly the same values of ε. That is why, two wind tunnel ε values are given for each measurement level. In the main portion of the CBL, discrepancies between the dissipation rates, derived alternatively by Eq. (5) and by Eq. (6), are much smaller than the scatter of data from other sources represented in the plot. The smallest dissipation rate values originate from the Schmidt and Schumann [48] LES study of a pure shear-free CBL. Larger ε values in the lower portion of the water tank CBL of Deardorff and Willis [9] are presumably a result of the TKE overproduction due to the previously discussed horizontal variations of the bottom heat flux in their experiments. Dissipation rates measured in the ocean and in the atmosphere are apparently affected by nonbuoyant contributions to the TKE production. Comparatively large ε values in the wind tunnel CBL result from the combined influence of shear, cellular circulations, and transition effects. All mentioned factors have already been considered earlier in the present paper.

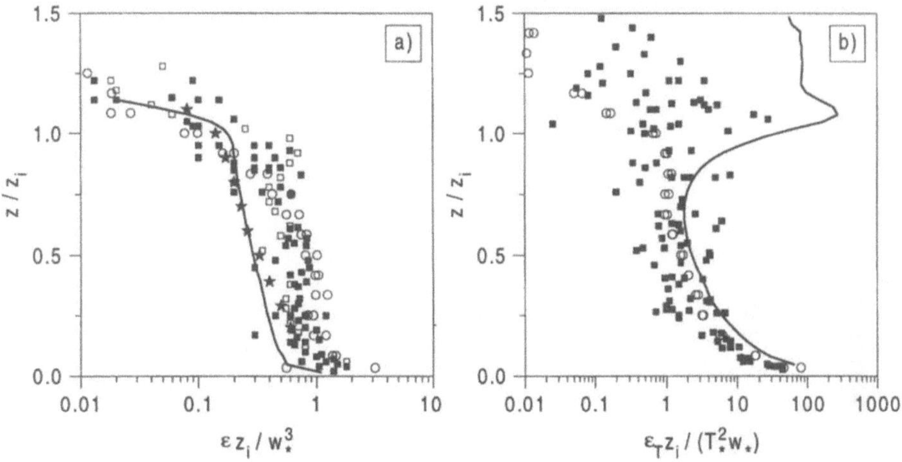

Figure 28. TKE dissipation rate (a), and destruction rate of temperature fluctuations (b) in the CBL. The wind tunnel data (circles) are compared with the atmospheric data from Caughey and Palmer [5], filled squares in (a), and Sorbjan [50], filled squares in (b); oceanic data of Shay and Gregg [49], empty squares; water tank model data of Deardorff and Willis [9], stars; and LES data (courtesy of U. Schumann), lines.

The temperature fluctuation destruction rate ε_T, (we define it here as destruction of the temperature half-variance), is obtained from the inertial-subrange temperature spectra using the Obukhov [35] law

$$P_T(k) = \beta \varepsilon_T \varepsilon^{-1/3} k^{-5/3}, \tag{7}$$

where β=0.8 is a dimensionless coefficient (Andreas [2], Schmidt and Schumann [48]). In Eq. (7), we have the choice of using ε determined either from Eq. (5) or from Eq. (6). It is clear, however, from Fig. 28b that the difference between the resulting ε_T profiles is small. This may be expected because $P_T(k)$ in Eq. (7) depends rather weakly on ε. In the lower half of the CBL, the discrepancies between ε_T from the three datasets represented in Fig. 28b are not very large. The drastic divergence of the ε_T profiles from different studies is observed in the upper portion of the CBL, where the turbulence regime is essentially governed by capping inversion effects. The wind tunnel model with weak inversion gives the smallest values of ε_T. The ε_T maxima in the atmospheric and LES profiles result from enhanced temperature fluctuations in the inversion layer with a strong stable temperature gradient (large $\mathrm{Ri}_{\Delta T}$ numbers).

3.4. EFFECTS OF ELEVATED SHEAR

To study the effects of elevated shear on the CBL turbulence regime, the initial flow configuration shown in Fig. 29 was employed. The experimental setup parameters were as follows: the mean-flow temperature and velocity, of 20 K and 1 m/s, respectively, were kept height-constant within the two lower layers of the tunnel; at z=0.3 m, between the second and the third layers, a velocity increment of 0.5 m/s and a temperature increment of 30 K were imposed; throughout the upper layers the velocity was kept height-constant at 1.5 m/s, and the temperature increased vertically at a rate of 67 K/m. The heat flux through the bottom was the same as in the basic case studied (see section 3.1.1).

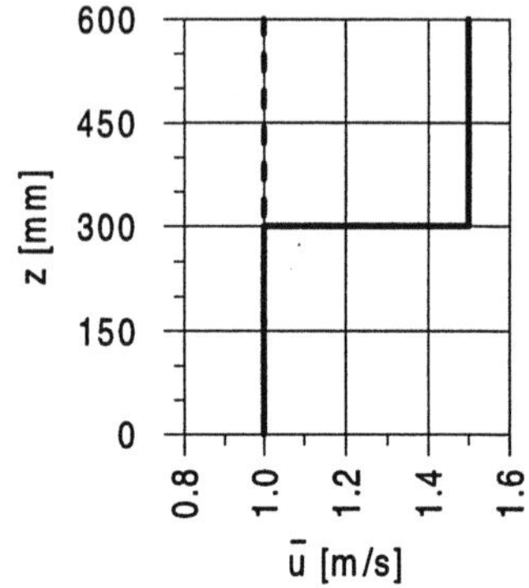

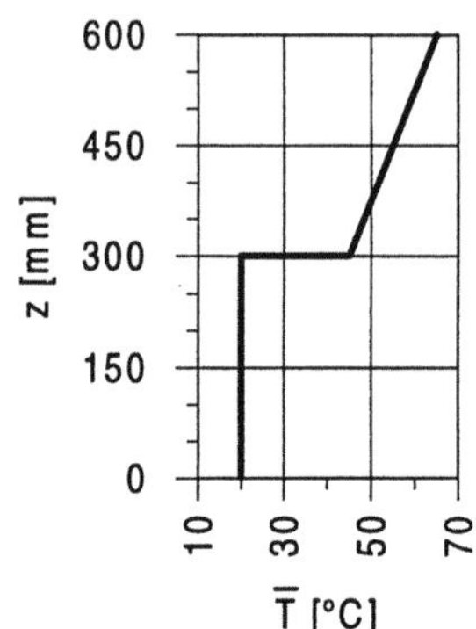

Figure 29. Velocity and temperature profiles at the test section inlet in the elevated-shear case study.

Results from the experiments with elevated shear are shown below in the non-normalized form. In the presence of elevated shear, the Deardorff [8] scaling, originally designed for the shear-free CBL driven by purely buoyant bottom-up forcing, is no longer relevant. We know of no scaling suggested for the CBL which appropriately accounts for effects of both the surface and elevated shears.

Modification of the TKE budget in the wind tunnel flow by elevated shear can be seen in Fig. 30. The main effect of the shear is an enhancement of the mechanical production of turbulence in the upper portion of the CBL. This production is balanced by the overall increase of the TKE dissipation rate. The thermal (buoyant) production of turbulence does not undergo considerable changes. It is possible to notice, however, that the buoyant production minimum, clearly seen near the CBL top in the case without elevated shear, decreases practically to zero in the presence of the shear and also shifts downwards. Our previous analysis has shown that the buoyant production minimum at the top of the wind tunnel CBL is primarily associated with cool updrafts penetrating aloft from the mixed layer, and to some extent also with the transport of heat down from the outer flow region by warm downdrafts (see section 3.2). Apparently, the elevated shear hampers the upward and downward motions at the CBL top and thus suppresses the heat exchange across the inversion layer. Under such conditions, the CBL growth associated with the penetration and entrainment processes, is considerably slowed down.

Such an impeded entrainment is clearly revealed by turbulence statistics measured at three subsequent stages of the CBL development, see Fig. 31. The capping inversion heights defined as the levels of most negative heat flux (Fig. 31a) progressively increase in the CBL without elevated shear. Once shear is applied, the growth stops. A similar effect may be observed in the evolution of the position of the temperature variance maximum in Fig. 31b. In the inversion-capped CBL, this maximum approximately corresponds to the middle of the inversion layer (Fedorovich *et al.* [15]). Figure 31b also shows that the elevated shear reduces the magnitude of the temperature fluctuation in the inversion layer.

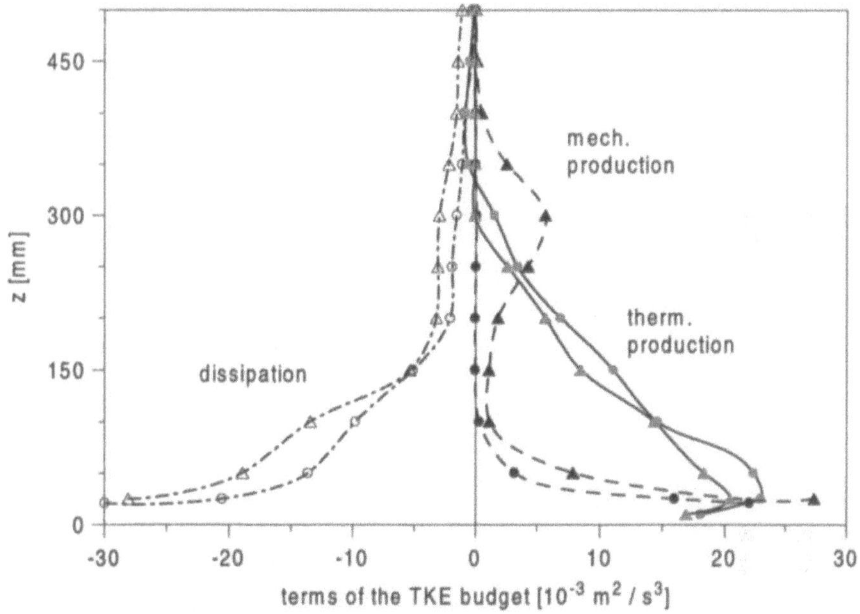

Figure 30. Main components of the TKE budget in the wind tunnel CBL with elevated shear (triangles) and without elevated shear (circles) at x=7.28 m.

Variances of the longitudinal and vertical velocities in Figs. 31c and 31d indicate that the turbulence production due to elevated shear is mainly attributed to the horizontal velocity fluctuations in the upper portion of the CBL, while the increase in the vertical velocity variance is relatively small.

Based on the data presented in Fig. 31, the following description of turbulence transformations in the sheared inversion layer may be outlined.

The imposed elevated shear deforms and partially dissolves the large-scale turbulence structures (up- and downdrafts) responsible for the vertical heat transport at the CBL top. This results in the enhancement of horizontal velocity fluctuations (see Fig. 31c) and leads to an energy redistribution between the larger-scale and smaller-scale vertical motions. The energy of the former decreases, and the energy of the latter increases. However, the integral of the vertical velocity spectrum (and thus also the vertical velocity variance) retains approximately the same value as in the shear-free inversion layer (Fig. 31d). Inside the entrainment layer, the smaller-scale vertical motions locally mix up the air from the destroyed convective structures. This can be an explanation for the smaller temperature variances at the CBL top as compared to the situation without elevated shear (Fig. 31b).

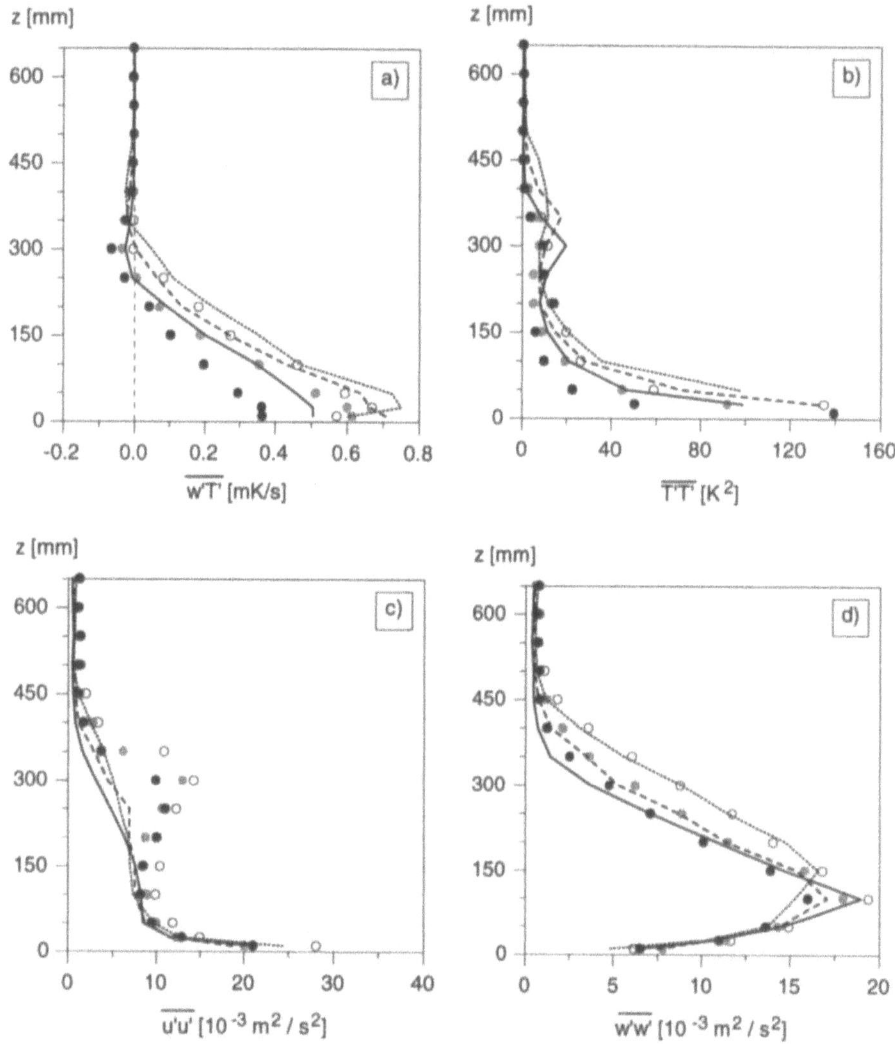

Figure 31. Vertical heat flux(d), and variances of (b) temperature, (c) longitudinal velocity, and (d) vertical velocity in the CBL with the elevated shear (points) and without the elevated shear (lines). Data refer to x=3.98 m (solid lines and black symbols), x=5.63 m (dashed lines and gray symbols), and to x=7.28 m (dotted lines and empty symbols).

A more profound insight into the turbulence mechanics within the sheared inversion layer could be provided by the spectral analysis of turbulence in this flow region. Such analysis is anticipated in the future program of wind tunnel experiments.

4. Summary and Conclusions

The presented results from the wind tunnel model study of the atmospheric CBL allow to conclude that the IHW thermally stratified wind tunnel adequately reproduces a number of important features of the atmospheric boundary-layer convective turbulence. Compared to water tank models of the CBL, the wind tunnel provides an advantageous opportunity to simulate the effects of combined buoyant and shear forcings in the atmospheric CBL.

The comparison of turbulence statistics and spectra, derived in the wind tunnel model by temporal averaging, with spatial averages from atmospheric, water tank and numerical CBL studies suggests that turbulence regimes in the nonsteady and horizontally evolving CBLs are similar to each other.

The Richardson numbers $\mathrm{Ri}_{\Delta T}$ and Ri_N in the performed experiments corresponded to convective conditions in the atmosphere when both capping inversion and hydrostatic stability of the upper air are weak. Under such conditions, the negative heat flux of entrainment and the magnitude of the temperature fluctuations within the inversion layer are rather small. Also, the horizontal velocity variance decays almost monotonously throughout the inversion layer, with merely a slight bend in the profile.

We found that wind shears at the surface and across the inversion layer can essentially modify the regime of turbulence production in the CBL. The elevated shear turned out to be an inhibitor of entrainment at the CBL top. With the bottom-up buoyancy to shear production ratio u_* / w_* of 0.35 to 0.43, the turbulence variances in the lower portion of the CBL are noticeably larger than in the shear-free CBL.

Another mechanism of turbulence enhancement which mostly affects the wind tunnel CBL at the early stages of its development is associated with the transition from the nonmixed unstable convective layer in the neutral environment to the well-mixed CBL capped by a temperature inversion. The vertical velocity fluctuations in the middle part of the simulated CBL are additionally enhanced by the quasi-steady helical circulations.

Velocity spectra measured in the bulk of the CBL show that the buoyancy is the dominant factor of turbulence production at smaller wavenumbers, while the shear contribution increases with larger wavenumbers. The combined effect of buoyant and shear forcings in the simulated CBL could be a reason for the elongation and flatness of the production ranges of the spectra measured.

The vertical to horizontal velocity spectral ratio in the tunnel is slightly higher than the local-isotropy value of 4/3. The characteristic value of the turbulent Reynolds number Re_l in the wind tunnel CBL is $3 \cdot 10^3$. In the mixed core and in the entrainment zone of the simulated CBL, the log-spectra of velocity and temperature feature apparent linear -5/3 subranges at $kz_i > 10$. Nevertheless, the typical extension of the inertial subranges in the wind tunnel CBL is quite short. It is of the order of kz_i magnitude. The inertial-subrange theory relationships have been used for the evaluation of the turbulence dissipation rates from the velocity spectra.

The TKE dissipation rates in the simulated sheared CBL are comparatively large over the entire depth of the layer. The temperature fluctuation destruction rates from the wind tunnel model are close to the atmospheric values in the lower portion of the CBL.

Within the inversion layer, the wind tunnel ε_T values are much smaller due to the weaker damping effect of the capping inversion.

In order to obtain information about the turbulence structure in the regions of the wind tunnel flow which are not accessible by the presently employed measurement technique, a parallel numerical study (LES) of the turbulence regime in the horizontally evolving atmospheric CBL is planned for the near future.

References

1. Adrian, R. J., Ferreira, R. T. D. S., and Boberg, T. (1986) Turbulent thermal convection in wide horizontal fluid layers, *Experiments in Fluids 4*, Springer-Verlag, 121-141.
2. Andreas, E. L. (1987) Spectral measurements in a disturbed boundary layer over snow, *J. Atmos. Sci.*, **44**, 1912-1939.
3. Bendat, J. S. and Piersol, A. G. (1986) *Random Data. Analysis and Measurement Procedures,* John Wiley & Sons.
4. Bruun, H. H. (1995) *Hot-Wire Anemometry*, Oxford University Press.
5. Caughey, S. J. and Palmer, S. G. (1979) Some aspects of turbulence structure through the depth of the convective boundary layer, *Quart. J. Roy. Meteor. Soc.*, **105**, 811-827.
6. Cenedese, A. and Querzoli, G. (1994) A laboratory model of turbulent convection in the atmospheric boundary layer, *Atmos. Environ.*, **28**, 1901-1914.
7. Cooley, J. W. and Tukey, J. W. (1965) An algorithm for the machine calculation of complex Fourier series, *Mathematics of Computation*, **19**, 297-301.
8. Deardorff, J. W. (1970) Convective velocity and temperature scales for the unstable planetary boundary layer and for Raleigh convection, *J. Atmos. Sci.*, **27**, 1211-1213.
9. Deardorff, J. W. and Willis, G. E. (1985) Further results from a laboratory model of the convective planetary boundary layer, *Bound.-Layer Meteor.*, **32**, 205-236.
10. Deardorff, J. W., Willis, G. E., and Lilly, D. K. (1969) Laboratory investigation of non-steady penetrative convection, *J. Fluid Mech.*, **35**, 7-31.
11. Deardorff, J. W., Willis, G. E., and Stockton, B. H. (1980) Laboratory studies of the entrainment zone of a convectively mixed layer, *J. Fluid Mech.*, **100**, 41-64.
12. Domnik, J. (1991) Application of LDA in gas flows, *Advanced Laser-Doppler Anemometry*. UMIST, University of Manchester, 1-25.
13. Drain, L. E. (1980) *The Laser Doppler Technique*, John Wiley & Sons.
14. Durst, F., Melling, A., and Whitelaw, J. H. (1987) *Theorie und Praxis der Laser-Doppler-Anemometrie*, Verlag G. Braun.
15. Fedorovich, E., Kaiser, R., Rau, M., and Plate, E. (1996) Wind tunnel study of turbulent flow structure in the convective boundary layer capped by a temperature inversion, *J. Atmos. Sci.*, **53**, 1273-1289.
16. Fedorovich, E. E. and Mironov, D. V. (1995) A model for shear-free convective boundary layer with parameterized capping inversion structure, *J. Atmos. Sci.*, **52**, 83-95.
17. Gisina, F. A. (1966) The effect of mean velocity and temperature gradients on the spectral characteristics of turbulence, *Izv. Acad. Sci. U. S. S. R., Atmos Oceanic Phys.*, **2**, 804-813 (Engl. Transl., 487-491.)
18. Greenhut, G. K. and Khalsa, S. J. S. (1987) Convective elements in the marine atmospheric boundary layer. Part I: Conditional sampling statistics, *J. Clim. Appl. Meteor.*, **26**, 813-822.
19. Holtslag, A. A. M. and Nieuwstadt, F. T. M. (1986) Scaling the atmospheric boundary layer, *Bound.-Layer Meteor.*, **36**, 201-209.
20. Hooper, W. P. and Eloranta, E. W. (1986) Lidar measurements of wind in the planetary boundary layer: the method, accuracy and results from joint measurements with radiosonde and kytoon, *J. Climate Appl. Meteor.*, **25**, 990-1001.
21. Jensen, N. O. and Busch, N. E. (1982) Atmospheric turbulence, in E. J. Plate (ed.), *Engineering Meteorology*, Elsevier, pp. 179-231.
22. Kaimal, J. C., Wyngaard, J. C., Haugen, D. A., Coté, O. R., Izumi, Y., Caughey, S. J., and Readings, C. J. (1976) Turbulence structure in a convective boundary layer, *J. Atmos. Sci.*, **33**, 2152-2169.
23. Kaiser, R. and Fedorovich, E. (1998) Turbulence spectra and dissipation rates in a wind tunnel model of the atmospheric convective boundary layer, *J. Atmos. Sci.*, **55**.

24. Kolmogorov, A. N. (1941) Local structure of turbulence in the incompressible fluid at very high Reynolds numbers, *Dokl. Akad. Nauk S.S.S.R.*, **30**, 299-303.
25. Kumar, R. and Adrian, R. J. (1986) Higher order moments in the entrainment zone of turbulent penetrative thermal convection, *J. Heat Trans.*, **108**, 323-329.
26. Lenschow, D. (1998) Observations of clear and cloud-capped convective boundary layers and techniques for probing them. See this volume.
27. Lenschow, D. H. and Stephens, P. L. (1980) The role of thermals in the convective boundary layer, *Bound.-Layer Meteor.*, **19**, 509-532.
28. Lenschow, D. H., Wyngaard, J. C., and Pennel, W. T. (1980) Mean-field and second-momentum budgets in a baroclinic, convective boundary layer, *J. Atmos. Sci.*, **37**, 1313-1326.
29. Lomas, C. G. (1986) *Fundamentals of Hot Wire Anemometry*, Cambridge University Press.
30. Mahrt, L. and Paumier, J. (1984) Heat transport in the atmospheric boundary layer, *J. Atmos. Sci.*, **41**, 3061-3075.
31. Meroney, R. N. (1998) Wind tunnel simulation of convective boundary layer phenomena. Simulation criteria and operating ranges of laboratory facilities. See this volume.
32. Meroney, R. N. and Melbourne, W. H. (1992) Operating ranges of meteorological wind tunnels for the simulation of convective boundary layer (CBL) phenomena, *Bound.-Layer Meteor.*, **61**, 145-174.
33. Mestayer, P. (1982) Local isotropy and anisotropy in a high-Reynolds-number turbulent boundary layer, *J. Fluid. Mech.*, **125**, 475-503.
34. Nelson, E., Stull, R., and Eloranta, E. (1989) A prognostic relationship for entrainment zone thickness, *J. Appl. Meteorol.*, **28**, 885-903.
35. Obukhov, A. M. (1949) Structure of the temperature field in turbulent flows, *Izvestiya Akad. Nauk S.S.S.R., Ser. Geogr. i Geofiz.*, **13**, 58-69.
36. Ogawa, Y., Diosey, P. G., Uehara, K., and Ueda, H. (1981) A wind tunnel for studying the effects of thermal stratification in the atmosphere, *Atmos. Environ.*, **15**, 807-821.
37. Panofsky, H. A., and Dutton, J. A. (1984) *Atmospheric turbulence*, John Wiley & Sons.
38. Paranthoen, P., Petit, C., and Lecordier, J. C. (1982) The effect of the thermal prong-wire interaction on the response of a cold wire in gaseous flows (air, argon and helium), *J. Fluid. Mech.*, **124**, 457-473.
39. Perry, A. E. (1982) *Hot-Wire Anemometry*, Clarendon Press, Oxford.
40. Plate, E. J. (1998) Convective boundary layer: a historical introduction. See this volume.
41. Poreh, M. and Cermak, J. (1984) Wind tunnel simulation of diffusion in a convective boundary layer, *Bound.-Layer Meteor.*, **30**, 431-455.
42. Poreh, M., Rau, M., and Plate, E. (1991) Design considerations for wind tunnel simulations of diffusion within the convective boundary layer, *Atmos. Environ.*, **25A**, 1250-1257.
43. Rau, M., Bächlin, W., and Plate, E. (1991) Detailed design features of a new wind tunnel for studying the effects of thermal stratification, *Atmos. Environ.*, **25A**, 1258-1263.
44. Rau, M. and Plate, E. (1995) Wind tunnel modelling of convective boundary layers, in J. E. Cermak, A. G. Davenport, E. J. Plate and D. X. Viegas (eds.), *Wind Climate in Cities*, Kluwer, pp. 431-456.
45. Rey, C., Schon, J. P., and Mathieu, J. (1979) Buoyancy effects in a wind tunnel simulation of the atmospheric boundary layer, *Phys. Fluids*, **22**, 1020-1028.
46. Ruck, B. (1987) *Laser-Doppler-Anemometrie*, AT-Fachverlag.
47. Sandborn, V. A. (1972) *Resistance Temperature Transducers*. Metrology Press, Fort Collins, Colorado.
48. Schmidt, H. and Schumann, U. (1989) Coherent structure of the convective boundary layer derived from large-eddy simulations. *J. Fluid. Mech.*, **200**, 511-562.
49. Shay, T. J. and Gregg, M. C. (1986) Convectively driven turbulent mixing in the upper ocean, *J. Phys. Oceanogr.*, **16**, 1777-1798.
50. Sorbjan, Z. (1991) Evaluation of local similarity functions in the convective boundary layer, *J. Appl. Meteor.*, **30**, 1565-1583.
51. Strickert, H. (1974) *Hitzdraht- und Hitzfilmanemometrie*, VEB Verlag Technik, Berlin.
52. Stull, R. B. (1988) *An Introduction to Boundary Layer Meteorology*, Kluwer Academic Publishers, Dordrecht.
53. Tschen, C. M. (1953) On the spectrum of energy in turbulent shear flow, *J. Res. NBS*, **50**, 51-62.
54. Tennekes, H., and Lumley, J. L. (1982) *A First Course in Turbulence*, The MIT Press.
55. Wiedemann, J. (1984) *Laser-Doppler-Anemometrie*, Springer-Verlag, Berlin.
56. Willis, G. E. and Deardorff, J. W. (1974) A laboratory model of the unstable planetary boundary layer, *J. Atmos. Sci.*, **31**, 1297-1307.
57. Willis, G. E. and Deardorff, J. W. (1976) On the use of Taylor's translation hypothesis for diffusion in the mixed layer, *Quart. J. Roy. Meteor. Soc.*, **102**, 817-822.
58. Wyngaard, J. C. (1992) Atmospheric turbulence, *Ann. Rev. Fluid Mech.*, **24**, 205-233.

59. Young, G. S. (1988) Turbulence structure of the convective boundary layer, *J. Atmos. Sci.*, **45**, 727-735.
60. Zilitinkevich, S. S. (1991) *Turbulent Penetrative Convection*, Avebury Technical, Aldershot.

REVIEW OF DIFFUSION PROCESSES IN THE CONVECTIVE BOUNDARY LAYER

F.T.M. NIEUWSTADT
J.M. Burgers Centre, Delft University of Technology
Rotterdamseweg 145, 2628 AL Delft, the Netherlands

1. Introduction

The Convective Boundary Layer (CBL) is one of the prototypes of the atmospheric boundary layer over a homogeneous surface, the other being the neutral and stable boundary layer. For a review of the meteorological characteristics of the three prototypes we refer to Nieuwstadt and Duynkerke [10]. Here we shall concentrate on the CBL. Convection or rather the term free convection is used to describe the process whereby turbulence is produced by buoyancy. In the atmospheric boundary layer, buoyant air generally results from the heating of the atmosphere at the surface which in case of a land surface is heated by solar radiation. Therefore the CBL is frequently identified with sunny and clear meteorological conditions. However, even small temperature variations are already sufficient to cause buoyant air motions. So the CBL is not restricted to only clear and sunny skies.

The turbulence structure of the CBL is well established as a result of atmospheric and laboratory experiments but also as a result of numerical simulation (see e.g. Wyngaard [15]). As argued above convection results from air heated at the surface and due to buoyancy this warm air starts to rise. By a process of self-organization the rising air is collected in large-scale columns of vertically rising air also known as thermals. Slowly descending air motions, called downdrafts, surround these thermals. The combination thermals and downdrafts forms a circulation system characteristic for almost all free convection flows above a horizontal surface.

The driving force of free convection is the heat flux, Q_o [W/m^2], at the surface which can be expressed as

$$Q_o = \rho c_p < w'\theta' >_o \tag{1}$$

E.J. Plate et al. (eds.), Buoyant Convection in Geophysical Flows, 371–399.

where ρ and c_p are the density and specific heat at constant pressure, respectively. The $<w'\theta'>_o$ is called the mean temperature flux at the surface where the angular brackets denote an appropriately defined ensemble average, a notation which we shall use in the following. The temperature flux should be interpreted as the mean correlation between the temperature, θ', and vertical velocity, w', fluctuations through which heat is transported by turbulence in the CBL. Note for instance that both for a thermal and downdraft $w'\theta' > 0$, i.e. heat is transported upwards, away from the surface. Another parameter which plays an important role in the CBL is the boundary-layer height h. One interpretation of h is the vertical distance to which a thermal rises on the average. Namely, the vertical motion of thermal is obstructed by the stable free atmosphere which forms a temperature inversion, a free surface so to speak, on top the CBL. Therefore, the boundary-layer height is sometimes also denoted as z_i (the height of the inversion).

With help of $<w'\theta'>_o$ and h we next introduce a velocity defined as

$$w_* = \left(\frac{g}{T_o} <w'\theta'>_o h\right)^{1/3}, \tag{2}$$

where g/T_o is called the buoyancy parameter. It is the acceleration per degree of temperature difference of a fluid particle with its surroundings. The w_* is known as the convective velocity scale and can be used to scale the turbulent velocities in a CBL. It is for instance representative for the vertical velocities within a thermal.

The w_*, $<w'\theta'>_o$ and h can be also used to scale other variables in the CBL and this is known as convective or mixed-layer scaling. For more details and results on the use of this scaling we refer again to Nieuwstadt and Duynkerke [10] and also to Wyngaard [15]. Here, we shall use mixed-layer scaling to express the results from our dispersion simulations without further explanation.

One point, however, remains to be discussed and that is how the strength of convection can be quantified. For this we must realize that turbulence in the atmospheric boundary layer can be also produced by other processes than convection. The most common alternative is surface friction which can be characterized by the surface shear stress τ_o. Based on this surface shear stress we can define the friction velocity u_* as

$$u_* = \sqrt{\frac{\tau_o}{\rho}}. \tag{3}$$

Like w_*, the u_* can be used to scale turbulence motions but in this case the turbulence produced by surface friction. The ratio w_*/u_* can now be interpreted as a measure of the strength of convection in the CBL (in comparison with the shear production). When $w_*/u_* \to \infty$ the convection is strong and weak convection results for $w_*/u_* \to 0$.

The objective of the present paper is to consider dispersion of pollutants in the atmospheric boundary layer. The three prototypes of the boundary layer mentioned above have each their own dispersion characteristics and for some discussion we refer to Nieuwstadt and Duynkerke [10], Nieuwstadt and Meeder [11] and the references mentioned therein. To describe dispersion in the atmospheric boundary layer, we may in principle use the same scaling approach in terms of characteristic parameters as we have introduced above. Here, we mainly restrict ourselves to a discussion of dispersion and diffusion processes in the CBL. For this we shall primarily make use of the results obtained by numerical simulation of the atmospheric boundary layer. Namely, the details that can be obtained from such simulations facilitate the interpretation of the results and their explanation in terms of physical processes. However, where possible, we will compare the numerical results with data obtained from atmospheric and laboratory experiments.

This review is organized as follows. First, we summarize in the next section, the main points of numerical simulation of atmospheric turbulence with emphasis on the specific problems that are connected to the simulation of dispersion. The section thereafter is devoted to a review of diffusion and dispersion results. First, we consider the results for the mean concentration and, second, the results for the concentration fluctuations.

2. Large-Eddy Simulation of Atmospheric Turbulence

The strong rise of computer power over the last decades has made it feasible to perform so-called numerical simulation of turbulence. This means that the governing equations for a turbulent flow, i.e. the conservation equations for mass, momentum and energy, are discretized by numerical methods. The resulting algebraic equations are then solved with help of a computer. When performing such simulation, one immediately encounters the limitation set by the resolution where resolution is defined here as the ratio of the largest to the smallest flow scale that one wants to simulate. It is an established fact of turbulence theory that this ratio is proportional to the Reynolds number[1] of the flow. For atmospheric turbulence the value of the characteristic Reynolds number is very large, say $O(10^9)$. For a simulation in

[1]The Reynolds number is formally defined as

$$Re = \frac{u\ell}{\nu}$$

where ν is the kinematic viscosity and where ℓ and u are the relevant scales for the largest scales of turbulent motion (for the CBL these would be w_* and h, respectively). It then which all flow scales are to be resolved, this would require a computer that is far beyond any present-day system or even any machine of the foreseeable future. Therefore, another technique is needed to simulate atmospheric turbulence. This is Large-Eddy Simulation (LES).

In a LES only those flow scales are numerically simulated of which the resolution can be handled by the computer that we use. These are the largest flow scales and in principle the ratio of the largest to smallest "large" scale that can be resolved, is given by the ratio of the flow domain to the grid size that is used. Let us say that twice the grid size is the smallest scale that we can resolve. All smaller flow scales can thus not be computed and these are therefore denoted as subgrid scales.

To obtain equations for the resolved scales in which the subgrid scales are removed, we apply a filter operation to the equations of motion. This is usually a spatial filter where in view of the requirement for the smallest resolved scales mentioned above, the filter length is determined by the grid size. The effect of the subgrid scale on the flow can not be removed by the filter operation. Namely, the small scales are essential for the dynamics of turbulence and their effect on the resolved scales appears as an additional term in the equations for the resolved motions. For instance, in the equations for the resolved motions a so-called subgrid-stress term appears. An additional model, the subgrid model, must be prescribed before the equations for the large-scales can be solved. The main premise of LES is that the details of this subgrid model are not very important for the simulation of the large-scale motions and, therefore, possible errors in the subgrid modelling are of no consequence.

The discussion presented above gives only a qualitative introduction of the principles of LES. For more quantitative details we refer to the literature, in particular Galperin and Orzag [4] for a general review of LES and to Mason [8] for a review of the atmospheric applications of LES.

LES is particularly well suited to simulate the CBL because this type of boundary layer is dominated by large-scale flow motions, the thermals and the downdrafts. In a comparison study between several LES codes (Nieuwstadt *et al.* [9]) it was shown that LES of the CBL is quite robust with respect to changes in the numerics and the subgrid-model details. This is in agreement the premise of LES mentioned above. The LES data were also found to agree quite reasonably with laboratory and atmospheric data. An example of such agreement is given in Fig. 1 where profiles of the horizontal and vertical velocity fluctuations, σ_u and σ_w, in a CBL are shown in terms of mixed-layer scaling. It is clear that the agreement between the LES and other data is quite good for the vertical velocity fluctuations, σ_w.

follows from turbulence theory that $Re^{3/4}$ can be interpreted as the ratio of the largest scale of the flow determined by inertia to the smallest scale determined by viscosity.

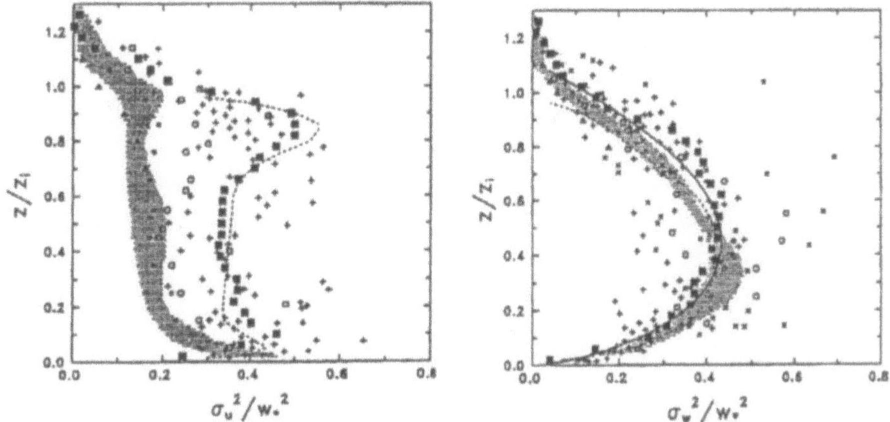

Figure 1. The rms of horizontal and vertical velocity fluctuations, σ_u (left) and σ_w (right), in a CBL in terms of mixed-layer scaling; the dotted band denotes the range of LES data, the symbols denote data from laboratory and atmospheric experiments and the various lines are empirical expressions fitted to these data (Hibberd and Sawford [5]).

The difference between experiments and simulation data for the horizontal fluctuations, σ_u, is believed to be caused by large-scale horizontal motions which are not resolved by the LES. Nevertheless, it is generally accepted that the LES of the convective boundary layer produces a quite realistic turbulent flow field. This will be our point of departure to discuss the simulation of dispersion processes.

3. Numerical Simulation of Dispersion Processes

Given the simulated turbulent velocity field as function of space and time, the next step is to extend the simulation with the dispersion and diffusion processes. For this we introduce a passive[2] contaminant with a concentration c. The concentration satisfies a conservation law given by

$$\frac{\partial c}{\partial t} = -u_j \frac{\partial c}{\partial x_j} + \kappa \frac{\partial^2 c}{\partial x_j^2}, \tag{4}$$

where u_j is the instantaneous velocity field and κ the molecular diffusion coefficient. We have written (4) in a form that the change of c as a function of time can be attributed to the two processes on the right-hand side of the equation. The first term on this right-hand side is the transport of the contaminant by the instantaneous turbulent velocity field and this process

[2]Passive in this context means that the contaminant does not influence the turbulent velocity field so that the simulation of the velocity and concentration field is decoupled.

is called dispersion. The second term is caused by molecular exchange of concentration and this is generally called diffusion.

Both processes mentioned above in connection with (4) involve all scales of the velocity and concentration field. In our LES only the large flow scales are considered and the large-scale velocity field is given by $\overline{u}_j$. The overbar denotes here the filter operation by which the subgrid scales are removed from the flow field. To bring (4) in an equivalent form so that it describes only the large scales, we must apply to it the filter operation. The result reads

$$\frac{\partial \overline{c}}{\partial t} = -\overline{u}_j \frac{\partial \overline{c}}{\partial x_j} + \frac{\partial f_j}{\partial x_j}, \tag{5}$$

where the f_j is given by

$$f_j = -\left(\overline{cu_j} - \overline{c}\,\overline{u}_j\right). \tag{6}$$

The first term on the right-hand side of (5) is the transport of $\overline{c}$ by the resolved large scales of flow motion and this term can thus be explicitly computed by our simulation. The second term describes the transport by the subgrid scales. It involves the filter applied on the product of c and u_j which is in general not equal to the product of the filtered c and u_j. The value of $\overline{cu_j}$ depends on how c and u_j are correlated with each other at the small scales which, within the context of our LES, we do not know. This is in principle the same problem which we encounter when we perform Reynolds averaging to obtain the equations of motion for the mean flow. The correlation $\overline{cu_j}$ is thus a new unknown in our problem and we need a model to express it in term of known quantities. This is what we have called the subgrid model.

We also observe that the molecular diffusion term has been omitted from (5). The background of this simplification is the fact that within an atmospheric LES the Reynolds number based on the smallest resolved scale is still very large[3] and this implies that molecular effects are negligible even on the smallest resolved scales. In other words, the subgrid term takes over the role of the diffusion term in (4).

Before we can consider the numerical methods to solve the (5), we will first consider the requirements that such a solution method must satisfy. The numerical method to solve $\overline{c}$ should have the following desirable properties:

1. *Conservation of mass*
   No mass of the contaminant c must either be added or removed from

[3]The smallest grid size in an atmospheric LES is usually $O(10\text{m})$.

the computational domain by numerical error. Otherwise, the computed value of $\bar{c}$ can not be interpreted as a concentration value due to sources explicitly introduced in our problem.

2. *Non-diffusive and non-dispersive*
   In order compute the dispersion and diffusion of the contaminant as accurately as possible it is clear that the numerical scheme should not introduce any additional diffusion and dispersion.
3. *Positive definite*
   By definition a concentration is always positive. In some case this requirement is essential to keep the computation stable. An example is the simulation of dispersion in connection with chemical reactions because the chemical rate equations require the concentration to be positive for a realizable reaction.

Given these requirements there are two methods to solve the equation for the resolved concentration $\bar{c}$. These are the *Eulerian* and *Lagrangian* method.

3.1. EULERIAN METHOD

In this method, we solve for the concentration $\bar{c}$ as a function of time and space in the computational domain. The main advantage of using this method is that in this case it simple to combine the simulation of the dispersion with other processes such as chemical reactions, plume rise, etc. A disadvantage is that in general it is far from trivial to satisfy the three requirements that we have given above. Conservation of mass can only be satisfied exactly if the equation for $\bar{c}$ is discretized in a so-called conservative form by applying a finite volume method. The second requirement depends on the details of the numerical scheme that is applied. In general first-order accurate numerical schemes are highly diffusive so that they have to be ruled out whereas second-order (linear) schemes[4] are in general dispersive and thus also not applicable without modification. Therefore, special (non-linear) schemes have to be developed to solve (5). The third requirement can be usually related to the dispersive property of a numerical scheme.

The consequence is that the Eulerian method requires complex numerical schemes in order to give a reasonably accurate simulation of the concentration field. Nevertheless, this method has been applied to a number of dispersion problem with success and we refer to the following references for some examples: Wyngaard and Brost [16], Haren and Nieuwstadt [5], Schumann [13], Sykes and Henn [14].

A typical computational geometry for an Eulerian simulation is shown in Fig. 2. The LES for the velocity field is carried out in a box as shown.

[4]It can be proven that all linear second-order schemes are dispersive.

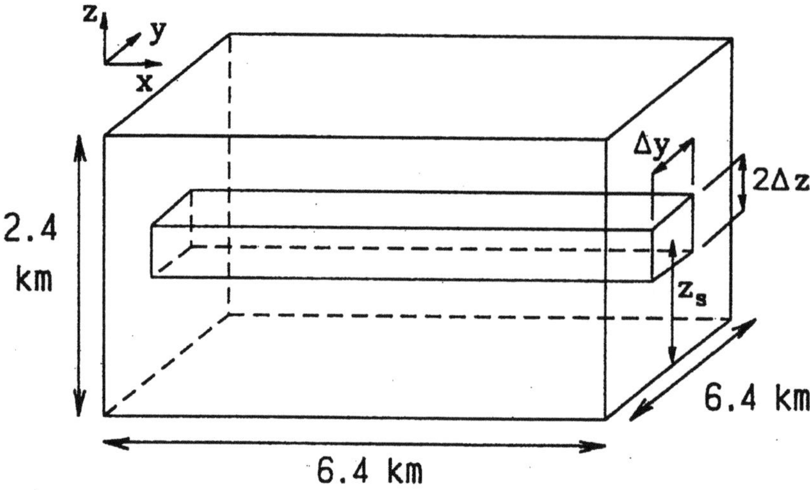

Figure 2. Computational domain for the simulation of a line source using the Eulerian method.

When the simulation has resulted in a fully developed state of convective turbulence, we introduce a line source as indicated in the figure by setting the concentration equal to some value, say c_o, at the grid points which form the line source. Simultaneously with the LES of the velocity field, we compute the development of the line source by solving (5) as a function of time and space. The result is a (large-scale) concentration distribution from which the statistics of the dispersion process can be computed.

Although an instantaneous line source seems to be at first sight a rather artificial source geometry, it can nevertheless be related to the more practical dispersion case of a continuous point source. This can be shown by applying Taylor's translation hypothesis. This hypothesis states that under the assumption that the streamwise dispersion is negligible with respect to the transport by a mean advection velocity U, the dispersion of the line source as function of t can be interpreted as the dispersion of a continuous point source as a function of downstream distance x. The distance x and time t are then related by the transformation

$$x = Ut \tag{7}$$

where the translation or advection velocity U should be independent of time and space coordinates.

We end this discussion on the Eulerian method with a few remarks on the subgrid model for f_j that is needed to solve (5). One usually adopts a model that is closely related to the model for the subgrid stresses. One of the most commonly used models is based on the gradient transfer hypothesis where the subgrid term is expressed in terms of the gradient of the resolved quantity. This model which can be formulated in various forms, has lead to quite acceptable results when applied to the CBL (see e.g. Nieuwstadt *et al.* [9]). For the subgrid diffusion term the gradient transfer hypothesis leads to

$$f_j = K_c \frac{\partial \bar{c}}{\partial x_j}, \tag{8}$$

where the exchange coefficient K_c is related to the grid size and the resolved velocity field.

3.2. LAGRANGIAN METHOD

This method can be best explained if we rewrite (5) in the following form

$$\frac{D\bar{c}}{Dt} = \frac{\partial f_j}{\partial x_j}.$$

The left-hand side of this equation can be interpreted as the change of the concentration when travelling with a fluid particle that moves with the resolved flow. The equation then implies that this concentration change is only due to the subgrid diffusion which the particles encounters on its way through the flow. In view of this interpretation it seems natural to compute the dispersion process by following fluid particles (with a given concentration) through the flow. This is generally known as the Lagrangian method.

The advantage of this method is that it satisfies the three requirements stated above almost trivially. The disadvantage is that it seems only suited to describe the dispersion of passive contaminants and its application with active contaminants in which other processes play a role, leads in general to complications. Nevertheless it has been used with success in a number of dispersion studies with as examples: Lamb [6] and Mason [7]. In the following section we present some new results on dispersion in the CBL that we have obtained with this method.

A typical geometry for a Lagrangian simulation of dispersion is illustrated in Fig. 3. Again we show the computational box in which the LES (of e.g. the CBL) is carried out by solving the filtered equations of motion for the velocity field. Let us assume that the turbulence is again fully developed. At the time $t = t_o$ we introduce a particle pair as shown in Fig. 3 with a vertical separation equal to $2\Delta z_o$ and centered around the release height z_s. The particle trajectories are then computed with a simple explicit Euler

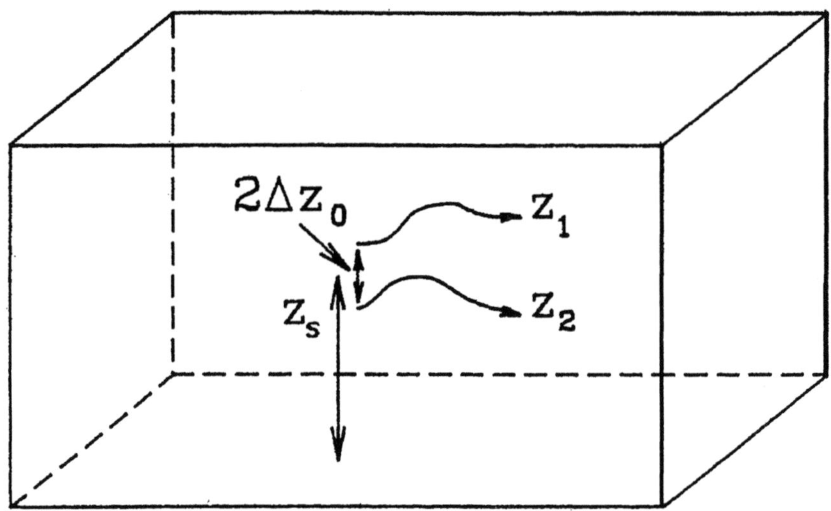

Figure 3. Computational domain for the simulation of particle trajectories used in the Lagrangian method of simulating dispersion.

scheme as a function of time[5]. At each release height we introduce 1500 particle pairs which are randomly distributed over the horizontal plane at $z = z_s$. Whenever a particle leaves the computational domain through the horizontal walls, the velocity field is extended to outside the domain by using the periodic lateral boundary conditions. The horizontal distance that a particle can travel is thus in principle unlimited.

The subgrid model for the Lagrangian method is not as well established as for the Eulerian method. Therefore, we shall discuss the model that we have used for the subgrid contribution to the particle trajectories in some detail. We assume that the contribution by subgrid motions can be described by a random walk process. This implies that the subgrid contribution to the particle trajectory is given by randomly distributed and uncorrelated displacements where the displacements have been chosen to satisfy a given diffusion coefficient. In an Eulerian terminology this is equivalent to the gradient diffusion process given by (8). Without further discussion we give here the resulting expression for the displacement that we have added to the particle trajectory computed for the resolved motion. It reads

$$\Delta x_i^{(\mathrm{sub})} = \frac{dK_c}{dx_i}\, dt + (2K_c)\, d\omega_t, \tag{9}$$

[5]We have also experimented with second-order numerical schemes, e.g. the predictor-corrector scheme, but no differences could be observed in the computed particle statistics.

where K_c is the subgrid diffusion coefficient which we have already encountered in (8) and $d\omega_t$ is the so-called Wiener process which is a random process with the following properties

$$\begin{aligned} < d\omega_t > &= 0 \\ < d\omega_t^2 > &= dt. \end{aligned}$$

4. Dispersion in the Convective Boundary Layer

The fact that the CBL consists of large-scale structures such as thermals and downdrafts, has a large influence on the dispersion characteristics of this boundary layer. In the thermals we have a strong upward velocity which is compensated by the downward motion in the downdrafts which are in general weaker. Continuity demands that the average velocity, e.g. averaged over a horizontal plane, is zero. The strong upward motions must thus be compensated by a larger number of weak downward motions. The consequence is that the probability density function of the vertical velocity fluctuations is positively skewed. The effect of this positive skewness will become apparent in the following subsections where we will discuss the general characteristics of dispersion in the CBL.

In our discussion we distinguish between the mean concentration and the concentration fluctuations which each have their own characteristics. In addition we shall subdivide our discussion of the results in those obtained with the Eulerian method and those obtained with the Lagrangian method.

4.1. MEAN CONCENTRATION

4.1.1. *Eulerian Method*

In the Eulerian method we compute the dispersion by solving the equations for the resolved concentration distribution $\bar{c}$ as a function of space (x, y, z), and time t. An example of a simulation result is shown in Fig. 4. Here, we present the instantaneous concentration patterns on a vertical plane resulting from a line source (see Fig. 2) released at $t = 0$. To facilitate interpretation of the results we show in the same figure an isoline plot of the vertical velocity fluctuations in which the large scale thermal motions can be clearly recognized.

Figure 4 shows that the concentration patterns are strongly influenced by the thermals, i.e. within a thermal the contaminant rises directly to the top of the boundary layer whereas in the surrounding downdrafts the concentration patterns move toward the surface. At the latest time instant shown in Fig. 4, we see that the concentration patterns fill the whole boundary layer with parts of the line source impinging on the surface and parts

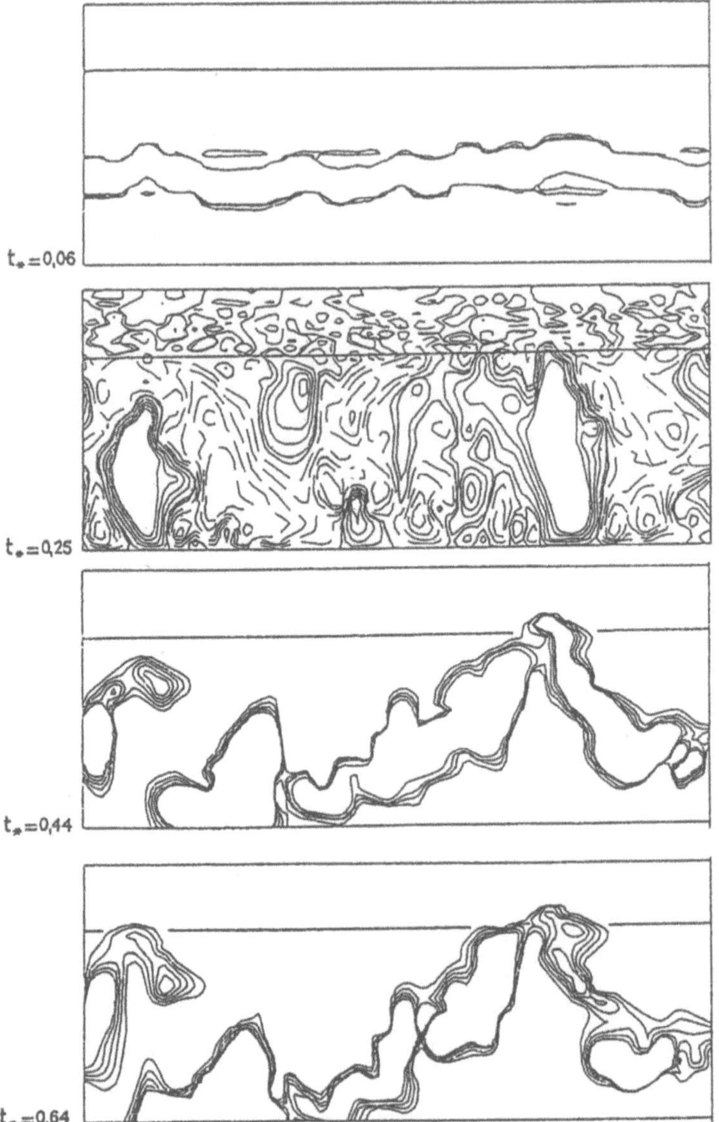

Figure 4. Isoline plot on a vertical plane through the computational domain of the instantaneous resolved concentration distribution of a line source at various times $t/t_* \equiv X = xw_*/(Uh)$ after the release of the line source; the second frame gives the isoline plot of the vertical velocity fluctuations where the dashed isolines mean negative or downward velocities and solid isolines positive or upward velocities, the horizontal line denotes the mean boundary-layer height.

near the top of the boundary layer. In dispersion terminology this is called looping plume behaviour and it can be recognized when one observes plume dispersion in the atmosphere during convective conditions.

With help of the simulation results we can compute the statistics of the concentration field. One of these statistics is the mean concentration which we shall denote as $<c>$. The angular brackets imply again an appropriate ensemble average. In the case of our LES with periodic horizontal boundary conditions, an ensemble average can be computed by taking a horizontal average over horizontal plane at height z. Therefore the mean concentration is only a function of height z.

In Fig. 4 we have seen that the line source behavior during dispersion is strongly influenced by the presence of large structures in the CBL. It seems clear that this influence will also appear in the behavior of the mean concentration statistics. This can be demonstrated by performing a LES of two dispersion problems which have been suggested by Wyngaard and Brost [16]. They propose to simulate the dispersion for the following source configurations. The first is an area source, i.e. a source which is uniformly distributed across a horizontal plane, at the surface and the second an area source at the top of the boundary layer. These two cases are denoted as top-down and bottom-up dispersion. Wyngaard and Brost [16]) found that the dispersion characteristics for both cases are very different. This can perhaps be best illustrates by considering the diffusion coefficient K which is defined as

$$K = -\frac{<\overline{w}'\overline{c}'>}{d<c>/dz}, \tag{10}$$

where $<\overline{w}'\overline{c}'>$ is the vertical (resolved) concentration flux due to turbulence and $d<c>/dz$ the vertical gradient of the mean concentration[6].

The results for K for the top-down case, denoted as K_t, and for the bottom-up case as K_b are shown in Fig. 5. We note that the values for K_t and K_b are clearly different with $K_b > K_t$. This implies that in a CBL the turbulent diffusion coefficient is dependent on the source location. This is in contradiction with the concept of gradient diffusion such as given by (10). Namely, for gradient transfer to be valid the exchange coefficient can only be a function of local conditions and this is clearly not the case here. An important conclusion is therefore that gradient transfer can not be applied to dispersion in the CBL. The physical reason for this result is the presence of the large-scale boundary-layer filling motions which as we have seen influence dispersion directly.

The strong effect of the large-scale motions on dispersion in the CBL can be illustrated by another example. Let us consider a line source at two source heights z_s, i.e. near the surface and at mid-boundary layer (see

[6]Note that by definition $<c>\equiv<\overline{c}>$

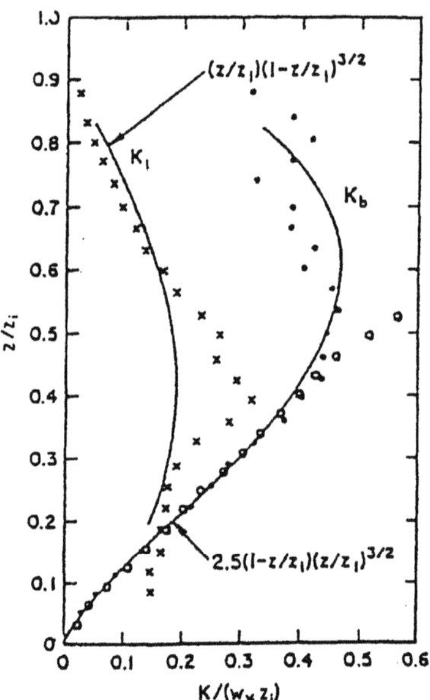

Figure 5. Diffusion coefficients for the top-down dispersion case (K_t) and for the bottom-up dispersion case (K_b) computed from a LES of these two dispersion cases in the CBL (Schumann [14]).

Fig. 2). The mean concentration patterns resulting from these two sources are shown in Fig. 6 where the data were obtained both from laboratory data and LES. We find that the experimental and computational data agree well. The figure also illustrates that the dispersion from a surface source leads to a completely different behaviour than the dispersion from an elevated source in the middle of the boundary layer. For the surface source the concentration rises directly to the top of the boundary layer and forms there an elevated maximum. On the other hand for the elevated source the plume seems to descend and impinge on the surface.

Both, perhaps rather surprising results, can be explained in terms of the presence of thermals in the CBL. For the surface source the thermals transport the contaminant very efficiently and quickly to the top of the boundary layer. The behavior of the elevated source can be explained in terms of the skewness of the probability density which, as we have seen, is also the consequence of the thermals. This skewness implies that a vertical downward velocity has a higher probability than an upward velocity. The result is a (on the average) downward moving plume.

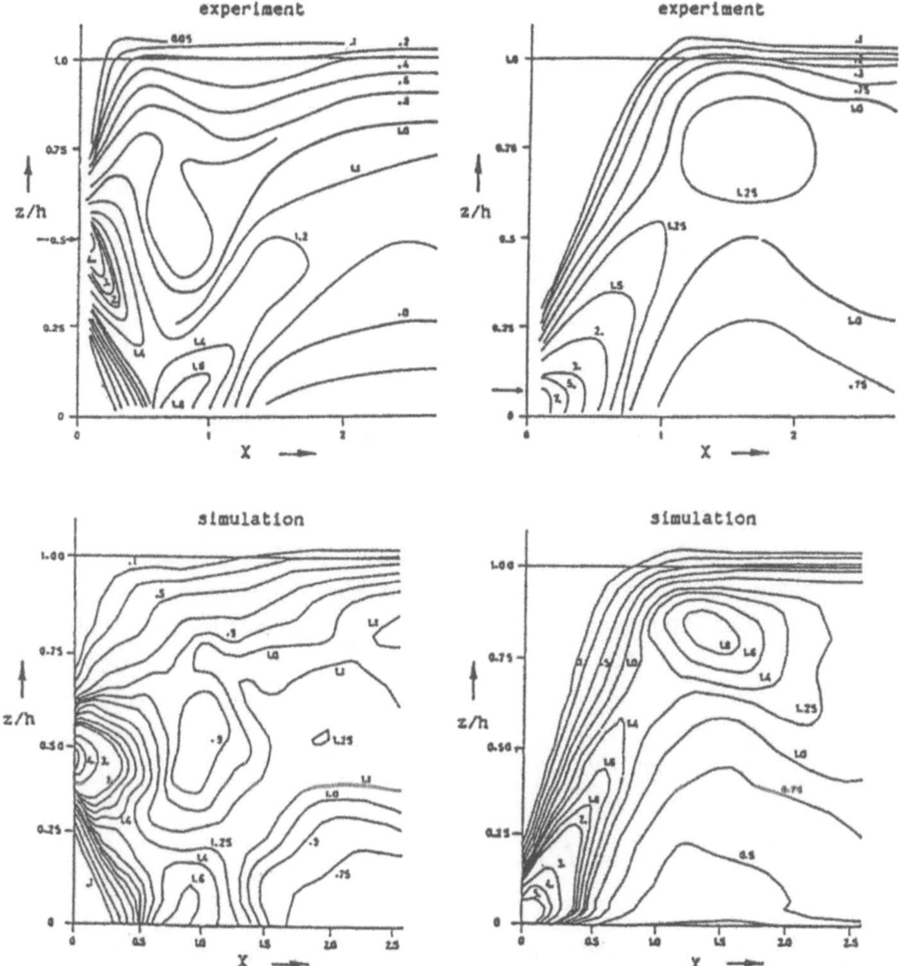

Figure 6. Mean concentration patterns for an instantaneous line source as a function of t/t_* for a near-surface source (left figures) and an elevated source (right figures) in the middle of the boundary layer; the two upper figures have been obtained from laboratory data and the two lower figures have been obtained from LES.

4.1.2. *Lagrangian Method*

In terms of a Lagrangian method, the mean concentration can be obtained from so-called single-particle statistics. To show this, let us consider a single particle released at time t_o from position $\boldsymbol{x}_s \equiv (x_s, y_s, z_s)$. The probability that a particle arrives at a position $\boldsymbol{x}$ at time t is given by $p(\boldsymbol{x}, t; z_s, t_o)$[7].

[7]Because of horizontal homogeneity of the turbulence field the probability density will depend only on $x - x_s$ or $y - y_s$.

Let now the initial distribution of the particle concentration be given by $c_o(\boldsymbol{x}_s)$, it then follows that the concentration at a later time is given by

$$<c(\boldsymbol{x},t)>=\int\int\int p\,(\boldsymbol{x},t;z_s,t_o)\,c_o(\boldsymbol{x}_s)\,d\boldsymbol{x}_s. \tag{11}$$

In the case of a point source with unit strength at $\boldsymbol{x}_o$ the initial distribution is given by $c_o(\boldsymbol{x}_s) = \delta\,(\boldsymbol{x}_s - \boldsymbol{x}_o)$ where δ is the Dirac delta function. Substitution into (11) then leads to

$$<c(\boldsymbol{x},t)>=p\,(\boldsymbol{x},t;z_o,t_o) \tag{12}$$

which proves that the mean concentration is equivalent to the single-particle probability density. The $p\,(\boldsymbol{x},t;z_s,t_o)$ can be computed by releasing a large number of particles and sampling at each time after the release the number of particles in a given volume around $\boldsymbol{x}$.

Here, we shall not consider the probability density determined by the Lagrangian method directly because we have already considered the mean concentration distribution in the section on the Eulerian method. Instead we shall consider the moments of the concentration distributions which can be conveniently computed from the particle trajectories by computing the appropriate moment of each particle trajectory and the averaging over all particle trajectories.

Let us first consider the first moment given by

$$<z>\,(z_s,t) = \int\int\int zp\,(\boldsymbol{x},t;z_s)\;d\boldsymbol{x} = \frac{1}{N_p}\sum_{j=1}^{j=N_p} z_j \tag{13}$$

where the integration should be extended over the whole computation domain in the horizontal and over the boundary-layer depth ($0 < z < h$). The z_j denotes the height of the jth particle and N_p the number of particles.

In Fig. 7 we show some results for $<z>$ for various source heights in terms of mixed-layer scaling where t_* is defined as h/w_*. In the left-hand frame of Fig. 7 we show the LES results with and without subgrid effects. It is clear that the influence of the subgrid model is negligible or in other words the mean height is primarily determined by the large-scale resolved eddies. In the following, we shall see that this remark applies to all single-particle statistics. In the right-hand frame of Fig. 7 we show the LES results of $<z>$ together with some experimental data denoted by symbols.

The results shown in Fig. 7 can be interpreted as follows. Initially the mean height $<z>$ is equal to the source height, z_s. At large diffusion times, i.e. $t/t_* \to \infty$ when the concentration is uniformly distributed across the boundary layer the value of $<z>$ approaches $0.5h$ which can be readily

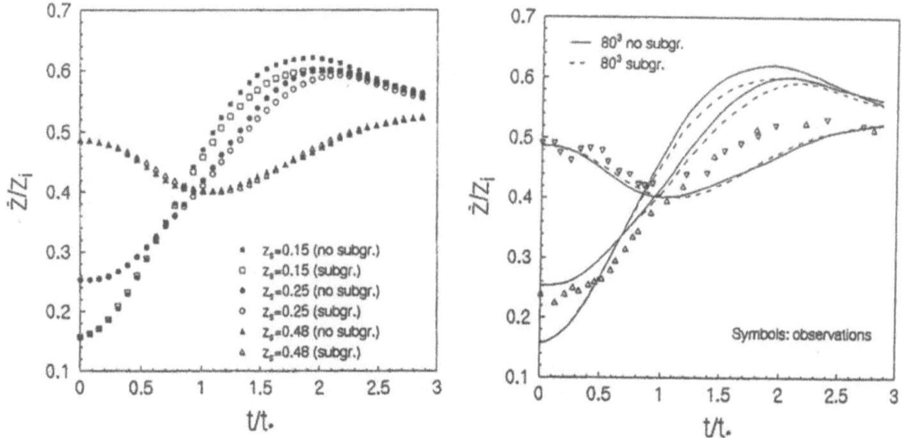

Figure 7. The mean height of particles released from the sources heights $z_s = 0.15, 0.25$ and $0.5h$ as a function of non-dimensional diffusion time; left-hand frame: the closed symbols are the trajectory computations without subgrid contributions and the closed symbols are the trajectory computations with subgrid contribution; right-hand frame: lines are the LES results (with and without subgrid contribution) and the symbols are observations.

derived from (13) by substituting a uniform probability density distribution $p(\boldsymbol{x}, t) = 1/h$. For the low source height, the z_s increases strongly as a function of diffusion time and overshoots its asymptotic value of $0.5h$. This behavior is equivalent with the strong rise of the surface plume which we have observed in Fig. 6. For the elevated source at mid-boundary layer the mean height first decreases below the asymptotic value which is again equivalent to the downward motion of the elevated plume shown in Fig. 6.

Next we consider the second moment of the probability or concentration distribution which is a measure for the spread of the particles with respect to their initial position. For that reason the second moment is also denoted as the dispersion parameter. For the horizontal and vertical direction, the dispersion parameters are denoted as σ_y and σ_z respectively and they are defined as

$$\sigma_y^2 = \int\int\int y^2 p(\boldsymbol{x}, t)\, d\boldsymbol{x} = \frac{1}{N_p} \sum_{j=1}^{j=N_p} y_j^2, \tag{14}$$

$$\sigma_z^2 = \int\int\int (z - <z>)^2 p(\boldsymbol{x}, t)\, d\boldsymbol{x} = \frac{1}{N_p} \sum_{j=1}^{j=N_p} \left(z_j^2 - <z>\right)^2, \tag{15}$$

where without any loss of generality the $< y >$ has been assumed to be equal to zero.

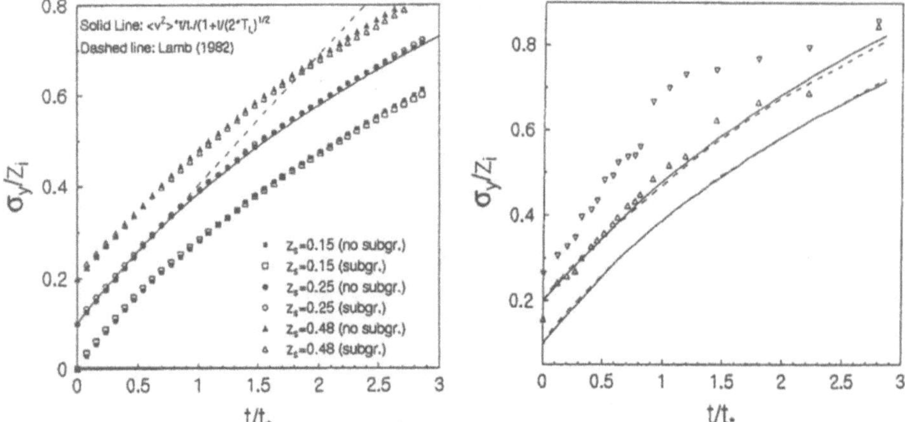

Figure 8. The horizontal dispersion parameter of particles released from the sources heights $z_s = 0.15, 0.25$ and $0.5h$ as a function of non-dimensional diffusion time (the results for the three source heights are shown displaced vertically by a distance $0.1h$ in order to avoid overlap); left-hand frame: the closed symbols are the trajectory computations without subgrid contributions and the closed symbols are the trajectory computations with subgrid contribution, the solid curve is the equation discussed in the text; right-hand frame: lines are the LES results (with and without subgrid contribution) and the symbols are observations.

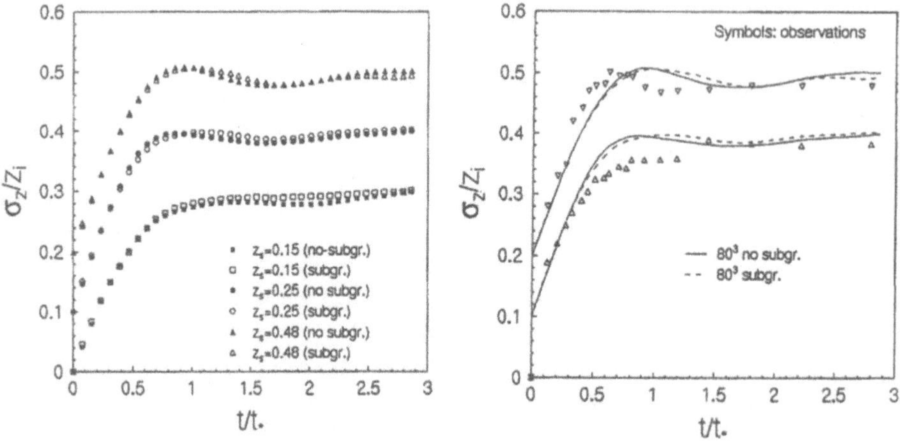

Figure 9. The vertical dispersion parameter of particles released from the sources heights $z_s = 0.15, 0.25$ and $0.5h$ as a function of non-dimensional diffusion time (the results for the three source heights are shown displaced vertically by a distance $0.1h$ in order to avoid overlap); left-hand frame: the closed symbols are the trajectory computations without subgrid contributions and the closed symbols are the trajectory computations with subgrid contribution; right-hand frame: lines are the LES results (with and without subgrid contribution) and the symbols are observations.

The computational results for σ_y and σ_z are shown in Figs. 8 and 9, respectively (note that in both figures the results of σ_y and σ_z for each source height have been displaced vertically by a constant distance to avoid overlap). For both σ_y and σ_z we find that the contribution of the subgrid motion to the statistics is negligible. Let us now consider both dispersion parameters in some more detail.

For the horizontal dispersion coefficient we find only minor differences as function of source height. Initially, the σ_y increases linearly with time and later on it changes to a parabolic dependence, i.e. $\sigma_y \propto t^{1/2}$. This is in agreement with the asymptotic limits which follow from Taylor's diffusion theorem. An expression which satisfies these two asymptotic limits and which at the same time fits well to the data, is given by

$$\frac{\sigma_y}{h} = \frac{< v'^2 > t/t_*}{\sqrt{1 + t/(2T_L)}}. \tag{16}$$

Here, $< v'^2 >$ is the variance of the horizontal velocity fluctuations which we obtain from our LES data and the T_L is the Lagrangian time scale which, as we shall see later, is equal to $0.7t_*$. In the right-hand frame of Fig. 8 we show the LES results for σ_y together with some experimental data. We find that the LES results underestimate the observational data by about a factor two. This result is consistent with the general underestimation of the horizontal velocity variance by LES data as we have already noticed in Fig. 1.

For the vertical dispersion coefficient shown in Fig. 9 we see that for all source heights the σ_z increases strongly to reach its asymptotic limit for a uniform concentration distribution which is $\sigma = h/\sqrt{12}$. In this case agreement with observational data shown in the right-hand frame of Fig. 9 is quite reasonable.

The Lagrangian computation method has the advantage that it allows us to obtain statistics which can not be easily retrieved from Eulerian data. One of these statistics is the Lagrangian autocorrelation function which for the horizontal velocity component reads

$$\rho_v(\tau; z_s) = \frac{1}{\sigma_v(t; z_s)\sigma_v(t_o : z_s)} \frac{1}{N_p} \sum_{j=1}^{j=N_p} v_j(t_o; z_s) v_j(t; z_s), \tag{17}$$

where $\tau = t - t_0$ and where we have assume that the particle starts at $t = t_o$ at $z = z_s$. A similar definition can be given for the vertical autocorrelation function ρ_w. The results for the horizontal and vertical correlation coefficient are shown in Fig. 10.

For the horizontal correlation function (Fig. 10, left) we find that it is independent of the source height. As a function of time it can be described

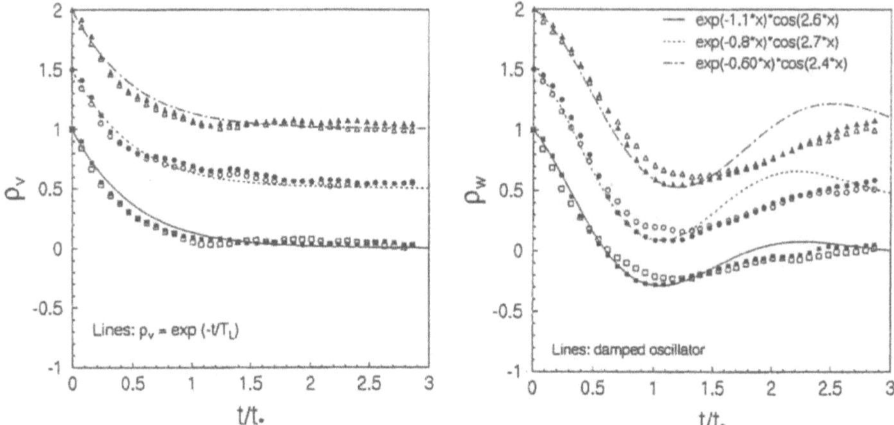

Figure 10. The horizontal autocorrelation function (ρ_v, left frame) and the vertical autocorrelation function (ρ_w, right frame) computed from particle trajectories released from the sources heights $z_s = 0.15, 0.25$ and $0.5h$ as a function of non-dimensional time (the results for the three source heights are shown displaced vertically by a constant distance in order to avoid overlap); the closed symbols are the trajectory computations without subgrid contributions and the closed symbols are the trajectory computations with subgrid contribution.

very well by an exponential expression which is also shown in this figure. From the exponential function we find that the Lagrangian time scale is equal to $0.7t_*$ which is the value that we have already used in (16). The vertical correlation function (Fig. 10, right) shows a more complex behavior which can be interpreted as the combination of an exponential decrease together with an oscillation as shown by the curves drawn in the figure. The oscillating behavior is the result of the constraints imposed on the vertical motion by the surface and the boundary-layer top. For instance, a particle that moves upward to the boundary-layer top is reflected downward again and this causes the negative correlation of velocity.

4.2. CONCENTRATION FLUCTUATIONS

The mean concentration of a contaminant is sufficient in many cases to estimate or predict the impact of air pollution. However, there are a number of problems for which higher-order statistics are needed. Examples are for instance problems related to odor where the mean concentration is not very important but rather when the concentration exceeds a certain level. The same is also true for problems such as the release of toxic or explosive gases. Finally, when one considers dispersion in combination with chemical reactions, it is also not sufficient to consider the mean concentration only. In all the cases mentioned above, one needs to take into account the higher-

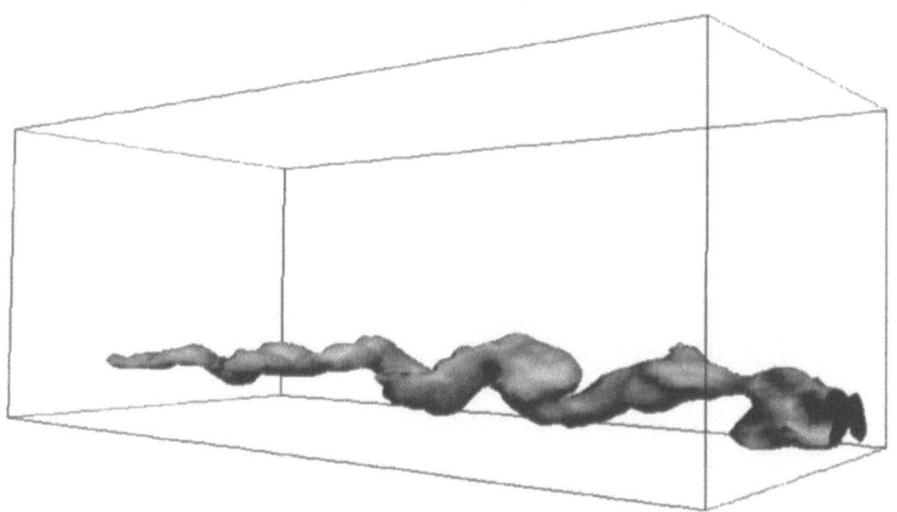

Figure 11. Instantaneous concentration contribution resulting from a point source in the neutral atmospheric boundary layer.

order statistics of the concentration field and in particular the second-order moments which are the concentration fluctuations. This will be the topic of the present section.

4.2.1. *Eulerian Method*

In the Eulerian method the concentration fluctuations can be formally computed as the second moment of the instantaneous concentration field according to

$$< c'^2 > = < (c - < c >)^2 > . \tag{18}$$

To understand the processes that contribute to the concentration fluctuations, let us consider an instantaneous plume resulting from a point source as illustrated in Fig. 11. On the average the plume is transported downwind by the mean velocity. On top of this downstream transport, the plume also fluctuates around the average plume axis. This fact will result in concentration fluctuations because, if one would measure at a single point near the perimeter of the plume, one would measure intermittently a zero concentration level and a non-zero concentration. It is clear that such a signal leads to a non-zero value of concentration fluctuations. This source of $< c'^2 >$ is denoted as macro-mixing or intermittency.

However, there is also another source of concentration fluctuations. To introduce this second source, we assume that near the centerline of the plume the intermittency is one or in other words, the concentration is here always non-zero. In that case there are nevertheless concentration fluctuations which are due to turbulent stirring of the concentration inside the instantaneous plume. Let us illustrate this mechanism by writing down the equation for the resolved concentration fluctuations which can be obtained from (5) by multiplying it with $\overline{c}$ and ensemble averaging. The result reads

$$\frac{D\frac{1}{2}<\overline{c}'^2>}{Dt} = \underbrace{<\overline{u}'_j\overline{c}'>\frac{\partial<\overline{c}>}{\partial x_j}}_{1} - \underbrace{\frac{\partial}{\partial x_j}\left(<\overline{u}'_j\overline{c}'^2> + <\overline{c}'f_j>\right)}_{2} - \underbrace{<f_j\frac{\partial\overline{c}}{\partial x_j}>}_{3}. \tag{19}$$

This equation shows that the change of concentration fluctuations following a (resolved) fluid element is caused by the processes represented by the three terms on the right-hand side of the equation. These processes are called: 1. *Production*, 2. *Transport* and 3. *Subgrid destruction.* Let us assume that the time derivative and the transport term are negligible. The latter is a reasonable assumption in particular in the initial phase of plume development. In that case a balance is established between production and destruction. We see that the production is connected to the mean concentration gradient. Its interpretation is that stirring by turbulent motions with a scale no larger than the instantaneous plume width leads in combination with a mean concentration gradient to concentration fluctuations. This production is sometimes denoted as micro-mixing. However, the (subgrid) destruction term prohibits that these concentration fluctuations grow without bound and they are limited by the interaction with the subgrid scales. This means that the subgrid term plays an essential role in the dynamics of concentration fluctuation in contrast to its role for the mean concentration for which we have seen that effects of the subgrid terms are frequently negligible.

To illustrate our discussion on concentration fluctuations, we consider the results of a LES of plume dispersion in comparison with observations. For the CBL only very little experimental data is available. Therefore, we have performed a simulation for plume dispersion in the neutral boundary layer, i.e. a boundary layer in which turbulence is solely produced by shear. Without going into detail of the computations, we want to point out that the instantaneous plume shown in Fig. 11 is the same as used in this particular computation.

The computational results for the neutral boundary layer have been compared with the carefully carried out experiments by Fackrell and Robbins [2] who performed dispersion experiments in a neutral boundary layer generated on a flat plate in a wind tunnel. We have rescaled their data to an atmospheric boundary layer for which we have performed the LES. A comparison of the LES results and the wind tunnel data is shown in Fig. 12 where we show the vertical profiles of the mean concentration and of the variance of concentration fluctuations at several distances from the source. The profiles have been normalized with the maximum value so that only

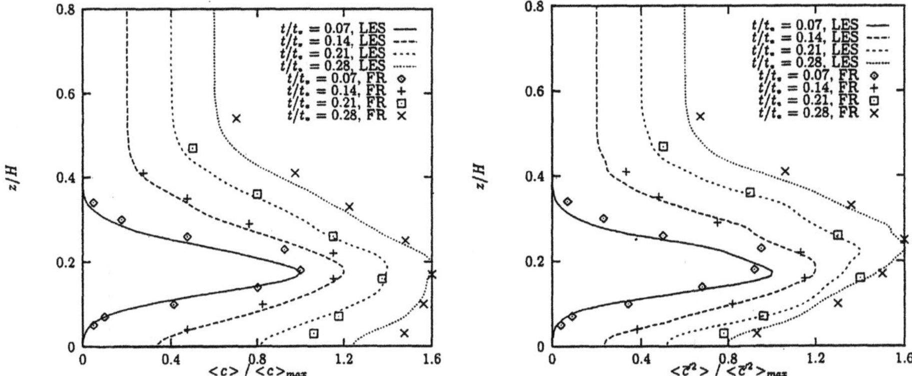

Figure 12. LES results for the dispersion of a point source in a neutral boundary layer (solid lines) compared with wind tunnel data (symbols): (left) the mean concentration and (right) the concentration fluctuations.

the shape of the profiles can be considered. The results indicate that the agreement between LES and experiments is quite acceptable both for the mean concentration and the concentration fluctuations.

It remains to be shown how well the values of the actual concentration fluctuations compare between simulations and experiments. The ensemble averaged concentration fluctuations can be written as

$$<c'^2>=<\overline{c}'^2> + <\overline{c'^2}>, \tag{20}$$

where $<\overline{c}'^2>$ denotes the contribution to the fluctuations by the resolved motions and $<\overline{c'^2}>$ the contribution by the subgrid scales. To compute this latter contribution we must again introduce a subgrid model which in our case reads

$$\overline{c'^2} = 3.74\,\frac{\overline{u_i'c'}^2}{e}, \tag{21}$$

where e is the subgrid kinetic energy. The background of this model is a balance between the production and destruction in the equation of the subgrid concentration fluctuations. The comparison between LES data and experiments shown in Fig. 13 indicates clearly that the subgrid contributions to the concentration fluctuations is non-negligible and essential for a good agreement with experiments. This result can be generalized to the statement that for all higher-order statistics the subgrid contribution can not be neglected. Therefore, a reliable simulation of these higher-order statistics of the concentration field is only possible when a good estimate of the subgrid contribution is available.

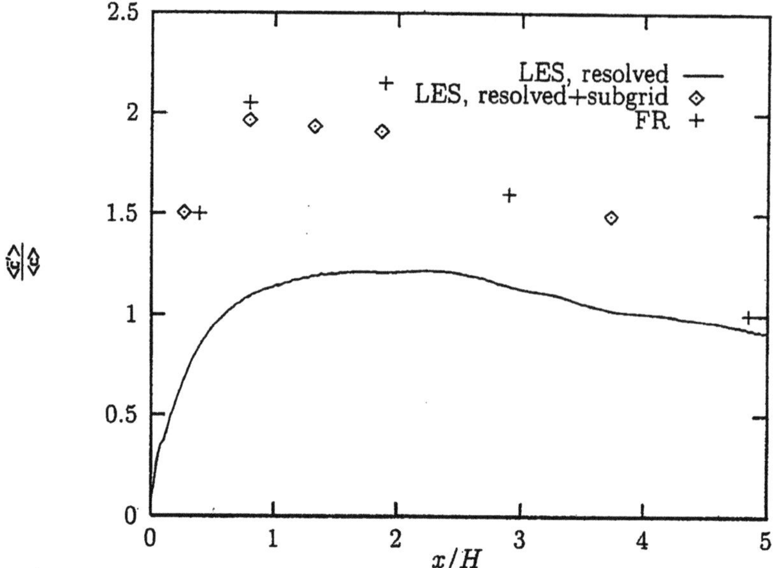

Figure 13. Magnitude of the concentrations fluctuations measured along the plume centerline as a function of distance from the source.

4.2.2. *Lagrangian Method*

In order to compute concentration fluctuations following the Lagrangian method by simulation of particle trajectories, we must look at two-particle statistics which can be obtained from the particle pair results that we have considered before (see Fig. 3). From the trajectories of the particle pairs we can compute the so-called two-particle probability density which is formally defined as

$$p\,(\boldsymbol{x}_1, \boldsymbol{x}_2, t_1, t_2; z_s - \Delta z_o, z_s + \Delta z_o, t_o)\,.$$

This function can be interpreted as the probability that a particle (1) released at $z_s - \Delta z_o$ at t_o, arrives at $\boldsymbol{x}_1$ while a second particle (2) released at $z_s + \Delta z_o$ at t_o, arrives at $\boldsymbol{x}_2$. Given an initial distribution for the concentration of particle pairs $C_o(z_s, \Delta z_o, t_o)$ we can write for the concentration variance

$$<\bar{c}^2> (\boldsymbol{x}, t) = \int\!\!\int p\,(\boldsymbol{x}, \boldsymbol{x}, t, t; z_s - \Delta z_o, z_s + \Delta z_o, t_o)\; C_o(z_s, \Delta z_o, t_o) dz_s d\Delta z_o. \tag{22}$$

When the two particles are independent, it follows that

$$p\,(\boldsymbol{x}_1, \boldsymbol{x}_2, t_1, t_2; z_s - \Delta z_o, z_s + \Delta z_o, t_o) = p\,(\boldsymbol{x}_1 t_1; z_s - \Delta z_o, t_o) \;\cdot\; p\,(\boldsymbol{x}_2 t_2; z_s + \Delta z_o, t_o) \tag{23}$$

and we find in that case for the (23)

$$<\bar{c}^2> (\boldsymbol{x}, t) = <\bar{c}>^2 (\boldsymbol{x}, t).$$

The variance of the concentration is equal to the square of the mean concentration or in other words the variance of concentration fluctuations: $<\bar{c}'^2> \equiv (<\bar{c}^2> - <\bar{c}>)^2$ is equal to zero. Therefore, we must allow for a dependence between the two trajectories of the particle pair in order to estimate the concentration fluctuations. In other words we need two-particle statistics to compute concentration fluctuations and thus one particle statistics is not sufficient.

Let us consider an example to illustrate the points made above. We assume an initial concentration distribution equal to a point source. In the section on mean concentration, we have seen that the probability of a single particle from this point source arriving at a given position is identical to the mean concentration $<c>$ at this position. If we assume that the particle is not diluted and keeps its the initial concentration c_o, we can also obtain the intermittency I from this one-particle statistic as

$$I = \frac{<c>}{c_o}.$$

This implies that one-particle statistics is sufficient to compute the process of macro-mixing and thus also the concentration fluctuations resulting from this process. However, when micro-mixing is taking into account, i.e. when a fluid particle mixes with its environment, the dependency of nearby particles trajectories must be considered in order to determine this mixing. Thus we may conclude that for the total concentration fluctuations resulting from both macro- and micro-mixing two-particle statistics are needed.

In the following, we shall instead of computing the two-particle probability density consider a related quantity, namely the distance between the two particles of a particle pair. The advantage of the latter quantity is that it can be directly computed from the particle-pair trajectories. Here we shall present some results obtained from a LES of the CBL. Let us first define the vertical component of the relative distance between the two particles. This is given by

$$<\Delta z^2> = \frac{1}{4} \sum_{j=1}^{N_p/2} \left(z_j^{(1)} - z_j^{(2)} \right)^2, \tag{24}$$

where $z^{(1)}$ denotes the first particle of the pair and $z^{(2)}$ the second. The summation runs over all $N_p/2$ particle pairs. The initial separation is $<\Delta z^2> = \Delta z_o^2$.

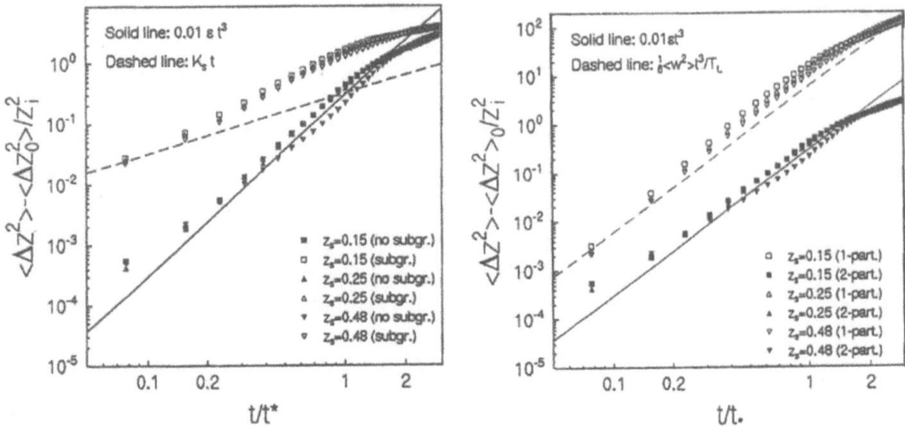

Figure 14. Statistics of relative distance between the two particles of a particle pair as function of dimensionless dispersion time for three source heights z_s: (left) relative distances computed with (open symbols) and without (closed symbols) subgrid contribution; (right) comparison of the relative distance between two particle computed from single particle statistics (open symbols) and from two-particle statistics (closed symbols).

In Fig. 14 (left) we show the results for $< \Delta z^2 > - \Delta z_o^2$ computed from the resolved motions only and also from the resolved motions plus subgrid contributions. The results show hardly any dependence on source height. We find that the contribution of the subgrid motions to the relative distance is not negligible. This is especially the case for small diffusion times when the particles are still close to each other. In particular we find that as $t \to 0$ the result for the total inter-particle distance computed as the sum of resolved plus subgrid, behaves as for a diffusion process, i.e. $< \Delta z^2 > - \Delta z_o^2 \simeq K_s t$. This is in accordance with the subgrid model (9) that we have applied for the Lagrangian trajectory calculation. In view of the uncertainty with respect to the subgrid contribution to $< \Delta z^2 >$ we shall turn our attention further only on the results obtained with the resolved motions. This means that we assume (artificial) particles that do not respond to subgrid fluctuations.

The results of Fig. 14 (left) indicate that the vertical relative dispersion of the two particles is independent of source height and follows a power law behavior with $< \Delta z^2 > \simeq \epsilon t^3$ where ϵ is the mean dissipation of turbulent kinetic energy which we have computed from our LES data. This is an interesting result because it is identical to the well-known result of particle separation in the inertial subrange, known as the Richardson's law. However, one may wonder whether such a result which in principle applies to dispersion caused by inertial subrange turbulence, can be used for these LES data for which the inertial subrange is hardly resolved. Therefore, we offer here another explanation. In a CBL the production by turbulence is dominated by the surface heat flux so that turbulence is governed by a balance by buoyant production and viscous dissipation. This implies

$$\epsilon \sim \frac{g}{T_o} < w'\theta' >_o .$$

The result found in Fig. 14 (left) can then be written as

$$< \Delta z^2 > - \Delta z_o^2 \simeq \frac{g}{T_o} < w'\theta' >_o t^3 ,$$

which implies that the relative distance follows free convective scaling. Free convection scaling leads thus to a power-law expression with the same exponent as Richardson's law.

Let us now consider another result which is illustrated in Fig. 14 (right). In this figure, we illustrate the mean distance between the two particles computed from the resolved motions together with a theoretical result. The theory, for the details of which we refer to Novikov [12], is based on the assumption that this separation can be computed from one-particle statistics. One assumes that the separation between two particles can be approximated by the separation of a single particle from a virtual point that travels in a straight line with the initial velocity of the particle, i.e the velocity at $t = t_o$. The result obtained with this theory is also shown in Fig. 14 (right). We see that the theoretical model greatly overestimates the two-particle results. The explanation is that at their release the two particles do not immediately separate but rather travel together without separation for some time until they reach a region of strong strain where they are forced to move apart. In other words the separation of the two particles occurs intermittently, i.e. only in regions of large strain. This explanation was first proposed by Fung *et al.* [3] and it is indeed observed by considering visualizations of individual particle pairs.

An estimate of this intermittent straining can be obtained by considering the following model (Fung *et al.*, [3]). Let us assume that the particle separation Δ is given by

$$\frac{d\Delta}{dt} \simeq f_c S, \tag{25}$$

where S is a measure for the strain rate that a particle pair experiences and f_c is an intermittency factor which gives the probability of encountering a strain rate which is able to separate the two particles. An estimate for the strain rate can be obtained by realizing that the most effective strain rate to separate a particle pair must have a size proportional[8] Δ and furthermore

[8]This is also the argument to derive the Richardson's law.

that the eddies of size Δ fall in the free convection regime. It then follows that

$$S = \left(\frac{g}{T_o} < w'\theta' >_o \Delta\right)^{\frac{1}{3}}.$$

After substitution of this expression in (25) we find for the solution

$$\Delta^2 \simeq f_c^3 \frac{g}{T_o} < w'\theta >_o t^3, \tag{26}$$

which explains not only the t^3 behavior but also that the intermittency has a large effect on the separation. Namely, f_c appears to the third power in (26). So even a value of f_c close to unity will lead to a substantial reduction of Δ, where one should also realize that for the theoretical one-particle model the f_c is by definition equal to 1.

5. Summary and Conclusions

We have considered here the dispersion of passive contaminants in the Convective Boundary Layer (CBL). The results discussed have been primarily obtained with the help of Large-Eddy Simulation (LES) which is a very suitable technique to simulate convective turbulence. The simulations of dispersion have been performed by means of two techniques: the Eulerian and Lagrangian computation method. In particular results for the dispersion from a point source have been discussed. The large-scale structures found in a convective boundary layer cause a non-standard dispersion behaviour.

References

1. Deardorff, J.W. and Willis, G.E (1984) Groundlevel concentration fluctuations from a buoyant and non-buoyant source within the convectively mixed layer, *Atmos. Environ.*, **18**, 1297-1309.
2. Fackrell, J.E. and Robbins, A.G. (1982) oncentration fluctuations and fluxes in plumes from point sources in a turbulent boundary layer, *J. Fluid Mech.*, **117**, 1-26.
3. Fung, J.C.H., Hunt, J.C.R., Malik N.A. and Perkins, R.J. (1990) Kinematic simulations of homogeneous turbulent flows generated by unsteady random fourier modes, *J. Fluid Mech.*, **236**, 281-318.
4. Galperin, B. and Orzag, S.A. (1993) *Large-eddy simulation of complex engineering and geophysical flows*, Cambridge Univ. Press.
5. Haren, L. van and Nieuwstadt F.T.M. (1989) The behaviour of passive and buoyant plumes in a convective boundary layer, as simulated with a large eddy model, *J. Applied Meteor.*, **28**, 818-832.
6. Lamb, R.G. (1978) A numerical simulation of dispersion from an elevated point source in the convective boundary layer. *Atmos. Env.*, **12**, 1297-1304.
7. Mason, P.J. (1992) Large-eddy simulation of dispersion in convective boundary layers with wind shear, *Atmos. Env.*, **26A**, 1561-1572.
8. Mason, P.J. (1994) Large-eddy simulation: a critical review of the technique, *Quart. J. Roy. Meteorol. Soc.*, **120**, 1-26.

9. Nieuwstadt, F.T.M., Mason, P.J., Moeng, C.-H. and Schumann, U. (1992) Large-eddy simulation of the convective boundary layer: a comparison of four computer codes. *Turbulent Shear Flows 8*, Springer-Verlag, 343-367.
10. Nieuwstadt, F.T.M. and Duynkerke, P.G., (1996) Turbulence in the atmospheric boundary layer. *Atmos. Research*, **40**, 111-142.
11. Nieuwstadt, F.T.M. and Meeder, J.P. (1997) Large-eddy simulation of air pollution dispersion: a review. *New Tools in Turbulence Modelling. Les Houches School*, May 21-31 1996, (Eds. O. Metais and J.O. Ferziger), Springer-Berlin, 265-279.
12. Novikov, E.A. (1989) Two-particle description of turbulence, Markov property and intermittency, *Phys. Fluids*, A, **1**, 326-330.
13. Schumann, U. (1989) Large-eddy simulation of turbulent diffusion with chemical reactions in the convective boundary layer, *Atmos. Env*, **23**, 1713-1727.
14. Sykes, R.I. and Henn, D.S. (1992) Large-eddy simulation of dispersion in the convective boundary layer, *Atmos. Env.*, **26A**, 3145-3159.
15. Wyngaard, J.C. (1992) Atmospheric Turbulence. *Annual Review of Fluid Mech.*, **24**, 205-233.
16. Wyngaard, J.C. and Brost, R.A. (1984) Top-down and bottom-up diffusion of a scalar in the convective boundary layer, *J. Atmos. Sci.*, **41**, 102-112.

CONVECTIVE PROCESSES IN FOREST FIRES

D. X. VIEGAS
Department of Mechanical Engineering
University of Coimbra
Apartado 10131, 3030 Coimbra, Portugal

Abstract

Buoyancy induced convective flows play a very important role in forest fires. The main properties of the convective column above the fire front are presented. The case of a linear fire front propagating in a uniform slope is used to demonstrate the importance of buoyancy induced convection in the spread of a fire. The case of a fire propagating in a canyon is presented as an example of the complex interaction of a background wind, topography and the fire.

1. Introduction

Forest fires constitute one of the major natural disasters that occur in various parts of the World, namely in Southern Europe. Undesired fires may have a very destructive impact on the social and economical activities, and on the natural environment at various temporal and spatial scales. As in some other natural disasters convective processes, namely those induced by buoyancy, play a very important role in the mechanisms of event start and development.

Once a forest fire is started and settled it will consume the vegetation in the close vicinity of its origin, which acts as a fuel of the porous solid type. As the quantity of available fuel is limited, the fire must propagate to surrounding vegetation in order to sustain itself. According to the vegetation layers that are involved in the propagation mechanisms, we will have ground fires, surface fires or crown fire regimes. The first propagation regime, which is usually accompanied by non-flaming combustion, is not very relevant in terms of heat release to the atmosphere and will not be considered here. Surface fires, particularly crown fires, can produce large amounts of thermal energy, of which only a fraction will contribute to fire spread. This energy will be released to the atmosphere in the form of a convection column. This column transports products of combustion mixed with ambient air entrained in it. The convection column above the fire is visible even at large distances from the fire by smoke particles, and constitute a sort of signature of the fire in the atmosphere.

If a wind flow is present, a complex interaction between the topography, the vegetation and the convection column will occur. A sort of interactive process will develop between the fire and the convective flow around it which determines the way the fire will spread. Vertical velocities induced by buoyancy in the combustion area

E.J. Plate et al. (eds.), Buoyant Convection in Geophysical Flows, 401–420.

within the vegetation layer may bring fuel particles aloft. These burning embers may be transported by wind to some distance from the fire before landing on the ground and eventually initiating new fire spots.

When a fire is burning fuel that is situated on the ground surface below tree canopies, the hot gases released from a ground fire below a tree can ignite the leaves, causing a crown fire. If the conditions permit it, this crown fire will involve much more fuel and propagate much faster than the initial surface fire.

The above mentioned phenomena constitute a sample of the complex situations that can be encountered in forest fires in which buoyancy induced mechanisms play a very important role. A quite complete review of current research on physical aspects of forest fires is given in [2]. Due to the number of factors involved, and to the difficulty of making experiments or even observations of these phenomena, the existing knowledge about many of them is still quite scarce. Specialised research on forest fires is usually aimed at understanding the mechanisms and at solving the problems that exist at various levels of fire development and spread, namely the heat transfer mechanisms at the fire front and the overall spread of a surface fire.

In this lecture we shall present an overview of the main mechanisms related to convection that occur in forest fires. It is by no means a complete review of the various roles that these mechanisms can play in the complex phenomena associated with forest fires.

First we shall describe the convection column that is produced by the fire in which strong vertical movements due to buoyancy occur. The interaction between the convection column and the existing wind field is also addressed. At a lower level, in the combustion zone, radiation from the flames is usually assumed to play a major role in the heat transfer process that causes fire spread. The very important influence of convection in defining the flame geometry and therefore the fire spread properties is analysed for the simple case of a fire line propagating in an inclined plane. The non-permanent spread of the fire line due to buoyancy generated convective flows near the flame front is demonstrated. This mechanism is then extended for wind driven fires. A canyon shaped geometry is used to demonstrate the effect of terrain configuration on the wind field, on the development of the thermal plume, and on fire spread.

2. Convective Column Above a Forest Fire

2.1. GENERAL PROPERTIES

One of the distinctive features of forest fires is the existence of a convective column above the fireline. This column can have a height ranging from some hundreds to several thousands of meters, with large scale vertical movements that induce a strong interaction between the fire on the ground and the lower layers of the atmosphere across a wide area.

In very intense fires, the convection column may not be significantly disturbed by ambient wind and we can distinguish the following zones in it, following [7]:

i. *Fuel Bed Zone* - Extending from the ground to the top of the vegetation layer involved in the combustion process.

ii. *Combustion Zone* - Consists of the actively burning zone, within and above the fuel bed.
iii. *Transition Zone* - Lies between the combustion or flaming zone and the more organised flow of the main convection column. Its vertical extension is usually less than 30 to 60m above the combustion zone.
iv. *Fire Convection Zone* - This is the main part of the convection column, extending from the transition zone up to the base of the column cap, its vertical extent can be in the range from 300m to more than 5000m. Condensation of water vapour released by combustion may occur in this zone.
v. *Smoke Fallout Zone* - This is a relatively thin layer of smoke spreading out from the convection column at the base of its cap.
vi. *Condensation Convection Zone* - In this zone heat from condensation is the main source of energy to feed the convection activity. Its light colour is derived from condensed water vapour or ice crystals. This zone is only formed in very intense fires with particular air mass properties.

If wind is present, the properties of the convection column may be changed according to the relative magnitude of the buoyancy induced velocities and the horizontal wind velocity at the various levels.

In the case of a fire that has spread along a large surface of non-homogeneous terrain, parts of the fire line may exhibit a behaviour similar to linear plumes, with localised columns as the ones described above and similar to thermal plumes originating from isolated sources.

2.2. THERMAL PLUME

The heat released by the combustion of forest vegetation induces a vertical motion of the air in the vicinity of the fire. If the vertical extension of the thermal plume that is formed is much larger than some characteristic horizontal dimension of the combustion zone, and if wind velocity is low, an axisymmetric turbulent thermal plume originating from a virtual point source can be assumed to occur. The properties of this flow were described and analysed by several authors, namely [6], [26], [4] and [5]. A brief summary based on these references will be given here.

2.2.1. Neutral Environment

The total buoyancy force $\rho_0 F_0$ produced by the fire per unit of time is the driving force of the flow. In the absence of density gradients in the ambient air (constant ρ_0) and of other energy exchanges, this force must be invariant along the vertical distance z. The plume will entrain ambient air and extend laterally with a typical transverse length $b \propto z$. At large distance from the origin a similarity assumption can be made in order to obtain the general properties of the plume development. Assuming a 'top-hat' profile for the transverse distribution of the flow properties, the following results can be obtained:

$$b = \frac{6}{5}\alpha z \tag{1}$$

$$\overline{w} = \frac{5}{6\alpha}\left(\frac{9}{10}\alpha\frac{F_0}{\pi}\right)^{\frac{1}{3}}.z^{-\frac{1}{3}} \tag{2}$$

$$g' = \frac{g.(\rho_0 - \rho)}{\rho_1} = \frac{5F_0}{6\alpha}\left(\frac{9}{10}7\alpha\frac{F_0}{\pi}\right)^{-\frac{1}{3}}.z^{-\frac{5}{3}}. \tag{3}$$

The reference density ρ_l corresponds to the ambient air density at some standard height, and the entrainment coefficient α, determined experimentally, has a value around 0.10.

In the case of a fire with a heat power release per unit area Q_a, the total buoyancy F_0 is given by:

$$F_0 = \frac{g.Q_a.A}{C_v\rho_a T_a}. \tag{4}$$

Penner *et al.* [21] propose a value of Q_a of 84W/m^2 for a city fire; for suburban and for forested areas the corresponding figures given are 14 and 2.3 respectively. This last value seems to be quite low as it is based on the assumption that a given large forested area takes twelve hours to burn. Based on data provided in [23] and [24] and much reduced values for the duration of the combustion reaction at a given area, values of Q_a between 8 and 30 W/m^2 were found for surface fires, while for crown fires the corresponding figure is of the order of 60 W/m^2. This value is close to that for a city centre.

2.2.2. *Unstable Environment*

In the case of unstable environment with a density gradient of the type:

$$\frac{g}{\rho_1}\left(\frac{d\rho_0}{dz}\right) = Cz^p, \tag{5}$$

in which C is a positive constant and p a negative exponent, p >-8/3. The 'top-hat' solution is given by:

$$\overline{w} = C^{\frac{1}{2}}(4+p)^{-\frac{1}{2}}\left(4+\frac{3}{2}p\right)^{-\frac{1}{2}}z^{\left(1+\frac{1}{2}p\right)} \tag{6}$$

$$g' = C\left(4+\frac{3}{2}p\right)^{-1}z^{(1+p)} \tag{7}$$

$$b = 2\alpha\left(3+\frac{1}{2}p\right)^{-1}z. \tag{8}$$

This general solution is not valid for the case of p=-8/3, corresponding to the uniform environment, discussed above, in which case it is necessary to specify a buoyancy flux to generate the plume. In the case of an unstable environment the properties of the plume are specified by the local density gradient given by (5).

It is interesting to notice that the exponent n of decay of velocity with height ($w \propto z^n$) and the rate of lateral expansion k of the plume ($b=k\alpha z$) change with p. For increasing values of p (increasing instability), n increases, even becoming positive, and k decreases. This means that the plume velocity will not decrease so quickly with height, and also that the lateral spread of the convection column will be less than in the uniform environment case.

As was said before, condensation may occur in the convection column. This situation can be tackled with a general solution using the appropriate values of the exponent in order to cope with the process. Based on the chemical reactions of combustion and on the properties of the atmosphere (cf. [9] and [10]), we can estimate that at the base of the convection column water vapour can be produced in a proportion of 1 to 10g per kg of air. Ambient air entrained in the plume will also carry water vapour that will be added to that of the combustion process and will be diluted in the plume. The increase of absolute humidity may be sufficient in some parts of the flow for condensation to occur. Such phenomena has been reported in large experiments or wildfires, for example in [29] and by [24].

2.2.3. *Stable Environment*

In the case of a stable atmosphere, the buoyancy flux will decrease with altitude and the ascending plume will come to rest at some finite height. In this situation no similarity solution exists and the flow equations must be solved for the entire domain. We can specify the initial buoyancy flux $\rho_1 F_0$, and a constant density gradient G, given by:

$$G = N^2 = -\frac{g}{\rho_1}\frac{d\rho_0}{dz}. \tag{9}$$

Turner [26] presents a numerical solution of the differential equations of the plume, that show that the density gradient and the vertical velocity become null for the following values of ζ, 2.13 and 2.8, respectively. The variable ζ is a non-dimensional function describing the vertical distance z above the ground, defined by:

$$\zeta = 0.345\alpha^{\frac{1}{2}} F_0^{-\frac{1}{4}} N^{\frac{3}{4}}.z. \tag{10}$$

The maximum height of the thermal plume is given by:

$$z_{max} = 5F_0 G^{-\frac{1}{4}}. \tag{11}$$

2.2.4. *Penetration in Inversion Layer*

If we have an inversion layer above the height z_i, with a global inversion gradient $\Delta\theta_i$, defined by [5]:

$$\Delta\theta_i = \int \frac{\partial\theta}{\partial z} dz \,, \tag{12}$$

the inversion strength is given by:

$$B_i = \frac{g}{\theta_i} \; \Delta\theta_i \;\; . \tag{13}$$

The loss of buoyancy of the thermal plume through the inversion layer will be $\Delta F = B_i V_v$, in which V_v is the volume flux of the plume.

The possibility of penetration is given by the condition:

$$F = \pi F_b \; > \; \Delta F \; = \; B_i \, V_v. \tag{14}$$

In the case of neutral atmosphere, with a = 0.08

$$V_v = 0.04 \; F_b^{1/2} z^{5/3}. \tag{15}$$

The minimum inversion base height for penetration to occur is then:

$$z_i < \frac{6.9 F_b^{2/5}}{B_i^{3/5}} \; . \tag{16}$$

2.2.5 *Non-Steady Effects*

Forest fires are not a permanent phenomenon. At a given place, due to the limited quantity of available fuel, the fire will either extinguish itself or propagate to adjacent areas over a period of time. Therefore, the thermal plume induced by a forest fire will in general be of the starting plume kind, or at least it will exhibit non-steady or pulsating behaviour. In many cases the fire will burn the available fuel at a given place before a fully developed convection column is formed.

2.2.6. *Cross Wind*

The previous analysis considered that the plume was developing in still air, which is not usually the case in nature. However, in many instances the vertical flow induced by the fire can be dominant in the surrounding flow field, even in the presence of a light to moderate wind field.

Rothermel [24] proposed a convection parameter defined as the ratio of the vertical impulse flux of the fire and the power due to the horizontal wind, in order to distinguish between buoyancy dominated and wind driven fires.

Briggs [4] presents a general model for the prediction of the rise of a thermal plume. Its application to the situation of a localised plume in a uniform environment in a cross wind of constant velocity predicts a growth of the height of the plume with the 2/3 power of linear distance x from the source:

$$z = 1.6\left(\frac{F_0}{\pi}\right)^{\frac{1}{3}} U^{-1} x^{\frac{2}{3}}. \tag{17}$$

2.2.7. *Linear Source*
Some parts of the fire front may have a two dimensional character and the corresponding thermal plume may be treated as a line plume. In the case of a uniform environment, and if we consider a buoyancy flux $\rho_0 F'_0$ per unit length and assuming Gaussian transverse profiles, [26] presented the following solutions:

$$w = 1.8 F'^{\frac{1}{3}}_0 \, exp\left(-32\frac{x^2}{z^2}\right) \tag{18}$$

$$g' = 2.6 F'^{\frac{2}{3}}_0 \, exp\left(-41\frac{x^2}{z^2}\right) \tag{19}$$

As can be observed, the centreline velocity remains constant along the plume. This prediction was validated in several experiments, namely in the laboratory measurements presented in [13].

In forest fires these line plumes will not be so intense as three-dimensional plumes, because they will tend to form in regions of the fire front that are propagating against wind or slope effects, and therefore they will not have a very high energy release rate. As will be shown later, forward spreading fires tend to have a non-stable fire front. They will easily form local heads of very high intensity fire that will burn for some time in a localised area, forming a convection column of the type described above.

Mercer and Weber [20] present a numerical model for linear plumes above a fire in a cross wind that can take into account the variability of wind velocity with height. At a large distance from the source, the height of the centreline of the plume increases linearly with distance. Temperature and velocity profiles for a neutral atmosphere are shown and the results are interpreted in terms of the possible damage to vegetation above the fire line.

Penner *et al.* [21] developed a numerical model for the estimation of the smoke plume dispersion above large scale fires, such as those from two well documented fires. The model can consider either two or three dimensional situations, and it takes into account the vertical structure of the atmosphere, the background wind, and the eventual condensation of water vapour. The model was applied to simulate the smoke plume development in three situations considered to be typical: fire in a city centre, fire in a suburban area, and fire in a forested area, for both two- and three-dimensional configurations.

2.3. EXPERIMENTAL DATA

Systematic studies of the convection column above large forest fires require experimental data to validate existing models. This is not easy task, given the unpredictable occurrence of a fire and the inherent difficulty of having the required personnel and equipment at the right place and time. Compounding this, is the fact that forest fires change their properties and location continuously therefore posing a serious threat for observers.

Ground, aerial and space observations have provided valuable information on smoke column development, giving hints to interpreting the complex interaction between the fire and the lower atmosphere. Examples of such studies are [11] and [25].

A promising method of observation is provided by the use of Doppler lidar and radar techniques, as described by [3]. These authors report observations obtained at two forest fires. Lidar and radar scans provided a detailed quantitative picture of the three-dimensional flow and smoke concentration in the surroundings of the fire.

3. Convective Effects at the Fire Front

3.1. ROLE OF CONVECTION IN FIRE PROPAGATION

In the following discussion we shall deal with a section of the fire front, and assume that the properties are uniform in the axis direction. We shall also assume that the fuelbed properties are uniform everywhere in the unburned region and that the ambient conditions (air temperature, wind velocity, and topography gradient) are also constant in the period of time and in the space considered in the analysis. A section of a fire front in such conditions is shown schematically in Fig. 1.

In flaming fires, radiation plays a major role in the heat transfer process from the reaction zone to the unburned fuel that drives fire propagation. Porosity of the fuelbed in forest areas, the radiation through the fuelbed itself, and the convection of cold or hot gases, induced by the heat release, make a significant contribution in the propagation mechanism. Radiation from the flame surface may be even more important depending on the size and shape of the flame.

In the absence of wind and in a horizontal fuelbed, convective motions will be induced by the buoyancy generated at the reaction zone, both inside the fuelbed and in the flaming zone. In principle, the flame front will be vertical. Taking into account the heat release from the burned area at the left of the flame front (Fig. 1) and its contribution to the updraft of ambient air, the flame front will tend to be inclined towards the burned zone. In these conditions the view factor from the flame will be relatively small and the radiation from the flame will not be very important. Natural convection induced inside the unburned fuel will have a cooling effect on it, and therefore will not contribute to fire spread. Radiation from the reaction zone through the fuelbed will therefore play a major role in this situation.

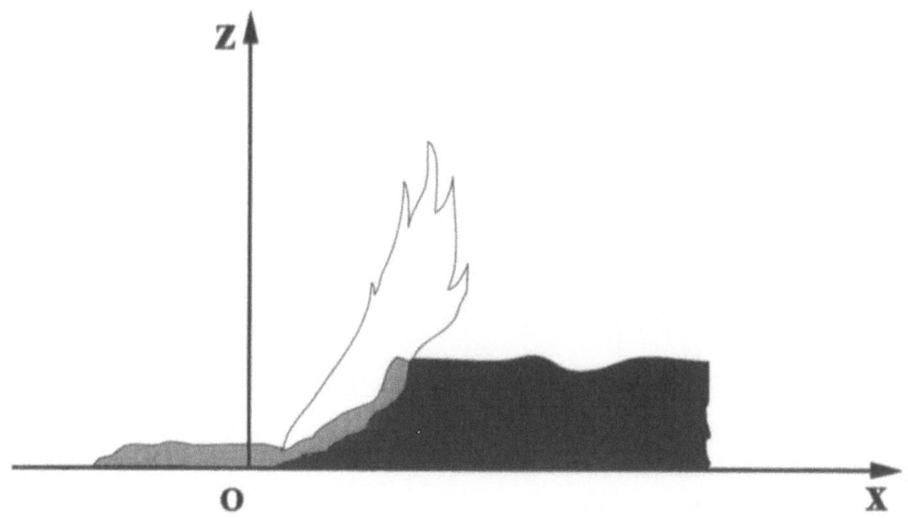

Figure 1. Schematic view of a two dimensional fire front in a porous fuelbed.

A contrary wind (negative *x*-axis direction) will enhance the above described mechanisms of heat transfer. The flame front will be even more inclined towards the burned area and its contribution will be even less. As the reaction zone geometry is not affected much by this contrary wind, the rate of spread remains practically unchanged in relation to the previous case. A negative slope, *i.e.* one in which the fire front is going downslope, has the same effect as a contrary wind.

A favourable wind brings several modifications to the scene. First of all, the chemical reaction is usually enhanced and the heat release from the burning area is much increased, compared with the no-wind or contrary wind cases. As a consequence, the flame size is generally much bigger in this case. Secondly, hot gases are convected to the unburned area preheating the fuel particles and accelerating their pyrolisis and ignition processes. Thirdly, the horizontal flow inclines the flame front, bringing it to the vicinity of the unburned fuel surface and increasing the radiation to the fuelbed drastically. In extreme cases the flame gases can even directly contact the fuel particles, in which case the convection from the flame to the fuel particles becomes a very important process. Positive slope will have similar effects. Finally, burning particles may be transported by the wind flow and released on the fuelbed ahead of the flame front, creating new fire spots, and contributing to a much faster spread of the fire.

Observations show that wind effects are very much dependent on the porosity of the fuelbed, increasing with the porosity, as could be expected. Slope effects are not as dependent on the fuelbed porosity, due to the weaker convective flows inside the fuelbed.

In order to characterise the wind field interaction with the flame front, the author [27] proposed the use of the wall shear stress produced by the flow over the fuel bed. As known, this parameter is related to the flow field near the surface, namely to the velocity profile and to heat and mass transfer processes at the interface, therefore it provides significant correlation with the wind induced rate of spread.

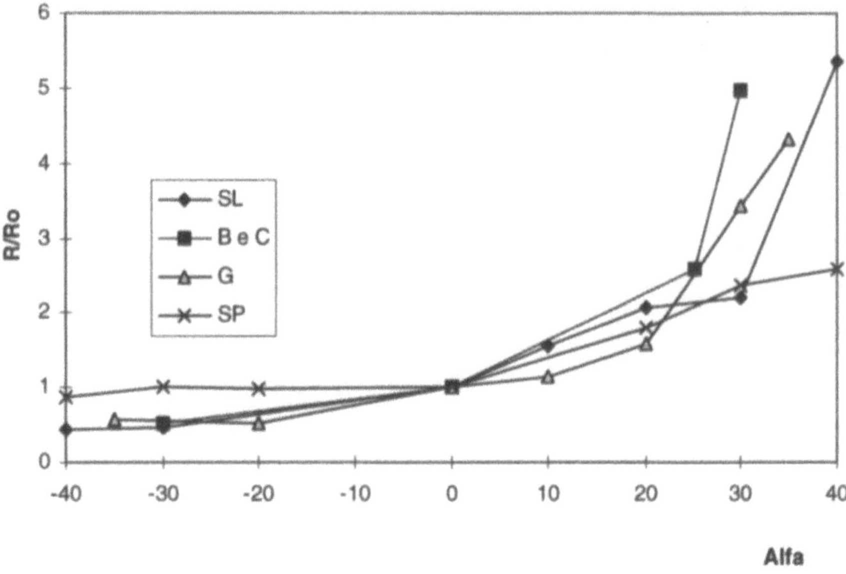

Figure 2. Variation of the rate of spread as a function of slope angle. Experiments with reference SP were performed with a point ignition fire; all others were performed with a linear ignition fire.

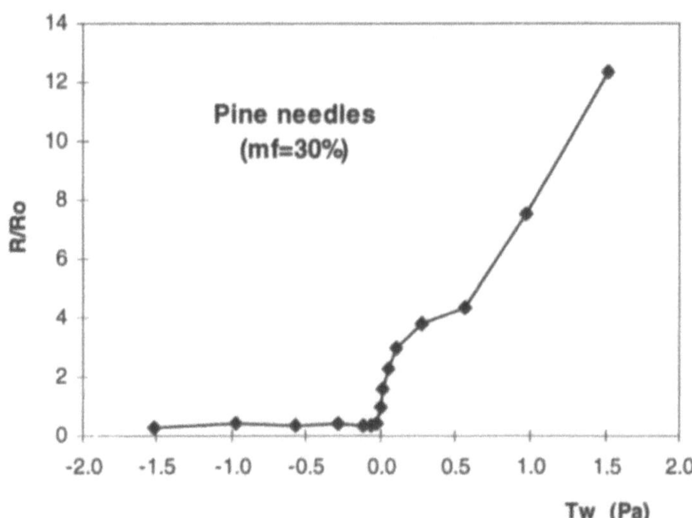

Figure 3. Variation of the rate of spread as a function of flow shear stress. Laboratory experiments performed with *Pinus pinaster* needles; R_0=0.44cm/s.

As a result of the above described mechanisms, the variation of the rate of spread of a linear fire front as a function of slope and wind velocity, determined in the laboratory, is shown in Figs. 2 and 3, respectively.

3.2. CASE OF FIRE PROPAGATING ON A SLOPE

In the previous considerations we assumed that the rate of spread vector, the wind velocity and the terrain gradient, when existing, were all aligned, although not necessarily in the same direction. In order to illustrate the relevance of convective effects in fire propagation we shall now consider the particular case of a linear fire front propagating in a slope with an arbitrary inclination, *i.e.* a case in which the nominal rate of spread vector is not aligned with the terrain gradient.

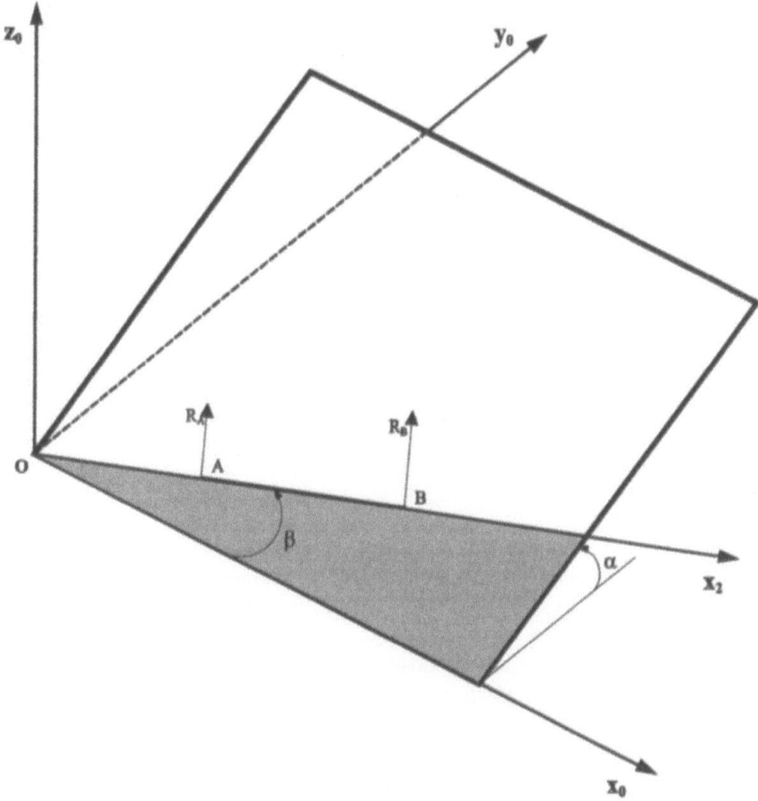

Figure 4. Schematic view of an inclined linear fire front on a slope.

Let us consider (see Fig. 4), a segment of a fire front on a plane of uniform slope α, that at some initial time can be considered to be linear and with an inclination β, in relation to the horizontal direction. It can be easily seen that the velocity vector of natural convection flow induced by the flame front has a component parallel to the fire line direction. This component of the flow transports heat across the fire front and therefore it induces non-uniform conditions at a fire line that was initially uniform. After a period of transition, the burning conditions at the fire front do not remain uniform across. The fire front and the shape of the flame becomes more like a trapeze instead of the initial rectangular form. These two conjugate effects contribute to a larger propagation velocity of the fire front on the upper segment of the fire line. As a consequence of this the movement of the fire line in the time interval Δt will be composed of a translation of average velocity R_m and a rotation of average velocity ω.

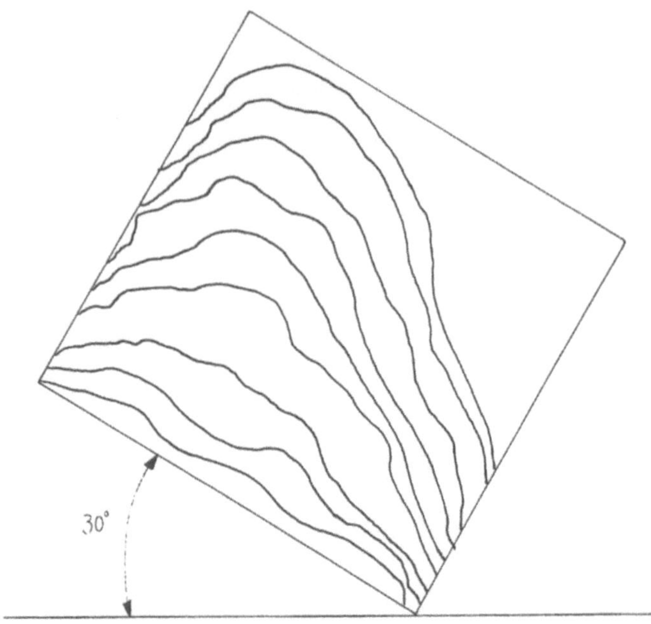

Figure 5. Location of the fire front at different time steps for a flame with initial inclination β=30° on a slope α=30°. The size of the burning table was 1.6x1.6m^2; time step between lines is 15s.

The above described effects can be observed in the field under real life conditions. However, it is quite difficult to make precise measurements of these phenomena due to the involvement of a large number of uncontrollable factors. A laboratory study was therefore set up to analyse this effect. A combustion table was built which could have an arbitrary slope α of $\pm$40°, with 5° steps, in which a linear fire line of 1.6m could be formed with any inclination β from 0 to 180°. A photographic camera placed above the table with its optical axis perpendicular to the surface of the table, ensured that still pictures of the fire front could be taken at desired time steps. The fuelbed was made of a uniform layer of pine needles with a load of 1kg/m^2. In each test a linear fire was ignited almost instantaneously and a set of pictures was taken at intervals that varied from 10 to 30 seconds, according to the expected spread of the fire. More details on the experimental procedure can be found in [28].

From these photos the location of the fire front was then obtained and analysed. The set of lines obtained in a test for the case in which α=30° and β=30° is shown in Fig. 5. As can be observed in that figure part of the fire line rotates tending to become parallel to the slope gradient and the other part rotates in the opposite direction - given the finite dimension of the fireline in the laboratory experiment, - tending to become parallel to the horizontal direction.

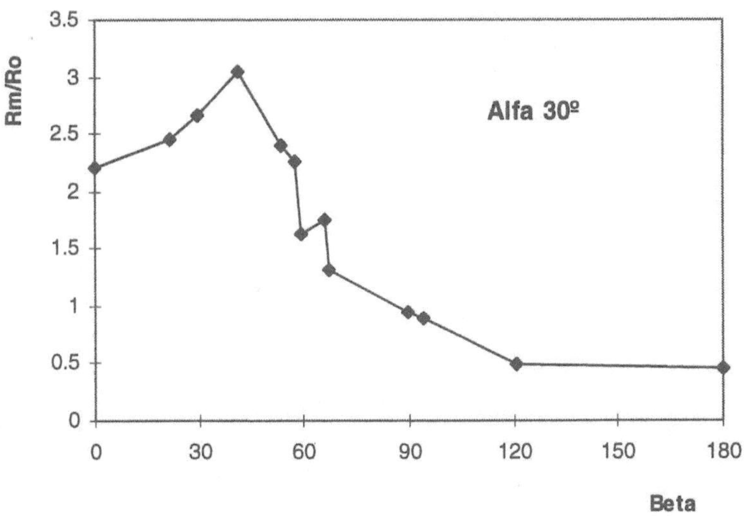

Figure 6. Average translation velocity of a linear fire front on a 30° slope, as a function of its inclination angle β, in relation to the horizontal.

The average translational and rotational speed of the fireline segment during its evolution can be measured in each test from the sequence of photos, as a function of the instantaneous inclination angle β. The average translation velocity R_m was made non-dimensional by dividing it by the basic rate of spread R_0, for the same fuelbed in no-slope and no-wind conditions. With this procedure results from tests made with slightly different values of R_0 can be presented as a single function that is independent of R_0. For the same reason the average rotational velocity ω was also divided by R_0, although a dimensional parameter is obtained in this case. The results obtained for α=30° are shown in Figs. 6 and 7. As can be seen for $0° < \beta < 90°$ the rotational velocity is non-zero, with a maximum value for $\beta \approx 50°$, in this case. This shows that a linear fire front inclined in relation to the slope gradient direction will not propagate in a stable manner, but will always change its direction of spread, tending to become parallel to the terrain gradient. This is essentially an effect due to natural convection induced by the fire front. Forced convection induced by wind will have a similar effect on the stability of the fire front.

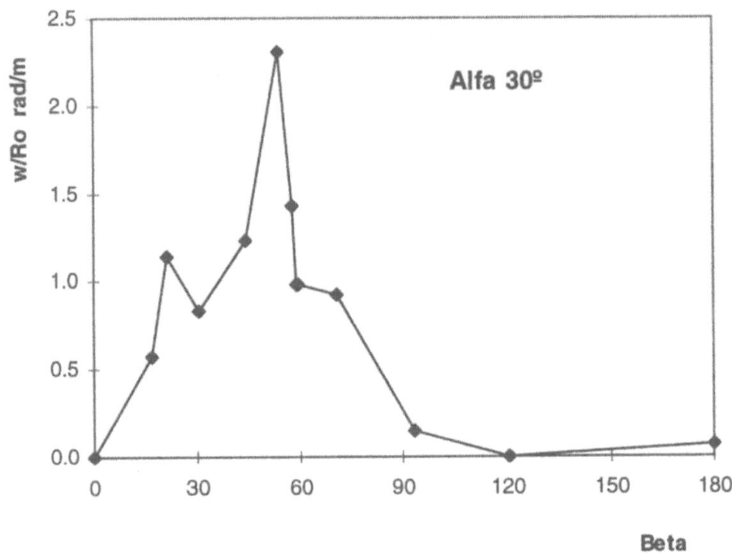

Figure 7. Average rotational velocity of a linear fire front on a 30° slope, as a function of its inclination angle β, in relation to the horizontal.

This mechanism can be used to interpret the evolution of a fire started by a single ignition point on a slope. Initially the fire spreads like a circle, but after a period of time its shape tends to become elongated, with faster spread in the upward direction than in the other directions. Assuming that each segment of the fire front interacts with its neighbours in a similar way, as if it belonged to a linear fire front with the same inclination angle, it is possible to predict the evolution of the fire front over time.

3.3. CASE OF WIND DRIVEN FIRE

Wind has a similar effect on fire propagation as terrain slope, but it can be even more intense. Stable fire propagation can be achieved only if the angle β between the rate of spread and wind velocity vectors is in the range: $\beta = 0°$ or $90° < \beta < 180°$. For β values in the range $0° < \beta < 90°$, the propagation of a fire line element will be unstable and the fire line element will always rotate during its movement, changing its spread properties even in nominally permanent conditions. The rotational velocity value will be a function of both β and the absolute value of wind velocity U (or shear stress τ_W).

Due to the importance of convective effects in fire spread, the problem of estimating the propagation of a fireline can be reduced to three simpler ones:

i) - the determination of the maximum spread direction, due to wind, slope or both, and the corresponding rate of spread value;

ii) - the determination of the rate of spread of a fireline in the same fuelbed without slope or wind;

iii) - the determination of the evolution of the parts of the fireline for which the angle β between the outward normal and that of the maximum spread direction is in the range 0°< β < 90°.

If slope and wind act simultaneously, as is quite common, their conjugate effect will depend on the relative magnitude and direction. There is not a general solution for this problem. Rothermel [23] proposed a vector sum of the rate of spread increase induced by slope and wind effects considered separately, in to order determine the maximum rate of spread direction. The general validity of this assumption is however not proved.

In complex topography situations numerical models which include thermal effects are required in order to cope with the interaction between the background wind, the topography, and the convective flow induced by the fire itself. The works of [8] and [14] give a valid contribution to this goal.

4. Case Study: Canyon Fire

4.1. GEOMETRICAL CONFIGURATION

An interesting case study on the role of convective flows induced by thermal effects on the ground due to a forest fire and the ambient wind flow is the one analysed by [19], for a canyon shaped terrain. This topographical configuration is associated with a large number of fatal accidents caused by forest fires in the past and it is therefore regarded as a potentially dangerous situation.

A simplified configuration of the canyon is shown in Fig. 8. It consists of a horizontal plane *XOY* and two inclined planes that intersect each other along a line that exists in the vertical plane *XOZ*. This is a symmetry plane of the canyon. The slope of each plane is θ, and the angle between two horizontal lines in these planes is α. The inclination angle ψ, of the intersection line in relation to the horizontal plane is determined by the previous parameters. Two different sets of geometrical configuration were tested, in order to assess the relative effect of the canyon shape in comparison to a simple inclined plane with the same slope θ or ψ. The definition of these sets is given in Table 1.

TABLE 1. Parameters of the geometrical configurations.

Set 1				Set 2			
Ref.	θ	α	ψ	Ref.	θ	α	ψ
a	30	60	16.10	e	39.23	60	22.21
b	30	90	22.21	f	30	90	22.21
c	30	120	26.57	g	25.24	120	•22.21
d	30	180	30	h	22.21	180	22.21

As can be seen in Table 1, Set 1 refers to a constant slope angle θ=30°, with the increase of α corresponding to an evolution from a closed canyon to a simple slope

(α=180°). Set 2 is characterised by a constant value ψ=22.21° of the inclination of the intersection line.

A numerical model to solve the full Navier-Stokes equations, including thermal effects for the turbulent flow using a boundary fitted grid, was developed, validated and applied to this situation. Details of the model can be found in [15], [16] and [17].

4.2. THERMAL EFFECTS

A constant heat flux was assumed to be generated at a portion of the soil surface near the base of the canyon and its influence on the oncoming flow was analysed for the various configurations. A set of 32 cells of 5.5x11m^2 was assumed to have a constant temperature of 450°C, creating a thermal plume that interacted with the oncoming wind flow. The incident wind flow was of a boundary layer type with a free stream velocity of 7m/s. The hot cells were placed at three locations: (1) just prior to the interception of the three planes, (2) 33 m ahead and (3) 66 m ahead. The numerical experiments were carried out only for configuration *c* (cf. Table 1).

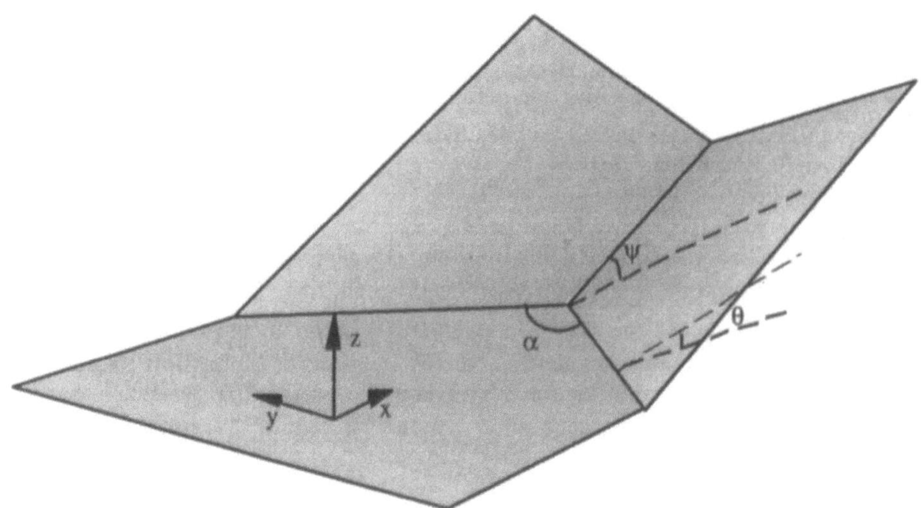

Figure 8. Schematic configuration of the canyon.

The inclination of the thermal plume in the symmetry plane in relation to the horizontal was 72°, 58° and 51° for cases (1), (2) and (3) respectively. The velocity distribution along a line of the symmetry plane 6m above the ground showed marked differences in comparison with the isothermal case. In case (1) there was an acceleration and a stagnation before the hot area was reached and a reverse flow immediately afterwards; in cases (2) and (3) there was an acceleration of the flow up to values of the order of 10m/s in both cases. Vertical temperature profiles at a distance of 67m from the hot area presented much higher temperatures near the ground for cases (2) and (3), while for case (1) the peak temperature was observed some 80m above the ground. This example shows the influence of the surrounding geometry in the development of a thermal plume. In the particular configuration of a canyon, it shows also that if the thermal effect is produced outside the stagnation region that exists at the base of the canyon, the induced velocities can be substantially enhanced the concave layout of the terrain surface.

4.3. FIRE SPREAD

A fire propagation algorithm was incorporated in the flow model in order to estimate the propagation of a point ignition fire at the base of the canyon.

The surface of the ground was assumed to be covered by uniform vegetation, a sort of herbaceous (fuel A) or a sort of shrub (fuel B), corresponding to fuel models 2 and 9, respectively, in the designation of [22]. The first one, being more porous, is more sensitive to wind effects in relation to the other. This can be judged by the fact that the rate of spread increases by a factor of 50 for fuel A if the wind velocity is increased from 0 to 5m/s, while it increases only 20 times for fuel B in the same conditions.

The rate of spread along the maximum spread direction is determined using Rothermel's model [23]. The direction of maximum spread at each location was found taking into account the conjugate effects of wind and slope according to [23]. The rate of spread along the other directions was found assuming an elliptical propagation of the fire from each cell, the form of the ellipse being given by the equations from [1]. A Dijkstra's algorithm (cf. [12]) was used to evaluate the spread of the fire by contagion to the surrounding cells.

The model was run for the various geometrical configurations described in Table 1, with an incident wind profile similar to the one considered in the previous case. Two basic approaches were made in the evaluation of fire spread:

(i) - *Static model* - the wind flow was considered not to be affected by the presence of the fire. i. e. all computations were made assuming the same wind flow over the ground that was evaluated for isothermal conditions.

(ii) - *Dynamic model* - the interaction between the fire and the wind flow was evaluated in an iterative process, in each time step, taking into account the actual location of the fire and the heat flux released by the burning and already burned zones. More details on the models can be found in [15], [18] and [19].

TABLE 2. Ratio of time to burn $2500m^2$ for extreme configurations.

	Set 1 *a/d*	Set 2 *e/h*
Fuel A	0.58	0.47
Fuel B	0.63	0.55

For both Sets 1 and 2 of the geometrical configurations, it was found that the rate of fire spread - assessed by the time evolution of the burned area in each case - showed a consistent increase for the canyon shape, in comparison to the slope with the same inclination (configurations *d* and *h*). The ratio of the time necessary to burn a given area of $2500m^2$ in configuration *a* in relation to configuration *d*, for Set 1, as well as the same relation for configurations *e* and *h*, for Set 2, is given in Table 2 for both fuels.

The static method is the one proposed originally by Rothermel, and is widely applied. It was found that the dynamic model predicted much higher rates of spread than the static one. Unfortunately no data are available to assess the validity of these results but they show the importance of taking into account the buoyancy convection induced by the fire and its influence on the wind field.

5. Conclusion

Convective processes play a very important role in forest fire physics and have a decisive influence on the propagation conditions of the fire and on its impact on the environment. The smoke plume above the fire is dominated by buoyancy generated convection that may interact with the lower part of the troposphere, and therefore with mesoscale meteorological phenomena. The heat power released by the fire and the vertical structure of the atmosphere will determine the development of this column and its overall effect on the fire spread conditions. The presence of wind will modify the shape of the thermal plume in a way that will depend on the relative intensity of wind velocity at various heights and on the vertical impulse of the fire. The transport of smoke for medium to short distances from the fire is a cause of concern for the safety and comfort of people and also for its impact on the quality of air.

At the fire front itself, in the close vicinity of the flaming zone convective processes induced by buoyancy or by wind for a given fuel and topography, determine the properties of the flame front, namely the radiation flux distribution to the unburned fuel. It was shown that whenever the convective flow is not aligned with the direction of spread, a non-stable situation will occur. If the angle between the convective flow and the direction of spread is in the range of $90° < \beta < 180°$, a practically stable propagation of the fire front will occur.

In the case of a canyon geometry it was shown that the overall terrain configuration modifies the wind flow and the thermal flow induced by heat sources near the ground. It was also shown that fire spread can be significantly faster in a canyon than on a slope with the same inclination. The interaction between the buoyancy flux induced by the fire and the wind flow increases the magnitude of the wind velocity above the fire and consequently causes an even faster spread of the fire.

Much more research is necessary to clarify the various mechanisms and processes involving thermal and mechanical convection in forest fires and their overall influence on the conditions of fire and smoke spread.

Acknowledgements

The author wishes to acknowledge the European Union (DG. XII and DG. XIII), for the support given to the research projects on forest fire physics which constituted the majority of the herein presented research. The support given by the Portuguese National Science Foundation for the same purpose is also acknowledged.

References

1. Anderson, H. E. (1983) Predicting Wind-driven Fire Size and Shape, USDA-FS, Ogden UT, Res. Pap. INT-305
2. André, J. C. S., Lopes, A. G. and Viegas, D. X. (1992) *A Broad Synthesis of Research on Physical Aspects of Forest Fires*, Cadernos Científicos sobre Incêndios Florestais, **2**, Coimbra.
3. Banta, R. M., L. D. Olivier, E. T. Holloway, R. A. Kropfli, B. W. Bartram, R. E. Cupp and M. J. Post (1992) Smoke-Column Observations from two Forest Fires Using Doppler Lidar and Doppler Radar, *J. of Applied Meteorology*, **31**, 1328-1349.
4. Briggs, G. A. (1975) Plume Rise Prediction - American Meteorological Society - *Workshop on Meteorology and Environment Assessment.*
5. Briggs, G. A. (1984) Plume Rise and Buoyancy Effects - *Atmospheric Science and Power Production*, Ed. By D. Anderson, U. S. Dep. Of Energy, chap. 8, 327-366.
6. Byram, G. and Nelson, R. M. (1974) Buoyancy characteristics of a Fire Heat Source - *Fire Technology*, **10**, 68-79.
7. Chandler, C. P., Cheney, P. Thomas, Trabaud, L. and Williams, D. (1983) *Fire in Forestry* **1**, John Wiley
8. Croba, D. and Tryfonopoulos, D.(1996) A Wind Modelling Code Capable of Coupling with Forest Fire Effects, Proc. *Workshop on Forest Fire Behaviour Modelling and Testing*, Ed. By D. X. Viegas, Coimbra, Portugal.
9. Drysdale, D. (1992) *An Introduction to Fire Dynamics*, John Wiley and Sons, Chichester, UK, 424 pp.
10. Eskinazi, S. (1975) *Fluid Mechanics and Thermodynamics of Our Environment*, Academic Press, New York, 422 pp.
11. Haines, D. A. and Smith, M. C. (1987) Three Types of Horizontal Vortices Observed in Wildland Mass and Crown Fires, *J. of Climate and Applied Meteorology*, **26**, 1624-1637.
12. Kourtz, P.H. and O'Regan, W.G. (1971) A Model for a Small Forest Fire ... to Simulate Burned and Burning Areas for Use in a Detection Model, *Forest Science*, Vol. 17 N.2.
13. Lee, S. L. and Emmons, H. W. (1961) A Study of Natural Convection above a Line Fire - J. *Fluid Mechanics*, **11**, part 3, 353-368.
14. Lockwood, F. C. and Croba, D. (1996) A Fundamentally Based Model for the Prediction of Large Fire Behaviour, Proc. *Workshop on Forest Fire Behaviour Modelling and Testing*, in D. X. Viegas, (ed) Coimbra, Portugal.
15. Lopes, A. M. G. (1993) *Modelação Numérica e Experimental do Escoamento Turbulento Tridimensional em Topografia Complexa: Aplicação ao Caso de um Desfiladeiro* - Ph. D. Thesis, Dep. Mechanical Engineering, University of Coimbra.
16. Lopes, A. M. G., Sousa, A. C. M. and Viegas, D. X. (1993a) Numerical simulation of three-dimensional turbulent Flow in Mountain Ridges Using a Boundary Fitted Coordinate System – *1st. European African Regional Conference*, Guernesey, Channel Islands.
17. Lopes, A. M. G., Sousa, A. C. M. and Viegas, D. X. (1993 b) Numerical simulation of Non-isothermal Turbulent Flow over 3-D Mountain Ridges - *Proc. Numerical Methods in Thermal Problems*, Swansea.
18. Lopes, A. M. G., A. C. M., Sousa and Viegas, D. X. (1994) Prediction of Turbulent Mixed Convection Induced by Localised Heat Sources on Ridges - *10th. Int. Heat and Mass Transfer Conf.*, Brighton, England.
19. Lopes, A. M. G., Sousa, A. C. M. and Viegas, D. X. (1995) Numerical Simulation of Turbulent flow and Fire Propagation in Complex Topography - *Numerical Heat Transfer Journal*, Part A, **27**, 229-253.
20. Mercer, G. N. and Weber , R. O. (1994) Plumes Above Line Fires in a Cross Wind - *Int. J. Wildland Fire* 4 (**4**): 201-207.
21. Penner, J. E., L. C. Haselman, Jr. and Edwards, L. L. (1986) Smoke-Plume Distributions above Large-Scale Fires: Implications for Simulations of "Nuclear Winter", *J. of Climate and Applied Meteorology*, **25**, 1434-1444.
22. Rothermel, R. (1972) *A Mathematical Model for Predicting Fire Spread in Wildland Fuels* - USDA-FS, Ogden UT, Res. Pap. INT-115.
23. Rothermel, R. (1983) *How to Predict the Spread and Intensity of Forest and Range Fires* - USDA-FS, Nat. Wildfire Coord. G., Tech, Rep. INT-143.
24. Rothermel, R. (1991) *Predicting the Behaviour and Size of Crown Fires in the Northern Rocky Mountains* - USDA-FS, Res. Pap. INT-359.
25. Small, R. D. and Larson, D. A. (1969) Velocity Fields Generated by Large Fires - Pacific Sierra Res. Corp.
26. Turner, J. S. 1973 - *Buoyancy Effects in Fluids*, Cambridge University Press, 368 pp.

27. Viegas, D. X. and Neto, L. P. C. (1992) Wall shear stress as a parameter to correlate the rate of spread of a wind induced forest fire - *Int. J. Wildland Fire*, **2** (4):69-86.
28. Viegas, D. X., Varela, V. and Borges, C. M. (1994) On the Evolution of a Linear Fire Front on a Slope - *Proc. II Int. Conference on Forest Fire Research*, Coimbra.
29. Vines, R. G. (1973) Air Movements Above Large Bush fires - *Proc. Tall Timber Fire Ecol. Conf.* **13**:295-301.

STRATOCUMULUS-TOPPED ATMOSPHERIC PLANETARY BOUNDARY LAYER

C.-H. MOENG
National Center for Atmospheric Research
P.O.Box 3000, Boulder, CO 80307 USA

1. Introduction

Marine stratocumulus often occurs over the subtropical oceans off the west coasts of the major continents throughout most of the summertime season when the large-scale subsiding motion is strong and the ocean surface is cold. This large-scale condition leads to a very moist, shallow planetary boundary layer (PBL). The lifting condensation level is below the top of the mixed layer, so an uniform layer of cloud can form in the upper part of the PBL.

The major climate impact of the marine stratocumulus regime is its effect on the global radiation budget. Marine stratocumulus reflects much more solar radiation back to space than the ocean surface does, while it emits about the same amount of longwave radiation as its underlying ocean surface. As a result, the cloud system can drastically decrease the net radiation input into the Earth. Randall *et al.* [14] estimated that a mere 4% increase in the area coverage of marine stratocumulus regime can produce a 2-3 K cooling in the global mean temperature. Slingo[18] using a 3-D general circulation model also showed large sensitivity of the Earth's radiation budget to changes in low clouds (e.g., cloud amount, liquid water path, and mean drop radius). Because cloud albedo depends sensitively on a) the cloud fraction and liquid water amount (which we are still unable to predict) and b) the dropsize distribution at the cloud top (whose relationship to large-scale variables is poorly understood), the marine stratocumulus cloud regime poses a big challenge for climate change studies.

The stratocumulus-topped PBL (STBL) can be driven by many forcings, such as surface heating, cloud-top longwave radiative and evaporative cooling, condensation, and wind shear. But the most typical STBL is dominated by cloud-top radiative cooling. The STBL is much more complicated

E.J. Plate et al. (eds.), Buoyant Convection in Geophysical Flows, 421–440.

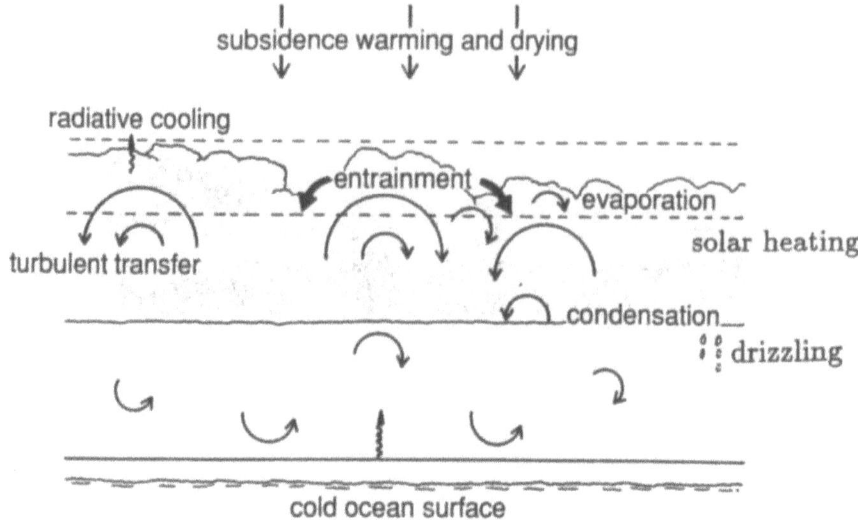

Figure 1. A sketch showing the various physical processes involved in the stratocumulus-topped PBL.

than the clear convective PBL because it has many more physical processes involved, as shown in Fig. 1 and discussed in the next section. The major complication comes from the fact that the buoyancy source, which drives turbulence, is located where entrainment takes place. Just how and how much entrainment affects this turbulence-generation process is poorly understood.

The mathematical description of the STBL is more complicated than that of the clear PBL because there are more thermodynamic variables to deal with. In the STBL the virtual potential temperature, which is a measure of buoyancy, is no longer a conserved variable. Temperature variables that are (approximately) conserved in a moist adiabatic process include the equivalent potential temperature θ_e and the liquid potential temperature θ_l. The moisture variable that is conserved in a moist adiabatic process (with no drizzle) is the total water mixing ratio q_T, i.e., the sum of water vapor q_v and liquid water q_l mixing ratio. The linearized relationships among these thermodynamic variables are:

$$\theta_v \approx \theta(1 + 0.61 q_v - q_l) \approx \theta + 0.61\theta_0 q_v - \theta_0 q_l, \tag{1}$$

$$\theta_e \approx \theta + \frac{L}{c_p} q_v, \tag{2}$$

$$\theta_l \approx \theta - \frac{L}{c_p} q_l, \tag{3}$$

where θ is the potential temperature, θ_0 is potential temperature in a reference state, L is the latent heat of condensation, and c_p is the specific heat of air at constant pressure. (More detailed description of these thermodynamic variables can be found in Stull [20].)

The buoyancy flux can be related to the fluxes of the conserved variables like liquid potential temperature θ_l and total water mixing ratio. The flux relationships are, obviously, different in cloudy (saturated air) and clear (unsaturated air) regions. For clear air such relationship can be easily dervied from (1) and (3),

$$\overline{w\theta_v} \approx \overline{w\theta_l} + 0.61\theta_0 \overline{wq_T} \quad \text{for clear air.} \tag{4}$$

For cloudy air, q_v can be approximated as q_* the saturation mixing ratio. Since $q_*(T,P) \approx q_*(T_0,P) + (\frac{\partial q_*}{\partial T})_p (T - T_0)$, then $q_v' \approx q_*(T,P) - q_*(T_0,P) \approx (\frac{\partial q_*}{\partial T})\theta' \approx \frac{c_p \gamma}{L}\theta'$, where $\gamma \equiv \frac{L}{c_p}(\frac{\partial q_*}{\partial T})_p$. Using this $q_v' - \theta'$ relationship along with (1) and (3), we can relate the buoyancy flux to the fluxes of the liquid potential temperature and total water mixing ratio as

$$\overline{w\theta_v} \approx A_1 \overline{w\theta_l} + B_1 \theta_0 \overline{wq_T} \quad \text{for cloudy air,} \tag{5}$$

where $A_1 = (1 + 1.61\gamma\epsilon)/(1+\gamma)$, $B_1 = (L/c_p)(1 - \epsilon + 0.61\gamma\epsilon)/(1+\gamma)$, and $\epsilon \equiv (c_p T_0 / L)$. Here the overline represents an ensemble average.

A problem occurs when the averaging is over a partly cloudy region where saturated and unsaturated air parcels both exist, such as a horizontal average over a bumpy cloud top level. Existing formulae for the averaged buoyancy flux over a partly cloudy region disagree on how the flux depends on fractional cloudiness, as discussed in Randall [13].

The flux profiles in the STBL are quite different from those in the clear convective PBL. In the clear PBL, the fluxes of conserved thermodynamic quantities are approximately linear with height under a quasi-steady state condition. Thus, the buoyancy flux remains nearly linear in the clear convective PBL. For the STBL, there exist additional heat sources/sinks due to radiation and condensation/evaporation. Using the liquid potential temperature and total mixing ratio as conserved variables (for the no-drizzle case), we can write the governing equations for their ensemble means (Θ_l and Q_T, respectively) as

$$\frac{\partial \Theta_l}{\partial t} = -\frac{\partial \overline{w\theta_l}}{\partial z} - \frac{1}{\rho c_p}\frac{\partial F_{rad}}{\partial z}, \tag{6}$$

and

$$\frac{\partial Q_T}{\partial t} = -\frac{\partial \overline{wq_T}}{\partial z}, \tag{7}$$

where F_{rad} is the radiative flux. Taking z derivative of (6) and (7), we have

$$\frac{\partial(\partial\Theta_l/\partial z)}{\partial t} = -\frac{\partial^2\overline{w\theta_l}}{\partial z^2} - \frac{1}{\rho c_p}\frac{\partial^2 F_{rad}}{\partial z^2}, \tag{8}$$

and

$$\frac{\partial(\partial Q_T/\partial z)}{\partial t} = -\frac{\partial^2\overline{wq_T}}{\partial z^2}. \tag{9}$$

A quasi-steady state of turbulence is usually referred to as a state where the vertical distributions of the means are constant in time, i.e., the left-hand sides of (8) and (9) equal zero. Therefore the total heat flux, $H \equiv \overline{w\theta_l} + \frac{1}{\rho c_p}F_{rad}$, and the total water flux, $\overline{wq_T}$, should remain linear with height. Because F_{rad}, the longwave radiative flux (or longwave plus solar radiative flux when the sun is present), is not linear with height, the liquid-potential-temperature flux $\overline{w\theta_l}$ alone is not linear with height either. Buoyancy flux, which can be calculated from (4) and (5), is also not a linear profile; it increases sharply at the cloud base level due to condensation and decreases sharply at the cloud top due to radiation and entrainment. Unlike the clear convective PBL, the profiles of turbulent fluxes within the STBL vary from case to case depending on the relative contribution of cloud-top cooling and surface heating to buoyancy forcing.

In this lecture, I will describe the physical processes in the STBL in section 2 and the cloud-top entrainment instability in section 3. In section 4, I will talk about an international effort in studying boundary-layer clouds by the Global Energy and Water-cycle Experiment (GEWEX) Cloud System Study (GCSS) Boundary-Layer Cloud Working Group.

2. Physical Processes

The most fundamental physical processes within the STBL are longwave radiation, turbulent mixing, entrainment, and cloud microphysics; they exist in all stratocumulus PBLs. In addition, solar radiation can be important in the daytime, and drizzle is often observed and may also be important in maintaining/dissipating the cloud layer.

2.1. LONGWAVE RADIATION COOLING AT THE CLOUD TOP

Longwave radiation cooling at cloud top exists due to the different radiative properties of water vapor and droplets. In the longwave radiation spectrum, there is a "window region" where water vapor is transparent to longwave radiation. But droplets can emit all longwave frequencies, i.e., they are black-body emitters. Thus, the downward longwave flux just above the cloud top, which comes from the emission of the overlying water vapor, is smaller than the flux coming out from the cloud layer. This leads to a sharp decrease in the downward flux just across the cloud top level. The upward flux, however, remains quite uniform with height, with only a slight change at the cloud base due to the slight difference in temperature at the surface and at the cloud base. The net flux thus has a sharp increase right at the cloud top, and this strong flux divergence leads to a strong cooling ($\approx$ several degrees per hour) over a very thin layer ($\approx$ 50m) at the cloud top, compared to the typical clear-air longwave radiation cooling of about 1$\sim$2 degrees per day.

When a PBL is cooled from the top, downdrafts that carry cooled (heavy) air parcels are negatively buoyant and hence are accelerated downward. This can drive turbulence, just like updrafts that carry the surface-heated (light) air parcels in the clear convective PBL (CBL) case. If the cloud top is a rigid lid, like the Earth's surface, then cooling from top should generate turbulence in an exact upside down sense as heating from bottom in the convective PBL. But the cloud top is a free surface; it is an interface that separates turbulent from nonturbulent air, and entrainment occurs there. This results in a subtle difference between the cooling-from-top and heating-from-surface PBLs (Moeng and Schumann [10]).

Nevertheless, some similar features were observed between the cooling-from-above and heating-from-below PBLs. For example, Nicholls [7] observed cellular patterns near the cloud tops, with broad updrafts in the centers and narrow downdrafts around the edges. This feature was also seen in a large eddy simulation (Fig. 2) as reported in Shen and Moeng [15].

2.2. ENTRAINMENT

Entrainment brings warm and dry air from above into the turbulent layer, and consequently modifies the cloudy air parcels that are being cooled radiatively. Thus, before those radiatively cooled air parcels descend and buoyantly "disturb" the flow field, they are modified by entrainment, as sketched in Fig. 3. This means that, unlike the CBL, the efficiency that cloud-top radiation cooling drives buoyancy (and hence TKE) depends on entrainment.

On the other hand, entrainment depends on the turbulence intensity. The stronger the turbulence, the stronger the entrainment rate. Without the condensation/evaporation effect (such as smoke cloud cases), the interactions between longwave radiation cooling and entrainment are as follows: Stronge entrainment brings in warm inversion air. When this warm air mixes in with the radiatively cooled air, these mixtures near the cloud top become less heavy and hence the efficiency of longwave cooling in driv-

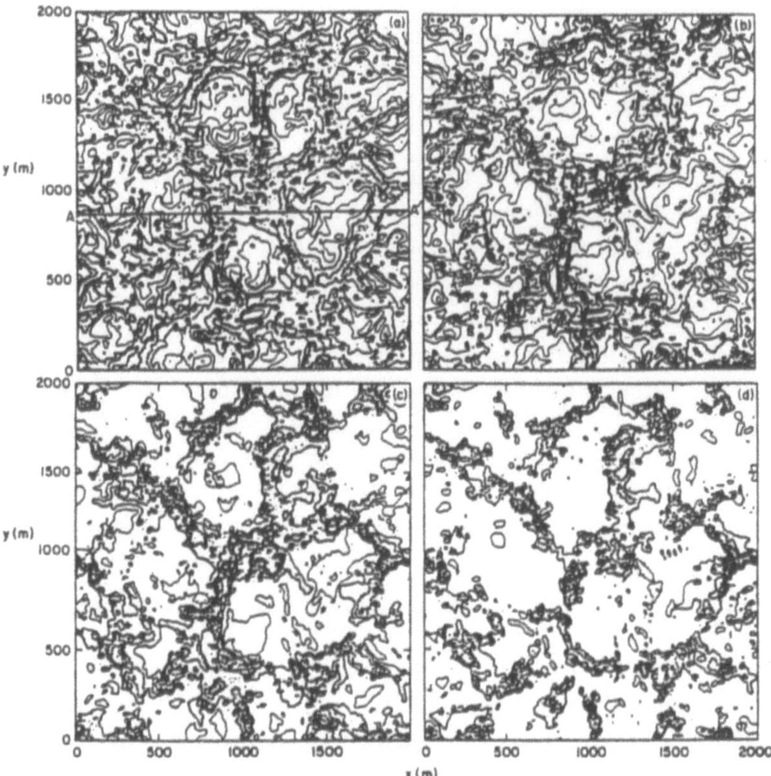

Figure 2. Horizontal cross sections of fluctuations of a) vertical velocity, b) virtual potential temperature, c) total moisture mixing ratio, and d) liquid water in the mid-PBL. The solid and dashed contour lines represent positive and negative values, respectively. (From Shen and Moeng [15])

ing turbulence is reduced. Weaker turbulence reduces entrainment. This is a negative feedback. But when cloud-top evaporative cooling dominates, stronger entrainment results in more evaporative cooling, and as a consequent enhances turbulence and gives stronger entrainment. This positive feedback between entrainment and cloud-top evaporative cooling is called cloud-top entrainment instability, which will be described in section 3.

2.3. CLOUD MICROPHYSICS

Cloud microphysical processes, such as condensation/evaporation and drizzle (light precipitation is often found in the STBL), further complicate the STBL structure. Condensation occurs in the cloud layer so there typically exists a sharp increase of buoyancy flux at the cloud base, as we will see in section 4. Evaporation occurs mostly near the cloud top where entrainment keeps bringing in warm and dry inversion air so the mixtures of the

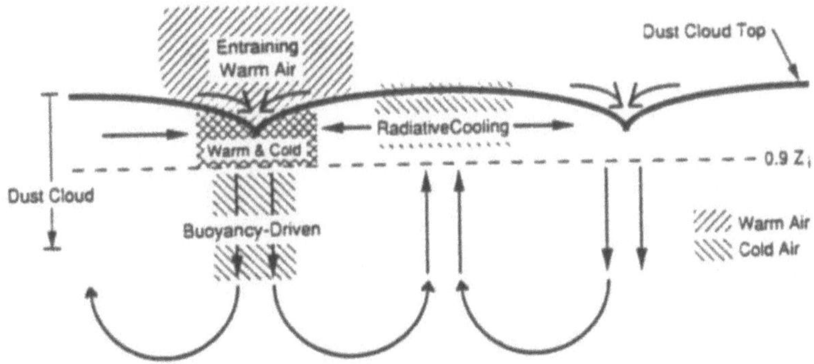

Figure 3. A sketch showing the idealized flow field (in arrows) and temperature fluctuations near the STBL top. Different shading signifies different temperature fluctuations. This sketch emphasizes only the locations where longwave radiative cooling is strongest and where the warm air entrainment is largest. (From Moeng and Schumann [10])

cloudy and clear air become unsaturated. Like radiative cooling, cloud-top evaporative cooling can also drive the turbulent motion within the STBL.

When droplets fall out of the cloud layer, they evaporate and result in cooling of the surrounding air. When the subcloud layer becomes colder than the cloud layer, due to such evaporation cooling, a thin stably stratified layer may occur in the subcloud layer, and lead to decoupling of cloud layer from its underlying surface (see Lenschow [5]). Other significant effects of drizzle include: decreasing the liquid water path, decreasing the entrainment rate, reducing the buoyancy production and hence lowering the turbulent kinetic energy, as found in an LES study by Stevens *et al.* [19].

As mentioned in section 1 the dropsize distribution near the cloud top is important in determining the amount of solar radiation reflected back to space. In this aspect cloud microphysics play a crucial role in modifying the impact of STBL on the global radiation budget.

2.4. SOLAR RADIATION

When we consider stratocumulus over the ocean, the influence of solar radiation comes from the solar absorption inside the cloud layer, rather than by providing surface heating. Unlike longwave radiation, solar heating inside the cloud is more uniformly distributed throughout the cloud layer. The part of solar heating near the cloud top can offset longwave radiation cooling in generating turbulence, while the solar heating inside the cloud layer rises the in-cloud temperature and thus the saturation mixing ratio. These two effects together can result in a diurnal cycle of marine stratocumulus even though the ocean surface temperature remains the same. That is, the cloud thins or even disappears in the daytime and thickens or reforms at night.

When the whole cloud layer is warmed by solar heating, the cloud layer may become warmer than the subcloud layer and again can lead to a thin stably stratified layer at the cloud base. This can "decouple" the cloud layer from the influence of the surface and cutoff the moisture supply from the surface into the cloud layer.

3. Cloud Top Entrainment Instability

Cloud-Top Entrainment Instability (CTEI) has been a popular topic in the field of marine stratocumuls cloud during the past decade mainly because it has been believed to be an important mechanism that can break up the PBL stratocumulus. The CTEI effect in breaking up stratocumulus was first suggested by Lilly [6]; he argued that when the air above the cloud layer has a smaller equivalent potential temperature than the cloudy air, i.e., $\Delta\Theta_e = \Theta_e(B^+) - \Theta_e(B^-) < 0$, where B^+ and B^- are levels just above and below the cloud top, then any air parcel that contains a mixture of air from above and air inside the cloud will remain colder than its surrounding unmixed cloudy air. So the mixture can accelerate downward, i.e., it becomes negatively buoyant downdrafts, and sink deeper into the cloud layer, as sketched in Fig. 4, and the cloud layer can eventually break up. This instability is the so-called CTEI.

Randall [12] and Deardorff [3] revised this CTEI criterion by adding the effects of liquid water on the buoyancy. From the virtual potential temperature formula shown in (1), we know that q_l can modify the buoyancy variable θ_v. Based on this definition and other relationships between different thermodynamic variables, Deardorff constructed a buoyancy mixing diagram and used it to demonstrate the CTEI mechanism. Randall, on the other hand, used the flux relationship between buoyancy and conserved variables to derive a criterion for CTEI. They used different methods but came up with the same criterion, which says that $\Delta\Theta_e$ has to be much smaller than zero in order for CTEI to occur.

I will use Randall's derivation for illustration. Assuming that air at the B^- level is totally saturated, similar to (4) and (5) we can relate the buoyancy flux to the fluxes of the equivalent potential temperature and total water mixing ratio for cloudy regions as follows:

$$\overline{w\theta_v}(B^-) = A_2\overline{w\theta_e}(B^-) - B_2\overline{wq_T}(B^-), \tag{10}$$

where A_2 and B_2 are again thermodynamic coefficients, which are different from A_1 and B_1 in (5) because now the equivalent potential temperature

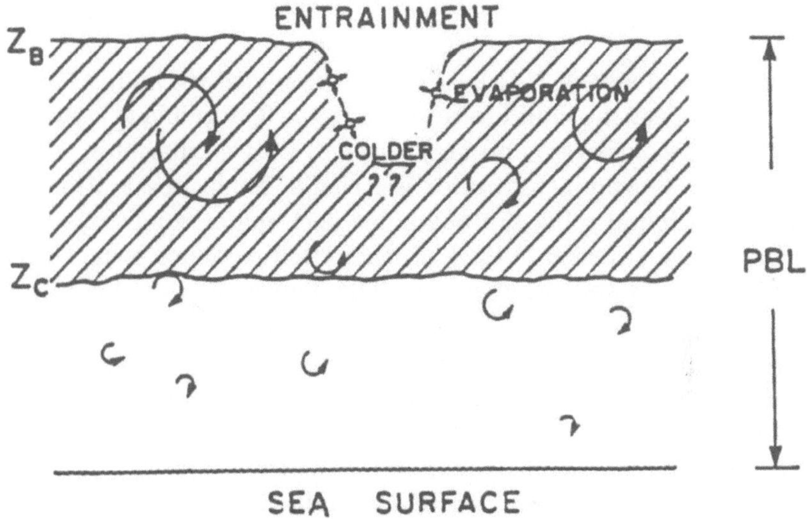

Figure 4. A sketch showing the involved mechanisms in the cloud-top entrainment instability process. (From Moeng and Arakawa [9])

is used instead of liquid potential temperature as the conserved thermodynamic variable. The jump conditions for the conserved variables are

$$\overline{w\theta_e}(B^-) = -w_e \Delta\Theta_e, \tag{11}$$

and

$$\overline{wq_T}(B^-) = -w_e \Delta Q_T, \tag{12}$$

where w_e is the entrainment rate and again Δ represents difference of the values between B^+ and B^- levels. Applying (11) and (12) in (10) yields

$$\overline{w\theta_v}(B^-) = w_e[-A_2\Delta\Theta_e + B_2\Delta Q_T]. \tag{13}$$

For the instability to take place, i.e., for downdrafts ($w' < 0$) to be negatively buoyant ($\theta_v' < 0$), buoyancy flux near the cloud top has to remain positive, i.e., $\overline{w\theta_v}(B^-) > 0$. Thus, the right-hand side of (13) has to be positive for downdrafts to accelerate downward, i.e.,

$$\Delta\Theta_e < \frac{B_2}{A_2}\Delta Q_T. \tag{14}$$

The right-hand side of (14) is almost always negative and is about -1 ~ -3K. If we neglect the liquid water term in the θ_v formula (1), the coefficient B_2 in (10) becomes zero and (14) becomes Lilly's criterion.

The above instability criterion was challenged by observations, however, which often found solid stratocumulus deck under conditions that satisfy the instability criterion (14). The most cited evidence is shown in Fig. 5, taken from Kuo and Schubert [4]. This figure shows that stratocumulus can exist when $\Delta\Theta_e < 0$ (Lilly's CTEI condition) and $\Delta\Theta_e < \kappa\frac{L}{c_p}\Delta Q_T$, where $\kappa = (B_2/A_2)(c_p/L)$ (Randall-Deardorff CTEI condition). (However, I would like to remind you that in practice it is difficult to define or measure the jump conditons, $\Delta\Theta_e$ or ΔQ_T. Soundings often show large variations in the thermodynamic fields just above the cloud top.)

Because of the controversy, many studies (e.g., Kuo and Schubert [4]; Siems *et al.*, [17]) began to look into more detail at the entrainment mixing and evaporation processes, using the following thermodynamical analysis. Consider a mixture that consists of χ mass units of warm air that is entrained from above and (1-χ) mass units of cool moist air within the cloud layer. χ is the fractional amount of inversion air in the mixture. Its equivalent potential temperature and total mixing ratio are

$$\theta_e = \chi\theta_{ea} + (1-\chi)\theta_{eb}, \tag{15}$$

$$q_T = \chi q_{Ta} + (1-\chi)q_{Tb}, \tag{16}$$

where the subscript a represents the above-cloud air and b represents the cloudy air, assuming no heat source/sink like radiative heating/cooling. Using the definitions of θ_e, θ_v, and

$$q_l = q_T - q_* \quad \text{when} \quad q_T > q_*, \tag{17}$$

$$q_l = 0 \quad \text{when} \quad q_T < q_*, \tag{18}$$

we can then iteratively solve for θ_v as a funtion of χ for given θ_{ea}, θ_{eb}, q_{Ta}, and q_{Tb}.

Figure 6 shows examples of θ_v of a parcel after mixing vs. χ for four different large-scale conditions that have θ_{ea}, θ_{eb}, q_{Ta}, and q_{Tb} located at points A, B, C, and D in Fig. 5, respectively. The kinks (i.e., minimum

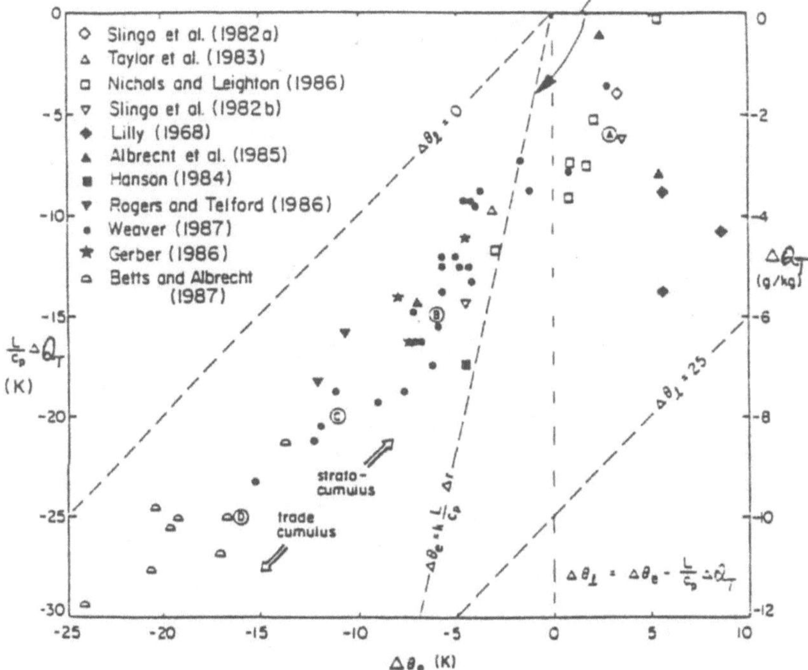

Figure 5. **A cloud-top jump condition diagram showing the Lilly and Randall-Deardorff instability criteria and observational data. Data with open symbols were taken from mid-latitude cases and those with solid symbols were from subtropical cases. (From Kuo and Schubert [4])**

buoyancy locations) in these four curves occur at $\chi = \chi_*$ where the mixture just becomes saturated.

Figure 6 shows that when the jump condition is on the stable side of the Lilly-Deardorff-Randall curve, such as at location A, all mixtures, with any χ amount, remain warmer than the unmixed cloudy environment. When the jump condition satisfies the Lilly-Deardorff-Randall criterion, some mixtures indeed become colder than their unmixed cloudy environment, and thus are negatively buoyant (i.e., accelerate downward). However, not all the mixtures are negatively buoyant; those with larger χ are still warmer than the cloudy air. Thus, Kuo and Schubert concluded that the CTEI criterion proposed by Deardorff and Randall is only a necessary, not sufficient, condition for cloud breakup. They also suggested, based on a two-dimensional numerical simualtion, that "surface evaporation and up-

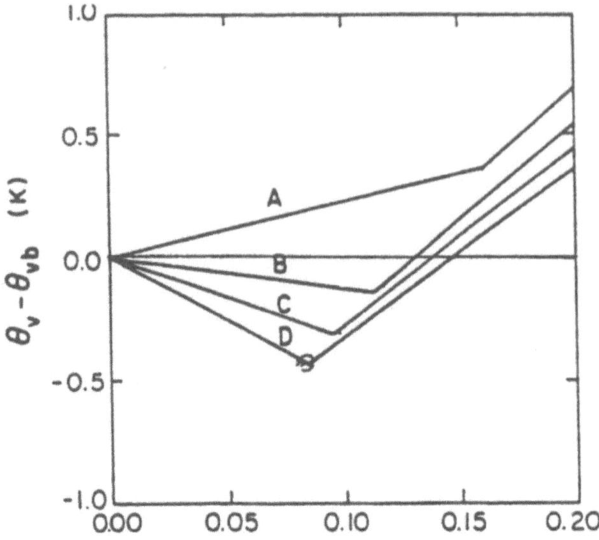

Figure 6. The θ_v difference against the fraction of the unsaturated air in the mixture for cases A, B, C, and D indicated in Fig. 5. The mixing ratio of liquid water content in the cloudy parcel is 0.38 g/kg. (From Kuo and Schubert [4])

ward transport processes can apparently provide enough water vapour to compensate for the evaporative effects associated with the weak evaporative instability".

Siems *et al.* [17] used the *buoyancy reversal* concept to interpret the CTEI; buoyancy reversal is a condition that after mixing parcels become denser than either of the unmixed fluids, and cannot return to their original place. Parcels experiencing buoyancy reversal are those having $\theta_v - \theta_{vb}$ smaller than zero in Fig. 6. Siems *et al.* defined a D value that is the ratio of $\theta_v - \theta_{vb}$ at χ_* and the buoyancy difference between the two unmixed layers, i.e.,

$$D = \frac{\theta_v(\chi_*) - \theta_{vb}}{\theta_{vb} - \theta_{va}}. \tag{19}$$

With this definition, we can show that Randall-Deardorff's instability criterion is $D > 0$, i.e., the minimum buoyancy of all mixtures, $\theta_v(\chi_*)$, becomes smaller than the cloudy-air environment θ_{vb}. Note that the denominator is always negative. Based on Shy and Breidenthal [16] laboratory experiments in a convective tank described next, and a two-dimensional numerical simulation, Siems *et al.* concluded that the D value has to be much larger

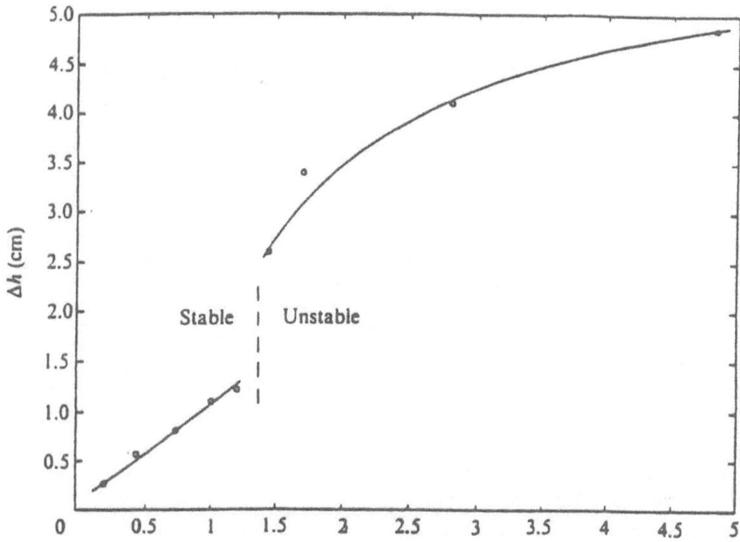

Figure 7. The entrained fluid as a function of buoyancy reversal parameter for large box experiment. (From Shy and Breidenthal [16])

than zero in order for sustained vigorous mixing (i.e., instability) to occur, a condition they believe is unlikely to occur in nature.

The Shy and Breidenthal [16] lab tank consists of two layers of fluids that can mix nonlinearly. These two fuilds are initially separated by a plate. The lower fluid is denser than the upper one, and after a sudden removal of the plate, these two fluids mix nonlinearly and produce mixtures that are denser than either of the two initial fluids. This mixing phenomenon is similar to the mixing of clear and cloudy air; after clear air is entrained into the cloud layer, evaporation occurs, and the mixtures can become denser than either of the unmixed clear or cloudy air. The laboratory results showed that when the buoyancy reversal is relatively large, $D > 1.3$, the whole system experiences "explosive entrainment" and becomes unstable, in the sense that "the perturbed interface did not relax back to horizontal, but developed a strong tilt." Because of the explosive entrainment, the depth of the lower fluid, Δh, increases significantly when $D > 1.3$, as shown in Fig. 7, taken from Shy and Breidenthal.

There is still debate going on as whether the CTEI phenomenon ever occurs in nature. This problem is difficult to answer from field experiments because the jump conditions of the thermodynamic variables appearing in (14) are very difficult to measure. In nature there are many other possible factors that can destroy the cloud layer; for example at daytime solar heating inside the cloud can thin or dissipate the cloud, and when decoupling occurs the moisture supply from the surface, which is needed to maintain the cloud moisture, can be totally cut off (ref. Lenschow [5]).

4. GCSS Boundary Layer Cloud Studies

Marine stratocumulus has been recognized as one of the crucial components in the cloud and climate issue. To study this cloud regime and to develop cloud parameterization schemes for use in climate models, an international orgnization, the GEWEX Cloud System Study (GCSS), formed a Boundary Layer Cloud Working Group. The idea is to use LES or cloud resolving models to generate a numerical database. To do that, the first step is naturally to examine the accuracy of these numerically generated data in representing cloud statistics in the context of developing or calibrating cloud parameterization schemes for climate models. Since there are very few complete observational datasets with which to evaluate the simulation performance, an alternative method to check the accuracy of the numerical database is to intercompare simulation results from different numerical codes to see if results are sensitive to numerical schemes or parameterizations of radiation, condensation/evaporation, and subgrid-scale turbulent motions.

A series of such intercomparison studies to cover different STBL types have taken place over the past few years. The first study took place at National Center for Atmospheric Research (NCAR) in August 1994. In that study, we chose an idealized STBL: horizontally homogeneous, nearly solid cloud deck, no solar heating, no decoupling, no drizzle, plus negligible amounts of surface heating and wind shear. Ten LES groups (shown in Tables 1 and 2, from Moeng *et al.* [11]; for the full names of the abbreviated institutes shown in the tables, refer to Moeng *et al.*), four 2D cloud-resolving modelers (CRMs), and six 1D cloud scheme modelers participated in that intercomparison workshop. We see from these Tables, the numerics and parameterizations of subgrid-scale turbulence, condensation, and longwave radiation are different among most of the LES codes. These numerics and parameterizations treatments raise uncertainties in LESs.

The results of that first intercomparison study of LES and 2D CRM were reported in Moeng *et al.* [11] and those of 1D models were given in Bechtold *et al.* [1]. Here I will talk about the LES intercomparion result only. Before we discuss the results, let me first show in Fig. 8 the major buoyancy forcing of this STBL case, i.e., cloud-top radiative cooling, which can only be internally determined in the LESs.

The maximum cooling at the cloud top shown in Fig. 8 varies from −3 to −6 K h^{-1}. KNMI (the Royal Netherlands Meteorological Institute) and

TABLE 1. The large-eddy simulation codes: scientists and numerics

LES	Scientist	Advection	Time Scheme	Time Step (sec)
KNMI	Pier Siebesma	2nd-order FD	Leapfrog	4
UKMO	Malcolm MacVean	2nd & 3rd FD	Leapfrog	≈0.46
UMIST	Jacques Pasquier	2nd & 3rd FD	Leapfrog	0.6
UW	Hugh Rand	2nd-order FD	Leapfrog	5
MPI	Andreas Chlond	2nd & Bott FD	AB 2nd	2
NCAR	Chin-Hoh Moeng	mixed spectral-FD	AB 2nd	0.5
UOK	Marat Khairoutdinov	5th-order FD in x,y	AB 3rd	2
ARAP	Ian Sykes	2nd-order FD	Leapfrog	≈2
WVU	Steve Lewellen	2nd-order FD	Leapfrog	≈2
CSU	Bjorn Stevens	6th-order FD	Leapfrog	2

TABLE 2. The SGS turbulence, condensation, and longwave radiation schemes used in the LES codes

LES	SGS-turbulence	SGS-condensation	IR scheme
KNMI	Deardorff TKE eq.	0 or 100%	Stephens and Welch et al.
UKMO	Smagorinsky-Lilly	0 or 100%	grey-body, q_l only
UMIST	Smagorinsky-Lilly	0 or 100%	mixed emissivity
UW	Smagorinsky-Lilly	0 or 100%	Roach-Slingo, q_l only
MPI	Deardorff TKE eq.	S-D SGS condensation	Cox-Stephens, q_l only
NCAR	Deardorff TKE eq.	0 or 100%	mixed emissivity
UOK	Deardorff TKE eq.	0 or 100%	mixed emissivity
ARAP	ARAP TKE eq.	S-D SGS condensation	grey-body, q_l only
WVU	ARAP TKE eq.	S-D SGS condensation	grey-body, q_l only
CSU	Smagorinsky-Lilly	explicit microphysics	mixed emissivity

UW (the University of Washington) produce the largest cloud-top cooling, and as we will see later, these two LESs produce the largest buoyancy flux within the cloud layer. Unlike the intercomparison study of the clear convective PBL performed by Nieuwstadt *et al.* [8] where the surface buoyancy flux forcing is given, we cannot just prescribe this cloud-top buoyancy flux forcing in the STBL and this makes the model intercomparison of the STBL more difficult.

Figures 9–12 show the vertical distributions of mean fields, fluxes, velocity variances, and the TKE budgets calculated from the sceond hour of simulation from the ten LES codes. Although they differ by magnitude,

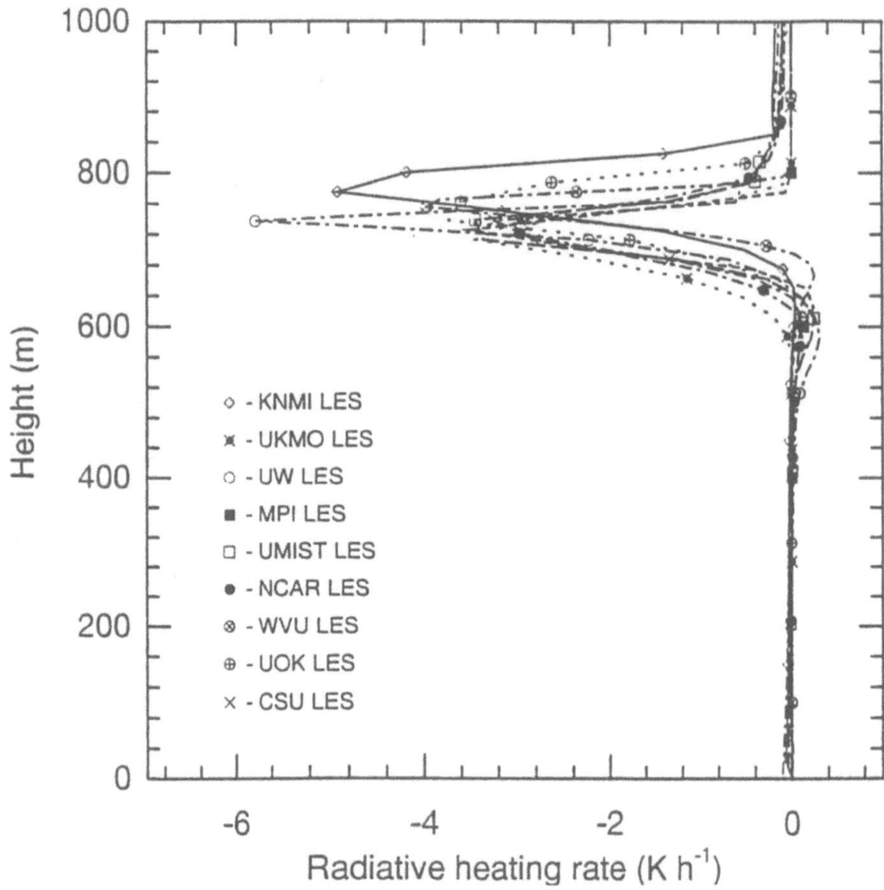

Figure 8. Vertical profiles of longwave radiative heating rate averaged over the second hour of simulation. (From Moeng *et al.* [11])

these statistics are quite similar in shape despite the code variations. Most of the magnitude differences can be explained by the difference of the cloud-top radiative cooling forcing. For example, the UW and KNMI, which generated the largest cloud-top forcing, simulated the largest turbulent fluxes and velocity variances as well; on the other hand, ARAP and WVU, which had the smallest amount of cloud-top cooling, simulated the smallest fluxes and velocity variances. The large variation in the mean liquid water content is worrying; this variation is partly due to the different saturation mixing ratio algorithm used in different models and partly due to the variation of the predicted entrainment rate. The large variation in the total water flux is also due to the variation of the entrainment-rate prediction.

The entrainment rate was found to differ significantly among the LESs,

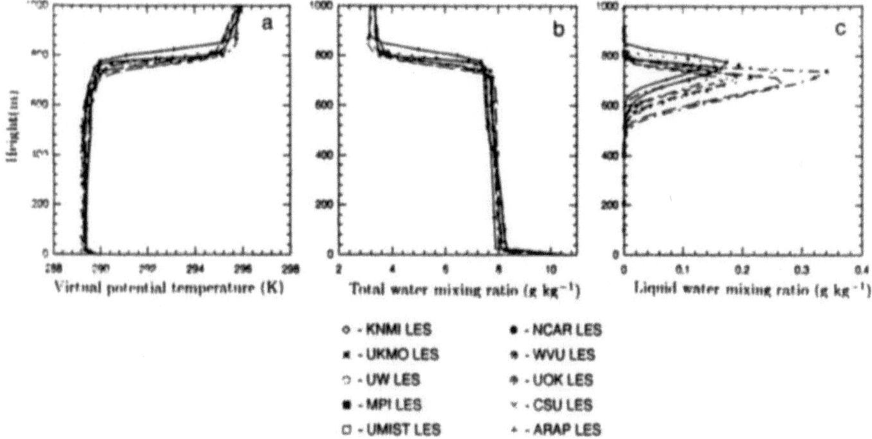

Figure 9. Vertical profiles of a) virtual potential temperature, b) total water mixing ratio, and c) liquid water mixing ratio averaged over the second hour of simulation from the ten LES codes. (From Moeng *et al.* [11])

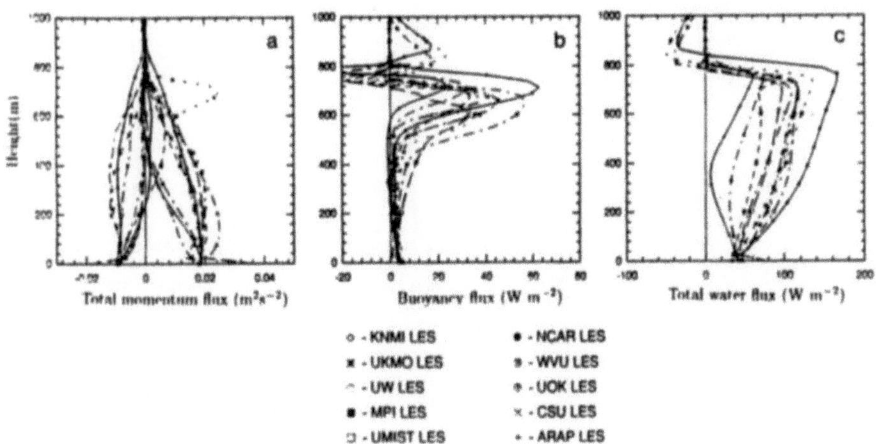

Figure 10. Same as Fig. 9 but for a) momentum fluxes of u and v, b) buoyancy flux, and c) total water flux. (From Moeng *et al.* [11])

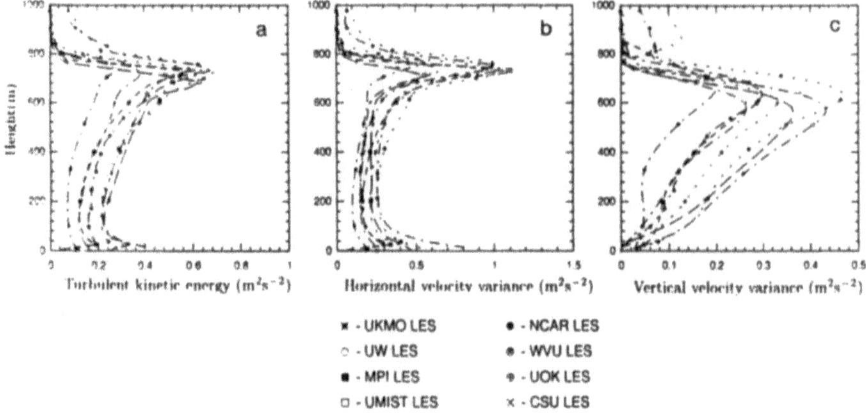

Figure 11. Same as Fig. 9 but for a) total TKE, b) resolved-scale horizontal velocity variance, and c) resolved-scale vertical velocity variance. (From Moeng *et al.* [11])

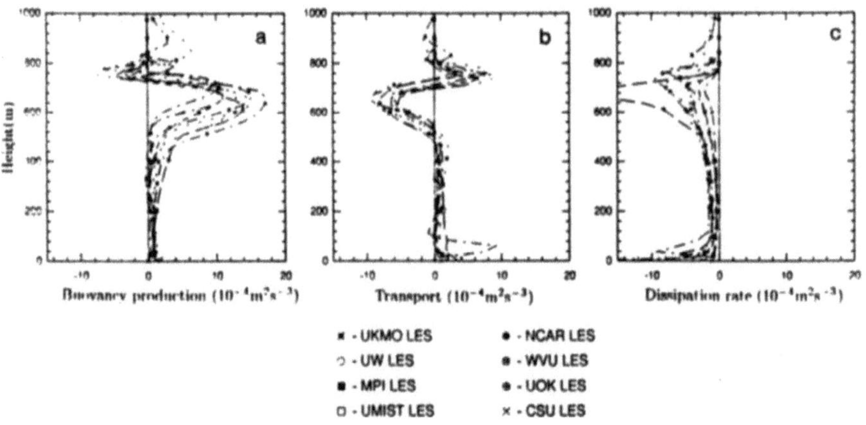

Figure 12. Same as Fig. 9 but for a) buoyancy production, b) turbulent transport, and c) molecular dissipation rate of the TKE budget. (From Moeng *et al.* [11])

TABLE 3. The averaged growth rate of the cloud-top height during the second hour of simulation

LES	averaged dz_i/dt (cm s^{-1})
KNMI	1.68
UOK	1.37
UMIST	0.94
NCAR	0.906
UKMO	0.856
CSU	0.686
UW	0.512
MPI	0.441
WVU	0.278
ARAP	0.236

as indicated in Table 3 (from Moeng *et al.* [11]). Among the model uncertainties due to numerics, subgrid-scale turbulence, subgrid-scale condensation, and radiation, none could be identified to explain such differences.

This led to the second intercomparison study which focused on the entrainment-rate prediction performance. This workshop was held in 1995 hosted by the KNMI of the Netherlands. To study just entrainment we chose an idealized smoke cloud case in which the condensation/evaporation effect can be neglected. In addition the same formula was given to calculate the longwave radiative cooling at the cloud top. This way, the variations on subgrid-scale condensation and radiative forcing were eliminated (or minimized) among the different LES codes. The result of this study was summerized in Bretherton *et al.* [2].

The third intercomparison workshop in 1996 focused on a simulation of an observed case, namely an ASTEX (Atlantic Stratocumulus Transition Experiment) case. This study also examined the effect of drizzle in STBL structure. This intercomparison study is now being analyzed and summarized by Peter Duynkerke (at Utrecht University, Netherlands).

References

1. Bechtold, P., Krueger, S.K., Lewellen, W.S., van Meijgaard, E., Moeng, C.-H., Randall, D.A., van Ulden, A., and Wang, S. (1996) Modeling a stratocumulus-topped PBL: Intercomparison among different one-dimensional codes and with large eddy simulation. *Bull. Amer. Met. Soc*, **77**, 2033–2042.
2. Bretherton, C.S., MacVean, M.K., Bechtold, P., Chlond, A., Cotton, W.R., Cuxart,

J., Cuijpers, H., Khairoutdinov, M., Kosovic, B., Lewellen, D., Moeng, C.-H., Siebesma, P., Stevens, B., Stevens, D.E., Sykes, I., and Wyant, M.C. (1998) An intercomparison of radiatively-driven entrainment and turbulence in a smoke cloud, as simulated by different numerical models. *Quart. J. Roy. Meteor. Soc.*, submitted.

3. Deardorff, J.W. (1980) Cloud-top entrainment instability. *J. Atmos. Sci.*, **37**, 131–147.
4. Kuo, H and Schubert, W.H. (1988) Stability of cloud-topped boundary layers. *Quart. J. Roy. Meteor. Soc.*, **114**, 887–916.
5. Lenschow, D.H. (1998) Observations of clear and cloud-capped convective boundary layers and techniques for probing them. See this volume.
6. Lilly, D.K. (1968) Models of cloud-topped mixed layers under a strong inversion. *Quart. J. Roy. Meteor. Soc.*, **94**, 292–309
7. Nicholls, S. (1989) The structure of radiatively driven convection in stratocumulus. *Quart. J. Roy. Meteor. Soc.*, **115**, 487–511.
8. Nieuwstadt, F.T.M., Mason, P., Moeng, C.-H. and Schumann, U. (1993) Large-eddy simulation of the convective boundary layer: A comparison of four computer codes. *Turbulent Shear Flows 8*, Springer-Verlag, 343–367.
9. Moeng, C.-H. and Arakawa, A. (1980) A numerical study of a marine subtropical stratus cloud layer and its stability. *J. Atmos. Sci.*, **37**, 2661–2676.
10. Moeng, C.-H. and Schumann, U. (1991) Composite structure of plumes in stratus-topped boundary layer. *J. Atmos. Sci.* **48**, 2280–2291
11. Moeng, C.-H., Cotton, W.R., Bretherton, C., Chlond, A., Khairoutdinov, M., Krueger, S., Lewellen, W.S., MacVean, M.K., Pasquier, J.R.M., Rand, H.A., Siebesma, A.P., Stevens, B. and Sykes, R.I., (1996) Simulation of a stratocumulus-topped planetary boundary layer: Intercomparison among different numerical codes. *Bull. Amer. Meteor. Soc.*, **77**, 261–278.
12. Randall, D.A. (1980) Conditional instability of the first kind upside-down. *J. Atmos. Sci.*, **37**, 125–130.
13. Randall, D.A. (1987) Turbulent fluxes of liquid water and buoyancy in partly cloudy layers. *J. Atmos. Sci.*, **44**, 850–858.
14. Randall, D.A., Coakley, J. A., Fairall, C.W., Kropfli, R.A., Lenschow, D.H. (1984) Outlook for research on subtropical marine stratiform clouds. *Bull. Amer. Meteor. Soc.*, **65**, 1290–1301
15. Shen, S. and Moeng, C.-H. (1993) Comparison of a computer-simulated stratus-topped boundary layer with aircraft observations. *Boundary Layer Meteorol.*, **65**, 29–53
16. Shy, S.S. and Breidenthal, R.E. (1990) Laboratory experiments on the cloud-top entrainment instability. *J. Fluid Mech.*, **214**, 1–15
17. Siems, S.T., Bretherton, C.S., Baker, M.B., Shy, S.S., and Breidenthal, R.E. (1990) Buoyancy reversal and cloud-top entrainment instability. *Quart. J. Roy. Meteor. Soc.*, **116**, 705–739
18. Slingo, T. (1990) Sensitivity of the Earth's radiation budget to changes in low clouds. *Nature*, **343**, 49–51
19. Stevens, B., Cotton, W.R., Feingold, G. and Moeng, C.-H. (1998) Large-eddy simulations of strongly precipitating, shallow, stratocumulus-topped boundary layer. *J. Atmos. Sci.*, submitted.
20. Stull, R.B. (1988) *An Introduction to Boundary Layer Meteorology*. Kluwer Academic Publishers. 666pp.

SHALLOW CUMULUS CONVECTION

A. P. SIEBESMA
Royal Netherlands Meteorological Institute (KNMI)
Postbox 201, 3730 AE De Bilt, The Netherlands

1. Introduction

Shallow cumulus convection plays a crucial role in determining the vertical thermodynamic structure of the atmosphere and influences the large-scale circulation significantly in both tropics and mid-latitudes. This is most clearly demonstrated in the Hadley circulation over the tropical oceans, such as schematically illustrated in Fig. 1. In the subtropical belts, the surface evaporation from the ocean increases significantly due to the enhanced vertical transport of heat and moisture by shallow cumulus clouds, which are usually referred to as trade-wind cumuli. This enhanced moisture, collected in the trade-wind boundary layer, is transported by the trade winds towards the intertropical convergence zone (ITCZ), where it is finally released as latent heat in deep convective precipitating cumulus towers. Since this latent heat release is an important engine of the Hadley circulation, the enhanced surface evaporation upstream in the trade-wind region can be regarded as a fuel supply for this circulation. It is in this context that the presence of shallow cumuli, though small in individual size and non-precipitating, intensifies the large-scale circulation. Locally, the vertical transport of heat and moisture by the shallow cumulus clouds is also important to counteract the drying and warming effects of the large-scale subsidence induced by the Hadley circulation. As a result, a steady thermodynamic state of the vertical structure of the trade-wind region can be maintained.

As already suggested by the title we will limit ourselves to *shallow* cumulus convection of clouds whose vertical extent is small enough to be non-precipitative. Therefore the word "shallow" should be interpreted as "non-precipitative". Consequently we will not touch upon any microphysical issues dealing with transitions of liquid water to ice and precipitation and the interaction of precipitation with the dynamics. It is exactly the feedback of the microphysics through the precipitation with the dynamics

E.J. Plate et al. (eds.), Buoyant Convection in Geophysical Flows, 441–486.

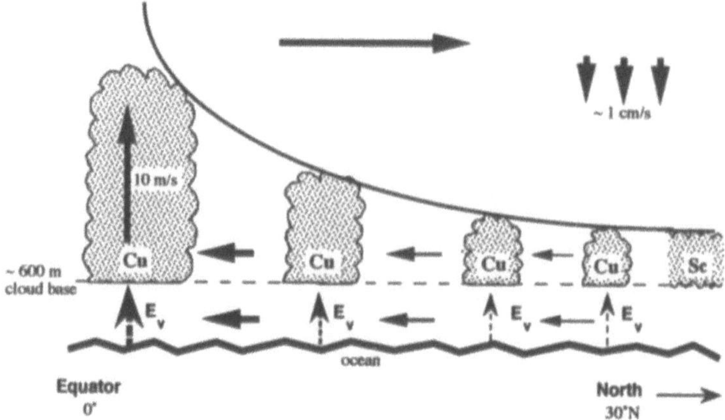

Figure 1. Schematic cross-section of one branch of the Hadley circulation, illustrating the enhancement of moisture convergence in the ITCZ due to the presence of trade-wind cumuli upstream (After Tiedtke[60]).

which makes precipitative *deep* cumulus convection fundamentally different and more complicated than non-precipitative *shallow* cumulus convection.

Main emphasis in this paper is placed on the vertical transport of moisture and heat. It turns out that the mixing of cumulus clouds with the environment is a key issue, both for the fundamental understanding and the more practical issue of parameterization of vertical transport by shallow cumulus convection. Although we are dealing with a relatively simple case of non-precipitative convection we shall see that the theoretical understanding of these mixing processes is still in a rather primitive state. This makes shallow cumulus convection such a challenging topic; while being the most simplest form of cumulus dynamics, apparently it is highly non-trivial and still awaiting better theoretical understanding.

The organization of this paper is as follows. As a prerequisite for the non-expert we give in section 2 a short overview of the essentials of thermodynamics for moist air. Section 3 reviews our current knowledge of *individual* shallow cumuli and their mixing with the environment. Section 4 deals with the dynamics of a cumulus ensemble and its interaction with the environment. After all, it is the interaction of a whole cloud ensemble with the large-scale flow that one needs to understand and to parameterize for large-scale weather and climate model purposes. Since there are only a few observational field experiments dealing with this more global issue, one is more and more relying on numerical cloud resolving models, such as large-eddy simulation (LES) models, for this task. Therefore we will discuss the state of the art of LES modeling for shallow cumulus convection in section 5. Finally in section 6 we will see how all this knowledge has accumulated in parameterizations of cumulus convection for large-scale models.

2. Essentials of Thermodynamics of Cloudy Air

2.1. INTRODUCTION

Everyone who reads a book on atmospheric thermodynamics runs the risk of getting overwhelmed by a seemingly endless list of temperatures. However, in cloud physics, one essentially needs only three temperatures: one for dry adiabatic processes, one for moist adiabatic processes and one for measuring buoyancy. The purpose of this section is to give a simple self-contained introduction into the thermodynamics of cloudy air. For a more complete overview the reader should consult the standard work of Iribarne and Godson [22] or chapter 4 of Emanuel's book on Atmospheric Convection [14] which is more up to date. The content of this chapter is neither rigorous nor complete. Its sole purpose is to offer the reader not familiar with this field, necessary operational knowledge required for understanding the rest of this paper.

2.2. BASIC THERMODYNAMICS OF CLOUDY AIR

We consider an isothermal parcel with volume V, temperature T and a total mass m. It contains a mass m_d of dry air, m_v of water vapor and m_l of liquid water. Since the masses are additive, we can write the total density ρ as the sum of the partial densities

$$\rho = \frac{m}{V} = \rho_d + \rho_v + \rho_l. \tag{1}$$

In order to take the effect of water vapor and liquid water into account for the equation of state we use the gas law for dry air and water vapor

$$p_d = \rho_d R_d T \quad , \quad p_v = \rho_v R_v T, \tag{2}$$

where R_d and R_v are the specific gas constants for dry air and water vapor. If we apply Dalton's law (additivity of partial pressures), assuming that the liquid water does not have an effect on the pressure, we can write the total pressure as

$$p = p_d + p_v, \tag{3}$$

and consequently, the equation of state of a moist parcel, containing dry air, water vapor and liquid water as

$$p = \rho R_d T \left[1 + (\varepsilon^{-1} - 1)\frac{\rho_v}{\rho} - \frac{\rho_l}{\rho} \right], \tag{4}$$

where $\varepsilon = R_d/R_v \cong 0.622$. This result motivates the use of the specific humidity for water vapor q_v and liquid water content q_ℓ defined as

$$q_v \equiv \frac{\rho_v}{\rho} \quad , \quad q_\ell \equiv \frac{\rho_l}{\rho}. \tag{5}$$

We can rewrite the equation of state (4) in a similar form as (2) if we define a virtual temperature T_v

$$p = \rho R_d T_v \quad , \quad T_v \equiv T\,[1 + 0.61 q_v - q_l]\,. \tag{6}$$

Note that T_v can be interpreted as the temperature that dry air must have in order to have the same density as the moist air under consideration. Since for a given pressure, T_v is inversely proportional to the density, it is a direct measure of the buoyancy of an air parcel.

Inspection of (6) shows immediately that moist air ($q_v > 0$) has a higher virtual temperature than dry air ($q_v = 0$) at the same pressure and temperature and hence a lower density. Presence of moisture therefore enhances the buoyancy of a parcel. Liquid water however works in the opposite direction; it increases the density of an air parcel. Since typically $q_v \sim O(10^{-2})$ and $q_\ell \sim O(10^{-3})$, the effect of humidity on T_v is less than 1%. Nevertheless, in cloud dynamics (see subsection 5.4) it turns out that the buoyancy of a cloud, *i.e.* the virtual temperature excess, is largely determined by these subtle humidity effects.

Water vapor condensates into liquid water if it exceeds the saturation specific humidity q_s, which is the maximum amount of water vapor an air parcel can contain before condensation takes place. It can be written using (2) and (3) in terms of the total pressure and the saturation vapor pressure $e_s \equiv \rho_s R_v T$

$$q_s \equiv \frac{\rho_s}{\rho} = \varepsilon \frac{e_s}{p + e_s(\varepsilon - 1)}. \tag{7}$$

The saturation curve e_s can be found by integrating the Clausius-Clapeyron relation [22]

$$\frac{d \ln e_s}{dT} = \frac{L}{R_v T^2}, \tag{8}$$

where L denotes the specific latent heat of vaporization. As a first approximation L may be considered constant with temperature and equal to its triple point value L_0. Integration of (8) from the triple point (e_0, T_0) then readily gives

$$e_s = e_0 \exp\left[-\frac{L_0}{R_v}(\frac{1}{T} - \frac{1}{T_0})\right], \tag{9}$$

with $e_0 = 6.107$ hPa, $T_0 = 273.15$ K, and $L_0 = 2.5 \times 10^6$ J/kg. More accurate solutions for the saturation curve can be found in [22, 14, 62]. The important observation to be made here is that the saturation specific humidity q_s, as defined by (7) is a function of p and T only, and is therefore a state variable.

2.3. CONSERVED VARIABLES

If one wants to study turbulent mixing of heat and moisture it is convenient to eliminate all internal sources and sinks. This can be done by absorbing these sinks and sources into the heat and moisture variables. Working with such "conserved" variables has the advantage that a mixture of two parcels can be easily desribed in terms of a linear combination of the properties of the two individual parcels. We will make explicitly use of this property in section 3.

In the absence of precipitation and the ice phase, condensational effects are the only sinks and sources of the humidity variables q_v and q_l. It is then easy to realise that the sum of these variables, the total water specific humidity

$$q_t = q_v + q_\ell, \tag{10}$$

is a conserved moisture variable, invariant for condensational processes and pressure changes.

For heat there are two sources and sinks that we want to eliminate by including them into a variable: 1) temperature changes due to adiabatic compression and expansion and 2) temperature changes due to condensational effects. We will introduce two temperatures: one that is conserved under "dry" adiabatic processes and another one that is also conserved under "moist" adiabatic processes, *i.e.* adiabatic processes including condensation. Combining the first and second principle of thermodynamics and again neglecting the ice phase and transitions of liquid water to precipitation, one can find for the specific entropy s of a parcel [14, 22, 62]

$$ds(p, T, q_\ell) = c_{pm} d\ln T - R_m d\ln p - \frac{L}{T} dq_\ell, \tag{11}$$

where $c_{pm} = c_{pd} + q_v c_{pv} + q_\ell c_{p\ell}$, and $R_m = R_d + q_v R_v$ are the mean parcel values for respectively the specific heat capacity and the specific gas constant (see Appendix B for precise values of the used constants). Note that the last term of (11) represents condensational effects. No approximations are involved in the derivation of (11). However, for the purpose of shallow cumulus convection it is sufficient to work with an approximate form. We use the dry air values for the specific heat and the specific gas constant (*i.e.* $c_{pm} \sim c_{pd}$ and $R_m \sim R_d$) in (11) and furthermore approximate the last term in (11) by $d(L_0 q_\ell/T)$. For a discussion on the errors made using the above mentioned approximations we refer to Tripoli and Cotton [62].

In the absence of liquid water ($q_\ell = 0$) the last term in (11) vanishes and the remaining terms define the potential temperature θ as $ds \equiv c_{pd} d \ln \theta$. Integrating (11) from a reference state, usually defined by a reference pressure $p_0 = 1000\text{hPa}$ gives

$$\theta = T \left(\frac{p_0}{p} \right)^{\kappa} \equiv T \pi^{-1}, \tag{12}$$

where $\kappa = R_d/c_{pd}$ and π is the Exner function. The potential temperature θ can be interpreted as the temperature an air parcel would obtain if compressed or expanded adiabatically to a pressure of 1000 hPa. It is therefore conserved during adiabatic processes ($ds = 0$) without phase changes ($dq_\ell = 0$). We will refer to such processes as dry adiabatic processes.

If we include phase changes, we can define the potential liquid water temperature θ_l by $ds = c_{pd} d \ln \theta_l$ [6]. Integration of the approximate form of (11) from a reference state defined by $p_0 = 1000\text{hPa}$ and $q_l = 0$ gives

$$\theta_\ell = \theta \ \exp \left(-\frac{L_0 q_\ell}{c_{pd} T} \right). \tag{13}$$

Within the approximations mentioned, θ_l is conserved under moist adiabatic processes ($ds = 0$), including phase changes.

The liquid water potential temperature θ_ℓ has advantages above the still widely used equivalent potential temperature θ_e. Firstly, θ_ℓ is especially useful as a model variable since it reduces to the (dry) potential temperature in the absence of liquid water. Secondly, since $q_\ell \sim O(10^{-3})$ in clouds, the term in the exponent of (13) is always much smaller than 1 so one can use a linearized version of θ_ℓ by employing a Taylor expansion

$$\theta_\ell \approx \theta - \frac{L_0}{c_{pd} \pi} q_\ell. \tag{14}$$

As an alternative for θ and θ_ℓ one can also use the so-called static energies, which in the next section will be proven to be useful for finding lapse rates. They are easily derived from (11), assuming hydrostatic equilibrium $dp = -\rho g dz$ and the gas law (6)

$$dh(T, z, q_\ell) = c_{pm} dT + g dz - L dq_\ell. \tag{15}$$

Although this can be integrated exactly [14], we will approximate the specific heat again by its dry air value ($c_{pm} \sim c_{pd}$) and L by its triple point value L_0. Integration gives for the unsaturated case (*i.e.* $dq_\ell = 0$) the *dry* static energy,

$$h_d = c_{pd}T + gz, \tag{16}$$

and for the saturated case the *liquid* water static energy

$$h_\ell = c_{pd}T + gz - L_0 q_\ell. \tag{17}$$

The static energies h_d and h_ℓ are closely related to θ and θ_ℓ. The dry static energy h_d is (as θ) conserved under unsaturated adiabatic transformations whereas the liquid water static energy h_l (as θ_ℓ) is also conserved under saturated adiabatic transformations. The only additional assumption that went into the derivation of the static energies is that the pressure change should be strictly hydrostatic.

2.4. LAPSE RATES

If a dry air parcel is lifted dry adiabatically it cools due to expansion. The rate at which the temperature is decreasing is given by the dry adiabatic lapse rate Γ_d. Since the dry static energy h_d is constant for such adiabatic vertical displacements, (16) immediately gives for the dry adiabatic lapse rate

$$\Gamma_d \equiv -\left(\frac{dT}{dz}\right)_{parcel} = -\left(\frac{\partial T}{\partial z}\right)_{h_d} = \frac{g}{c_{pd}} \cong 9.76\text{K/km}. \tag{18}$$

If a saturated parcel is lifted adiabatically it will condensate and, due to the resulting latent heat release, the temperature will not fall as rapidly as in the unsaturated (dry) case. To calculate this moist adiabatic lapse rate we substitute $q_\ell = q_t - q_{sat}(p, T)$ in (17). This way h_ℓ is only a function of z, p and T. Using hydrostatic equilibrium plus the fact that h_ℓ and q_t remain constant along a moist adiabat, we find for a saturated parcel, using (17), (7) and (9)

$$\Gamma_m \equiv -\left(\frac{dT}{dz}\right)_{parcel} = -\left(\frac{\partial T}{\partial z}\right)_{h_\ell, q_t} = \frac{g}{c_{pd}} \frac{\left(1 + \frac{Lq_s}{R_d T}\right)}{\left(1 + \frac{L}{c_p}\frac{\partial q_s}{\partial T}\right)} < \Gamma_d. \tag{19}$$

The moist adiabatic lapse rate Γ_m is always smaller than the dry adiabatic lapse rate Γ_d. Its actual value however depends strongly on temperature. For $T = 290$ K and $p = 1000$ hP we find $\Gamma_m = 4.5$ K/km, much smaller than Γ_d. For lower temperatures, however, the difference between Γ_m and Γ_d becomes progressively smaller.

2.5. LOCAL STABILITY OF THE ATMOSPHERE

The presence and intensity of convection is largely determined by the thermodynamic stability of the atmosphere. This stability is usually defined by the sign of the buoyancy of a vertically displaced atmospheric parcel. More precisely, given average profiles of $\overline{T}$ and $\overline{q_t}$, we consider a parcel at a height z_0 with the average atmospheric properties at that height and displace it vertically an infinitesimal distance δz. The resulting vertical velocity w is then given by Newton's second law of motion

$$\frac{dw}{dt} = \frac{d^2\delta z}{dt^2} = B\left(z + \delta z\right) \cong \frac{dB}{dz}\delta z, \tag{20}$$

where the buoyancy force per unit mass B working on the displaced parcel is given by

$$B = -g\frac{\rho_p - \overline{\rho}}{\rho_p} = g\frac{T_{v,p} - \overline{T_v}}{\overline{T_v}}, \tag{21}$$

where the subscript p denotes parcel properties. In the last step of (21) we used the equation of state (6) and assumed that the parcel's pressure adjusts instantaneously to the environmental pressure, *i.e.* $p_p = \overline{p}$. In (20) we recognize the equation of the harmonic oscillator whose solution is either oscillating or exponentially growing, depending on the sign of dB/dz. Therefore, it is natural to define the local stability of the atmosphere at a height z_0 as

$$\begin{aligned} \frac{d\overline{T}_v}{dz} > \frac{dT_{v,p}}{dz} &\Longleftrightarrow \text{stable} \quad , \\ \frac{d\overline{T}_v}{dz} < \frac{dT_{v,p}}{dz} &\Longleftrightarrow \text{unstable} \quad , \end{aligned} \tag{22}$$

which means that one has to compare the lapse rate of T_v of a parcel with the vertical gradient of the atmospheric profile of T_v. The virtual temperature lapse rate can be calculated along the same lines as the temperature lapse rates in the previous subsection. Ignoring for simplicity effects due to the vertical gradient of $\overline{q}_t$ it is easy to see that for an unsaturated atmosphere at z_0, the borderline between stable and unstable is simply given by the *dry* adiabatic lapse rate Γ_d, while for a saturated atmosphere it is the *moist* adiabatic lapse rate Γ_m.

This property allows the introduction of an important concept of conditional instability. Suppose that locally the atmosphere has a vertical temperature gradient in between a moist and dry adiabat such as schematically illustrated in Fig. 2. This implies that the atmosphere at such a point is stable for unsaturated parcels but unstable for saturated parcels. In that case we speak of conditional instability which can be formulated as

$$\begin{aligned} (-\partial\overline{T}/\partial z) > \Gamma_d \quad &\Longleftrightarrow \quad \textit{Absolute instability} \\ \Gamma_m < (-\partial\overline{T}/\partial z) < \Gamma_d \quad &\Longleftrightarrow \quad \textit{Conditional instability} \\ (-\partial\overline{T}/\partial z) < \Gamma_m \quad &\Longleftrightarrow \quad \textit{Absolute stability.} \end{aligned} \tag{23}$$

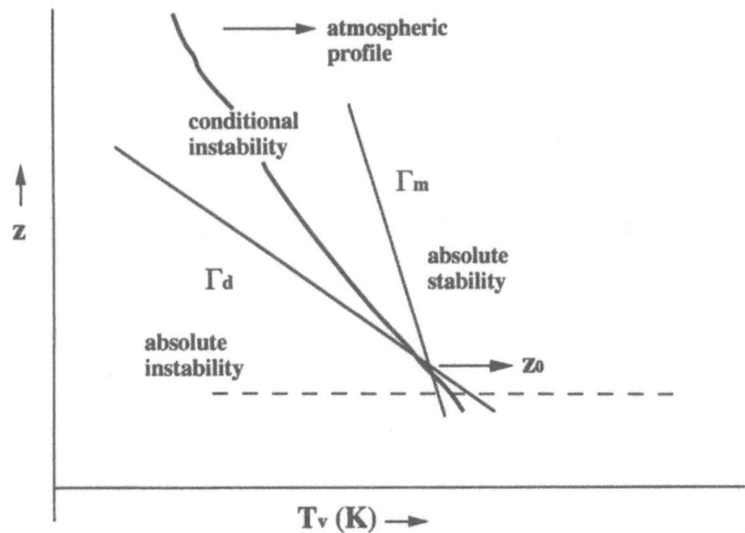

Figure 2. A conditional unstable profile together with a dry and a moist adiabat.

Although conditional instability is a purely mathematical concept, it is a typical property of the atmosphere in the cloud layer; while saturated cloud parcels are unstable with respect to vertical pertubations, unsaturated parcels are stable. The concept of local conditional instability should not be mistaken for conditional instability of atmospheric layers in which finite-amplitude displacements are considered. This is the topic of the next subsection.

2.6. THE PARCEL METHOD AS A POOR MAN'S CLOUD MODEL

We will now proceed to investigate finite-amplitude displacements of parcels and see that the concept of conditional instability has special consequences for moist convection. This is most easily illustrated by considering a typical vertical profile of the virtual potential temperature for a trade-wind-cumulus-topped boundary layer. Such a profile is schematically illustrated in Fig. 3. Near the surface there is a thin absolutely unstable layer, the surface layer, above which lies a dry convective boundary layer. On top of this subcloud layer there is a conditionally unstable cloud layer in which cumulus clouds are formed. Finally, an absolutely stable inversion layer lies on top.

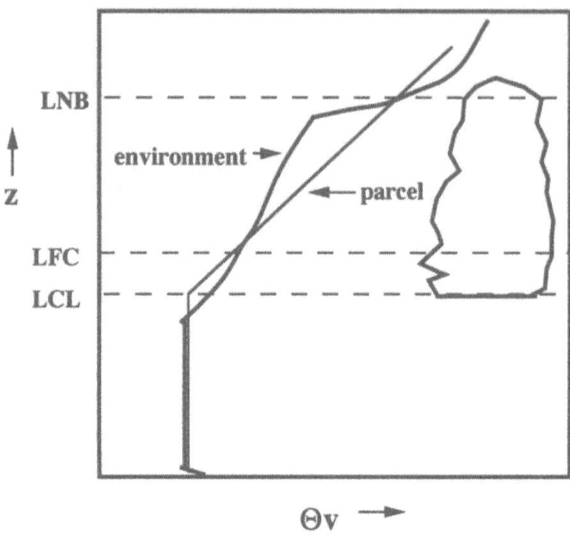

Figure 3. A schematic environmental profile of $\overline{\theta}_v$ for a cumulus topped boundary layer along with a θ_v profile for a parcel that is released from the surface layer.

As a metaphore for a rising thermal, we release a parcel from the surface layer with thermodynamic properties $\overline{T}_0$ and $\overline{q}_{t,0}$ of that layer. As the parcel rises it will cool with a lapse rate of 9.8 K/km until it gets oversaturated at the *lifting condensation level* (LCL). Note that at this point the parcel is negatively buoyant, and would simply fall back unless it has enough kinetic energy to continue. If that is the case it will continue rising but with a moist adiabatic lapse rate as given by (19), since it has a "condensation engine" aboard. Because the cloud layer is conditionally unstable, the moist adiabat describing the parcel's temperature must intersect the atmospheric profile. This occurs at the so-called *level of free convection* (LFC) where the parcel becomes positively buoyant with respect to the environment. From this point on the parcel can rise freely until the moist adiabat crosses the profile again in the inversion layer at the *level of neutral buoyancy* (LNB). Possibly, due to accumulated kinetic energy, the parcel can penetrate slightly further into the inversion (overshoot). Viewing a parcel as a rising thermal, it can serve as a poor man's cloud model and the LCL and the LNB can be used as estimates for cloud base and cloud top, respectively.

This example shows that for finite-displacements the stability of a parcel depends on the origin z_0 of the parcel. Therefore, as a sort of non-local integrated stability function, one can define the *convective available potential energy* (CAPE) as the buoyancy of a parcel, integrated from the height z_0 where it is released to its to its actual position

$$CAPE\,(z, z_0) \equiv \int_{z_0}^{z} B dz = R_d \int_{p(z)}^{p(z_0)} (T_{v,p} - \overline{T}_v) d \ln p, \tag{24}$$

which is directly related to the area between the two curves drawn in Fig. 3. CAPE can be considered as a measure of how much energy there is potentially available for moist convection, but it does not provide a clue on the actual release of this potential energy.

One may wonder why there is positive CAPE in the cloud layer anyhow. This is due to the fact that the region between the LCL and the LFC acts as a potential energy barrier for a parcel lifted from near the surface. This prevents a parcel to rise from the dry boundary layer into the cloud layer, *unless* it has enough kinetic energy to overcome this barrier. So due to this potential barrier, positive CAPE accumulates in the cloud layer as it is not directly consumed and transformed into kinetic energy by cumulus clouds. This leaves us with the physical picture of instabilities diagnosed by CAPE being build up slowly by radiative cooling and surface heating. The CAPE is occasionally consumed if a strong thermal overcomes the barrier and activates a cloud. In that case, the potential energy is transformed into kinetic energy by the cloud dynamics.

The mechanism that activates cumulus convection is called the trigger function [27] and is still a matter of debate, especially for the case of deep convection. In the case of shallow convection where CAPE in general is not so large, *i.e.* the potential barrier is quite small, the trigger is a local phenomenon caused by small perturbations such as local thermals that are strong enough to overcome the barrier.

It is the presence of the potential barrier that makes cumulus convection so much more intermittent in space and time than turbulence in the dry convective boundary layer; cloud updrafts occur in bursts in an otherwise quiet cloud layer. Another consequence is that the dynamics of vertical motions is quite different. Strong saturated updrafts in narrow unstable cloud channels and relatively weak unsaturated downdrafts in the stable environment give a highly positively skewed vertical velocity distribution which should be contrasted with the more classical Kolmogorov turbulence in the dry boundary layer. Figure 4 sums up these differences in a cartoonlike fashion.

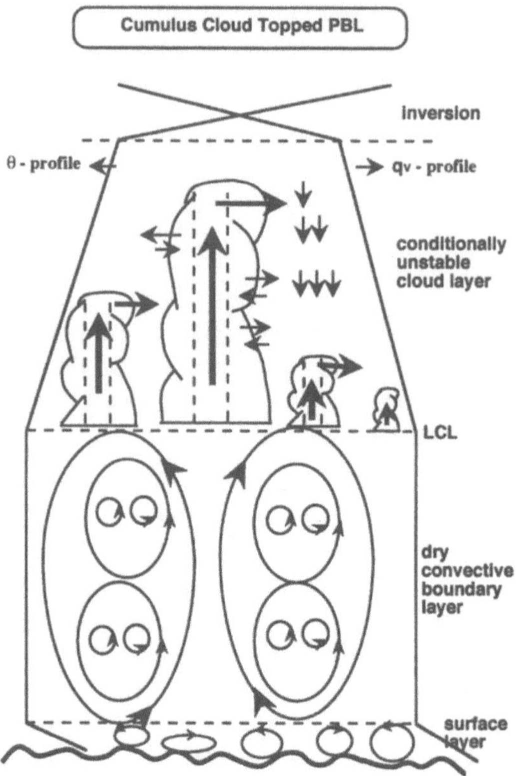

Figure 4. Cartoonlike view of the dynamics of the cumulus topped boundary layer.

3. Mixing and Dynamics of Individual Clouds

3.1. INTRODUCTION

If one observes a cumulus cloud, it is easy to recognize various stages of its life cycle such as sketched schematically in Fig. 5. First, a cloud may be born due to a strong rising thermal in the convective boundary layer that reaches the LCL. If during this first stage, the ascending cloud top is between the LCL and LFC we speak of a *forced cloud* [56] because it is still negatively buoyant and mechanically forced by subcloud thermals below. If the kinetic energy is large enough, the cloud will overcome the potential barrier and reach the LFC after which it becomes positively buoyant. The cloud is then in its *active phase* where rapidly growing turrets with sharp interfaces are visible. This development continues until the cloud attains its mature stage when it reaches its maximum height around or slightly above its neutral buoyancy level (LNB). If finally the fueling of moist air from the subcloud layer below ceases (possibly due to downdrafts induced by the cloud itself), the cloud enters its *passive stage*. During this decaying phase the original cloud base disappears, the cloud interface becomes fuzzy and the cloud will dissolve by evaporation.

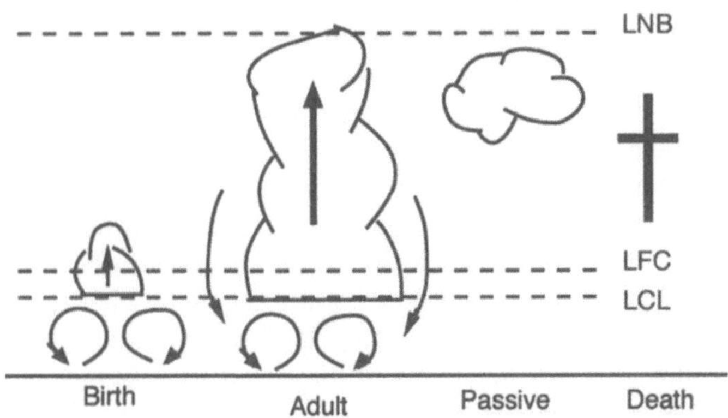

Figure 5. **Schematic life cycle of a cumulus cloud.**

A cloud could simply be described as a rising parcel if, during its active phase, it behaved simply as a burst of subcloud air that is injected into the conditionally unstable layer *without* mixing with its environment and finally detraining by evaporation at the LNB. In that case, modeling of vertical transport of moisture and heat would be straightforward, using the parcel concept introduced in section 2.6.

However, in reality matters are more complicated. Since the work of Stommel [55] it is known that clouds continuously mix air with their environment. More specifically, it was shown by Warner [65] that the liquid water content q_l, measured during many penetrations of cumulus clouds at various heights, is substantially lower than its adiabatic value based on the parcel method. Moreover, it was found that the ratio of the measured liquid water to its adiabatic value decreases systematically with height. This classic result has been confirmed by other authors, at least qualitatively [40, 25], and is an indication that clouds continuously dilute by turbulent mixing with their environment. This process which is usually referred to as *entrainment* and which will be made mathematically precise in section 6, directly affects cloud top height, vertical velocity and in-cloud fields. It is therefore crucial to understand the dynamics and mixing mechanisms of cumulus clouds with their environment. This will be the main theme of the remainder of this section. A review on entrainment in cumulus clouds from a more experimental point of view has been given recently by Blyth [8].

3.2. LATERAL MIXING

The first cloud models, developed in the sixties [54, 47, 48, 49], were all based on models of plumes and thermals [63]. The essence of all these models is that the picture of a rising parcel is extended to include *lateral* mixing of the parcel with its environment during the ascent.

Rather than giving the classic plume similarity arguments to calculate the effects of lateral entrainment, we proceed here by giving some simple hand-waving arguments. In this way we avoid the discussion to what extent plume models can be used for cloud modeling. Consider a cloud with upward rising air with a velocity w_c, which is entraining air laterally from the environment (see Fig. 6a). Further we make the assumption that the entrained environmental air homogenizes instantaneously with the cloudy air. As a result we can use the so-called "top-hat" approximation for thermodynamic variables $\phi = \{\theta_\ell, q_t\}$. This implies that at a height z, ϕ can take only two values, ϕ_c in the cloud and ϕ_e in the environment. Since the only way a conserved variable ϕ_c can change is through lateral mixing, we can write the Lagrangian time derivative of a rising parcel in the general form

$$\frac{d\phi_c}{dt} = F_{mixing}, \tag{25}$$

where we are searching an expression for F_{mixing}. The only parameters available are ϕ_c, ϕ_e and a time scale τ. Since we have no additional knowledge, we choose the simplest form for F_{mixing} and assume that the cloud dissolves by lateral mixing in a typical cloud life-time τ

$$F_{mixing} = -\frac{\phi_c(z) - \phi_e(z)}{\tau}. \tag{26}$$

If furthermore we assume steady-state and substitute (26), we can write (25) in the more familiar form

$$\frac{\partial \phi_c}{\partial z} = -\frac{1}{w_c \tau}(\phi_c - \phi_e) \equiv -\varepsilon(\phi_c - \phi_e), \tag{27}$$

where ε is usually referred to as the fractional entrainment rate. A more fundamental derivation of (27) is given in section 6.4. The result (27) allows a simple physical interpretation for ε. Realizing that a shallow cumulus cloud typically makes one eddy turnover, we can interpret the cloud lifetime τ as the eddy turnover time. In that case, ε^{-1} is proportional to the vertical size h_c of the cloud

$$\varepsilon = \frac{1}{w_c \tau} \sim \frac{1}{h_c}. \tag{28}$$

We can use (28) to make an order of magnitude estimate for ε. If we take $\tau = 10^3$s and $w_c = 1\text{ms}^{-1}$ we find

$$\varepsilon \sim 10^{-3}\text{m}^{-1}. \quad (29)$$

Similarity arguments for plumes give the same result (27), but additionally with a quantitative estimate for the fractional entrainment rate [63]

$$\varepsilon = \frac{2\alpha}{R}, \quad (30)$$

where R is the radius of the plume and α is a dimensionless constant close to 0.1, a value determined from plumes in laboratory experiments. The use of this result in cloud models is, though still being common practice, highly questionable [43].

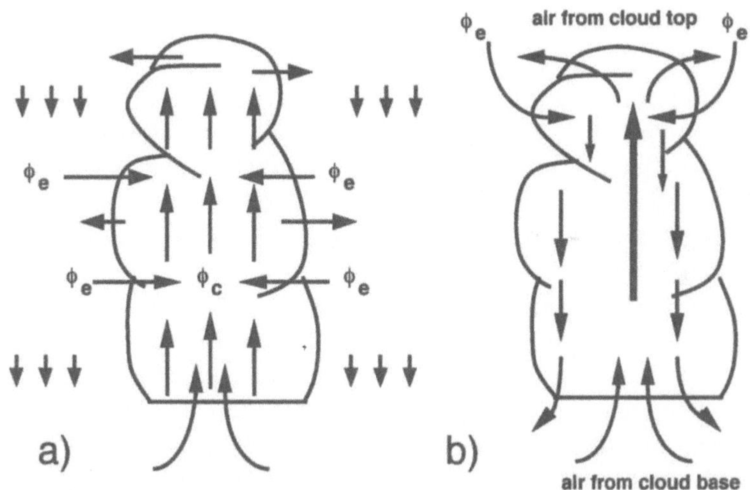

Figure 6. Schematic sketch of two extreme mixing mechanisms: a) *lateral mixing* where environmental air is continuously entrained into upward rising cloud air and b) *vertical mixing* where environmental air mixes with cloud air only near cloud base and cloud top.

3.3. VERTICAL MIXING

The concept of lateral entrainment has received considerable criticism over the last 25 years. A fundamental problem was pointed out by Warner [66]. Comparison of a lateral entraining cloud model with cloud measurements showed that (27) can not simultaneously produce a realistic liquid water content and a cloud top height. Adjusting increasing ε in order to get a

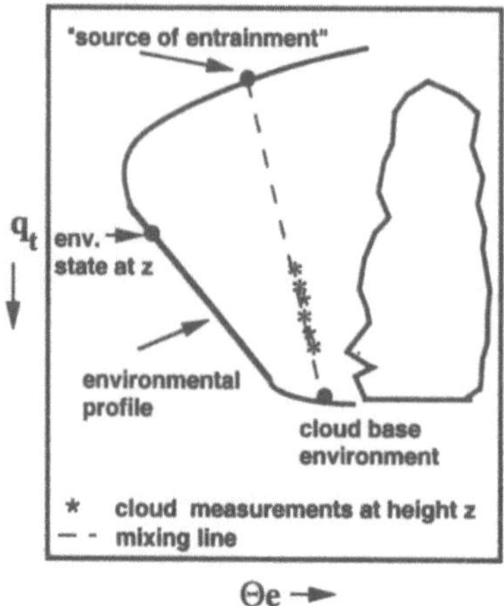

Figure 7. Schematic conserved variable diagram with a environmental sounding and in-cloud data points (∗) measured at height z that fall on a straight mixing line. The upper intersection point of the mixing line with the sounding is usually interpreted as the source of entrainment.

realistic liquid water content predicted a too low cloud top height (*i.e.* LNB). Vice versa, reduction of ε gave a correct cloud top but a too high liquid water content. A second riddle was the observation of strong random fluctuations of liquid water in the cloud with no systematic variations from cloud edges towards the middle, which is hard to explain with a simple lateral entraining cloud model.

This gave rise to the idea, already suggested by Squires [53], that vertical rather than lateral mixing is the principal mechanism for diluting cumulus clouds. In this view a cloud consists of relatively undiluted updrafts that entrain environmental air only at cloud base and at the zero buoyancy level near the maximum cloud top. Subsequently, evaporation will induce negatively buoyant *penetrative downdrafts* in the cloud through cloud top instability (Fig. 6b).

Empirical thermodynamical evidence for such a vertical mixing mechanism was presented by Paluch [38]. The idea is presented in Fig. 7 which shows an idealized conserved variable diagram of a typical environmental sounding in terms of the conserved variables q_t and the equivalent potential temperature θ_e (which could have been θ_l equally well), defined by

$$\theta_e = \theta \, \exp\left(\frac{L_0 q_v}{c_{pd} T}\right). \tag{31}$$

Comparing (31) with the definition of θ_ℓ (13) shows that these two temperatures differ only by q_t in the exponent. Therefore, if θ_ℓ and q_t are conserved variables, so is θ_e.

A conserved variable diagram has the advantage that mixing of two different parcels results in a mixed parcel which falls on a *mixing line*, *i.e.* a straight line connecting the thermodynamic states of the two original parcels. The position of the final mixture on this mixing line is determined by the relative concentrations of the original parcels. Therefore, if cloud air would consists of a two-point mixture of cloud base air and cloud top air, as suggested by Squires [53], in-cloud measurements at a specific height would scatter on a straight mixing line which connects the environmental sounding at cloud base and cloud top (Fig. 7).

By applying this analysis to sailplane measurements inside developing cumuli congestus in Colorado, it was found indeed that cloud data scattered fairly well on a straight line between cloud base and a point (the so-called source of entrainment) several kilometers above the level of observation [38]. These results were interpreted as empirical evidence for vertical mixing of undiluted air from cloud base with environmental air near cloud top through penetrative downdrafts. This view, in some sense the opposite extreme from the lateral entraining plume, resolved the paradox raised by Warner [66] and also explained qualitatively the strong fluctuations of the liquid water which could be induced due to evaporation in the penetrative downdrafts.

3.4. ASCENDING CLOUD TOP ENTRAINMENT

The rather extreme picture of vertical mixing is nevertheless not without problems. Conceptually it is rather odd since it suggests that clouds would only entrain once they have reached their maximum cloud top. Since Paluch [38], numerous studies have been reported that used the same analysis to infer the source of entrainment, with rather ambiguous conclusions. Some studies claimed that the source of entrained air originated near the cloud top [29, 4, 23, 39], but also entrainment sources were reported near the observation level [41, 9, 7]. In most cases the source level was less than 1 km from the level of observation.

A more refined view was put forward by Blyth *et al.* [7] as a result of a study of more than 80 continental cumulus clouds above Montana. A compilation of 44 cases is shown in Fig. 8 and shows that the source of entrained air was only slightly above (~500 m) the level of observation. No dependence of the source level on the existence of up- and downdrafts was found. These findings favor a picture of a shedding thermal for cumulus convection such as sketched in Fig. 8. The top of the cloud consists of a undiluted rising core with a toroidal circulation. This advancing cloud top is inducing mechanically forced downdrafts of the environmental air that is mixed with the core slightly below the advancing top. The resulting mixed

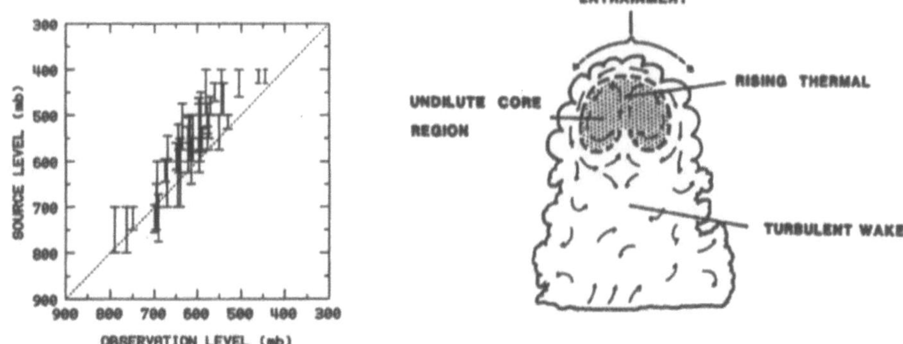

Figure 8. Left panel: A scatter plot of the height of the "source of entrainment" vs. the height of the observation level for 44 different cases using the "mixing line" analysis as explained in Fig. 7. Right panel: Interpretation of the scatter plot leading to a schematic view of a cloud as a shedding thermal. See text for further discussion [7].

parcels have a reduced buoyancy and are left behind in a trailing wake *and stay there.* With this mechanism the mixed air at each level consists of two-point mixture of cloud base air (the undiluted core) and environmental air slightly above the level of observation, in agreement with the observed mixing line.

The weak point of the proposed mechanism is that only undiluted cloud air from the cloud base mixes with the environmental air. Indeed it is true that undiluted air has been found at all levels within cumulus clouds [17, 23] but this air represents only a small fraction of the cloud. It is then difficult to understand how only this small undiluted core region participates in all mixing events with the environment.

3.5. PALUCH ANALYSIS REINTERPRETED; BACK TO AN ENTRAINING PLUME MODEL?

Although the interpretation of straight lines as two-point mixing is tempting due to its simplicity, one should be extremely cautious with it. As pointed out by Taylor and Baker [58] a rather linear distribution of in-cloud (θ_e, q_t) data points can equally well be interpreted as a result of any buoyancy sorting mechanism. Their point is illustrated by drawing a zero buoyancy line in a conserved variable diagram. Such a line consists of all possible combinations of θ_e and q_t that have the same θ_v as the environment at the level of observation. The point is that this zero buoyancy line lies quite close to a mixing line between cloud base and the level of observation. In fact, in-cloud measurents are not decisive enough to discriminate statistically between these two lines. Therefore an alternative interpretation of the straight mixing lines is simply that most cloud parcels are close to their zero buoyancy level. This picture favors the buoyancy sorting mechanism which assumes that all parcel mixtures are en route to their zero buoyancy level. Apparently, parcels with considerable buoyancy excess are so unstable that they are quickly transported to their zero buoyancy level and consequently are quite rarely observed.

Additional evidence for this latter interpretation has been given recently by Lin and Arakawa [30] who performed a Paluch analysis on the output of a 2-D cloud-resolving model. When the data points in a cloud are plotted in a conserved variable diagram they are distributed on a quasi-straight line that intersects the sounding close to the level of observation. Naively one might interpret this as a two-point mixing of cloud base air with environmental air near the observation level. However, by calculating the backward trajectories it was shown that the cloud air originated from *multiple* levels, all *below* the level of observation.

These results put both the extreme vertical mixing mechanism and the shedding thermal model of Blyth [7] in jeopardy since both views lean heavily on the two-point mixing interpretation. Since observational thermodynamic evidence is apparently not sufficient we discuss next some direct kinematic cloud observations.

Aircraft observations of trade wind cumulus bands off the coast of Hawaii were analyzed by Raga *et al.* [40]. For active clouds below the inversion they found that vertical velocities were almost exclusively positive. Only above the inversion equally strong downdrafts were observed. Similar results were found by Jonas [25] who studied small maritime cumulus clouds over the North sea. Traverses through active cumuli showed organized updrafts in the clouds with only a thin shell of downdrafts of around 2 m/s outside the clouds. The values of θ_ℓ and q_t in the downdrafts were not too different from the far field values at the observation heights. This led to the conclusion that the descent around the cloud edges is due to mechanical forcing rather than penetrative downdrafts driven by evaporative cooling. These results are supported by numerical simulations of Klaassen and Clark [28].

In view of these findings the following picture emerges. Following Blyth *et al.* [7], an active cumulus cloud can still be viewed as a rising thermal with a *lateral entraining ascending cloud top*. However, we can relax the condition that only undiluted air rises and mixes with the environment and that mixed air stays behind in a trailing wake. Instead, the rising thermal exists of different mixtures, ranging from the most buoyant undiluted parcels to nearly zero buoyant mixtures. Only when a mixture becomes negatively buoyant it will stay behind and eventually detrain by evaporation. This is essentially the same mechanism as proposed by Taylor and Baker [58].

It does not suffer from the Warner paradox [66] since the cloud top is determined by the undiluted parcels while other parcels dilute the cloud by lateral entrainment. It gives quasi-linear mixing lines [58] and is qualitatively in agreement with the observed kinematics. Ironically enough, this picture of what we will call an *intermittent entraining thermal* is not that different from the old classic plume models. The main difference is that in the present picture the lateral entrained air is assumed *not* to homogenize with the rest of the cloud air instantaneously.

3.6. MODELS FOR MIXING IN CUMULI

Since there is still no consensus on what is the principal mechanism of entrainment (and detrainment) in cumulus clouds, there coexist a wide range of models and schemes, each emphasising different aspects of the mixing mechanism. In order to categorize the various schemes in a comprehensible way it is convenient to introduce two time scales:

- τ_h as the typical time to homogenize lateral entrained air with cloud air, and
- τ_c as cloud life time, *i.e.* $\tau_c \sim h_c/w_c$ where h_c is the vertical size of the cloud and w_c the typical vertical velocity in the cloud (see (28)).

In the extreme case that $\tau_h \ll \tau_c$, lateral entrained air is immediately homogenized with cloudy air at that height. Then, it is sufficient to work only with cloud averaged values. This is the basic assumption of the classic entraining plume model introduced in section 3.2. On the other side of the spectrum we have the situation that $\tau_h \gg \tau_c$. This implies that the entrained air is mixing so slowly with the cloudy air, that it can only take place near the zero buoyancy level where the vertical velocity goes to zero. Therefore this limit covers the vertical mixing case, as described in section 3.3 where cloud air essentially mixes with environmental air only at the maximum cloud top after which the mixture sinks back to its zero buoyancy level. This idea has been translated into a simple model first by Telford [59].

Most models and schemes that have been proposed over the last 10 years are in between these two extremes. If $\tau_h \sim \tau_c$, entrained air can travel a considerable distance in the vertical before being homogenized with the rest of the cloud. Therefore it is necessary to use a probability density function of the cloud fields, or equivalently, an ensemble of rising and descending cloud parcels. We will refer to these class of models as *stochastic* [42] or *episodic* [13]. The essence of all these models is that the cloud is represented by an ensemble of parcels, each one having a different mixing fraction with environmental air (see Fig. 9). Each parcel is then sent to its zero buoyancy level whereafter it is assumed to detrain by evaporation. Since there is a

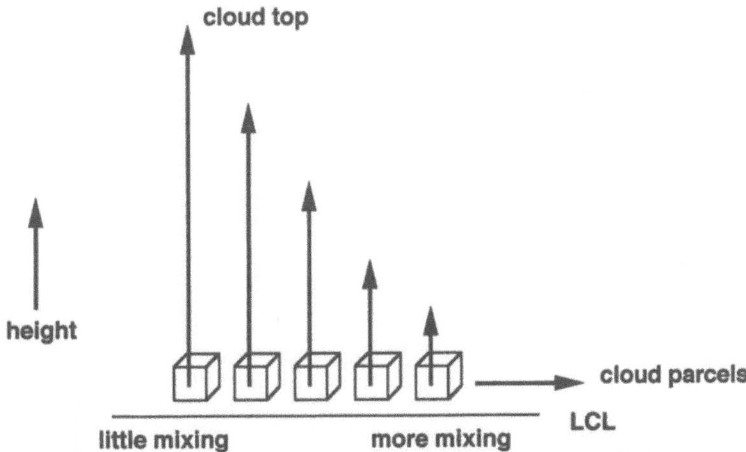

Figure 9. **Principle of stochastic mixing. The cloud is modeled as a ensemble of mixtures, each one having different mixing fractions with environmental air. See text for further discussion.**

lot of free parameters in this approach, many variations on this theme are possible and indeed have been formulated [13, 42, 26, 21].

A model that despite its simplicity captures some essential features of the intermittent entraining thermal described in section 3.5 is the cloud model proposed by Kain and Fritsch [26]. It essentially consists of a rising plume model where at each level equal amounts of cloudy and environmental air are used to make an ensemble of mixtures that is assumed to be Gaussian distributed. The negatively buoyant mixtures are assumed to detrain while the positive buoyant mixtures are mixed into the plume updraft.

4. Observational Large-Scale Budgets of Cloud Ensembles

Present state of the art global circulation models (GCM) have a typical resolution of 100 km. For shallow cumulus convection this implies that a suitable parameterization should take into account the statistical effect of a whole cloud ensemble rather than the dynamics of individual clouds. On the one hand, this is a complicating factor since various clouds may differ substantially, most importantly, in size. Dependent on the initial conditions at cloud base, clouds can reach quite different maximum heights. On the other hand, it facilitates parameterizations since one does not have to consider individual cloud life cycles.

Ideally, to study the effect of a cloud ensemble on the large-scale flow, one ought to analyze large-scale budgets based on observations. Unfortunately, only a few large-scale budget studies are reported. Notable examples are studies based on the undisturbed periods of field experiments such as BOMEX [18, 19, 36, 15], ATEX [3] and more recently TOGA COARE [24].

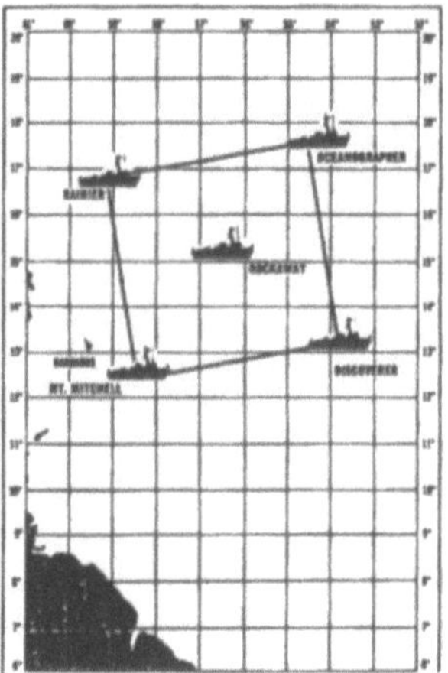

Figure 10. Map of the fixed ship array during BOMEX.

The usual strategy to derive large-scale budgets form such field experiments is still by means of rawinsonde data. As an example, consider the undisturbed period of the Barbados Oceanographic and Meteorological Experiment (BOMEX). Figure 10 shows the fixed ship locations from which soundings were taken each 1.5 hour during BOMEX. The large-scale vertical velocity, *i.e.* the subsidence, can be obtained from the area averaged continuity equation, which, after employing the divergence theorem, is

$$\oint \hat{\mathbf{n}} \cdot \mathbf{v} dl + \frac{\partial \overline{w}}{\partial z} = 0, \tag{32}$$

where $\mathbf{v}$ denotes the horizontal velocity vector, w the vertical velocity and bars indicate area averages. The contour integral is taken along the fixed ship array and $\hat{\mathbf{n}}$ denotes an outward pointing normal unit vector along the contour. Just for the sake of simplicity we assume a constant density. Since the contour integral can be estimated using the rawinsonde data, the large-scale subsidence $\overline{w}$ can be obtained as a residual.

The area averaged budget equations for heat and moisture are

$$\begin{aligned} \frac{\partial \overline{\theta}}{\partial t} + \overline{\mathbf{v}} \cdot \nabla \overline{\theta} + \overline{w} \frac{\partial \overline{\theta}}{\partial z} &= -\frac{\partial \overline{w'\theta'}}{\partial z} + \frac{\pi L}{c_p}(c - e) + Q_R \equiv Q_1, \\ \frac{\partial \overline{q_v}}{\partial t} + \overline{\mathbf{v}} \cdot \nabla \overline{q_v} + \overline{w} \frac{\partial \overline{q_v}}{\partial z} &= -\frac{\partial \overline{w'q_v'}}{\partial z} - (c - e) \equiv -Q_2, \end{aligned} \tag{33}$$

where primes denote deviations from the area averages, and Q_R is the heating rate due to radiation. All the terms on the left hand side of (33) can be obtained by rawinsonde data: the storage term (first term), the large-scale advection term (second term) by employing the divergence theorem and, most importantly, the subsidence term (third term) by using the already diagnosed subsidence $\overline{w}$. As a result, the so-called subgrid processes on the right hand side of (33) are obtained as an overall residual. Historically, these residuals are referred to as the apparent heat source Q_1 and the apparent moisture sink Q_2. They consist of a turbulent flux divergence term (first term) that describes the vertical transport of heat and moisture, a net condensation term $c - e$ (second term) and, in case of the heat equation, an additional radiative heating term Q_R.

Usually during field experiments such as BOMEX, measured surface fluxes of heat and moisture are also available. These can be related to Q_1 and Q_2 by a vertical integration of (33)

$$\begin{aligned} \int_0^\infty (Q_1 - Q_R)dz &= (\overline{w'\theta'})_{srf}, \\ \int_0^\infty Q_2 dz &= -(\overline{w'q'_v})_{srf}, \end{aligned} \tag{34}$$

and provide an additional integral check for the diagnosed subgrid processes.

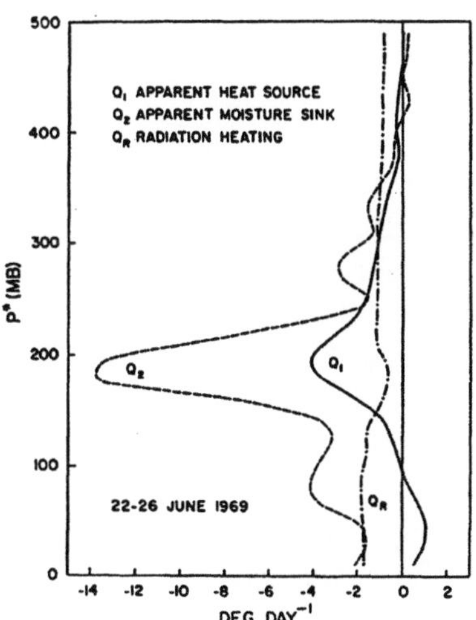

Figure 11. Observed Q_1, Q_2 and Q_r for the undisturbed BOMEX period (from Nitta and Esbensen [36]).

The above described analysis method is not without problems. The left hand side of (33) is in general dominated by the subsidence term, even for the case of a typical divergence of only 5×10^{-6} s^{-1} (corresponding to a subsidence of only 5×10^{-3} ms^{-1} at 1000 m). Such a small divergence can only be diagnosed by (32) if there is a persistent subsidence. This was the case during the undisturbed period of BOMEX where a 4-day stationary state was observed. A summary of the large-scale budget analysis is shown in Fig. 11. Due to subsidence a strong large-scale drying and heating was diagnosed. Since the storage term was small it can be concluded that this large-scale forcing was counteracted by a mixture of turbulent transport, condensational processes and radiative cooling such as diagnosed by Q_1 and Q_2.

In conclusion, observations give at best an estimate for the *overall* subgrid processes such as expressed by Q_1 and Q_2. This is useful information for evaluating models and parameterizations but certainly not enough for a fundamental understanding of the physical processes that are at play and which are needed to design a parameterization on a physically sound basis. Ideally, one would like to have information on how the subgrid processes are build in terms of turbulent fluxes and condensational heating. For this purpose we have to resort to numerical cloud resolving models such as large-eddy simulation (LES) models. This is the topic of the next section.

5. LES Modeling of Shallow Cumulus Convection

5.1. INTRODUCTION

The resolution of a large-eddy simulation (LES) model is such that the largest eddies of a three-dimensional turbulent field are explicitly resolved. Therefore, the bulk of the turbulent field is resolved by the model and the dynamics of the smaller subgrid eddies can be expressed realistically in terms of the resolved eddies, due to the well-known scaling behavior in the inertial subrange. This approach, pioneered by Deardorff [12], has been successfully applied to the dry convective boundary layer for which it was originally designed (see paper of Nieuwstadt [35] in this volume). Sommeria [51] was the first one to use an LES model to study a cumulus-topped boundary layer by including a condensation scheme. Various LES studies of shallow cumulus cases have been reported since then [52, 5, 10, 33, 11, 43]. In this section we will mainly discuss LES results based on BOMEX such as reported in [43] for the following reasons:

- BOMEX describes a shallow cumulus case that is *typical* for the whole trade-wind region, which makes it more universal than "just a case".
- During the undisturbed phase of BOMEX a well-defined steady state of the atmosphere was observed and realistic estimates of the large scale

forcing could be made. This provides a *critical* test for an LES model, since it has to produce a realistic cumulus cloud dynamics that precisely counteracts the prescribed large-scale forcing in order to maintain the steady state.
- Since the LES results indeed succeed in maintaining the stationary state such as observed, the additional advantage of long time averaging is allowed, which improves the output statistics.
- The BOMEX case, such as described in [43] has been subjected to an intercomparison of 11 LES codes [46]. Since the report of this intercomparison is in progress, here we show only results from the LES model used in [43]. However, it should be realized that all the main features we discuss in this section are supported by other LES codes.

5.2. CASE DESCRIPTION AND GENERAL RESULTS

In [43] a large-eddy simulation has been described based on observations during the undisturbed period of BOMEX. During this period, non-precipitating clouds were the only type of cumuli that were observed under steady state conditions. Initial profiles of the specific humidity q_v and the potential temperature θ are shown in Fig. 12 along with observations. Note the typical shape of the profiles, consisting of a well-mixed convective subcloud layer, a conditionally unstable cloud layer topped by a strong inversion.

The large-scale forcing, dominated by subsidence, is prescribed using results of budget studies from BOMEX [19, 43]. This subsidence causes strong drying and heating, especially in the inversion layer. Since this is a case with a low cloud cover it is not necessary to use an interactive radiation scheme but instead a clear sky radiative cooling is prescribed.

Using the prescribed large-scale forcings and the initial profiles as described above, an LES run of 7 hours has been performed. Figure 12 shows the initial profiles and the mean profiles of θ and q_v after 3 and 7 hours of simulation. From this we can see that the model is in a steady-state, in agreement with observations.

This implies that the latent heat surface flux (about 150 $\mathrm{W/m^2}$ in this case) has to be distributed by turbulent transport so as to counteract the large-scale drying by subsidence. In order to gain some insight in how this is achieved we show in Fig. 13 the various terms of the budget equations for q_v and q_ℓ

$$\frac{\partial \overline{q_v}}{\partial t} = -\frac{\partial \overline{w'q_v'}}{\partial z} - (c-e) + \left(\frac{\partial \overline{q_v}}{\partial t}\right)_{\text{forcing}},$$

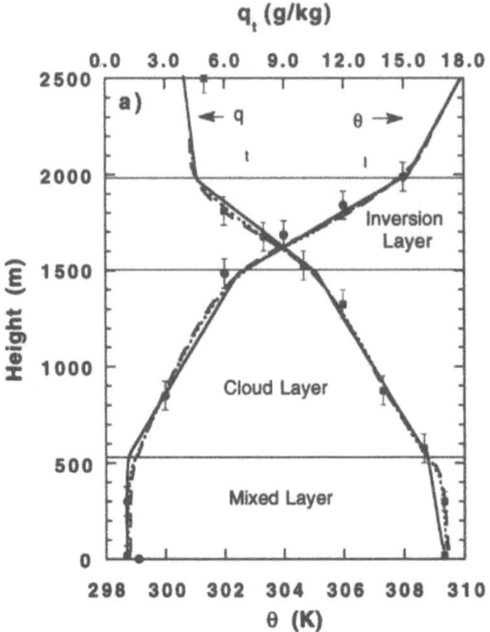

Figure 12. **Horizontally averaged profiles of θ and q_v at times t=0 hr, (full lines) t=3hr (dotted lines) and t=7hr (dashed lines). The circles and squares are the observed values.**

$$\frac{\partial \overline{q_\ell}}{\partial t} = -\frac{\partial \overline{w'q'_\ell}}{\partial z} + (c - e) + \left(\frac{\partial \overline{q_\ell}}{\partial t}\right)_{\text{forcing}}, \tag{35}$$

as determined by the LES model. The bars indicate an area average over the horizontal domain of the model. The large-scale advection and subsidence terms have been collected in one forcing term (*cf.* (33)). In order to improve the statistics, all terms in (35) are averaged over the last 4 hours of the simulation.

For liquid water we observe a strong positive condensation near cloud base around 500m, which is balanced by turbulent transport; the generated liquid water is transported upward, riding on its own generated buoyancy. Near the inversion where the clouds loose their buoyancy the upward transported liquid water is evaporated back into the water vapor. So after all is said and done there is no net liquid water production. The liquid water is only a temporary disguise of the moisture to generate the necessary buoyancy in order to penetrate through the conditionally unstable layer into the inversion layer and to deposit air from subcloud layer there. From the q_v budget we see that the liquid water is not the only moisture that is

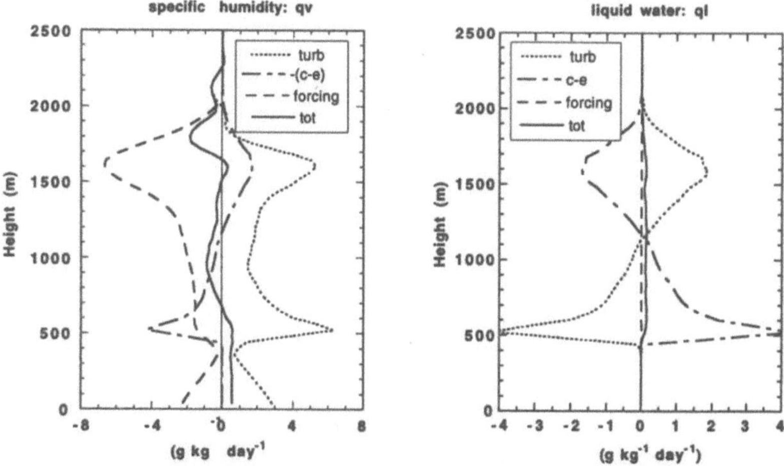

Figure 13. **The various terms from the q_ℓ (left panel) and the q_v (right panel) budget equation (35) integrated over the last 4 hrs of simulation. Terms are as follows: turb = turbulent flux divergence, c-e = condensation minus evaporation, forcing = large-scale forcing and tot: total tendency as a result of all processes.**

transported. Since the air is saturated in the buoyant cloud channels, most of the moisture transport is through the water vapor.

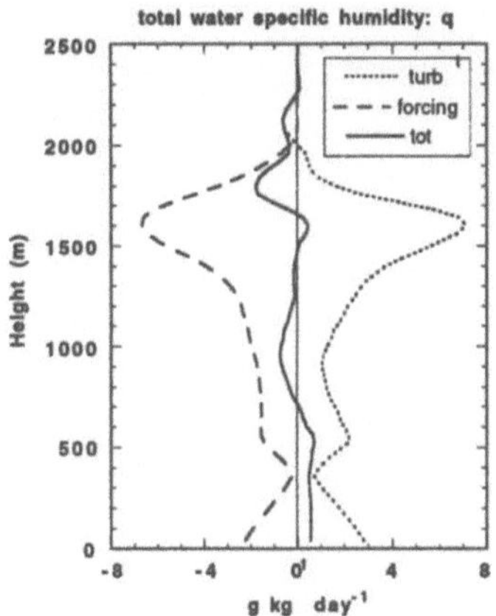

Figure 14. As Fig. 13 but now for the q_t budget equation (36).

The total moisture transport can be easily summarized by considering the sum of the two budget equations (35)

$$\frac{\partial \overline{q_t}}{\partial t} = -\frac{\partial \overline{w'q_t'}}{\partial z} + \left(\frac{\partial \overline{q_t}}{\partial t}\right)_{\text{forcing}} . \qquad (36)$$

Since the condensation term has dropped out we can now directly observe (see Fig. 14) a balance between the large-scale forcing and the moisture transport due to the cumulus convection. For a similar discussion on the heat budget equation we refer to Siebesma and Cuijpers [43].

5.3. CLOUD COVER AND MASS FLUX

Figure 15 shows a snapshot of the 3-D cloud field generated by the LES model. A striking feature is that clouds vary significantly in size. In fact only very few clouds actually do reach the inversion. This is illustrated more quantitatively in Fig. 16a where the fractional cloud cover a_c (see Appendix A for a precise definition) is shown as a function of height.

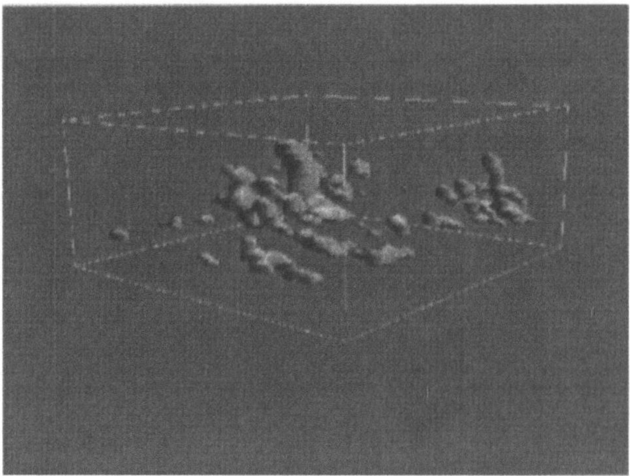

Figure 15. **Snapshot of the 3-D cloud field generated by the LES model**

A strong systematic decrease of a_c with height is apparent. Figure 16a also shows the part of the cloud fraction due to updrafts and downdrafts. From this we conclude that cloud downdrafts play only a minor role. Finally, the cloud core cover, defined as the fraction of cloudy grid points that are positively buoyant, is shown in Fig. 16a. Note that 50 % of the cloudy grid points are positively buoyant. The remaining part is either passive or forced.

The vertical mass transport through the cloud ensemble is measured by the mass flux which is defined as $M \equiv \rho a_c w_c$. Setting ρ equal to 1 for simplicity, it is essentially the product of the cloud cover a_c and the vertical velocity w_c of the cloud ensemble. The mass flux is an important quantity, since in many cumulus parameterizations, the turbulent flux of a field ϕ is estimated using the so-called mass flux approximation.

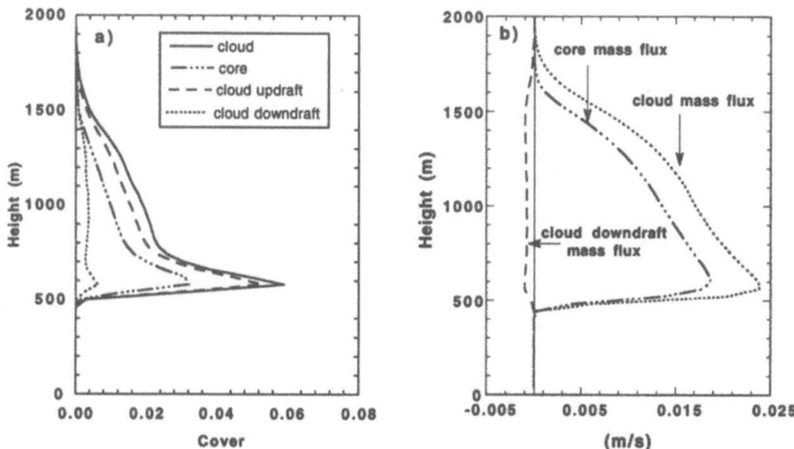

Figure 16. Fractional cover (a) and mass flux (b) of: clouds, cloud updrafts, cloud downdrafts and the cloud core as a function of height. See text for further discussion

$$\overline{w'\phi'} \approx M(\phi_c - \phi_e). \tag{37}$$

In Fig. 16b we show the mass flux for the whole cloud ensemble, the cloud core and the updrafts and downdrafts within the clouds. A few important conclusions can be drawn:

- There is a systematic decrease of the mass flux with height, which is due to the decrease of the cloud cover with height.
- Mass downdrafts in the clouds are unimportant. Apparently, penetrative evaporatively driven downdrafts are not present *within* the clouds.
- Most of the vertical mass transport is carried by the core of the clouds.

5.4. CONDITIONALLY SAMPLED THERMODYNAMIC FIELDS

We saw in section 3 that the in-cloud values of $q_{t,c}$ and $\theta_{l,c}$ are only constant with height, if clouds do *not* mix with the environment. In Fig. 17 we show the conditionally sampled cloud and core fields of q_t and θ_ℓ. The effect of mixing with the environment is clearly visible. Note that the cloud fields are now representing the whole cloud ensemble (see Appendix A for precise definitions) rather than individual clouds as in section 3. From Fig. 17 we can also conclude that the core has mixed less intensively with the environment than the whole cloud ensemble. This is due to the fact that the core points are mostly in the center of the clouds and are effectively shielded from the environment.

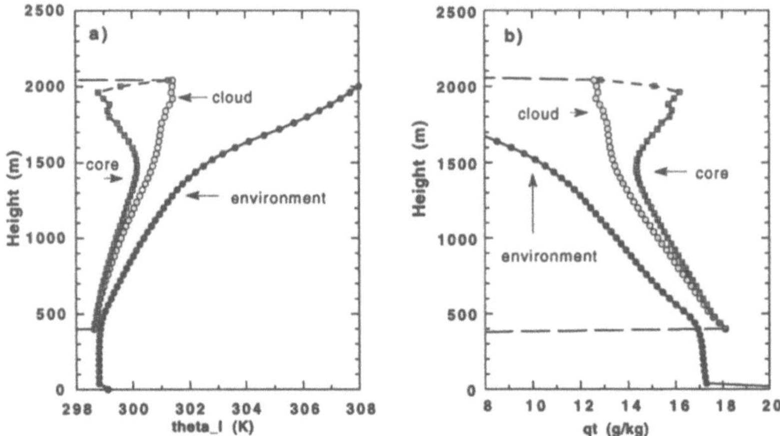

Figure 17. Conditionally sampled cloud and core fields of q_t and θ_ℓ along with the environmental profiles.

In Fig. 18 we show the environmental virtual potential temperature $\overline{\theta_v}$ along with the values of $\theta_{v,c}$ in the clouds and in the core. We find the rather striking result that the cloud ensemble is almost neutrally buoyant with respect to the environment in the cloud layer, while the core is only marginally positively buoyant with a virtual temperature excess of only a few tenths of a degree. Apparently, in the present shallow cumulus case there is only a small potential barrier, *i.e.* little negative CAPE between LCL and the LFC. Hence, there is little chance to build up CAPE, because it is almost directly consumed by the clouds and transferred into kinetic energy, thereby maintaining an almost neutrally buoyant state. This situation has to be contrasted with deep convection where a considerable CAPE can build up that is subsequently released in more violent and precipitating deep convective events. In this sense deep convection is more episodic than shallow cumulus convection.

5.5. ENTRAINMENT OF THE CUMULUS ENSEMBLE

In the previous section we saw how the temperature and moisture of the cloud ensemble are affected by entrainment. In order to quantify the entrainment we assume that the same lateral entrainment arguments for a single cloud (see section 3.2) can now be applied to a cloud *ensemble*. In that case equation (27) can be used and written as

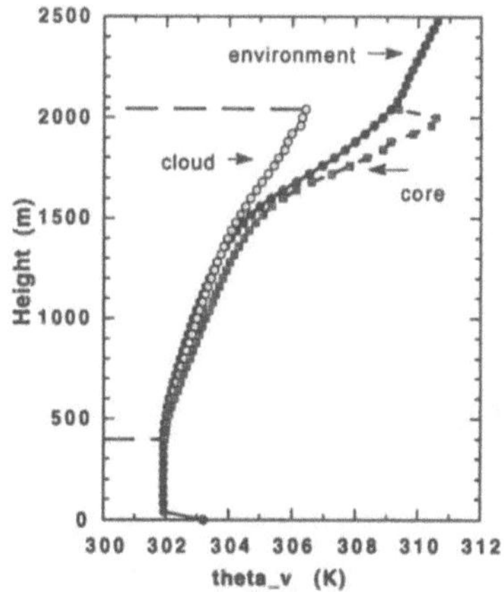

Figure 18. As in Fig. 17 but for θ_v.

$$\varepsilon = -\frac{\partial \phi_c / \partial z}{(\phi_c - \phi_e)}, \tag{38}$$

where the fractional entrainment rate ε should now be interpreted as representative of a whole cloud ensemble. Using the conditionally sampled fields we can therefore determine the fractional entrainment. Results are shown in Fig. 19 for the cloud ensemble (left panel) and the core ensemble (right panel). Each calculation is done for both $\phi = q_t$ and $\phi = \theta_\ell$. From this we can draw the following conclusions:

- We find a typical value of $\varepsilon \approx 3 \times 10^{-3}$ m^{-1} in agreement with the order of magnitude estimate (29).
- This value is much higher than the classic plume result (30): Using a typical cloud radius of 500 m in (30) gives $\varepsilon \approx 2 \times 10^{-4}$ m^{-1}, a factor of 10 smaller than the present LES results.
- As already noted in the previous section, entrainment in the core is systematically smaller than in the clouds.
- Above the cloud base we observe a systematic decrease of ε with height. This can be interpreted as evidence for (28). Assuming that mixing at height z is dominated by those cloud members of the ensemble that actually do reach that height (see Fig. 20), we find, using (28), for the fractional entrainment rate ε of the whole cloud ensemble

$$\varepsilon \sim \frac{1}{z - z_{lcl}}. \tag{39}$$

- Since ε is a purely turbulent mixing parameter, it should only depend on the velocity field. This implies that equation (38) should give the

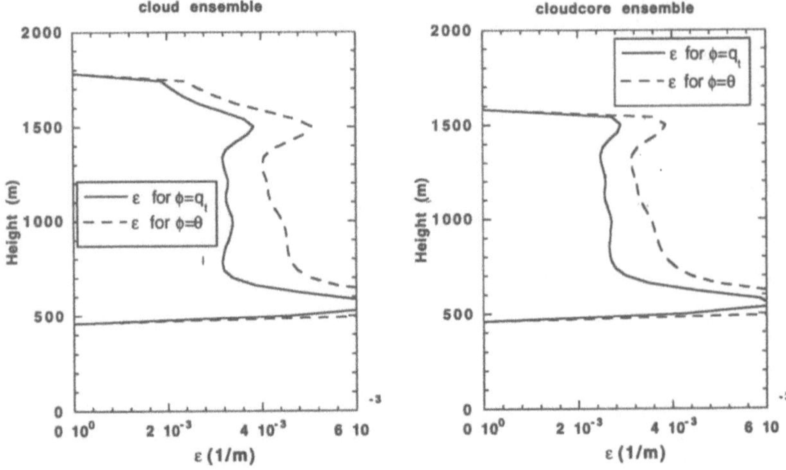

Figure 19. Fractional entrainment rate ε using (38) based on $\phi = \{\theta_\ell, q_t\}$ for both the cloud ensemble (left panel) and the core (right panel).

same results for q_t and θ_ℓ. This is true only approximately (see Fig. 19). Clearly, this is due to one or more of the numerous assumptions that went into (38). This point will be further investigated in the next section.

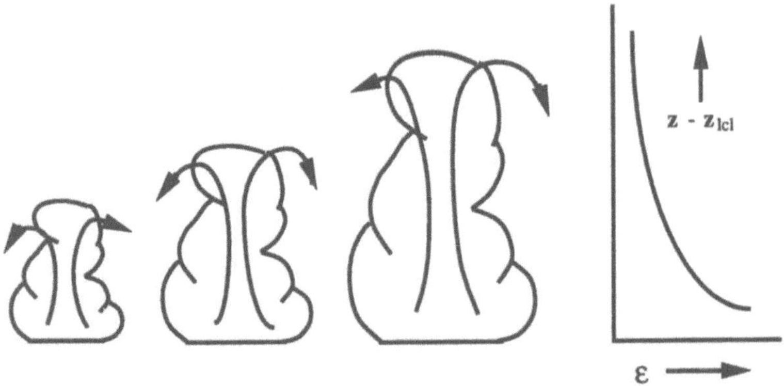

Figure 20. Schematic illustration of the arguments leading to (39). The most active eddies with respect to the turbulent mixing between clouds and the environment at height z are those eddies formed by clouds that actually do reach that height.

6. Parameterizations of Entrainment and Detrainment

6.1. INTRODUCTION

So far, entrainment has been used as a rather vague mixing concept. In fact the only quantitative result, (38), is based on hand-waving arguments. Also in the literature, the concepts of entrainment and detrainment often remain rather descriptive. Therefore, we will start in section 6.2 with precise definitions of entrainment and detrainment. The objective is then to invoke these definitions in separate budget equations for the cloudy part and the environmental part and determine the lateral mass exchange between these regions as residuals. However, in order to do so we have to make a mean field approximation for the lateral mass exchange and test its validity with LES (see section 6.3). Further simplifications made in section 6.4 lead finally to a mass flux parameterization for shallow cumulus. The most important free parameters in this parameterization are precisely the entrainment and detrainment rates. Typical values will be determined from LES. Ideally one would like to have a dynamical parameterization for these rates. An attempt in this direction is finally presented in section 6.5.

We saw in section 5.3 that most of the vertical transport takes place in the core of the clouds. Therefore to remain concrete, from now on the subscripts c and e should be interpreted as the core part and its complementary part. However, it should also be understood that the arguments hold for any arbitrary interface, so the derived equations can also be used for a cloud interface or, in case of the dry convective boundary layer, for an updraft-downdraft interface.

6.2. DEFINITIONS OF ENTRAINMENT AND DETRAINMENT

In Appendix A we derived the spatially averaged continuity equation for the cloudy region

$$\frac{\partial a_c}{\partial t} + \frac{1}{A}\oint_{\text{interface}} \hat{\mathbf{n}}\cdot(\mathbf{u}-\mathbf{u}_{\text{i}})dl + \frac{\partial a_c w_c}{\partial z} = 0, \tag{40}$$

where $a_c = A_c/A$ is the fractional cloud cover, $\hat{\mathbf{n}}$ is a unit vector perpendicular to the interface, $\mathbf{u}$ is the full 3D velocity vector of the mass flow at the interface and $\mathbf{u}_{\text{i}}$ is the velocity of the interface. In mass flux parameterizations [2] the entrainment and detrainment rates E and D are usually introduced as

$$\frac{\partial a_c}{\partial t} + (D-E) + \frac{\partial a_c w_c}{\partial z} = 0. \tag{41}$$

Comparing (40) and (41) shows that $D-E$ can be defined as

$$D - E \equiv \frac{1}{A}\oint_{\text{interface}} \hat{\mathbf{n}}\cdot(\mathbf{u}-\mathbf{u}_{\text{i}})dl, \tag{42}$$

which clearly defines $D - E$ as the net mass exchange across the interface. Let us emphasize that it is the mass velocity $\mathbf{u}$ *relative* to the interface velocity $\mathbf{u_i}$ that enters the definition (42). This way it is guaranteed that $D - E = 0$ if a cloud is simply advected by the mean wind.

Since entrainment deals with inflow and detrainment with outflow we proceed by postulating separate definitions for E and D

$$
\begin{aligned}
E &= -\frac{1}{A}\oint_{\hat{\mathbf{n}}\cdot(\mathbf{u}-\mathbf{u_i})<0} \hat{\mathbf{n}}\cdot(\mathbf{u}-\mathbf{u_i})dl, \\
D &= \frac{1}{A}\oint_{\hat{\mathbf{n}}\cdot(\mathbf{u}-\mathbf{u_i})>0} \hat{\mathbf{n}}\cdot(\mathbf{u}-\mathbf{u_i})dl,
\end{aligned}
\tag{43}
$$

so that entrainment is just defined as that part of the contour integral (42) where there is an inflow of mass into the cloudy part and detrainment as the complementary part where there is mass outflow into the environment (see Fig. 21).

Figure 21. Entrainment is the inflow of mass and detrainment outflow of mass across the interface.

6.3. MEAN FIELD APPROXIMATIONS

Matters become complicated more if we want to employ E and D, as defined in (43), in the budget equations of an arbitrary field ϕ for the core and its complementary, environmental part, (see Appendix A for a derivation),

$$
\begin{aligned}
\frac{\partial a_c\phi_c}{\partial t} &= -\frac{\partial a_c\overline{w\phi}^c}{\partial z} - \frac{1}{A}\oint_{\text{interface}} \hat{\mathbf{n}}\cdot(\mathbf{u}-\mathbf{u_i})\phi dl + a_cF_c, \\
\frac{\partial(1-a_c)\phi_e}{\partial t} &= -\frac{\partial(1-a_c)\overline{w\phi}^e}{\partial z} + \frac{1}{A}\oint_{\text{interface}} \hat{\mathbf{n}}\cdot(\mathbf{u}-\mathbf{u_i})\phi dl + (1-a_c)F_e,
\end{aligned}
\tag{44}
$$

where all possible sinks and sources are absorbed in the forcing term F. At this point it is impossible to infer E and D as defined in (43). In order to obtain budget equations useful for parameterizations we make the mean field approximation that entrainment transports *average* environmental properties into the core and, vice versa, *average* core properties are detrained into the environment, (see Fig. 22),

$$
\begin{aligned}
E_\phi \phi_e &\simeq -\frac{1}{A} \oint_{\hat{\mathbf{n}}\cdot(\mathbf{u}-\mathbf{u_i})<0} \hat{\mathbf{n}} \cdot (\mathbf{u} - \mathbf{u_i}) \phi dl, \\
D_\phi \phi_c &\simeq \frac{1}{A} \oint_{\hat{\mathbf{n}}\cdot(\mathbf{u}-\mathbf{u_i})>0} \hat{\mathbf{n}} \cdot (\mathbf{u} - \mathbf{u_i}) \phi dl.
\end{aligned} \tag{45}
$$

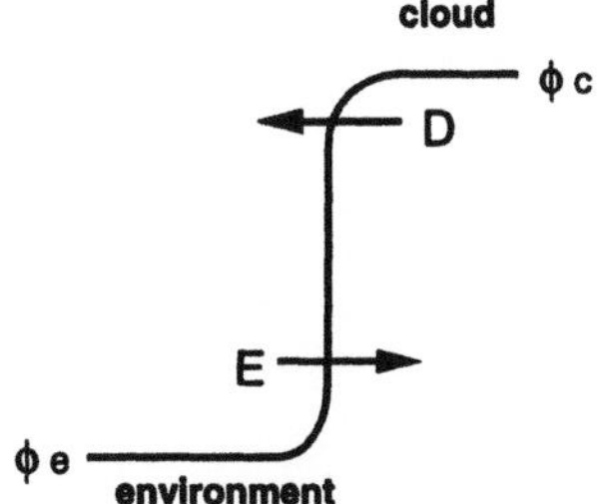

Figure 22. The mean field approximation (45) assumes a top-hat distribution for the field ϕ : the average environmental field ϕ_e is entrained into the clouds and the average cloud field ϕ_c is detrained into the environment.

We have indexed the "effective medium" exchange rates E_ϕ and D_ϕ to indicate that there might be a ϕ dependence. Let us stress however that (45) is a useful approximation only if $E_\phi = E$ and $D_\phi = D$. Accepting this strong assumption for the moment we can use (45) in (44):

$$
\begin{aligned}
\frac{\partial a_c \phi_c}{\partial t} &= E\phi_e - D\phi_c - \frac{\partial a_c \overline{w\phi}^c}{\partial z} + a_c F_c, \\
\frac{\partial (1-a_c)\phi_e}{\partial t} &= -E\phi_e + D\phi_c - \frac{\partial (1-a_c)\overline{w\phi}^e}{\partial z} + (1-a_c) F_e.
\end{aligned} \tag{46}
$$

The validity of the assumption (45) has been tested using LES results for BOMEX by calculating E and D as residuals of (46) for different fields $\phi = \{\theta_\ell, q_t\}$ [46]. The results, displayed in Fig. 23, do not show any ϕ-dependence. Although this is not a complete proof for $E = E_\phi$ and $D = D_\phi$, it does show that (46) gives consistent results so that we can indeed use these equations to investigate mixing between core and the environment.

Before we proceed with simplifying (46) even more to derive suitable parameterizations let us pause for a moment and ask the question whether

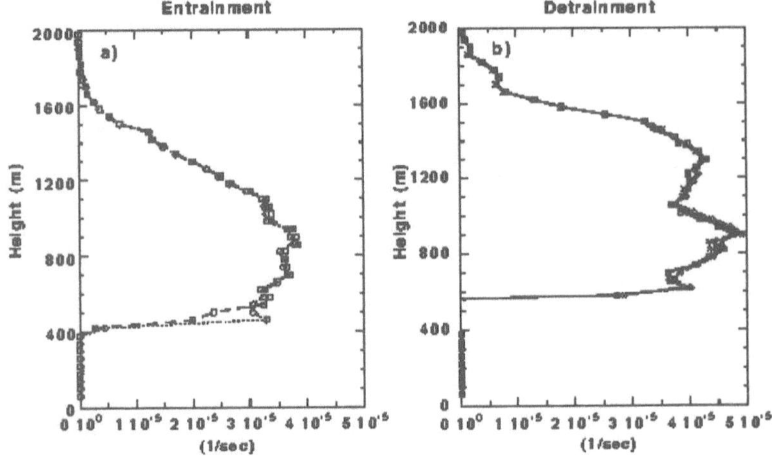

Figure 23. Entrainment and detrainment rates determined using LES data for different fields. In a) entrainment rates are obtained as a residual of (46) for $\phi = q_t$ (squares) and $\phi = \theta_\ell$ (circles). In b) as in a) but for detrainment. In b) we also plot the detrainment rate computed as a residual of the continuity equation (41). Note an extraordinary data collapse for all calculations.

assumption (45) already excludes stochastic mixing mechanisms such as introduced in section 3.6. In other words, do we already make a choice for lateral mixing when taking (46) as a starting point for parameterizations? The answer is partly yes since (46) assumes a top-hat distribution for the mixing between clouds and the environment, similar to the assumption that led to the lateral mixing result (27). On the other hand, by keeping the third term on the rhs of (46), there is still a freedom for variability of the fields within the cloudy and environmental regions. Therefore, by accepting (46), we are halfway to a lateral mixing model.

6.4. MASS FLUX PARAMETERIZATION WITH LATERAL MIXING

In order to formulate a practical parameterization for large-scale models, further simplifications of (46) are necessary. We therefore make three approximations:

- the mass flux approximation $\overline{w\phi}^c \simeq w_c\phi_c$,
- the core cover is much smaller than 1, *i.e.* $a_c \ll 1$, so that $\phi_e \approx \overline{\phi}$,
- steady state for the cloud ensemble $\partial a_c\phi_c/\partial t = 0$,

so that (41) and (46) can be simplified to

$$\frac{\partial M}{\partial z} = E - D, \tag{47}$$

$$\frac{\partial M\phi_c}{\partial z} = E\overline{\phi} - D\phi_c, \tag{48}$$

$$\frac{\partial \overline{\phi}}{\partial t} = -\frac{\partial M\left(\phi_c - \overline{\phi}\right)}{\partial z} + \overline{F}. \tag{49}$$

where $M = a_c w_c$. This set of equations is essentially the mass flux parameterization for cumulus convection as proposed by Tiedtke [61] .The first term on the rhs of (49) is precisely the turbulent flux divergence within the mass flux approximation (37) that needs to be parameterized. Hence, the parameterization is reduced to the determination of M and ϕ_c. These can be obtained from the steady state cloud model, as defined by (47) and (48), *provided* we know E and D plus the boundary conditions of M and ϕ_c at the cloud base. The latter, also known as the closure problem, is an interesting and important topic in itself but is outside the scope of the present paper. The entrainment and detrainment rates are usually parameterized in terms of the mass flux

$$E = \varepsilon M, \qquad\qquad D = \delta M, \tag{50}$$

where fractional entrainment and detrainment rates ε and δ are introduced. Within the present framework it is easy to show that ε as defined in (50) coincides with the one introduced in (27) and used in (38).

In most parameterization ε and δ are prescribed as constants with typical values $\varepsilon, \delta \sim 10^{-4}\mathrm{m}^{-1}$ [1, 16, 61], based on (30). Using LES output for BOMEX, typical values for the exchange rates have been determined [43]

$$\begin{aligned} \varepsilon &\approx 1.5 \sim 2.5 \times 10^{-3}\,\mathrm{m}^{-1}, \\ \delta &\approx 2.5 \sim 3\ \times 10^{-3}\,\mathrm{m}^{-1}, \end{aligned} \tag{51}$$

The spread in the results in (51) is mainly due to variation of the rates with height. Various sensitivity tests have been made by varying the resolution and domain size of the LES model. In all cases the fractional entrainment and detrainment rates remained within the range indicated by (51). Note that these values appear to be almost one order of magnitude larger than values suggested by (30) and used in many cumulus parameterizations. Furthermore, note that the detrainment is systematically larger than the entrainment consistent with the monotonic decrease of the mass flux with height (see section 5.3).

One may wonder how typical the results in (51) are for shallow convection in general. Recently we have made LES runs based of a North Sea cumulus case, as defined in [50] and of the Puerto Rico case, as defined in [11]. In both cases we found typical values in the range indicated by (51).

The parameterization (47-49) has been evaluated analytically for BOMEX [44] using constant prescribed values of ε and δ as suggested by the

LES results (51). The resulting turbulent fluxes and cloud fields compared well with the LES results indicating that the mass flux parameterization (47)-(49) is a sound approach *provided* the correct values for ε and δ are used.

We do not claim that the numerical values of ε and δ in (51) are universal. They are likely a function of the environmental conditions and will change from case to case. What is needed is a dynamical parameterization for these exchange rates based on physical concepts rather than using prescribed values. An attempt at that is made in the next subsection.

6.5. PARAMETERIZATION FOR ENTRAINMENT AND DETRAINMENT

Making similar approximations as in the last subsection, we can write the vertical momentum equation of the cloudcore region as [48, 46]

$$B - \varepsilon w_c^2 - w_c \frac{\partial w_c}{\partial z} = 0, \tag{52}$$

where the buoyancy B is given by

$$B = g \frac{\left(\theta_{v,c} - \overline{\theta}_v\right)}{\overline{\theta}_v}. \tag{53}$$

Equation (52) is a well-known and useful result since it relates the fractional entrainment to the buoyancy B and the vertical cloudcore velocity w_c. However, since w_c is an extra unknown variable it does not provide a solution for ε directly. We therefore consider the continuity equation (47) using (50) to arrive at

$$\frac{1}{a_c}\frac{\partial a_c}{\partial z} + \frac{1}{w_c}\frac{\partial w_c}{\partial z} = \varepsilon - \delta. \tag{54}$$

A proposal of Nordeng [37] is to link δ to the core cover divergence since detrainment is closely related to the evaporation of clouds. The remaining ε is then coupled to the vertical velocity divergence. This way one obtains two equations out of (54)

$$\frac{1}{w_c}\frac{\partial w_c}{\partial z} = \varepsilon, \tag{55}$$

$$\frac{1}{a_c}\frac{\partial a_c}{\partial z} = -\delta. \tag{56}$$

Combining (55) with (52) and integrating the vertical velocity from cloud base z_{lcl} to an arbitrary height z provides a parameterization for the fractional entrainment ε only in terms of buoyancy $B(z)$ and the cloudcore base vertical velocity

$$\varepsilon(z) = \frac{B(z)}{2\left[w_c^2(z_{lcl}) + \int_{z_{lcl}}^{z} B(z)dz\right]}. \tag{57}$$

Note that this parameterization has the same scaling behavior as (39) if $B(z)$ is height independent.

Given the boundary conditions, *i.e.* the cloud fields at the cloud base, this provides a closed set of equations; the cloud model (47-48) provides $B(z)$ if ε is known and, vice versa, (57) provides $\varepsilon(z)$ if $B(z)$ is known. Therefore all cloudcore fields ϕ_c, the buoyancy B and ε can be solved iteratively.

The final free parameter that remains to be determined is the fractional detrainment δ in order to determine the mass flux from (47). Since it seems that a typical feature of shallow cumulus convection is that the mass flux decreases monotonically with heigth, the simplest way to assure this is to parameterise the fractional detrainment rate δ as

$$\delta = \varepsilon \frac{1}{z_{lnb} - z} \qquad\qquad z_{lcl} \leq z \leq z_{lnb} \tag{58}$$

where z_{lnb} is the neutral buoyancy level, that is used as an estimate of the highest cloud top.

Tests with a one-column version of the ECMWF model with the entrainment/detrainment parameterization (57-58) in conjunction with the mass flux parameterization (47-49) has given promising results [45].

7. Conclusions and Perspectives

This paper has reviewed our current understanding of shallow cumulus convection with emphasis on the mixing between cumulus clouds and their environment. Entrainment of dry air into cumulus clouds and detrainment of cloudy air into the environment affects significantly the dynamics and their interaction with the environment. It determines where in the cloud layer the ventilated heat and moisture from the subcloud layer will be deposited.

The main controversy in the field is whether this mixing can be described by a relative simple lateral mixing principle, or whether one has to resort to the more elaborate buoyancy sorting models. Essential in the latter mechanism is that mixing of clouds with the environment takes place in discrete events whereafter both ascending positively buoyant and descending negatively buoyant cloud parcels are all en route to their zero buoyancy level where they finally detrain by evaporation.

As discussed in section 3.5, it is difficult to draw conclusions on the basis of conserved variable diagrams only. Recent observations of kinematics [25,

40], LES results [43] and 2-D CRM results [30] seem to suggest that cloud downdrafts, although present, do not play a dominant role in the dynamics of shallow cumulus clouds. This supports the picture of an intermittant rising thermal, existing of an ensemble of rising mixtures, as schematically sketched in Fig. 9. Such a model does not suffer from the Warner paradox [66] since the cloud top is determined by the most buoyant undiluted parcels while other parcels dilute the cloud sufficiently by lateral entrainment. This intermittant entraining thermal model is somewhat in the middle between the classic plume model and the more trendy episodic buoyancy sorting models [13, 42].

In the light of this one may wonder why a simple mass flux parameterization [61] is so well capable in reproducing the shallow cumulus dynamics such as observed during BOMEX, provided it is feeded with the correct lateral entrainment rates [44]. Probably this is the case because the dynamics is mainly determined by the active core, *i.e.* the most buoyant part of the cloud ensemble. This way even a single entraining plume model of the core can overcome the Warner paradox as suggested also in [31]. Nevertheless, the possibility for a need of a description of *multiple* rising mixtures should not be excluded. Analysis of LES data using conserved variable diagrams should give further insight into this topic.

Another important related issue that has not been addressed in this paper deals with the question of how much air is entrained from the dry convective boundary layer into the cloud layer. This determines the boundary layer ventilation, boundary layer equilibrium and the surface evaporation. In General Circulation Models (GCM's) it is still common practice that turbulent transport in the subcloud layer and in the cloud layer are parameterized separately. This implies that for a shallow cumulus parameterization the turbulent fluxes at the cloud base need to be specified, which is called the closure problem. Usually this is solved by assuming a rather arbitrary coupling on the latent heat flux at the surface [61] or by relating it to CAPE [16, 37]. The dry convective boundary layer parameterization scheme [20, 32], on the other hand, has entrainment at the top with no knowledge of the existence of clouds, which leads to possible overlaps between the schemes and double counting of transport mechanisms [34]. A physical coupling between the turbulent transport in the cloud and dry convective boundary layer and its parameterization needs to be established. It is clear that the boundary layer ventilation is determined by the strongest thermals in the subcloud layer that condensate at the lifting condensation level (LCL) and have enough kinetic energy to reach the level of free convection (LFC). An investigation of the dynamics of such strong updrafts in the dry convective boundary layer can be regarded as a first preliminary step in this direction [64].

8. Acknowledgements

I would like to thank Aad van Ulden for interesting discussions and suggestions and Harm Jonker for a critical reading of the manuscript. Special thanks are given to Vanda Grubišić for a careful and constructive review of the paper.

Appendix A: Budget Equations for Conditionally Sampled Fields

8.1. DEFINITIONS AND AVERAGING PROCEDURES

Consider an arbitrary field ϕ in a domain with an horizontal area A. We denote a horizontal spatial average of this field with an overbar

$$\overline{\phi}(z) \equiv \frac{1}{A}\int_0^{L_x}\int_0^{L_y}\phi(x,y,z)dxdy \quad \text{with} \quad A = L_xL_y. \tag{59}$$

Within this domain there exist possibly nonconnected cloudy parts that vary in time and space. We define cloud and environment averages as spatial averages over the corresponding parts

$$\overline{\phi}^c = \phi_c \equiv \frac{1}{A_c}\int\int_{\text{cloudy part}}\phi dxdy,$$

$$\overline{\phi}^e = \phi_e \equiv \frac{1}{A_e}\int\int_{\text{env part}}\phi dxdy, \tag{60}$$

where A_c and A_e denote the area over the cloudy and the environmental part. Clearly, ϕ_c and ϕ_e are trivially related

$$\overline{\phi} = a_c\phi_c + (1-a_c)\phi_e, \tag{61}$$

where $a_c = A_c/A$ is the fractional cloud cover. The vertical turbulent flux of ϕ is defined as

$$\overline{w'\phi'} \equiv \overline{w\phi} - \overline{w}\overline{\phi}. \tag{62}$$

Likewise we can define turbulent fluxes within the cloudy and environmental part

$$\begin{aligned} \overline{w'\phi'}^c &\equiv \overline{w\phi}^c - w_c\phi_c, \\ \overline{w'\phi'}^e &\equiv \overline{w\phi}^e - w_e\phi_e. \end{aligned} \tag{63}$$

Again the cloudy and environment fluxes are related

$$\overline{w'\phi'} = a_c\overline{w'\phi'}^c + (1-a_c)\overline{w'\phi'}^e + a_c(w_c - \overline{w})(\phi_c - \phi_e). \tag{64}$$

We will also need the average field ϕ_b and the turbulent flux at the cloud boundary

$$\overline{\phi}^b = \phi_b \equiv \frac{1}{L_b} \int_{\text{cloud interface}} \phi dl,$$

$$\overline{w'\phi'}^b \equiv \overline{w\phi}^b - w_b \phi_b, \tag{65}$$

where L_b denotes the total length of the cloud boundary.

BUDGET EQUATIONS

The prognostic equation for ϕ for a nondivergent velocity field can be written as

$$\frac{\partial \phi}{\partial t} + \nabla_h \cdot \mathbf{v}\phi + \frac{\partial w\phi}{\partial z} = F, \tag{66}$$

where $\mathbf{v}$ denotes the horizontal part of the velocity vector and w its vertical component. For the sake of simplicity we shall assume constant density but the equations can be easily generalized for the quasi-Boussinesq approximation. All possible source and sink terms are collected in F. The objective is to integrate all the terms of (66) over a horizontal cloudy area $A_c(z,t)$. Applying the Leibnitz theorem on the first and third term of the lhs of (66),

$$\frac{1}{A} \int_{A_c(z,t)} \frac{\partial \phi}{\partial t} dxdy = \frac{\partial a_c \phi_c}{\partial t} - \phi_b \frac{\partial a_c}{\partial t},$$

$$\frac{1}{A} \int_{A_c(z,t)} \frac{\partial w\phi}{\partial z} dxdy = \frac{\partial a_c \overline{w\phi}^c}{\partial z} - \overline{w\phi}^b \frac{\partial a_c}{\partial z}, \tag{67}$$

we find

$$\frac{\partial a_c \phi_c}{\partial t} - \phi_b \frac{\partial a_c}{\partial t} + a_c \overline{\nabla_h \cdot \mathbf{v}\phi}^c - \overline{w\phi}^b \frac{\partial a_c}{\partial z} + \frac{\partial a_c \overline{w\phi}^c}{\partial z} = a_c F_c. \tag{68}$$

For $\phi = 1$ and $F_c = 0$ we recover the continuity equation

$$\frac{\partial a_c}{\partial t} - \frac{\partial a_c}{\partial t} + a_c \overline{\nabla_h \cdot \mathbf{v}}^c - w_b \frac{\partial a_c}{\partial z} + \frac{\partial a_c w_c}{\partial z} = 0, \tag{69}$$

which can be put in a more transparent form by applying the divergence theorem

$$\frac{\partial a_c}{\partial t} + \frac{1}{A} \oint_{\text{interface}} \hat{\mathbf{n}} \cdot (\mathbf{u} - \mathbf{u}_i) dl + \frac{\partial a_c w_c}{\partial z} = 0, \tag{70}$$

where $\hat{\mathbf{n}}$ is an outward pointed unit normal vector at the interface at height z, $\mathbf{u}$ is the full 3D vector of the (mass) velocity at the interface and $\mathbf{u}_i$

is the velocity of the interface itself. Note that the normal vector does not necessary point into the horizontal direction. Likewise, using the divergence theorem, (68) can be rewritten as

$$\frac{\partial a_c \phi_c}{\partial t} + \frac{1}{A} \oint_{\text{interface}} \hat{\mathbf{n}} \cdot (\mathbf{u} - \mathbf{u}_i) \phi dl + \frac{\partial a_c \overline{w\phi}^c}{\partial z} = a_c F_c. \qquad (71)$$

Appendix B: List of Thermodynamic Constants

symbol	name	value
c_{pd}	specific heat capacity at constant pressure for dry air	1005 J kg^{-1} K^{-1}
c_{pv}	specific heat capacity at constant pressure for water vapor	1870 J kg^{-1} K^{-1}
c_l	specific heat capacity for liquid water	4190 J kg^{-1} K^{-1}
g	Acceleration due to gravity	9.81 m s^{-2}
L_0	Latent heat of vaporization at 273.15K	2.5 10^6 J kg^{-1}
p_0	reference pressure	1000 hPa
R_d	gas constant for dry air	287.05 J kg^{-1} K^{-1}
R_v	gas constant for water vapor	461.50 J kg^{-1} K^{-1}
ε	R_d/R_v	0.622
κ	R_d/c_{pd}	0.286

References

1. Anthes, R.A. (1977) A cumulus parameterization scheme utilizing a one-dimensional cloud model, *Mon. Wea. Rev.* **105**, 270-286.
2. Arakawa, A., and Schubert H. (1974) Interaction of a cumulus cloud ensemble with the large-scale environment. Part I. Theoretical formulation and sensitivity tests, *J. Atmos. Sci.* **31**, 674-701.
3. Augstein, E., Riehl H., Ostapoff F., and Wagner V. (1973) Mass and energy transports in an undisturbed trade-wind flow, *Mon. Wea. Rev.* **101**, 101-111.
4. Austin, P.H., Baker, M.B., Blyth, A.M. and Jensen, J.B. (1985) Small-scale variability in warm continental cumulus clouds, *J. Atmos. Sci.* **42**, 1123-1138.
5. Beniston, M.G., and Sommeria, G. (1981) Use of a detailed planetary boundary layer model for parameterization purposes, *J. Atmos. Sci.* **38**, 780-797.
6. Betts, A.K. (1973) Non-precipitating cumulus convection and its parameterization, *Quart. J. Roy. Meteor. Soc.* **99**, 178-196
7. Blyth, A.M., Cooper, W.A. and Jensen, J.B. (1988) A study of the source of entrained air in Montana cumuli, *J. Atmos. Sci.* **45**, 3944-3964.
8. Blyth, A.M. (1993) Entrainment in cumulus clouds, *J. App. Meteor.* **32**, 626-641.

9. Boatman, J.F. and Auer Jr., A.H. (1983) The role of cloud top entrainment in cumulus clouds, *J. Atmos. Sci.* **40**, 1517-1534.
10. Bougeault, Ph. (1981) Modeling the trade-wind cumulus boundary layer. Part I: Testing the ensemble cloud relations against numerical data, *J. Atmos. Sci.* **38**, 2414-2428.
11. Cuijpers, J.W.M, and Duynkerke, P.G. (1993) Large-eddy simulation of trade-wind cumulus clouds, *J. Atmos. Sci.* **50**, 3894-3908.
12. Deardorff, J.W., (1973) Three-Dimensional Numerical Modeling of the Planetary Boundary layer, in D.A Haugen (ed.), *Workshop on Micrometeorology*, Americ. Meteor. Soc., Boston, pp. 271-311.
13. Emanuel, K.A., (1991) A scheme for representing cumulus convection in large-scale models, *J. Atmos. Sci.* **48**, 2313-2335.
14. Emanuel, K.A., (1994) *Atmospheric Convection.* Oxford Univ. Press, Oxford.
15. Esbensen, S. (1978) Bulk thermodynamic effects and properties of small tropical cumuli, *J. Atmos. Sci.* **35**, 826-837.
16. Gregory, D. and Rowntree P. R. (1990) A mass flux scheme with representation of cloud ensemble characteristics and stability-dependent closure, *Mon. Wea. Rev.* **118**, 1483-1506.
17. Heymsfield, A.J., Johnson, P.N. and Dye, J.E. (1978) Observations of moist adiabatic ascent in northeast Colorado cumulus clouds, *J. Atmos. Sci.* **35**, 1689-1703.
18. Holland, J.Z., (1972) Comparative evaluation of some BOMEX measurements of sea surface evaporation, energy flux and stress, *J. Phys. Oceanogr.* **2**, 476-486.
19. Holland, J.Z., and Rasmusson, E.M. (1973) Measurement of atmospheric mass, energy and momentum budgets over a 500-kilometer square of tropical ocean. *Mon. Wea. Rev.* **101**, 44-55.
20. Holtslag, A.A.M. and Boville, B.A. (1993) Local versus nonlocal boundary-layer diffusion in a global climate model. *J. of Climate* 6, **6**, 1825-1842.
21. Hu, Q. (1997) A cumulus parameterization based on a cloud model of intermittently rising thermals, *J. Atmos. Sci.* **54**, 2292-2307.
22. Iribarne, J.V. and Godson, W.L. (1973) *Atmospheric Thermodynamics.* D. Reidel, Dordrecht.
23. Jensen, J.B., Austin P.H., Baker, M.B. and Blyth, A.M. (1985) Turbulent mixing, spectral evolution and dynamics in a warm cumulus cloud, *J. Atmos. Sci.* **42**, 173-192.
24. Johnson, R.H., and Lin, X. (1997) Episodic trade wind regimes over the western pacific warm pool, *J. Atmos. Sci.* **54**, 2020-2034.
25. Jonas, P.R. (1990) Observations of cumulus cloud entrainment, *Atmos. Res.* **25**, 105-107.
26. Kain, J.S. and Fritsch, J.M. (1990) A one-dimensional entraining/detraining plume model and its application in convective parameterization *J. Atmos. Sci.* **47**, 2784-2802.
27. Kain, J.S. and Fritsch, J.M. (1992) The role of the convective "trigger function" in numerical forecasts of mesoscale convective systems, *Meteorol. Atmos. Phys.* **49**, 93-106.
28. Klaassen, G.P., and Clark T.L. (1985) Dynamics of the cloud-environment interface and entrainment in small cumuli: Two-dimensional simulations in the absence of ambient shear. *J. Atmos. Sci.* **42**, 2621-2642.
29. Lamontagne, R.G., and Telford, J.W. (1983) Cloud top mixing in small cumuli, *J. Atmos. Sci.* **40**, 2148-2156.
30. Lin, C., and Arakawa, A. (1997) The macroscopic entrainment processes of simulated cumulus ensemble. Part I: Entrainment sources, *J. Atmos. Sci.* **54**, 1027-1043.
31. Lin, C., and Arakawa, A. (1997) The macroscopic entrainment processes of simulated cumulus ensemble. Part II: Testing the entraining plume model, *J. Atmos. Sci.* **54**, 1044-1053.
32. Louis, J-F., (1979) A parametric model of vertical eddy fluxes in the atmosphere, *Bound. Layer Meteor.* **17**, 187-202.
33. Nicholls, S., Lemone, M.A., and Sommeria, G. (1982) The simulation of a fair

weather marine boundary layer in GATE using a three-dimensional model, *Quart. J. Roy. Meteor. Soc.* **108**, 167-190.
34. Meijgaard, E. van (1996) Interaction of boundary-layer turbulence with shallow cumulus clouds over land. Proceedings on HIRLAM Workshop on Physical Parameterizations. Helsinki.
35. Nieuwstadt, F.T.M., (1998) Review of diffusion processes in the convective boundary layer. See this volume.
36. Nitta, T., and Esbensen, S. (1974) Heat and moisture budget analyses using BOMEX data, *Mon. Wea. Rev.* **102**, 17-28.
37. Nordeng, T.E. (1994) Extended versions of the convective parametrization scheme at ECMWF and their impact on the mean and transient activity of the model in tropics, Technical Memorandum No. 206, ECMWF, 41pp.
38. Paluch, I.R., (1979) The entrainment mechanism in Colorado cumuli, *J. Atmos. Sci.* **31**, 1028-1039.
39. Pontikis, C., Rigaud, A. and Hicks, E. (1987) Entrainment and mixing as related to the microphysical properties of shallow warm cumulus clouds, *J. Atmos. Sci.* **44**, 2150-2165.
40. Raga, G.B., Jensen, J.B., and Baker, M.B. (1990) Characteristics of cumulus band clouds off the coast of Hawaii, *J. Atmos. Sci.* **47**, 338-355.
41. Raymond, D.J., and Wilkening, M.H. (1982) Flow and mixing in New Mexico mountain cumuli, *J. Atmos. Sci.* **39**, 2211-2228.
42. Raymond, D.J., and Blyth, A.M. (1986) A stochastic model for nonprecipitating cumulus clouds, *J. Atmos. Sci.* **43**, 2708-2718.
43. Siebesma, A. P. and Cuijpers J. W. M. (1995) Evaluation of parametric assumptions for shallow cumulus convection, *J. Atmos. Sci.* **52**, 650-666.
44. Siebesma, A. P. and Holtslag A. A. M. (1996) Model impacts of entrainment and detrainment in shallow cumulus convection, *J. Atmos. Sci.* **53**, 2354-2364.
45. Siebesma, A.P. (1996) On the Mass Flux Approach for Atmospheric Convection. ECMWF Seminar Proceedings : New Insights and Approaches to Convective Parametrization. 34pp.
46. Siebesma, A.P., Brown, A., Cuxart, J., Sanchez, E., Moeng, C-H, Cotton, W.R., Jiang, H., Grubisic, V., Bretherton, C., Chlond, A., Khairoutdinov M., Krueger S., Lewellen, D., Macvean, M.K., Stevens, D. and Stevens, B. (1997) Intercomparison of 3d/2d cloud resolving models for the shallow-cumulus topped boundary layer, *In progress.*
47. Simpson, J., Simpson, R.H.D., Andrews, A. and Eaton, M.A. (1965) Experimental cumulus dynamics, *Rev. Geophys.* **3**, 387-431.
48. Simpson, J., and Wiggert, V. (1969) Models of precipitating cumulus towers, *Mon. Wea. Rev.* **97**, 471-489.
49. Simpson, J (1971) On cumulus entrainment and one-dimensional models, *J. Atmos. Sci.* **28**, 449-455.
50. Smith, S.A. and Jonas, P. (1995) Observations of the turbulent fluxes in fields of cumulus clouds, *Q. J. R. Meteorol. Soc.* **121**, 1185-1208.
51. Sommeria, G. (1976) Three-dimensional simulation of turbulent processes in an undisturbed trade-wind boundary layer, *J. Atmos. Sci.* **33**, 216-241.
52. Sommeria, G. and Lemone, M.A. (1978) Direct testing of a three-dimensional model of the planetary boundary layer against experimental data, *J. Atmos. Sci.* **35**, 25-39.
53. Squires, P. (1958) Penetrative downdraughts in cumuli, *Tellus* **10**, 381-389.
54. Squires, P. and J. S. Turner (1962) An entraining jet model for cumulonimbus updraughts, *Tellus* **14**, 422-434.
55. Stommel, H. (1947) Entrainment of air into a cumulus cloud, *J. Meteorol.* **4**, 91-94.
56. Stull, R.B. (1985) A fair-weather cumulus cloud classification for mixed-layer studies, *J. Clim. Appl. Meteorol.* **24**, 49-56.
57. Stull, R.B. (1988) *An Introduction to Boundary Layer Meteorology*, Kluwer Academic Publishers, 666pp.

58. Taylor, B.R. and Baker, M.B. (1991) Entrainment and detrainment in cumulus clouds, *J. Atmos. Sci.* **48**, 112-121.
59. Telford, J.W. (1975) Turbulence, entrainment and mixing in cloud dynamics, *Pure Appl Geophys.,* **113**, 1067-1084.
60. Tiedtke, M. (1987) The parametrization of moist processes. Part 2: The parametrization of cumulus convection. *ECMWF Lecture Series*, 56pp.
61. Tiedtke, M. (1989) A comprehensive mass flux scheme for cumulus parameterization in large-scale models, *Mon. Wea. Rev.* **177**, 1779-1800.
62. Tripoli, G.J. and Cotton, W.J. (1981) The use of ice-liquid water potential temperature as a thermodynamic variable in deep atmospheric models, *Mon. Weather Rev.* **109**, 1094-1102.
63. Turner, J.S. (1973) *Buoyancy effects in fluids* Cambridge University Press, 367 pp.
64. Van Ulden, A.P and Siebesma, A.P. (1997) A model for strong updrafts in the convective boundary layer, in Proceedings of: 12th Symposium on Boundary Layers and Turbulence, Vancouver, Canada, Americal Meteorological Society, 45 Beacon Street, Boston Massachusetts, MA02108-3693, USA. pp. 257-259.
65. Warner, J. (1955) The water content of cumuliform clouds, *Tellus* **7**, 449-457.
66. Warner, J. (1970) On steady-state one-dimensional models of cumulus convection, *J. Atmos. Sci.* **27**, 1035-1040.

Index

MIX
Papier aus verantwortungsvollen Quellen
Paper from responsible sources
FSC® C105338

If you have any concerns about our products,
you can contact us on
ProductSafety@springernature.com

In case Publisher is established outside the EU,
the EU authorized representative is:
Springer Nature Customer Service Center GmbH
Europaplatz 3, 69115 Heidelberg, Germany

Printed by Libri Plureos GmbH
in Hamburg, Germany